Intelligent Machine Vision

Springer-Verlag London Ltd.

Bruce Batchelor and Frederick Waltz

Intelligent Machine Vision

Techniques, Implementations and Applications

Springer

Bruce Batchelor, BSc, PhD, CEng, FIEE, FBCS
Department of Computer Science, Cardiff University,
PO Box 916, Cardiff CF24 3XF, UK

Frederick Waltz, BS, MS, PhD, FRSA
Consultant, Image Processing and Machine Vision, Minnesota, USA

British Library Cataloguing in Publication Data
Batchelor, Bruce G. (Bruce Godfrey), 1943-
Intelligent machine vision : techniques, implementations and applications
1. computer vision 2. Image processing - Digital techniques
I. Title II. Waltz, Frederick
006.3'7
ISBN 978-1-4471-1129-0

Library of Congress Cataloging-in-Publication Data
Batchelor, Bruce G.
Intelligent machine vision / Bruce Batchelor and Frederick Waltz.
p. cm.
Includes bibliographical references and index.
ISBN 978-1-4471-1129-0 ISBN 978-1-4471-0239-7 (eBook)
DOI 10.1007/978-1-4471-0239-7
1. Computer vision. 2. Image processing - Digital techniques. I. Waltz, Frederick M. II. Title
TA1634 .B35 2001
006.3'7—dc21
00-069840

ISBN 978-1-4471-1129-0

http://www.springer.co.uk

Originally published by Springer-Verlag London Limited in 2001
Softcover reprint of the hardcover 1st edition 2001

Typesetting: Electronic text files prepared by authors

34/3830-543210 Printed on acid-free paper SPIN 10648185

Rejoice in the Lord always.
I will say it again: Rejoice!

For my darling wife, Eleanor; my mother, Ingrid; my late Father, Ernest; my daughter, Helen; my son, David; my grandchildren Andrew, Susanna, Louisa, and Catherine.

Bruce

My wife Dorothy and son Bob have been very patient and helpful during the preparation of this book, for which I thank them sincerely. Unfortunately for them, I have at least one more book that I need to write before I can "really" retire. As the Preacher said long ago (*Ecclesiastes 12:12*),

> *Of making many books there is no end…*

Fred

Contents

Preface

Imagine that you are at a dinner party, sitting next to person who was introduced to you as a *Machine Vision Systems Engineer* (or *Vision Engineer* for short). Since you are not quite sure what that entails, you ask him to explain for the layman. He might reply thus:

"I study machines that can see. I use a video camera, connected to a computer, to find defects in industrial artifacts as they are being made."

This description is just short enough to avoid glazing of the eyes through boredom in people who are not trained in technology. However, it is very limited in its scope. For example, it does not encompass visual-sensory machines other than computers, or non-video image sensors. It also ignores the lighting, optics, and mechanical handling, which are all essential for a successful vision system. Compare the "dinner-party definition" given above with the following formal statement, which introduces the broader concept of *Artificial Vision*:

Artificial Vision is concerned with the analysis and design of opto-mechanical-electronic systems that can sense their environment by detecting spatio-temporal patterns of electromagnetic radiation and then process that data, in order to perform some useful practical function.

Machine Vision is the name given to the engineering of Artificial Vision systems. When browsing through the technical literature, a person soon encounters another term: *Computer Vision* (CV). He will quickly realize that *Machine Vision* (MV) and *Computer Vision* (CV) are used synonymously by many conference speakers and the authors of numerous technical articles and books. This is a point on which we strongly disagree. Indeed, this book is founded on our belief that these subjects are fundamentally different. There is a certain amount of academic support for our view, although it must be admitted that our most vociferous critics are to be found in universities and research institutions. On the other hand, most designers of industrial vision systems implicitly acknowledge this dichotomy, often by simply ignoring much of the academic research in CV, claiming that it is irrelevant to their immediate needs. In the ensuing pages, we will argue that *Machine Vision* should be recognized as a distinct academic and practical subject that is fundamentally different from Computer Vision and other related areas of study: *Artificial Intelligence* (AI), *Pattern Recognition* (PR), and *Digital Image Processing* (DIP). While Machine Vision makes use of numerous algorithmic and heuristic techniques that were first devised through research in these other fields, it concentrates on making them operate in a useful and practical way. This means that we have to consider all aspects of a vision system, not just techniques for representing, storing, and processing images inside a computer.

The problem of nomenclature arises because MV, CV, DIP, and sometimes AI and PR, are all concerned with the processing and analysis of pictures within electronic equipment. We did not mention computers explicitly in the preceding sentence, because Artificial Vision does not necessarily involve a device that is recognizable as a computer. Of course, it is implicit in the very name of CV that processing of images takes place inside a computer. On the other hand, MV does not impose such a limitation, although computer-based processing of images is encompassed by this term. MV also allows the image processing to take place in specialized digital networks, arrays of field-programmable gate arrays (FPGAs), multi-processor configurations, optical/opto-electronic computers, analog electronics, and various hybrid systems.

Although MV, CV, and DIP share a great many terms, concepts, and algorithmic techniques, they require a completely different set of priorities, attitudes, and mental skills. The division between MV and CV reflects the division that exists between engineering and science. Just as engineers have long struggled to establish the distinction between their subject and science, we are trying to do the same for MV and its scientific/mathematical counterpart: *Computer Vision.* The relationship between MV and CV is nicely illustrated by analogy with a jig-saw puzzle:

Position Code
0101, 1100, 1110, 0100
1101, 1000, 0110, 0010
1011, 1010, 0000, 0111
0011, 1111, 1001, 0001

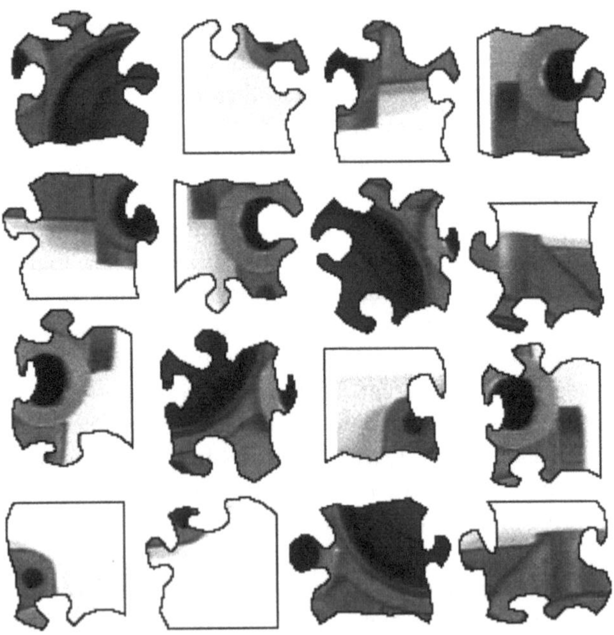

(The original image is shown in Plate 9(TR).) The position of each of the pieces within the puzzle can be defined by the 64-bit Position Code. (In this simple example, we do not allow rotation of the pieces.) From a theoretical point of view, this image, taken together with the 64-bit Position Code, represents a lossless transformation of the original image; together, they contain exactly the same amount of information as the original view. However, from a practical point of view, this transformation has a major effect; neither a machine nor a person can recognize this object nearly so easily. Although 99.902% of the information contained in the original (256*256 pixel) image has been retained within the jig-saw puzzle, to a vision engineer, or a psychologist, they are very different indeed. Clearly, a jig-saw puzzle is a very poor way to represent images for machines, or people, to analyze or recognize. Since Machine Vision is concerned with practical issues, a system must be able to recognize samples from a given class of objects, within acceptable limits of time and cost. Purely theoretical issues are not its concern. While finding loss-free transformations is important in certain applications, fast, cheap, reliable recognition of objects, or faults, is often of greater direct value, particularly for such tasks as industrial inspection and robot control. Obviously, the camera must view the appropriate face of the object being examined. This requires that we pay due attention to mechanical handling, as well as optimizing the lighting-optical subsystem, the choice of camera, the representation, storage, processing, and analysis of images. We cannot afford to concentrate on any one of these to the detriment of any other. This is the primary and inviolate rule for designing a successful Machine Vision system. We must always be aware that a module or subroutine that produces a minimal-loss transformation of information (like the jig-saw puzzle), may not be as useful in practice as one that produces a greater loss but expedites further processing. In Chapter 3, we will see several illustrations of this principle. While jig-saw puzzles do not have any serious practical application in Machine Vision, they typify the kind of artificial world that is often studied under the umbrella of Computer Vision.

During the 1970s and 1980s, Japan taught the rest of the world the importance of quality in manufacturing. The West learned the hard way: markets were lost to companies whose names were hitherto unknown. Many long-established and well-respected companies were unable to meet the challenge and failed to survive. Those that did were often faced with difficult years, as their share of the market shrank. Most companies in Europe and America have come to terms with this now and realize that quality has a vital role in establishing and maintaining customer loyalty. Hence, any technology that improves or simply guarantees product quality is welcome. Machine Vision has much to offer manufacturing industry in improving product quality and safety, as well as enhancing process efficiency and operational safety. Machine Vision owes its rising popularity to the fact that optical sensing is inherently clean, safe, and versatile. It is possible to do certain things using vision that no other known sensing method can achieve. Imagine trying to sense stains, rust, or surface corrosion by any other means!

The recent growth of interest in industrial applications of Machine Vision is due, in large part, to the falling cost of computing power. This has led to a proliferation of vision products and industrial installations. It has also enabled the development of cheaper and faster machines, with increased processing power. In many manufacturing companies, serious consideration is being given now to applying Machine

Vision to such tasks as inspecting, grading, sorting, counting, monitoring, controlling, and guiding. Automated Visual Inspection systems allow manufacturers to keep control of product quality, thus maintaining/enhancing their competitive position. Machine Vision has also been used to ensure greater safety and reliability of the manufacturing processes. The whole area of flexible automation is an important and growing one, and Machine Vision will continue to be an essential element in future *Flexible Manufacturing Systems.* There are numerous situations where on-line visual inspection was not possible until the advent of Machine Vision. This can lead to the development of entirely new products and processes. Many companies now realize that Machine Vision forms an integral and necessary part of their long-term plans for automation. This, combined with the legal implications of selling defective and dangerous products, highlights the case for using Machine Vision in automated inspection. A similar argument applies to the application of vision to robotics and automated assembly, where human operators are currently often exposed to dangerous unhealthy/unpleasant working conditions. Machine Vision is a valuable tool in helping to avoid this.

No Machine Vision system existing today, or among those planned for the foreseeable future, approaches the interpretative powers of a human being. However, current Machine Vision systems are better than people at some quantitative tasks, such as making measurements under tightly controlled conditions. These properties enable Machine Vision systems to out-perform people, in certain limited circumstances. Industrial vision systems can generally inspect simple, well-defined mass-produced products at very high speeds, whereas people have considerable difficulty making consistent inspection judgements in these circumstances. Machine Vision systems exist that can handle speeds as high as 3000 parts per minute, which is well beyond human ability. Experiments on typical industrial inspection tasks have shown that, at best, a human operator can expect to be only 70–80% efficient, even under ideal conditions. On many tasks, a Machine Vision system can improve efficiency substantially, compared with a human inspector. A machine can, theoretically, do this for 24 hours/day, 365 days/year. Machine Vision can be particularly useful in detecting gradual changes in continuous processes (e.g., tracking gradual color or texture variations in web materials). Gradual changes in shade, texture, or color are unlikely to be detected by a person.

Currently the main application areas for industrial vision systems occur in automated inspection and measurement and, to a lesser extent, robot vision. Automated visual inspection and measurement devices have, in the past, tended to develop in advance of robot vision systems. In fact, QC-related applications, such as inspection, gauging, and recognition, currently account for well over half of the industrial Machine Vision market. This has been achieved, in many cases, by retrofitting inspection systems onto existing production lines. There is a large capital investment involved in developing a completely new robotic work cell. Moreover, the extra uncertainty and risks involved in integrating two relatively new and complex technologies makes the development of robot vision system seem a daunting task for many companies and development has lagged behind that of inspection devices.

The technical difficulties involved in controlling flexible visually-guided robots have also limited the development. On the other hand, automated visual inspection systems now appear in every major industrial sector, including such areas as consumer goods, electronics, automobile, aerospace, food, mining, agriculture, pharmaceuticals, etc.

Machine Vision systems for industry first received serious attention in the mid-1970s. Throughout the early 1980s, the subject developed slowly, with a steady contribution being made by the academic research community, but with only limited industrial interest being shown. It seemed in the mid-1980s that there would be a major boost to progress, with serious interest being shown in vision systems by the major American automobile manufacturers. Then came a period of disillusionment in the USA, with a large number of small vision companies failing to survive. In the late 1980s and early 1990s, interest grew again, due largely to significant progress being made in making fast, dedicated image processing hardware. For many applications, it is possible now to provide sufficiently fast processing speed on a standard computing platform. Throughout the last 25 years, academic workers have demonstrated feasibility in a very wide range of products, representing all of the major branches of manufacturing industry.

We have watched this subject develop and blossom, from the first tentative steps taken in the mid-1970s to today's confident multi-million dollar industry. In those early days, industrial engineers and managers were mentally unprepared for the new technology. Although there is far greater acceptance of vision system technology nowadays, many customers are still unwilling to accept that installing and operating a vision system places a new intellectual responsibility on them. They still fail to understand that machines and people see things very differently. During the last 25 years, we have seen similar mistakes, based on a lack of appreciation of this point, being made many times over. Sadly, customer education is still lagging far behind technological developments in Machine Vision. The responsibility for making Machine Vision systems work properly does not lie solely on the shoulders of the vision engineer. The customer has a duty too: to listen, read, and observe, Making the most from the strengths and weaknesses of Machine Vision technology involves appreciating a lot of subtle points. This book discusses just a small proportion of these. We have had great fun developing and writing about this amazing technology and hope that some of our own excitement will be evident in these pages. Our clumsy attempts at building machines that can "see" simply emphasizes how wonderful and mysterious our own sense of sight is. It is humbling to realize that our supposedly sophisticated technology is no match for the gift of sight that our own Creator chose to bestow on us.

It is customary in many areas of writing to use so-called gender-neutral phrases, such as "*he/she*", "*s/he*", "*his/her*" etc. We regard these as being both clumsy and counter-productive. To improve clarity and to avoid placing women after men, we use the words, "*he*", "*his*" and "*him*" in the traditional way, to encompass both sexes, without granting precedence to either.

Acknowledgments

It is our pleasure to thank the following people who have contributed greatly to the preparation and development of ideas outlined in this book and who have kindly supplied us with materials included in these pages.

- **Dr. Paul Whelan** wrote Chapter 8 and was the joint author of Chapter 2 (with BGB). Dr. Whelan heads the team that developed the *JVision* software.
- **Dr. Andrew Jones** was responsible for the early stages of development of the *PIP* software. He also contributed to Chapter 6.
- **Mr. Michael Daley** wrote the revised *JIFIC* software. He also designed and built the *MMB* interface module.
- **Mr. Ralf Hack** wrote most of the *PIP* software.
- **Mr. Stephen Palmer** wrote additional *PIP* image processing software.
- **Miss Melanie Lewis** wrote the *WIP* software.
- **Mr. George Karantalis** was responsible for the initial development of *CIP* and wrote most of its image processing software.

The following MSc degree students wrote *CIP/JIFIC* software and provided material for Chapter 7:

Mr. Alan Christie
Mr. James Graham
Mr. Roger Harris
Mr. Richard Hitchell
Mr. Gareth Jones

The following BSc degree students wrote *CIP/JIFIC* software and provided material for Chapter 7:

Mr. Paul Goddard
Mr. Adam Gravell
Mr. Gareth Hunter
Mr. Geraint Pugh
Mr. Marc Samuel
Mr. Geraint Samuels

- **Mr. Robert Waltz** helped with the preparation and layout of this book.
- **Mr. Bruce Wright** created the cartoon shown in Fig. 1.1.
- **Mr. Rupert Besley** kindly gave permission to reprint the cartoon shown in Figure 1.3.
- **Dolan Jenner Inc.** gave us permission to use the photographs of the Micro-ALIS and ALIS 600 lighting systems shown in Figures 1.7 and 1.8.

Our gratitude is also extended to the following people, who have made significant contributions to the development of our ideas and understanding: Dr. Barry Marlow, Dr. David Mott, Dr. John Chan, Dr. John Miller, Dr. John Parks, Dr. Jonathon Brumfitt, Mr. Peter Heywood, and Dr. Mark Graves.

Glossary of terms, products, and acronyms

abc/N	A PIP operator which has an arity of N
abc/[M - N]	A PIP operator which has an arity of M to N inclusive
abc/[M,N]	A PIP operator which has an arity of M or N
&	Goal conjunction operator, alternative to ',' in Prolog
, (comma)	Goal conjunction operator, Prolog
. (period)	Terminator for a rule in Prolog
:-	Rule-definition operator, Prolog (Read as "Can be proved to be true if ...")
; (semicolon)	Goal union operator, Prolog
[]	Empty list in Prolog
[H|T]	Head (H) and tail (T) of a list, Prolog
⇐	Assignment operator
->	Conditional evaluation operator in Prolog
-->	Production-rule operator for Definite Clause Grammars, Prolog
Δ	Backtracking operator in Prolog
•	Goal repetition operator, infix
æ	Apple Event operator in PIP
AI	Artificial Intelligence
Algorithm	A prescription for solving a given computational problem (e.g. finding the roots of a quadratic equation). An algorithm has a defined outcome over a known application domain (cf. Heuristic).
ALIS	Lighting Prototyping System, Dolan Jenner, Inc.
Alternate image	One of the active image stores in PIP
Arity	Number of arguments of a Prolog predicate/PIP command
Autoview	Autoview Viking, a commercial IVS, British Robotics Systems Ltd. (no longer trading)
AVI	Automated Visual Inspection
AXI	Automated X-ray Inspection
BMVA	British Machine Vision Association
C	Computer language
C++	Computer language
Chain code	Method for describing the edge of a blob-like object in a binary image
CIP	Cyber Image Processing system, Authors G. Karantalis, Gareth Jones, Alan Christie, Roger Harris, Gareth Hunter, James Graham, Adam Gravell, Marc Samuel. Based on Java.
CPU	Central Processing Unit of a computer
CS	Computer Science
Current image	One of the active image stores in PIP
CV	Computer Vision

DIP	Digital Imaging Processing
Dyadic	Image processing operator which uses two input images
FIC	Flexible Inspection Cell, Cardiff University, Cardiff, Wales.
FPGA	Field Programmable Gate Array
GUI	Graphical User Interface
HCI	Human–Computer Interface
Heuristic	A computational procedure designed to provide a result close to that desired for a given application but which may sometimes fail to achieve a satisfactory result. A heuristic procedure may eventually be proved to be algorithmic, albeit over a restricted domain.
HMI	Human–machine interface
I/O	Input–Output
if	Rule-definition operator. Equivalent to ':-'
IVS	Interactive Vision System (prototyping system for designing a TVS)
Java	Computer language
JIFIC	Java Interface for the Flexible Inspection Cell, Cardiff University, Cardiff, Wales.
Journal	Window used in PIP to store image processing command sequences
JVision	Java Vision System, Dublin City University, Dublin, Ireland. Now called NeatVision
Local operator	Image processing operator which calculates the intensity of a given pixel by combining the intensities of several neighboring pixels in the input image
LPA	Logic Programming Associates Ltd., Studio 4, Royal Victoria Patriotic Bldg., Trinity Road, London SW18 3SX, England.
Macintosh OS	Operating system for the Macintosh family of computers
MacProlog	Version of Prolog for the Macintosh computer, Logic Programming Associates Ltd.
MMB	Mike's Magic Box, hardware interface module used to control the FIC
MMI	Machine–to–machine interface
Monadic	Image processing operator which uses only one input image
MV	Machine Vision
MVA	Machine Vision Association of the Society of Manufacturing Engineers
NeatVision	New name for JVision
N-tuple	Type of image filter/feature detector
PC	Personal Computer. The term usually refers to a computer based on the Windows operating system and Intel microprocessor. It is also used at times to include computers based on the Apple Macintosh and other operating systems.

1 Machine vision for industrial applications

Gwelediad yw sail pob gwybodaeth. Nid byd, byd heb wybodaeth.
(Vision is the basis of all knowledge. The world is no world without knowledge.)
Welsh proverbs

1.1 Natural and artificial vision

Vision is critical for the survival of almost all species of the higher animals, including fish, amphibians, reptiles, birds, and mammals. In addition, many lower animal phyla, including insects, arachnids, crustacea, and molluscs, possess well-developed optical sensors for locating food, shelter, a mate, or a potential predator. Even some unicellular organisms are photosensitive, even though they do not have organs that can form high-resolution images. Vision bestows great advantages on an organism. Looking for food, rather than chasing it blindly, is very efficient in terms of the energy expended. Animals that look into a crevice in a rock to check that it is safe to go in are more likely to survive than those that do not do so. Animals that use vision to identify a suitable mate are more likely to find a fit, healthy one than those that ignore the appearance of a possible partner. Animals that can see a predator approaching are more likely to be able to escape capture than those that cannot. Compared with other sensing methods based on smell, sound, and vibration, vision offers far greater sensitivity and resolution for discriminating the four essentials for survival listed above.

Of course, vision is just as important to *homo sapiens* as it is to any other animal species; life in both the forest and a modern industrial society would be impossible without the ability to see. Human society is organized around the highly refined techniques that people have developed to communicate visually. People dress in a way that signals their mood, social standing, and sexual availability. Commerce is dependent upon handwritten, printed, and electronic documents that all convey messages visually. Education relies heavily upon the students' ability to absorb information that is presented visually. Writers of technical books such as this exploit the reader's ability to understand complex abstract ideas through the visual medium provided in printed text, diagrams, and pictures. The leisure and entertainment industries rely on the fact that vision is the dominant human sense. So important is vision for our everyday lives that its loss is regarded by most people as one of the worst fates that could happen to a human being.

Manufacturing industry is critically dependent upon human beings' ability to analyze complex visual scenes. This is exploited in innumerable and highly variable tasks, ranging from the initial design process, through initial forming, machining, finishing, assembly, and quality control, to final packing. The value of vision as a safe way to sense many of the dangers that exist in a factory does not need explanation. Vision is hygienic and does not pollute the object under observation. With very few exceptions, the light levels needed to inspect an industrial artifact do not affect it in any way. Vision is potentially very fast; propagation delays from the object under inspection to sensor are typically just a few nanoseconds. (The sensor and its associated processing system will normally introduce much longer delays than this.) Vision provides a great deal of very detailed information about a wide variety of features of an object, including its shape, color, texture, surface finish, coating, surface contamination, etc. Vision is extremely versatile and we can employ a wide range of optical techniques to widen its range of application even further. For example, it is possible to use stroboscopic light sources, structured lighting, color filters, coherent illumination, polarized light, fluorescence, and many other cunning optical "tricks" to sense a very broad range of physical phenomena.

Human vision is truly remarkable in its ability to cope with both subtlety and huge variations in intensity, color, form, and texture. The dynamic range is enormous: about 10^{14}:1. People can see extremely subtle changes of color that are difficult to emulate in a machine. (This is why it is so difficult to match colors after car-body repair.) People can identity faces they know in a milling crowd, despite great changes in appearance that have taken place since the individuals concerned last met. On the other hand, they can register two matching pictures with great precision. A person can drive a car at high speed, yet a watchmaker can make very fine adjustments to a mechanism. A player can strike the cue ball in the game of billiards with exquisite accuracy. A painter can capture the subtlety of color in a woodland scene on his canvas, while an experienced foundry worker can identify the temperature of an incandescent ingot from its color very accurately. All of these visual skills are exercised for inspection, grading, sorting, counting, and a great many other tasks required for manufacturing industry.

Despite its obvious benefit to industry, human vision has its limitations. It is unreliable (Figure 1.1). Many factors contribute to this, most notably fatigue, discomfort, illness, distraction, boredom, and alcohol/drug ingestion. People cannot work at the speed required to provide 100% inspection of the products of many fast manufacturing processes. Boring repetitive tasks, such as watching high-speed conveyors carrying unstructured heaps of raw materials, are particularly prone to unreliable behavior. Many inspection tasks require precise dimensional/volumetric measurement, which people cannot perform accurately. For safety's sake, people must not be exposed to high temperatures, toxic/stifling atmospheres, high levels of smoke, dust, and fumes, biological hazards, risk of explosion, excessively high noise levels, gamma and x-rays. On the other hand, people can damage certain products by clumsy handling, and spread infection by coughing and shedding skin and hair. In addition, people cannot rigorously apply objective inspection criteria, particularly those relating to the aesthetic properties of highly variable items such as food products. For these reasons, engineers have long dreamed of

PIP	Prolog Image Processing. Authors A. C. Jones, R. Hack, S. C. Palmer, B. G. Batchelor, Cardiff University, Cardiff, Wales.
PR	Pattern Recognition
Predicate	Logical postulate expressed in Prolog whose truth value is to be determined, roughly similar to a procedure in a conventional language.
Prolog	Artificial Intelligence language (see Appendix B)
Prolog+	IVS consisting of Prolog controlling an image processor
PVR	Polar vector representation of a blob-like object in a binary image
RAM	Random Access Memory
ROI	Region of Interest. That part of an image that defines the spatial limit of operation of an image processing function
ROM	Read Only Memory
ROPE	Remotely Operated Prototyping Environment
SCIP	Scripting for CIP, Cardiff University, Cardiff, Wales, originally developed by G. E. Jones
Scratch	Image store in PIP
SKIPSM	Separated Kernel Image Processing using (finite) State Machines
SPIE	SPIE – The International Society for Optical Engineering
Structuring element	Kernel of a Mathematical Morphology operator
SUSIE	Southampton University System for Image Evaluation, 1974, authors J. R Brumfitt and B. G. Batchelor
System 77	Name for Autoview Viking when sold in USA, 3M Company
TVS	Target Vision System (a Machine Vision System used on the factory floor)
VCS	Vision Control System, a commercial IVS, Vision Dynamics
VDL	Vision Development Language, a commercial IVS, 3M Company
VSP	Very Simple Prolog+
Windows	Windows 95/98/NT operating systems. Microsoft Corporation, Inc.
WinProlog	Version of Prolog, intended for the Windows operating system
WIP	Windows Image Processing. Author Melanie Lewis.
WWW	World Wide Web

Software and hardware availability

PIP interactive prototyping software (Macintosh®)

A stand-alone version can be downloaded free of charge.

URL: *http://bruce.cs.cf.ac.uk/bruce/LA_how-to-get-it.html*

A source code version is available only to collaborating partners.

Requires MacProlog.

Contact Bruce Batchelor via e-mail: *bruce@cs.cf.ac.uk*

WIP interactive prototyping software (Windows®)

Available only to collaborating partners.

Contact Bruce Batchelor via e-mail: *bruce@cs.cf.ac.uk*

CIP interactive prototyping software (Java®)

Application and applet versions available.

URL: *http://www.cs.cf.ac.uk/User/G.Karantalis/default.html*

Lighting Advisor

Macintosh® version. Hypercard is required.

URL: *http://bruce.cs.cf.ac.uk/bruce/LA_how-to-get-it.html*

A preliminary on-line version is available.

URL: *http://www.cs.cf.ac.uk/User-bin/M.R.F.Lewis/lighting.html*

A superior on-line version is under development.

URL: *http://www.cs.cf.ac.uk/bruce/index.html*

JVision (now called NeatVision)

Contact Dr. Paul Whelan via e-mail: *whelanp@eeng.dcu.ie*

Also see URL *www.neatvision.com*

JIFIC control software

Contact Michael Daley via e-mail: *mike@cs.cf.ac.uk*

MMB interface module

Contact Michael Daley via e-mail: *mike@cs.cf.ac.uk*

SKIPSM software

Contact Frederick Waltz via e-mail: *fwaltz@isd.net*

building a machine that can "see." Such a machine would need to be able to sense its environment using a video camera, interfaced to an electronic information-processing system (i.e., a computer or a bigger system containing a computer).

Fig. 1.1 Human beings are easily distracted from boring tasks. Almost anything is more exciting to think about or look at than objects moving past on a conveyor belt.

1.2 Artificial vision

For the moment, we will use the phrase Artificial Vision, since we do not yet want to get into a detailed discussion about the precise meanings of the more commonly used terms Computer Vision and Machine Vision. The application of Artificial Vision systems to manufacturing industry motivates the work reported in this book. This is an area where machines are beginning to find application, in preference to people. Artificial Vision does not necessarily attempt to emulate human or animal vision systems. Indeed, the requirement that industrial vision systems must be fast, cheap, and reliable, together with our rather limited knowledge about how people and animals actually see, together make this approach unprofitable. When the author of this chapter is asked by a non-technical person what he does for a living, he replies that, as an academic, he studies *machines that consist of a television (more properly, video) camera connected to a computer and uses them to inspect objects as they being made in a factory* (Figure 1.2). Such a definition is acceptable around the dinner table but not for a technical book such as this. In fact, designing a vision system that is useful to industry in practice requires a multidisciplinary approach, encompassing aspects of all of the following technologies:

- Sample preparation
- Sample presentation to the camera
- Illumination
- Optics
- Image sensing
- Analog signal processing
- Digital information processing
- Digital systems architecture
- Software
- Interfacing vision systems to other machines
- Networking
- Interfacing vision systems to people
- Existing industrial work and Quality Assurance practices

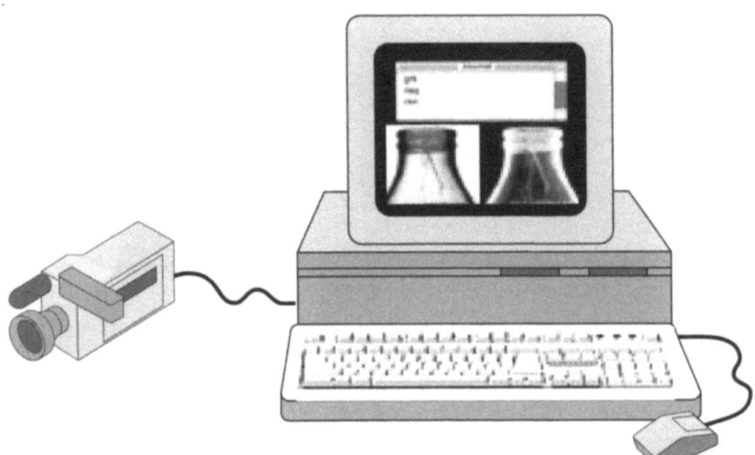

Fig. 1.2 A vision system? The usefulness of this diagram is limited to introducing total novices to image processing. This is not a complete Machine Vision system.

Figure 1.3 shows that many of these points must also be considered when a human being performs the inspection task. To justify the remaining points, in the context of Artificial Vision, the reader is referred to Plates 5 to 16, which illustrate some typical industrial vision applications. Lighting is obviously of critical importance when inspecting transparent, glossy, and glinting artifacts, such as glass bottles, coins, and electric light bulbs. Specialized lighting and viewing methods are needed to examine inside deep holes (e.g., the two hydraulics components). In some situations, we may need to use specialized sensing techniques, such as x-rays, to detect details that we could not otherwise see. The safety implications of using ultraviolet and penetrating radiation are of crucial importance, since even the most technically successful project will be judged a failure if the equipment does not

Fig. 1.3 A typical industrial inspection task! (Original caption: *Checking the spelling of sea-side rock.* Rock is a hard-candy cylinder that is distinguished by having the name of the holiday resort where it was sold spelled out by lettering, which runs internally along its whole length. Reproduced by kind permission of the artist.)

meet certain minimum legal requirements. Mechanical, rather than human, handling of the parts to be inspected is vital when the process would place the operator in danger; for example when X-ray sensing is used, or inspecting 25 ton slabs of red hot steel. We may need to apply a high level of intelligent interpretation to analyze images of plants, since they are totally unpredictable in form. Cake decoration patterns also require intelligent analysis, because they are highly variable, as are the shapes of baked products, such as loaves, croissants, etc. Intelligence is also needed to tell the time from an analog clock. Complex analytical methods of a different type are needed to analyze the surface texture of a slice of bread, or cork floor tiles. If we do not pay due attention to the application requirements, certain tasks may apparently require complicated solutions. However, these tasks may be quite straightforward if we constrain them in some way. The electrical connector in

Plate 9 is difficult to inspect if it is presented haphazardly to the camera. On the other hand, if it is delivered to the camera in a well-controlled manner (i.e., its position and orientation are both fixed), the inspection task becomes almost trivial. In some applications, defining the inspection task unambiguously requires a sophisticated user interface. For example, to define the criteria for inspecting the hydraulics manifold (Plate 9), we would need to tell the inspection machine which holes to measure, how large they should be, whether they are tapped and chamfered, and how they should relate geometrically to one another. This requires close interaction between the user and the inspection machine.

In our present state of knowledge about this subject, it is absolutely essential that we tailor each system to the specific demands of the application. In this respect, the design process is driven by the applications requirements in the broadest sense, not by the computational procedures alone. We cannot, for example, design a "universal" vision system, nor could we afford the cost if we did. Neither can we devise a system that can tolerate uncontrolled lighting, ranging from moonlight through standard room lighting to sunlight. We must control the light falling on the scene that is to be examined. We do this in two ways:

1. Use curtains, boards, or other rigid sheeting to block ambient light.
2. Apply light from stable lamps driven by regulated power supplies. Feedback control of the light levels is highly desirable.

1.3 Machine Vision is not Computer Vision

During a recent conversation, one of the authors (BGB) discussed the meanings of the terms Machine Vision (MV) and Computer Vision (CV) with another researcher, who is a very able and active worker in the mathematical and algorithmic aspects of Computer Science. That is, he works in what we maintain is Computer Vision. He regards the terms *Machine Vision* and *Computer Vision* as being synonymous. In order to highlight the difference between them, BGB challenged him to design a vision system that is able to decide whether a bright "silver" coin is lying with its obverse or reverse side uppermost. He could not do so! The reason for this is simple. This very intelligent man lacks training in an appropriate experimental subject (e.g., engineering or physics). Since the starting point for this challenge is a coin, not a computer file, the lighting and optical subsystems must be designed before any work on image processing algorithms is contemplated. We will present several further anecdotes to support our contention that MV requires skills and knowledge that CV ignores. We do not intend to belittle the intellectual component of Computer Vision. On the contrary, it is our belief that CV is intellectually "deep," while Machine Vision requires a broad range of skills. There should be mutual respect between the practitioners of these two subjects. Unfortunately, this is not always so. We will argue that MV should be regarded as a distinct academic and commercial subject, deserving its own institutions for organizing meetings, publishing results, honoring its leading exponents, and with its own funding program for research and advanced development.

There is a natural assumption among non-specialists that MV and CV are synonymous, because they both refer to Artificial Vision. In addition, many researchers blur the boundaries separating *Machine Vision* from the subjects of *Artificial Intelligence* (*AI*), *Digital Image Processing* (*DIP*), and *Pattern Recognition* (*PR*, alternatively called *Neural Networks*). Of these, DIP is most often confused with MV because it too is concerned with Artificial Vision. DIP is an essential component of Computer Vision, so we need not discuss it separately. PR and to a smaller extent AI are sometimes perceived as encompassing aspects of MV, although the distinction is greater than for CV. Several other terms need to be mentioned before we begin our discourse in earnest. *Automated Visual Inspection* (*AVI*), *Robot Vision*, and *Visual Process Control and Monitoring* are subsets of MV. However, the term *Robot Vision* is also used when referring to the theoretical aspects of robot control as a subset of AI.

All of the terms mentioned above are used in the technical literature and by speakers at numerous business meetings and conferences in an imprecise way. There is no doubt that the sloppy use of these terms has been the cause of a considerable amount of confusion. This has often resulted in unsatisfactory service being provided to would-be customers of vision systems. It has also hampered research, and has probably hindered career progression of engineering staff.

The problem of nomenclature arises because MV, CV, DIP, and sometimes AI and PR are all concerned with the processing and analysis of pictures within electronic equipment. Notice that we did not mention computers explicitly, since not all processing of pictures takes place within a device that is recognizable as a digital computer. Of course, CV assumes that these activities *do* take place inside a computer. On the other hand, MV does not impose such a limitation, although computer-based processing of images is encompassed by this term. MV allows the processing of images to take place in specialized digital networks, arrays of field-programmable gate arrays (FPGAs), multiprocessor configurations, optical/opto-electronic computers, analog electronics, or hybrid systems.

It must be understood that MV, CV and DIP share a great many terms, concepts, and algorithmic techniques but they have a completely different set of attitudes and priorities (see Table 1.1).

To summarize, the division between MV and CV reflects the separation that exists between engineering and science. Just as engineers have long struggled to establish the distinction between their studies and science, we are trying to do the same for MV and its scientific/mathematical counterpart (i.e., CV).

This confusion of terms is even manifest in the names of learned societies. For example, the *British Machine Vision Association* holds a series of conferences in which almost all of the papers are in what we maintain is CV. There is certainly a great difference between the articles included in the proceedings of the BMVA on the one hand and those organized by the *Machine Vision Association* of the *Society of Manufacturing Engineers*, or *SPIE – The International Society for Optical Engineering*. Indeed, it is doubtful whether many of the papers that are acceptable to a BMVA audience would be welcomed at an MVA conference, or vice versa. This fact

Table 1.1. Comparing Machine Vision and Computer Vision. Entries in the Machine Vision column relate to the factory-floor target machine (TVS), unless otherwise stated, in which case they refer to an interactive prototyping tool kit (IVS)

Feature	Machine Vision	Computer Vision
Motivation	Practical	Academic
Advanced in theoretical sense	Unlikely. (Practical issues are likely to dominate over academic matters.)	Yes. Academic papers often contain a lot of "deep" mathematics
Cost	Critical	Likely to be of secondary importance
Dedicated electronic hardware for high-speed processing	Very likely	No (by definition)
Designers willing to use non-algorithmic solutions to problems	Yes (e.g., systems are likely to benefit from careful lighting)	No
In situ programming	Possible	Unlikely
Input data	A piece of metal, plastic, glass, wood, etc.	Computer file
Knowledge of human vision influences system design	Most unlikely	Very likely
Most important criteria by which a vision system is judged	a) Ease of use. b) Cost-effectiveness. c) Consistent and reliable operation	Performance
Multidisciplinary	Yes	No
Nature of an acceptable solution	Satisfactory performance	Optimal performance
Nature of subject	Systems Engineering, practical	Computer Science, academic (i.e., theoretical)
Operates free standing	a) IVS interacts with human operator. b) TVS must be able to operate free-standing	May rely on human interaction
Operator skill level required	a) IVS medium/high. b) TVS must be able to cope with low skill level	May be very high
Output data	Simple signal, to control external equipment	Complex signal, for human being
Speed of processing	a) IVS must be fast enough for effective interaction. b) Real-time operation is very important for TVS	Not of prime importance
User interface	Critical feature for IVS and TVS	May be able to tolerate weak interface

does not in any way detract from the worth or intellectual content of either "camp." It merely reflects a difference of approach and priorities. In the case of Machine Vision, the latter are usually manifest in terms of the design criteria imposed by the application. See Table 1.2.

Table 1.2 Technologies needed to design a Machine Vision system

Technology	Remarks
Algorithms/heuristics for image processing	One of the principal subjects in this book. NB. This constitutes a common area of interest, shared with Computer Vision. However, the approach is very different.
Analog and video electronics	Normally used for preprocessing to reduce the data rate and improve signal/noise ratio.
Communications	• Network to computers and control systems. • Connect to other machines in the factory, PLCs, etc.
Digital electronics	Normally used for preprocessing to reduce the data rate and improve signal/noise ratio.
Industrial engineering	Whole system must be designed to be robust and provide reliable operation in a hostile factory environment.
Lighting	Lighting is a critical component for any Machine Vision system. (NB. Shrouding is needed to eliminate ambient light and is just as important as the applied lighting.)
Mechanical handling	• Essential for presenting objects to the camera in a controlled manner. (Defective parts must not jam the handling unit.) • Also needed for mounting lights and camera rigidly.
Optics	Numerous special effects can be provided by lighting and optics and can often convert a very difficult problem into a trivial one.
Quality control and production engineering	System should be designed to comply with current industrial, operational, and environmental working and quality-control practices.
Sensor	The imaging sensor may provide the ability to detect 1-D, 2-D, or 3-D spatial patterns in any one of the following physical media: chemical composition of surface/bulk material, gamma ray, infrared, magnetic field, microwave, neutron flux, pressure, surface profile, ultrasonic, ultraviolet, visible, X-ray, etc.
Software	Common area of interest shared with Computer Vision.
Systems architecture	Good overall "balance" throughout the system is essential.
User interface	System should be designed to provide a good ergonomic interface to the human operator.

1.3.1 Applications

We will restrict our discussion to industrial applications of vision system for such tasks as inspecting, measuring, verifying, identifying, recognizing, grading, sorting, counting, locating and manipulating parts, monitoring/controlling production processes, etc. Vision systems may be applied to raw or processed materials, piece parts, assemblies, packing, packaging materials and structures, continuous-flow products (webs, extrusions, etc.), simple and complex assemblies, or to tools and manufacturing processes. They can also be applied to complex systems (e.g., complicated physical structures, pipework, transport mechanisms, etc.) and to a wide range of factory-management tasks (e.g., monitoring material stocks, checking the floor for debris, controlling cleaning, ensuring safe working conditions for human operators, etc.).

Machine Vision is also applied to a variety of other areas having considerable significant commercial importance and academic interest:

- Aerial/satellite image analysis
- Agriculture
- Document processing
- Forensic science, including fingerprint recognition
- Health screening
- Medicine
- Military applications (target identification, missile guidance, vehicle location)
- Publishing
- Research, particularly in physics, biology, astronomy, materials engineering, etc.
- Security and surveillance
- Road traffic control

Of course, there are numerous other less-important applications areas that we do not have space to mention here. It would be possible to make similar remarks in relation to these non-industrial applications areas, in order to establish a broader horizon for Machine Vision. However, the scope of the present discussion is more limited, since this book concentrates on techniques and machines for industrial applications. Moreover, in the small amount of space available to us here, we cannot do complete justice to these additional application areas. We will therefore deliberately restrict our discussion by referring only to industrial applications. Nevertheless, we should point out that any application which places demands on the speed, cost, reliability, size, or safety of a vision system imposes constraints that must be taken into account by adopting good Systems Engineering design practices. The basic tenet of all work in MV is that it is unacceptable to limit our concern to the mathematics/algorithms, or any other single aspect of the system design task.

1.3.2 Does it matter what we call the subject?

The answer is most definitely "Yes!" There are several reasons why MV deserves to be identified as a subject in its own right:

- To ensure proper refereeing of academic papers
- To promote appropriate funding of research
- To protect customers from claims of expertise in MV by workers who have a narrow field of expertise and are therefore not competent to design a complete system
- To protect the careers of its practitioners

Most technical papers in Machine Vision are not intellectually "deep," as those in CV usually are. Papers on MV are usually much broader in their scope. It is usual, for example, to find a rather small amount of detailed mathematical analysis in a paper on MV. If it is present at all, then it is almost certainly directly relevant to the solution of a practical problem. On the other hand, many papers in CV seem to be written with the deliberate intention of including as much mathematics as possible, perhaps to impress the readers with the author's erudition. The author of this chapter has on several occasions received quite critical comments from referees of academic papers who are clearly unaware of the distinctive nature of MV, although the same papers were subsequently published in forums more sympathetic to "systems" issues. There is a distinct tendency for workers in CV to dismiss papers in MV as being trivial or shallow because they do not contain a lot of analysis. On the other hand, a well-written paper on Machine Vision is likely to cover a wide range of practical issues, such as object transportation, lighting, optics, system architecture, etc. Hence, papers that concentrate solely on the mathematics or algorithms of image processing and understanding may not be acceptable in a journal or conference on Machine Vision, because no connection is made to "real-world" problems.

In similar way, the assessment of applications for grants or promotion can be biased if the referee thinks his skills are appropriate when in fact they are not. For this reason, the distinction between MV and CV is crucial. A balanced impression of a person's total contribution to a project cannot be gained by someone who does not understand the full scope of his work. Just as it would be unsuitable for a person trained in, say, mathematics to judge a hardware engineer, so it is inappropriate for a CV *specialist* to make categorical judgements about an MV *generalist.* (The significant words are set in italics.)

There is good reason to believe that grant applications are sometimes rejected because the referee fails to understand the full significance of a new approach. Referees who look for significant new techniques in algorithms will not normally be impressed by grant applications that concentrate on "systems issues," such as the interrelationships between the choice of mechanical handling device, optics, camera, and computing system architecture. As a result, good MV grant applications are sometimes rejected. Of equal significance is the problem of inferior grant applications being accepted because the referee is unaware of broad "systems" issues, or new developments in MV thinking that make the proposed approach obsolete or less than optimal. For example, a new kind of camera may eliminate the need for a certain type of image processing operation. Or, a new type of lamp or optical filter may enhance image contrast to the point where only the very simplest type of processing is necessary.

Perhaps the most important single reason for needing to protect the identity of MV is to avoid the damage that can be done to its reputation by charlatans. Indeed, a great deal of harm has already been done by specialists in CV, DIP, PR, and to a lesser extent, AI *dabbling* in Machine Vision when they are not competent to do so. Ignoring the image acquisition is a sure way to generate an over-complicated image processing solution to a problem. Often, the requirements of an application could be satisfied better by a simple adjustment to the lighting or optics, rather than adding sophistication to the image processing. The classic problem of this type is that of thresholding an image. There are *many* ways to achieve high-contrast images, several of which do not require the use of conventional back illumination [BAT94a]. Ignoring such a possibility and then resorting to an over-complicated dynamic thresholding technique is simply poor engineering practice. There is no doubt that many installed vision systems use excessively complicated algorithms to compensate for poor lighting. From time to time, over-zealous and ill-informed salesmen have claim that their vision product is able to operate in uncontrolled ambient light. This is simply not true. Given the present state of our knowledge of the subject, claims of this kind are fraudulent! Many other vision systems have been less reliable than they might have been, had they been designed from the outset with a good appreciation of all of the "systems issues" involved.

To summarize, there is no excuse for fooling a customer who has no detailed knowledge of either CV or MV into thinking that a specialist in CV is competent to design a system for use in a factory. Unfortunately, intellectual honesty is not universal and the consequences of misleading customers like this can have a major impact on the wider long-term public perception of a technology such as MV. There are subtle ways in which erroneous views can be propagated and perpetuated. Every MV engineer knows that very subtle points can sometimes lead to the undoing of an otherwise successful application. For example, static electricity may make it very difficult to keep optical surfaces clean. This may reduce image contrast sufficiently to render a certain algorithmic approach unreliable. Another example is the health hazard associated with inspecting objects on a fast-moving conveyor using an area-scan camera and stroboscopic illumination. Using a line-scan camera avoids this altogether.

1.4 Four case studies

Lest the reader remain unconvinced, we will now consider four practical applications which highlight the essential identifying feature of Machine Vision – the need for a Systems Engineering approach to design. A paper presented at a recent conference,[BAT94c, BAT97], listed a series of "proverbs" in an attempt to encapsulate important lessons gleaned from a wide range of design exercises for industrial vision systems. This list has since been enlarged and can be viewed via the World Wide Web [PVB]. Five proverbs relate specifically to the need to define the intended function of a vision system:

- *The specification of the vision system is not simply what the customer wants.*
- *No machine vision system can solve the problem that the customer forgot to mention when placing the order.*
- *The required functionality and performance of a vision system should be indicated in the specification.*
- *Beware when the customer says "By the way!... We would like to be able to inspect some other objects as well."*
- *The customer said his widgets were made of brass. He did not think to state that they are always painted blue and are oily.*

Another proverb has gained particular resonance with the authors' colleagues working in MV and summarizes a danger that is peculiar to the design of vision systems:

- *Everybody (including the customer) thinks he is an expert on vision and will tell the vision engineer how to design the machine.*

Bear these points in mind as we trace the history of four very different industrial projects in which vision systems were commissioned:

Project 1. The project specification was drawn up without reference to what is technically possible or economically feasible. No experienced MV engineer was involved.

Project 2. The vision system was designed successfully by an experienced team working full time in MV. An unusual image-acquisition subsystem was designed in a series of experiments using very simple equipment. This greatly simplified the image processing requirements.

Project 3. The client specified the problem in a totally inappropriate way. Good engineering practice detected the problem before it was too late.

Project 4. The specification of an AVI system evolved during the preliminary contract negotiations and the final machine was very different from that first envisioned. This dialog would never have taken place without an experienced MV engineer being involved.

In each case, the engineers working in the client company had no previous experience with Machine Vision technology. They differed in their willingness to learn and accept the advice given by an experienced vision engineer. As a result, these four projects reached very different conclusions.

1.4.1 Doomed to failure

The first project epitomizes an approach that almost inevitably leads to failure and disillusionment with MV technology.

A certain company makes complex high-precision components for domestic goods, by a process involving several stages of pressing and punching sheet steel. Some of the products display a variety of faults, including both functional and cosmetic defects. The company, in association with a well-respected university department, set up a joint project. The investigator was a young graduate in Production Engineering with no prior experience of Machine Vision. In addition, his supervisors had very limited experience in this technology. Inexplicably, the university–company partnership succeeded in gaining funding for this project from both government and board-level company sources. The university also had an established link with a major electronics company that lead to a decision, quite early in the course of the project, to purchase a certain (old-fashioned) turn-key image processing system. This decision was made for political, rather than sound technical reasons. The specification of this system was far inferior to that available from a modern general-purpose computer costing about 25% of the price. At the stage when this decision was made, the lighting-optics-viewing arrangement had not been considered seriously and no image processing algorithm had been designed or selected. There was no evidence at all that this turn-key system would be able to implement the necessary image processing functions, since the requirements at this stage were simply not known. It is interesting to note that, during the middle phase of the project, two highly experienced MV engineers were given privileged access to the work of this group and counselled strongly but unsuccessfully against the approach being adopted. (Recall the proverb which states that everyone thinks that he is an expert in vision.) The system specification was drawn up without any reference whatsoever to what is possible in technical, financial, or human terms. The outcome in such a situation is inevitably failure and disillusionment with MV technology.

1.4.2 A successful design

In the mid-1980s, engineers in a certain company were in the late stages of building a fully automatic flexible-manufacturing system designed to produce over 50 different products, when they realized that they would be unable to complete the task without a non-contact parts-location system. The requirement was well-specified and was highly suitable for a vision system. The client company contacted a company specializing in designing and building Machine Vision systems. The latter company could not, at first, see a path leading towards a successful solution and, in turn, consulted a very experienced MV researcher. He devised an experiment involving a plastic meal tray, a dressmaker's tape measure, a desk lamp, and a standard surveillance-type video camera. (He needed to simulate a motorized turntable, a line-scan camera, and oblique lighting.) Acquiring each image took over an hour, with three people participating. Nevertheless, a series of test images was obtained and a suitable image processing algorithm devised, using an interactive Machine Vision system of the type described in later pages of this book. On the basis of these seemingly clumsy experiments, the vision company was able to design, build, and install a well-engineered target system. It worked very well, except that sunlight caused intermittent optical interference. However, a solution was soon found: a metal screen was placed around the camera and object to be inspected to block out ambient light. The final vision system was highly successful and worked reliably for several years. The important point to note is that consulting a suitably experienced Machine Vision engineer early in the project led to a highly successful solution. Although the resources used in the early experimental stages of this project were severely limited, they were appropriate for the purpose and a good system design ensued.

1.4.3 Moldy nuts

A highly experienced and well-respected MV engineer was commissioned to design a machine to inspect nuts (the edible variety), to detect patches of mold on the surface. He was supplied with a large bag of nuts to enable him to design the inspection machine. He expended a lot of effort on this exercise. When he had completed the task to his satisfaction, he contacted the client to arrange a demonstration prior to delivering and installing it in the food-processing factory. When the engineers from the client company arrived for the demonstration, they brought with them another large bag of nuts, hoping to test the vision system on previously unseen samples. (This idea is, of course, very sensible in principle.) The vision system did not respond well in these tests and failed to satisfy the client's expectations, for one very simple reason: the new sample of nuts had fresh mold which was green. The samples originally supplied to the vision engineer to help him in the design process had older brown mold. The client had omitted to tell him that the green mold quickly changes color to brown. This fact was obvious to them but they had not thought it important to tell the vision engineer. Since brown mold patches can be identified by eye, it was "obvious" to the client that the green mold could be detected just as easily by a Machine Vision system. It is a universal truth that, whenever the client tries to tell the vision engineer how to do his job, trouble is about to occur.

The lesson for us is that although the system was not designed appropriately in the first instance, the good engineering practice of testing the system on data independently of that used during the design highlighted a serious problem. The result was annoying but not disastrous, because it was fairly easy to redesign the system to recognize the colored mold. In fact, the revised design was considerably simpler than the first one!

1.4.4 A system that grew and grew

In this section we describe how one application, which was initially conceived as requiring a single-camera vision system, grew to much more massive proportions. As we will see, the original concept grew very considerably during the dialog between the customer and the vision engineer. Of course, this is the best time of all for the specification to change! In this project, the vision engineer was encouraged by the client to explain how Machine Vision could form part of a much larger system. In this case, the system that was eventually built was a complete parts-handling machine which just happened to implement Machine Vision, among several other functions. In the machine finally delivered to the customer, the vision sub-system was fully integrated into the larger machine and was no more, nor less, important to the overall function than that of the other equipment. In other words, the vision engine formed just one of many components of the larger system. The specification of the system was developed during several conversations involving the MV engineer and his client.

1. The MV systems engineer was invited to provide advice about the provision of an Automated Visual Inspection system for components used in the automotive industry. At first sight, the task seemed straightforward: the customer wanted to ensure that certain safety-critical components met their specification, but wanted to avoid the long and tedious tests then being performed manually. A significant speed increase (greater than 100%) was also expected. The initial specification was for a machine that could locate four holes at the corners of the component, which consists of a flat nearly-rectangular metal plate, with a central cylindrical spindle normal to the plate (Plate 9(CR)). In the first instance, measuring the positions of the centers of these holes appeared to be a straight-

forward task. However, the project became ever more complex as successive meetings took place between the MV engineer and the client.

2. The *Production Engineer* had ambitious ideas! He reasoned that, if the inspection system could perform this task, then why not others as well? So, he added the requirement that it should check that the holes are tapped. Clearly, the camera used to locate the holes could not be used to examine these female threads, because the latter task requires an oblique view and a different lighting arrangement.
3. The third requirement, also suggested by the Production Engineer, was for another camera with its associated image processing sub-system to check and measure the male thread at the end of the spindle.
4. Then, a requirement was drawn up for yet another camera, to check that the heat treatment that this part receives during its manufacture has been completed. (Heat treatment causes perceptible color change.) Up to now, the object to be examined could be inspected while it was static.
5. Since the spindle diameter is critically important, yet another requirement was introduced. However, this one was satisfied without the use of vision. Four swinging arms, each carrying a digital micrometer, were used to measure the spindle diameter. The result is more accurately than a simple vision system could achieve. Requirements 1 to 5 were all suggested by the Production Engineer.
6. Next, came the requirements defined by the *End-of-line Packer*: The parts are sprayed with an anti-corroding agent before being shipped out of the factory. Before they are sprayed, the parts are cleaned. This involves an all-over spray, with particular attention being paid to the four holes, which might contain swarf and other particulate debris. The cleaning process was to take place in an on-line washer, with the parts being transported on a continuously-moving conveyor belt. The cleaning process also imposed a need for a component drying station.
7. The *Quality Control Engineer* was the next person to participate in specifying the system requirements, by suggesting an additional function for the new machine. He pointed out that there is a need for a marking system, enabling defective components, along with the machine and shift responsible, to be identified later. Since the parts are painted shortly after they have been made, the marking must be engraved. In order to accommodate this function, the conveyor belt had to be lengthened, which had an unfortunate effect on product safety. Objects that have been rejected as faulty could be placed back on the *ACCEPT* region of an open conveyor. This has been referred to as the *Middle Management problem.* People who pick up objects from the production line to examine them by eye do not always replace them properly afterwards. Notice also that there are financial incentives for operators to move *REJECT* components deliberately onto the *ACCEPT* stream, since this *appears* to increase productivity. To prevent this, faulty parts should pass into a secure cabinet, accessible only to a trusted member of staff.
8. The *Line Supervisor* provided the last suggestion for embellishments to the planned machine, which by this time would be the size of a grand piano. With

only a little additional complication, the machine can be made totally automatic, with no operator intervention whatsoever. This required the provision of loading and ejection bays.

So far, we have not even begun to discuss the attitude of the accountants to all this! That is another story that need not concern us here. Let it suffice to say that the accountant did finally agree that the (enlarged) machine should be built. To a novice, this story may seem exaggerated. However, it is a real-life experience and, in general terms, should be regarded as typical rather than exceptional. Notice how much of the specification of this machine is concerned with non-imaging tasks. Of considerable importance in the negotiations is the improvement in confidence that results from the vision engineer being able to demonstrate each step in the developing inspection strategy. (We will return to this particular point many times in subsequent pages.)

1.5 Machine Vision is engineering, not science

The experience outlined in these four case studies is not unique. They all emphasize the importance of proper initial planning. One crucial part of this is reaching a clearly-defined agreement between the customer and the vision engineer about the real objectives of the exercise. The Machine Vision Association has long advocated the use of a *pro forma* specification which has been prepared and published specifically for this purpose [MVAa]. The use of a predefined structure for the system and contract specification should not be taken as an excuse to stifle dialog between the vision engineer and his customer. The specification for the system described in Section 1.4.4 evolved during several weeks of detailed discussion between these parties working closely together. Dialog is an essential feature of a successful design, since neither the vision engineer nor the client has access to all of the information necessary to complete the task, without the involvement of the other.

Another point to note is that an experienced vision engineer may not always give the answer that the customer would like to hear, but it may save a lot of problems later. One of the most valuable pieces of advice that a vision engineer can give is "*Don't use vision.*" Blunt words are sometimes necessary, to counter the prejudiced opinions of naive people with an over-optimistic view of what Machine Vision can accomplish, in practice.

Excessive optimism arises whenever people try to translate what they *think* they see into terms relating to a machine. When they first perceive the possibility of using MV technology, would-be customers often expect miracles because they think of themselves as being experts in vision. As a good working rule, it may be assumed that this is never the case, since introspection is a very unreliable method of determining how we perceive the world around us. Would-be customers are almost always totally unaware of both the limitations and potential of modern vision systems. Even if they have read about the subject, they are unaware of many of the subtle points of detail that can destroy a vision system's chances of ever working effectively and reliably. Very often, during the course of the initial discussions between the customer and vision engineer, various requirements not directly related to image processing become apparent and from then on form an integral part of the specification of a system with a far broader function than was at first imagined. While professional Computer Scientists are very much concerned with practical issues, such as data entry, the ergonomics of screen design, modes of system failure, etc., their academic colleagues tend to concentrate more on the details and performance of complex algorithms. Most of the work in Computer Vision takes place in academic departments; very few developments take place in a commercial environment. In other words, CV is principally an academic pursuit, even though it is motivated by serious applications in such areas as medicine and forensic science, where algorithm performance is of paramount importance. It is usual for a Computer Scientist to accept data for analysis without seriously questioning its quality. Indeed, CS training engenders this attitude! The starting point for most work in CV is a computer file. In such a situation, the term Computer Science is highly appropriate.

On the other hand, a Machine Vision engineer starts work, not with a computer file but a physical object, such as a machined metal component, a plastic molding, a sample of a web product, or a view of (part of) a complex machine. He will naturally be concerned with a very broad range of issues, and therefore needs to be a Systems Engineer, not a (Computer) scientist.

1.6 Structure, design, and use of machine vision systems

Figure 1.4 shows, in block-diagram form, the structure of a vision system intended for use on the factory floor. Figure 1.4(a) is perhaps the most important diagram in the whole book, because it emphasizes that the design of a Machine Vision system requires a multi-disciplinary approach, combining and harmonizing skills and knowledge of the following areas of engineering: mechanical, video, electrical, analog and digital electronic, illumination, optics, software (not visible in the diagram), ergonomics (human–computer interfacing), production, Quality Assurance and, of course, systems integration.

A vision system intended for the factory floor will be termed a *Target Vision System* (*TVS*). This name was chosen because a TVS is the end result of the design process that we will study in detail in later chapters. It cannot be claimed that Figure 1.4 represents all TVSs, since some systems may not require all of the modules depicted there. For example, when low-speed processing is acceptable, a standard computer may be able to perform those operations that might otherwise be expected to take place in the analog and digital preprocessing modules. In other instances, the intelligent processing module may not be needed. Figure 1.4 is comprehensive in its scope but requires some (minor) modification to enable it to model a system based on a line-scan camera or laser scanner. In broad terms, this diagram properly represents systems which employ microwave, infrared, ultraviolet, X-ray, and gamma-ray sensing. Furthermore, only minimal changes are required to encompass neutron-beam, electron-beam, and ultrasonic imaging systems. It also covers visible-light microscopy and systems which exploit both fluorescence and phosphorescence.

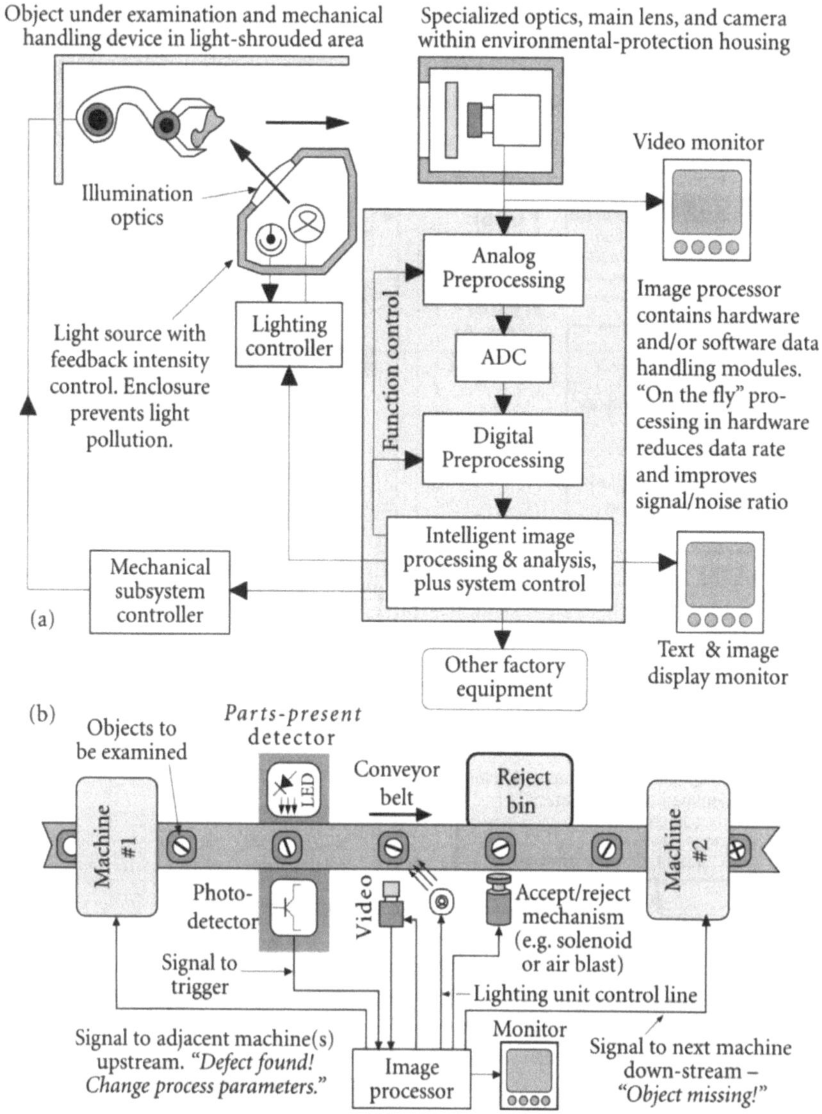

Fig. 1.4 (a) Archetypal Machine Vision system: general scheme, applicable to a wide range of industrial applications. (b) Typical on-line Automated Visual Inspection system. The image processor has a similar internal structure to that shown in the shaded box in (a).

1.6.1 Using an IVS for problem analysis

The design process for a TVS has been summarized in the form of a flow chart in Figure 1.5. It is not possible to design a vision algorithm as one would, for example, write a program to perform numerical analysis. The reason is exactly the same as that which prevents lay-people from becoming experts on vision by introspectively analyzing their own thought processes. A person writing a vision program needs to see the results of each step in the procedure. With very few exceptions, a person cannot simply look at a picture, mentally analyze it as a computer would, and then decide what algorithmic steps are needed to extract a given item of information from it. Interactive image processing was devised in the mid-1970s and was soon

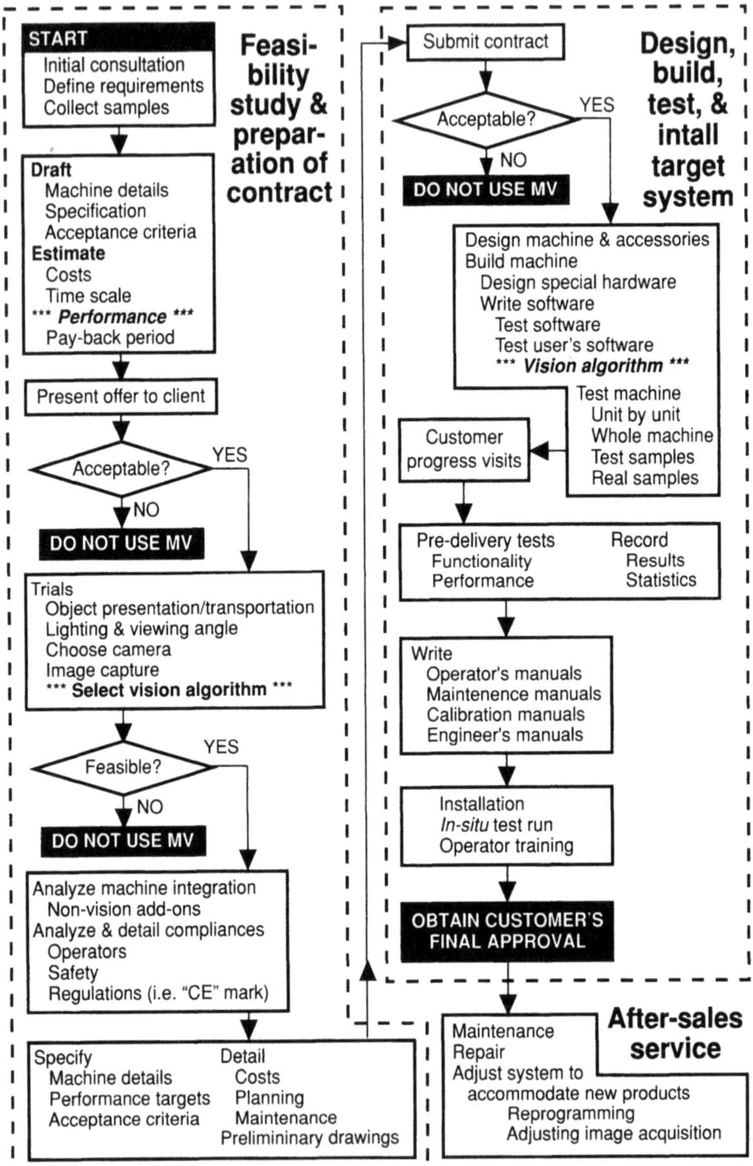

Fig. 1.5 Flow chart for the design of a Target Vision System.

found to provide a valuable contribution to the process of designing and/or choosing a vision algorithm. Table 1.3 provides a brief summary of the development of interactive vision systems. Their role in designing an industrial vision system is illustrated in the following anecdotes, which describe actual events.

Table 1.3 Interactive Vision Systems, historical perspective

Name	Organization	Year	Language	Remarks
SUSIE2	University of Southampton	1976	Fortran	The progenitor of all the other systems listed in this table
MAVIS	Transaction Security Ltd.	1979	Fortran	
Autoview Viking	British Robotic Systems Ltd.	1981	RTL	Sold in USA by 3M Vision Systems as System 77
VCS	Vision Dynamics Ltd.	1982	C	Runs under MS-DOS
VDL	3M Vision Systems	1986	C	Runs on 68000 processor
Supervision	Image Inspection Ltd.		C	Runs under MS-DOS
Intelligent Camera	Image Inspection Ltd.	1988		Self-contained hardware device based on bit-slice processor
Autoview Viking XA	British Robotic Systems Ltd.	1986	C and Prolog	Autoview Viking interfaced to Prolog
ProVision	Cardiff University /3M Co.	1989	Hardware and Prolog	Image processing is performed in hardware. Prolog is hosted on computer.
VSP	Cardiff University	1990	Hardware and Prolog	Image processing is performed in hardware. Prolog is hosted on computer.
Prolog+	Cardiff University	1991	Hardware and Prolog	Builds on top of VSP
PIP	Cardiff University	1995	C and Prolog	The most advanced system yet developed. Runs under Macintosh OS.
WIP	Cardiff University	1997	C and Prolog	Emulation of PIP running under Windows
CIP	Cardiff University	1997	Java	Remote operation via Internet

A certain company wished to build a system for checking the characteristics of an aerosol spray. Among other parameters, the spray-cone angle was to be measured. The particular spray material that interested the client would have been rather unpleasant to work with. Instead, he was prepared to study the application by using a less noxious material: furniture polish. A simple set-up was arranged, consisting of a light immediately above the spray cone. A black board, located behind the spray, was kept dark and acted as the visual background. Using this simple arrangement, several images were digitized and then processed, using the Autoview Viking interactive vision system [BAT82a]. The IVS was then used to explore various processing options. On this occasion, it took an experienced vision engineer only a few minutes to derive a suitable image-processing algorithm capable of estimating the spray-cone angle and deriving a measure of its symmetry [BAT83a]. On the basis of this quick test, a system was then designed and built for the application which involved measuring oil sprays.

In another project, a company wished to detect cracks in forged steel components for automobiles. In this instance, the client already employed a sophisticated method for highlighting cracks. This involves magnetizing the component, then immersing it in a suspension (within a kerosene base) of very fine ferro-magnetic particles that are also fluorescent. The components are then shaken to remove excess "ink" and are dried in hot air. They are then illuminated by ultraviolet light, but with no applied visible light. The cracks fluoresce (in the visible wave-band) because the fluorescent particles tend to adhere to the surface in regions where there is a high-remnant magnetic field. The cracks highlighted in this way were then viewed using a vidicon camera. The resulting digital images are of low quality, with a considerable amount of noise [BAT83b]. Again, the images were analyzed using the Autoview Viking prototyping vision system and an algorithm was discovered within a few hours. However, on this occasion an entirely new algorithm was developed: the PIP operator *crk/1*, which is explained in Chapter 2. It has subsequently been employed in many other industrial vision systems.

1.6.2 Structure of an IVS

As a starting point, we may take it that an IVS is organized according to the scheme outlined in Figure 1.2. The camera supplies a video signal to the computer via a plug-in card called a *frame store* or *digitizer*. (The former contains sufficient built-in memory to store at least one image. On the other hand, a digitizer is able to make use of the fact that the computer's RAM and input port are fast enough to enable the signal from the camera to be placed directly into the main memory.) Later, we will see that a IVS also requires quite sophisticated I/O and GUI interfacing. However, for the moment these can be ignored. The principal role of an IVS is as a prototyping tool kit for designing a TVS. In order to serve this function, the former need not be fast but it must be versatile. It must possess a rich set of commands and it should be possible to incorporate these into programs containing all of the usual features, e.g. looping, conditional statements, branching, simple I/O, etc.

1.6.3 Other uses for an IVS

A properly-designed interactive vision system operated by a skilled user has an important part to play in building customer confidence. It is also of great value in education and training. When attending a technical conference on Machine Vision, the author has often encountered a paper supposedly describing a "new" algorithm but which has in fact been known to him for some years. The reason is that he, like any other person who is experienced in using an IVS, has gained a considerable amount of unrecorded knowledge, which extends far beyond that published in technical articles and books. Another common experience at a conference is to discover a previously unknown algorithm which can be programmed very easily in the IVS control language. Three other situations also occur:

a. An algorithm described in the conference paper presents an entirely new idea, which leads to further enhancement of the IVS command repertoire.

b. The algorithm and/or application has previously been studied by the present author but was not published.
c. The idea lacks any real merit!

The reader is asked to overlook the seeming lack of modesty here and to understand that a powerful interactive vision system allows an experienced person to work very quickly and accomplish tasks that would otherwise take far longer.

1.7 Other design tools

Despite the rapid progress that has taken place in Machine Vision technology during the last decade, there remains an outstanding and pernicious problem: the very large amount of highly-skilled engineering effort needed to develop systems for each new application. It has been claimed many times, by many salesmen, authors, and conference speakers, that Machine Vision is a very flexible technology, ideal for sensing defects in a very wide range of industrial products and manufacturing processes. The fact that Machine Vision is potentially able to fulfil the needs of a very diverse range of industrial inspection, monitoring, and control tasks has already been established in numerous laboratory-based feasibility studies [BAT94b]. However, this oft-claimed versatility is merely an illusion if there are not enough people with the necessary systems-engineering skills to design and build machines that will continue to operate effectively and reliably for very long periods in the hostile working environment found on the factory floor. Bearing these ideas in mind, it becomes clear that a variety of computer-based design tools for Machine Vision would be valuable. If properly implemented, such tools would reduce the skill level for many of the sub-tasks in the design process. In this way, we hope one day to improve the overall effectiveness of the available pool of skilled labor.

The major tasks for which help is most needed are

- Overall planning of a vision system, given a range of general application constraints.
- Configuring the optical sub-system (i.e., placing the lighting, optical elements and camera).
- Calculating field-of-view, depth of focus, camera/image resolution, etc.
- Choosing an appropriate image sensor (camera).
- Designing the optical sub-system (i.e., choosing the lens/mirror configuration, coatings, mountings, detailed design of multi-element systems, etc.).
- Selecting and/or designing an appropriate image processing algorithm. (This relies very heavily on interactive vision systems, and is the main topic of this book.)
- Designing and programming a target system for use on the factory floor.

Of particular note are the following developments

- Lighting Advisor Expert System [PEN]
- Lighting Science Database [ITI]
- Lighting Advisor [BAT94a]

- Machine Vision Application Requirements Check List [MVAa]
- Machine Vision Lens Selector [MVAb]
- Opto*Sense [OPT]
- Tools for designing camera configurations [SNY92]

Work is continuing to combine the best features of such tools as these into a single integrated software package [BAT99].

1.7.1 Lighting Advisor

The *Lighting Advisor* is a catalog of almost 200 different lighting and viewing methods, and is in the form of a HyperCard stack. (HyperCard is a Macintosh-based program which closely resembles a browser in its mode of operation and actually predates the World Wide Web by about ten years. However, it is limited to operating on a stand-alone computer or via a Macintosh local-area network.) The Lighting Advisor displays both text and line drawings to explain each of the lighting/viewing methods. There is also provision to display sample images (Figure 1.6). It provides several different ways to navigate through the database. The emphasis in its design was to provide a useful tool to assist practicing vision system designers. Work is currently proceeding to integrate the Lighting Advisor with the PIP interactive vision system, described in Chapters 4 and 6. To this end, a new version of the Lighting Advisor has been written using the same version of Prolog that is used to implement PIP.

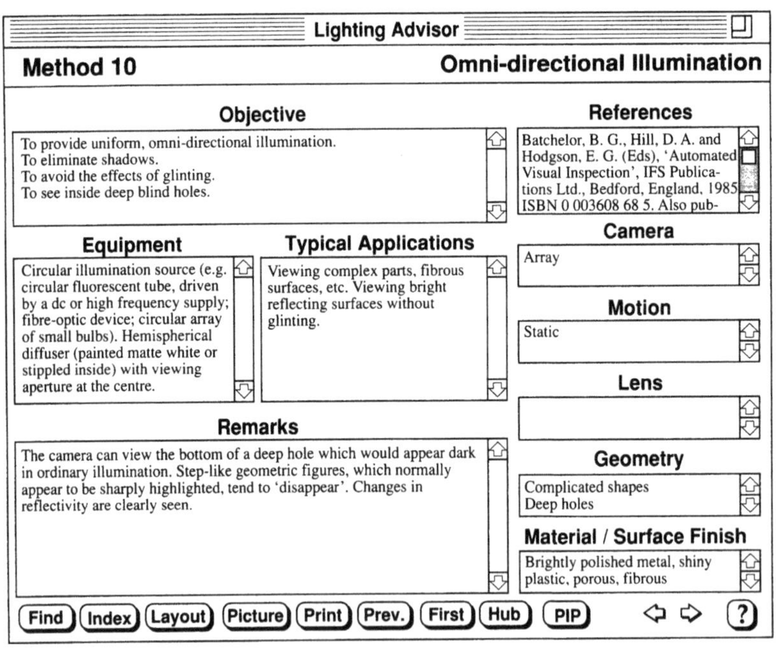

Fig. 1.6 (a) Lighting Advisor method description card.

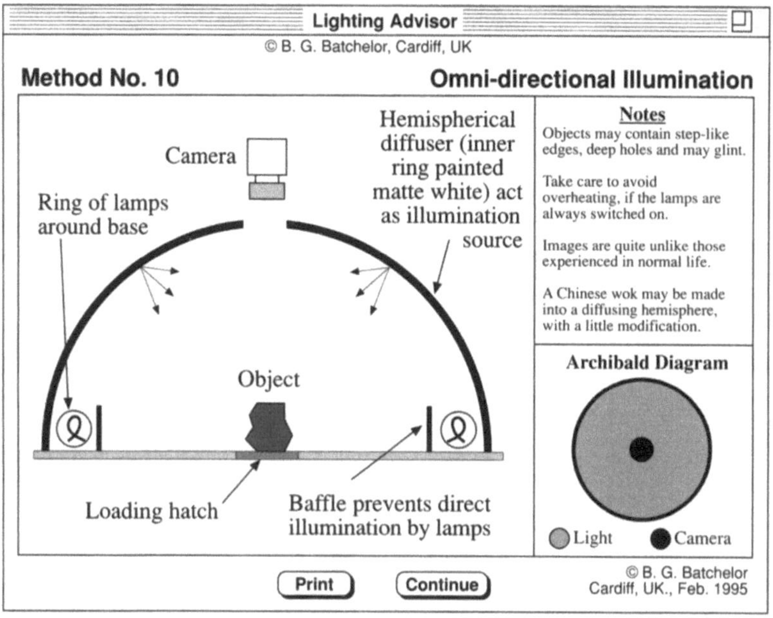

Fig. 1.6 (b) Lighting Advisor layout diagram.

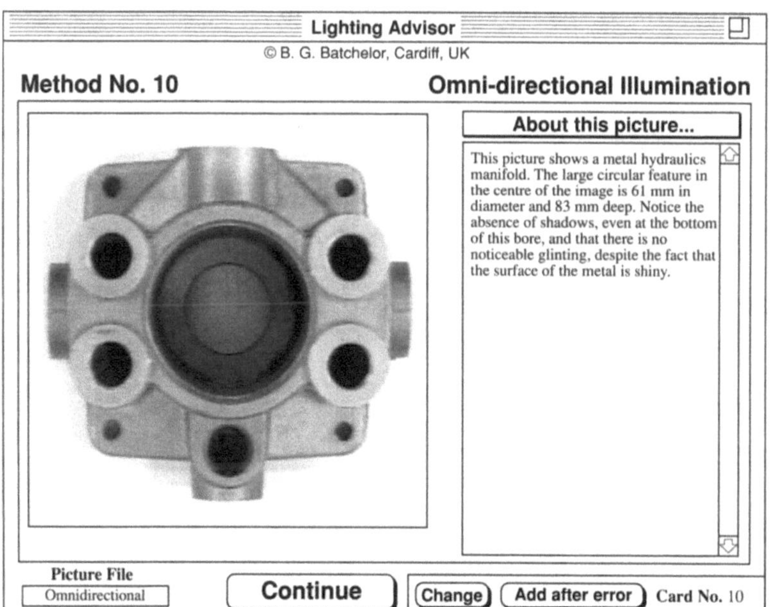

Fig. 1.6 (c) Lighting Advisor picture card, showing a typical image obtained with this method.

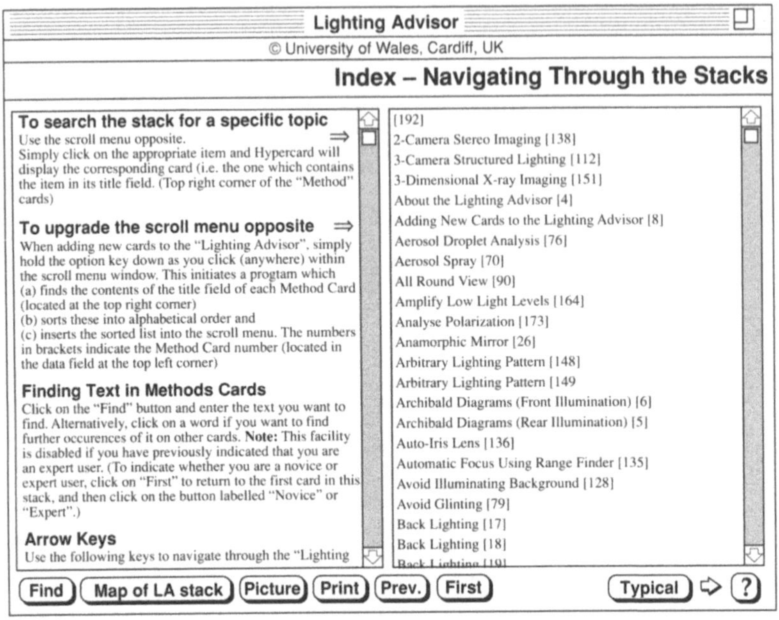

Fig. 1.6 (d) Lighting Advisor index card.

1.7.2 Prototyping kits for image acquisition

Another important development to assist in the design process are two prototyping systems for lighting and viewing. One of these, called Micro-ALIS, consists of an

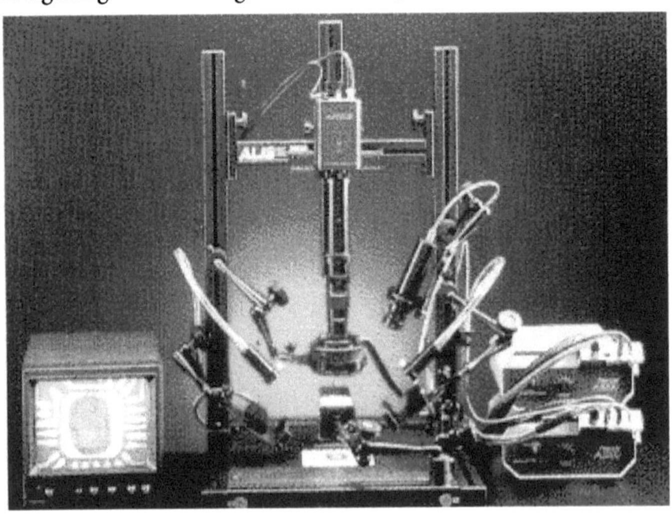

Fig. 1.7 Micro-ALIS (product of Dolan Jenner, Inc.).

optical bench and a range of high-precision optical-fibre illuminators, with rapid-release clamps to hold them *in situ* (Figure 1.7).

The other system, called ALIS 600, is a light-proof cabinet containing a range of different light sources, which are operated under computer control (Figure 1.8)[ALIS]. This closely resembles the *Flexible Inspection Cell* (*FIC*), which will be discussed in detail in Chapter 4. The latter consists of a metal frame holding a range of cameras and lamps. It also contains a flexible manipulator, consisting of an (X,Y,θ)-table and a pick-and-place arm. The Flexible Inspection Cell should properly be regarded as forming part of a larger integrated prototyping tool kit, called *ROPE* in Chapter 4. This also incorporates an interactive vision system (PIP, WIP, CIP, and JVision are all contenders) and other design tools such as the Lighting Advisor. Indeed, one of the main lessons of this book is that close integration among these and other design aids is needed if we are ever to realize the full potential of Machine Vision technology.

Fig. 1.8 ALIS 600 (product of Dolan Jenner, Inc.).

1.8 Outline of this book

There already exist several excellent books on Digital Image Processing [RUS95] that describe techniques for enhancing, transforming and analyzing pictures within a computer, although most concentrate solely on algorithmic techniques, totally ignoring implementation details. Some of the algorithms covered in this way are impractical for use within an industrial vision system because they are far too slow. A few books discuss software techniques for image processing, in some cases listing programs in such languages as C or Java [WHE00]. A few others describe hardware implementation. The approach that we will take is quite different. Greater emphasis will be placed in the following pages on implementation and broader issues relating to the bigger system of which the image processor is just a part. In Chapter 2, the basic algorithms of image representation and processing are first introduced. However, only those algorithms that are perceived as being practical for use in industrial vision systems are included. The approach is traditional and we define many of the basic commands implemented in the interactive and target vision systems described in later chapters. In Chapter 3 we demonstrate that there are many possible ways to implement certain image processing operations. In some cases, there may, in fact, be several different ways to perform a given calculation exactly. Some of these variations are made possible through the use of different image representations, while others rely on our representing the problem in a different mathematical form. Approximate methods are clearly important too, provided that bounds can be placed on the errors that they introduce. The reader is urged to question the fundamental nature of an application before deciding on the details of the algorithm. Reformulating the application task may, in some situations, yield a completely different way of approaching the problem, thereby making it more tractable. Problem formulation is an important part of designing an effective vision system and is therefore covered in considerable detail. Heuristics are introduced that can speed up certain algorithms. Heuristics that usually work (but cannot always be guaranteed to do so) can often be useful. It is emphasized that *statistical performance* is usually more important than being right every time, if the latter leads to an undue loss of speed. The nature of the mistakes that are made can be even more important than their total number. Hence, fast techniques which guarantee that any errors are "safe" are often more useful than slow optimal algorithms.

The features that distinguish an industrial vision system from a standard image processing package form the main topic discussed in Chapter 4. These include the broad aspects of the human–computer interface, controlling external devices (including lamps, video multiplexors, and various electromechanical devices), and operating a system remotely via the Internet. The role of an interactive vision system as a prototyping tool kit is discussed in detail. The need for effective user–machine interaction has a profound influence on the design of the TVS and particularly the IVS. As a result, it is recommended that a wide range of (non-vision) features be specially built into an IVS, to make it as user-friendly as possible. These include: scripting and programming, logical reasoning, the ability to build an expert system around the image processor, speech recognition, speech synthesis, natural language understanding, on-line help, and close interaction with other design aids. One particular implementation of an IVS, called *PIP*, is discussed in detail. This is based on Macintosh OS and uses the Artificial Intelligence language *Prolog* to provide the top-level control. (*PIP* derives its name from *Prolog Image Processing*.) Another system, based on the *Windows* operating system, called *WIP*, is also described, but in not quite so much detail. At the moment, *WIP* (*Windows Image Processing*) is not so well developed as PIP, which effectively defines the "end goal" for the designers of WIP. A third system, called *CIP* (*Cyber Image Processing*), is also discussed. This is written in Java and hence is able to operate via the Internet. CIP is also able to control the same external devices as PIP. However, CIP does not have such a well-developed command repertoire and currently lacks PIP's top-level control facility using Prolog.

Chapter 5 discusses various topics relating to hardware and fast software implementation of image processing operators. The reformulation of algorithms (Chapter 3) becomes critically important in designing fast implementations. Of particular note is the so-called *SKIPSM* method for implementing a large and important family of image processing functions, including many which fall under the umbrella of *mathematical morphology*. In the past, implementation of these algorithms has been hampered by the low speed of execution. However, the SKIPSM approach allows *very* large filters, which combine many picture points, to be executed in the same time as those that use just a few. This counter-intuitive situation represents a major advance for both hardware and software implementations. Surprisingly, the same implementation technique can be applied to a much larger group of operators, which one would not normally associate with the name mathematical morphology. The architecture and general principles of operation of pipeline, parallel, and concurrent processors are all discussed in this chapter.

Software implementations are discussed in Chapter 6 and 7. The former concentrates on PIP, with a smaller amount of detail being given for its close relative, WIP. The organization of CIP is described in Chapter 7. The latter is likely to be important in the long term, since Java is platform-independent and seems to be gaining ever more supporters. PIP, WIP, and CIP all follow the same general principles of operation. However, they differ in their implementation rather more than they do in their user interfaces. A completely different approach to the design of the user interface of an IVS is described in Chapter 8, where *JVision* is introduced. While the command repertoire is roughly similar to that provided in PIP, the user specifies program sequences using visual programming.

In Chapter 9, we discuss several applications to demonstrate how an IVS is used in practice.

Finally, Chapter 10 provides a summary of this book and an overview of the prospects for the use of intelligent machine vision.

2 Basic machine vision techniques

Image – An optical appearance or counterpart of an object, such as is produced by rays of light either reflected as from a mirror, refracted as through a lens, or falling on a surface after passing through a small aperture.
Oxford English Dictionary

The purpose of this chapter is to outline some of the basic techniques used in the development of industrial machine vision systems. These are discussed in sufficient detail to allow an understanding of the key ideas outlined elsewhere in this book. In the following discussion we will frequently indicate the equivalent PIP operators for the vision techniques described. PIP commands appear in square brackets. In certain cases, sequences of PIP commands are needed to perform an operation and these are similarly listed.

2.1 Representations of images

We will first consider the representation of *monochrome* (grey-scale) images. Let i and j denote two integers where $1 \leq i \leq m$ and $1 \leq j \leq n$. In addition, let $f(i,j)$ denote an integer function such that $0 \leq f(i,j) \leq W$. (W denotes the white level in a grey-scale image.) An array F will be called a *digital image*.

$$F = \begin{vmatrix} f(1,1) & f(1,2) & \dots & f(1,n) \\ f(2,1) & f(2,2) & \dots & f(2,n) \\ f(3,1) & f(3,2) & \dots & f(3,n) \\ \dots & \dots & \dots & \dots \\ f(m,1) & f(m,2) & \dots & f(m,n) \end{vmatrix}$$

An address (i,j) defines a position in F, called a *pixel*, *pel*, or *picture element*. The elements of F denote the intensities within a number of small rectangular regions within a real (i.e., optical) image (see Figure 2.1). Strictly speaking, $f(i,j)$ measures the intensity at a single point but if the corresponding rectangular region is small enough, the approximation will be accurate enough for most purposes. The array F contains a total of $m \times n$ elements and this product is called the *spatial resolution* of F. We may *arbitrarily* assign intensities according to the following scheme:

$f(i,j) = 0$	black
$0 \leq f(i,j) \leq 0.33W$	dark grey
$0.33W < f(i,j) \leq 0.67W$	mid-grey
$0.67W < f(i,j) < W$	light grey
$f(i,j) = W$	white

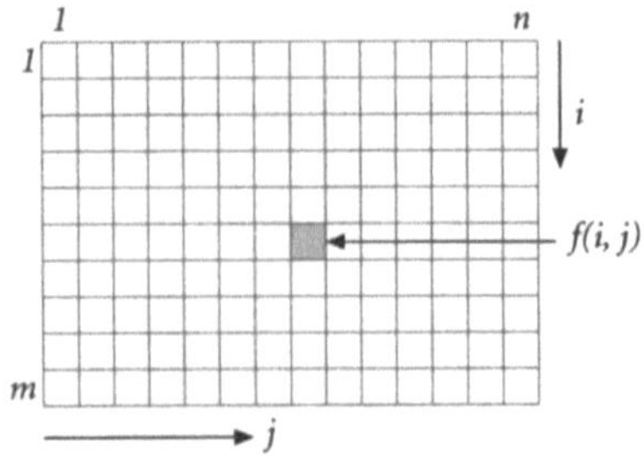

Fig. 2.1 A digital image consisting of an array of $m \times n$ pixels. The pixel in the ith row and the jth column has an intensity equal to $f(i,j)$.

Let us consider how much data is required to represent a grey-scale image in this form. Each pixel requires the storage of $\log_2(1+W)$ bits. This assumes that $(1+W)$ is an integer power of two. If it is not, then $\log_2(1+W)$ must be rounded up to the next integer. This rounding up can be represented using the ceiling function $<\ldots>$. Thus, a grey-scale image requires the storage of $<\log_2(1+W)>$ bits. Since there are $m \times n$ pixels, the total data storage for the entire digital image F is equal to $m \times n < \log_2(1+W)>$ bits. If $m = n \geq 128$, and $W \geq 64$, we can obtain a good image of a human face. Many of the industrial image-processing systems in use nowadays manipulate images in which $m = n = 512$ and $W = 255$. This leads to a storage requirement of 256 Kbytes/image. A binary image is one in which only two intensity levels, *black (0)* and *white (1)*, are permitted. This requires the storage of $m \times n$ bits/image.

An impression of color can be conveyed to the eye by superimposing *four* separate imprints. (Cyan, magenta, yellow, and black inks are often used in printing (Plates 19 and 20).) Ciné film operates in a similar way, except that when different colors of light, rather than ink, are added together, *three* components (red, green, and blue) suffice. Television operates in a similar way to film: the signal from a color television camera may be represented using three components: $R = \{r(i,j)\}$, $G = \{g(i,j)\}$, $B = \{b(i,j)\}$, where R, G and B are defined in a similar way to F. The vector $\{r(i,j), g(i,j), b(i,j)\}$ defines the intensity and color at the point (i,j) in the color image. (Color image analysis is discussed in more detail in Chapter 4.) Multispectral images can also be represented using several monochrome images. The total amount of data required to code a color image with r components is equal to $m \times n \times r<\log_2(1+W)>$ bits, where W is simply the maximum signal level on each of the channels.

In order to explain how moving scenes may be represented in digital form, ciné film and television will be referred to. A ciné film is, in effect, a time-sampled representation of the original moving scene. Each frame in the film is a standard monochrome or color image, and can be coded as such. Thus, a monochrome ciné film may be represented digitally as a sequence of two-dimensional arrays $[F_1, F_2, F_3, F_4, \ldots]$. Each F_i is an $m \times n$ array of integers, as we defined above when discussing the coding of grey-scale images. If the film is in color, then each of the F_i has three components. In the general case, when we have a spatial resolution of $m \times n$ pixels, each spectral channel permits $(1+W)$ intensity levels, there are r spectral channels, and p is the total number of "stills" in the image sequence, we require $m \times n \times p \times r \langle \log_2(1+W) \rangle$ bits/image sequence.

We have considered only those image representations which are relevant to the understanding of simple image processing and analysis functions. Many alternative methods of coding images are possible but these are not relevant to this discussion.

2.2 Elementary image-processing functions

The following notation will be used throughout this section, in which we will concentrate upon grey-scale images unless otherwise stated.

- i and j are row and column address variables and lie within the ranges $1 \leq i \leq m$ and $1 \leq j \leq n$ (Figure 2.1).
- $A = \{a(i,j)\}$, $B = \{b(i,j)\}$ and $C = \{c(i,j)\}$.
- W denotes the white level.
- $g(X)$ is a function of a single independent variable X.
- $h(X,Y)$ is a function of two independent variables, X and Y.
- The assignment operator "$\leftarrow$" will be used to define an operation that is performed upon one data element. The assignment operator "$\Leftarrow$" will be used to indicate that an operation is to be performed upon *all* pixels within an image.
- k, $k1$, $k2$, and $k3$ are constants.
- $N(i,j)$ is that set of pixels arranged around the pixel (i,j) in the following way:

$(i-1, j-1)$	$(i-1, j)$	$(i-1, j+1)$
$(i, j-1)$	(i, j)	$(i, j+1)$
$(i+1, j-1)$	$(i+1, j)$	$(i+1, j+1)$

Notice that $N(i,j)$ forms a 3×3 set of pixels and is referred to as the *3×3 neighborhood* of (i,j). In order to simplify some of the definitions, we will refer to the intensities of these pixels using the following notation:

A	B	C
D	E	F
G	H	I

Ambiguities over the dual use of A, B, and C should not be troublesome, because the context will make it clear which meaning is intended. The points $\{A, B, C, D, F, G, H, I\}$ are called the *8-neighbors* of E and are also said to be *8-connected* to E. The points $\{B, D, F, H\}$ are called the *4-neighbors* of E and are said to be *4-connected* to E.

2.2.1 Monadic pixel-by-pixel operators

These operators have a characteristic equation of the form

$c(i,j) \Leftarrow g(a(i,j))$ or $E \Leftarrow g(E)$

Such an operation is performed for all (i,j) in the range $[1,m],[1,n]$ (see Figure 2.2). Several examples will now be described.

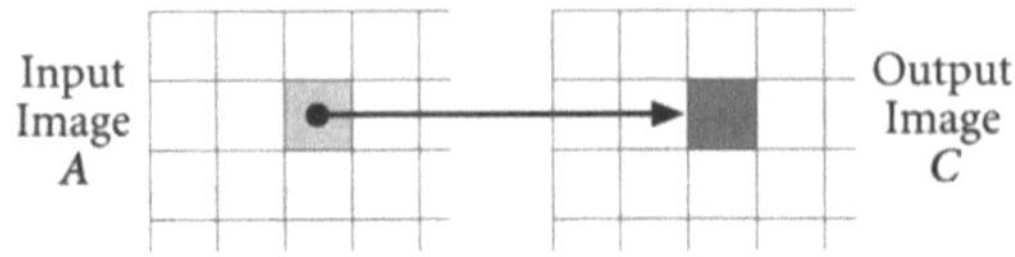

Fig. 2.2 Monadic pixel-by-pixel operator. The $(i,j)^{th}$ pixel in the input image has intensity $a(i,j)$. This value is used to calculate $c(i,j)$, the intensity of the corresponding pixel in the output image.

Intensity shift [acn]

$$c(i,j) \Leftarrow \begin{cases} 0 & a(i,j)+k<0 \\ a(i,j)+k & 0 \le a(i,j)+k \le W \\ W & W < a(i,j)+k \end{cases}$$

k is a constant, set by the system user. Notice that this definition was carefully designed to maintain c(i,j) within the same range as the input, *viz.* [0,W]. This is an example of a process referred to as *intensity normalization.* Normalization is important because it permits iterative processing by this and other operators in a machine having a limited precision for arithmetic (e.g., 8 bits). Normalization will be used frequently throughout this chapter.

Intensity multiply [mcn]

$$c(i,j) \Leftarrow \begin{cases} 0 & a(i,j)\times k<0 \\ a(i,j)+k & 0 \le a(i,j)\times k \le W \\ W & W < a(i,j)\times k \end{cases}$$

Logarithm [log]

$$c(i,j) \Leftarrow \begin{cases} 0 & a(i,j)=0 \\ W\dfrac{\log(a(i,j))}{\log(W)} & \text{otherwise} \end{cases}$$

This definition arbitrarily replaces the infinite value of log(0) by zero, and thereby avoids a difficult rescaling problem.

Antilogarithm (exponential) [exp] $c(i,j) \Leftarrow W\exp(a(i,j))/\exp(W)$

Negate [neg] $c(i,j) \Leftarrow W - a(i,j)$

Threshold [thr]

$$c(i,j) \Leftarrow \begin{cases} W & k1 \le a(i,j) < k2 \\ 0 & \text{otherwise} \end{cases}$$

This is an important function, which converts a grey-scale image to a binary format. Unfortunately, it is often difficult or even impossible to find satisfactory values for the parameters $k1$ and $k2$.

Highlight [hil]

$$c(i,j) \Leftarrow \begin{cases} k3 & k1 \le a(i,j) < k2 \\ a(i,j) & \text{otherwise} \end{cases}$$

Square [sqr] $c(i,j) \Leftarrow [a(i,j)]^2 / W$

2.2.2 Dyadic pixel-by-pixel operators

Dyadic operators have a characteristic equation of the form

$c(i,j) \Leftarrow h(a(i,j), b(i,j))$

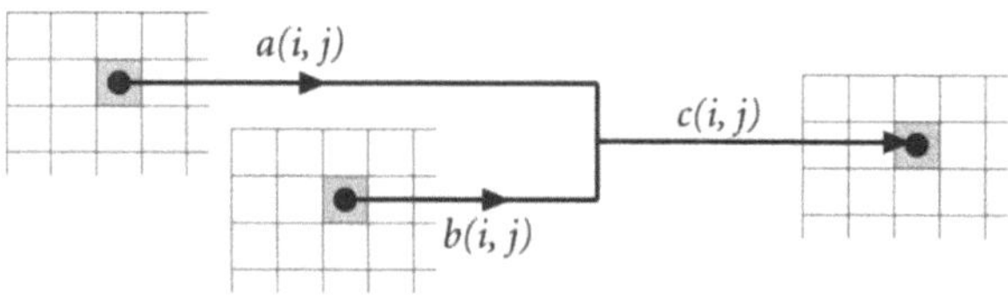

Fig. 2.3 Dyadic pixel-by-pixel operator. The intensities of the $(i,j)^{th}$ pixels in the two input images are combined to calculate the intensity $c(i,j)$ at the corresponding address in the output image.

There are two input images: $A = \{a(i,j)\}$ and $B = \{b(i,j)\}$ (Figure 2.3), while the output image is $C = \{c(i,j)\}$. It is important to realize that $c(i,j)$ depends upon only $a(i,j)$ and $b(i,j)$. Here are some examples of dyadic operators.

Add [add] $c(i,j) \Leftarrow [a(i,j) + b(i,j)] / 2$

Subtract [sub] $c(i,j) \Leftarrow [(a(i,j) - b(i,j)) + W] / 2$

Multiply [mul] $c(i,j) \Leftarrow [a(i,j) \times b(i,j)] / W$

Maximum [max] $c(i,j) \Leftarrow \text{MAX}\ [a(i,j), b(i,j)]$

When the maximum operator is applied to a pair of binary images, the *union* (OR function) of their white areas is computed. This function may also be used to *superimpose* white writing onto a grey-scale image.

Minimum [min] $c(i,j) \Leftarrow \text{MIN}\ [a(i,j), b(i,j)]$

When A and B are both binary, the *intersection* (AND function) of their white areas is calculated.

2.2.3 Local operators

Figure 2.4 illustrates the principle of the operation of local operators. Notice that the intensities of several pixels are combined in order to calculate the intensity of just one pixel. Among the simplest of the local operators are those using a set of nine pixels arranged in a 3×3 square. These 3×3 operators have a characteristic equation of the following form:

$$c(i,j) \Leftarrow g(a(i-1,j-1), a(i-1,j), a(i-1,j+1), a(i,j-1), a(i,j), a(i,j+1), a(i+1,j-1), a(i+1,j), a(i+1,j+1))$$

where $g(\bullet)$ is a function of nine variables. This local operator is said to have a 3×3 *processing window*. That is, it computes the value for one pixel on the basis of the intensities within a region containing 3×3 (= 9) pixels. Other local operators employ larger windows and we will discuss these briefly later. In the simplified notation which we introduced earlier, the above definition reduces to

$$E \Leftarrow g(A, B, C, D, E, F, G, H, I).$$

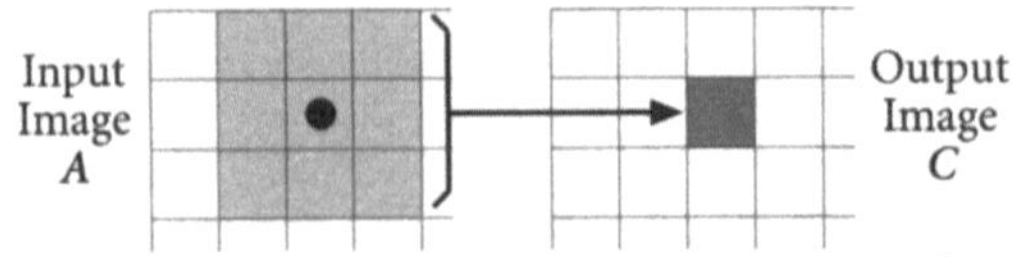

Fig. 2.4 Local operator. In this instance, the intensities of nine pixels arranged in a 3×3 window are combined using linear or nonlinear processes. Local operators may be defined which uses other, possibly larger, windows, which may or may not be square.

2.2.4 Linear local operators

An important subset of the local operators is that group which performs a linear weighted sum, and which are therefore known as *linear local operators.* For this group, the characteristic equation is:

$$E \Leftarrow k1(A \times W1 + B \times W2 + C \times W3 + D \times W4 + E \times W5 + F \times W6 + G \times W7 + H \times W8 + I \times W9) + k2$$

where $W1, W2, \ldots, W9$ are weights, which may be positive, negative, or zero. Values for the normalization constants $k1$ and $k2$ are given later. The matrix illustrated below is termed the *weight matrix,* and is important because it determines the properties of the linear local operator.

W1	W2	W3
W4	W5	W6
W7	W8	W9

The following rules summarize the behavior of this type of operator. They exclude the case where *all* the weights and normalization constants are zero, since this would result in a null image:

1. If all weights are either positive or zero, the operator will *blur* the input image. Blurring is referred to as *lowpass filtering.* Subtracting a blurred image from the original results in a highlighting of those points where the intensity is changing rapidly and is termed *highpass filtering.*
2. If $W1 = W2 = W3 = W7 = W8 = W9 = 0$ and $W4, W5, W6 > 0$, then the operator blurs along the rows of the image. Horizontal features, such as horizontal edges and streaks, are not affected.
3. If $W1 = W4 = W7 = W3 = W6 = W9 = 0$ and $W2, W5, W8 > 0$, then the operator blurs along the columns of the image. Vertical features are not affected.
4. If $W2 = W3 = W4 = W6 = W7 = W8 = 0$ and $W1, W5, W9 > 0$, then the operator blurs along the diagonal (top-left to bottom-right). There is no smearing along the orthogonal diagonal.

5. If the weight matrix can be reduced to a matrix product of the form *PQ*, where

$P =$

0	*0*	*0*
V4	*V5*	*V6*
0	*0*	*0*

and $Q =$

0	*V1*	*0*
0	*V2*	*0*
0	*V3*	*0*

the operator is said to be "separable." This is important because in such cases it is possible to apply two simpler operators in succession, with weight matrices P and Q, to obtain the same effect as that produced by the overall operator.

6. The successive application of certain linear local operators which use windows containing 3×3 pixels produces the same results as linear local operators with larger windows. For example, applying that operator which uses the weight shown below (left) twice in succession results in an image similar to that obtained from the 5×5 operator with the weight matrix shown below (center). (For simplicity, normalization has been ignored here.) Applying the same 3×3 operator three times in succession is equivalent to using the 7×7 operator shown below (right).

1	1	1
1	1	1
1	1	1

1	2	3	2	1
2	4	6	4	2
3	6	9	6	3
2	4	6	4	2
1	2	3	2	1

1	3	6	7	6	3	1
3	9	18	21	18	9	3
6	18	36	42	36	18	6
7	21	42	49	42	21	7
6	18	36	42	36	18	6
3	9	18	21	18	9	3
1	3	6	7	6	3	1

Notice that all of these operators are also separable. Hence it would be possible to replace the last-mentioned *7×7* operator with six simpler operators: *3×1, 3×1, 3×1, 1×3, 1×3, and 1×3*, applied in any order. Note, however, that it is not always possible to replace a given large-window operator with a succession of *3×3* operators. This becomes obvious when one considers, for example, that a *7×7* operator uses 49 weights and that three *3×3* operators provide only 27 degrees of freedom. Separation is often possible, however, when the larger operator has a weight matrix with some redundancy; for example, when it is symmetrical.

7. In order to perform normalization, the following values are used for *k1* and *k2*.

$$k1 \leftarrow 1 \Big/ \left(\sum_{p,q} |W_{p,q}| \right) \qquad k2 \leftarrow \left| 1 - \left[\left(\sum_{p,q} W_{p,q} \right) \Big/ \left(\sum_{p,q} |W_{p,q}| \right) \right] \right|$$

8. A filter using the following weight matrix performs a *local averaging function* over an *11×11* window *[raf(11,11)]*.

1	1	1	1	1	1	1	1	1	1	1
1	1	1	1	1	1	1	1	1	1	1
1	1	1	1	1	1	1	1	1	1	1
1	1	1	1	1	1	1	1	1	1	1
1	1	1	1	1	1	1	1	1	1	1
1	1	1	1	1	1	1	1	1	1	1
1	1	1	1	1	1	1	1	1	1	1
1	1	1	1	1	1	1	1	1	1	1
1	1	1	1	1	1	1	1	1	1	1
1	1	1	1	1	1	1	1	1	1	1

1	1	1	1	1	1	1	1	1	1	1

This produces quite a severe 2-directional blurring effect. Subtracting the effects of a blurring operation from the original image generates a picture in which spots, streaks, and intensity steps are all emphasized. On the other hand, large areas of constant or slowly-changing intensity become uniformly grey. This process is called *highpass filtering*, and produces an effect similar to unsharp masking, which is familiar to photographers.

2.2.5 Nonlinear local operators

Largest intensity neighborhood function [lnb]

$E \Leftarrow \mathrm{MAX}(A, B, C, D, E, F, G, H, I)$

This operator has the effect of spreading bright regions and contracting dark ones.

Edge detector [command sequence: ***lnb, sub***]

$E \Leftarrow \mathrm{MAX}(A, B, C, D, E, F, G, H, I) - E$

This operator is able to highlight edges (i.e., points where the intensity is changing rapidly).

Median filter [mdf(5)]

$E \Leftarrow \mathrm{FIFTH\_LARGEST}\ (A,B,C,D,E,F,G,H,I)$

This filter is particularly useful for reducing the level of noise in an image. Noise is generated from a range of sources, such as video cameras and x-ray detectors, and can be a nuisance if it is not eliminated by hardware or software filtering.

Crack detector. This operator is equivalent to applying the PIP sequence *[lnb, lnb, neg, lnb, lnb, neg]* and then subtracting the result from the original image. This detector is able to detect thin dark streaks and small dark spots in a grey-scale image; it ignores other features, such as bright spots and streaks, edges (intensity steps), and broad dark streaks. (This is an example of an operator that can be described far better using computer notation rather than mathematical notation.)

Roberts edge detector [red]. The Roberts gradient is calculated using a 2×2 mask. This will determine the edge gradient in two diagonal directions.

$$E \Leftarrow \sqrt{(A-E)^2 + (B-D)^2}$$

The following approximation to the Roberts gradient magnitude is called the *Modified Roberts operator.* This is simpler and faster to implement and it more precisely defines the PIP operator *red.* It is defined as

$$E \Leftarrow \{|A-E| + |B-D|\}/2$$

Sobel edge detector [sed]. This popular operator highlights the edges in an image. Points where the intensity gradient is high are indicated by bright pixels in the output image. The Sobel edge detector uses a 3×3 mask to determine the edge gradient.

$$E \Leftarrow \sqrt{[(A+2B+C)-(G+2H+I)]^2+[(A+2D+G)-(C+2F+I)]^2}$$

The following approximation is simpler to implement in software and hardware and more precisely defines the PIP operator *sed*:

$$E \Leftarrow \{|(A+2B+C)-(G+2H+I)|+|(A+2D+G)-(C+2F+I)|\}/6$$

See Figure 2.5 for a comparison of the Roberts and Sobel edge detector operators when applied to a sample monochrome image. Note that, while the Roberts operator produces thinner edges, these edges tend to break up in regions of high curvature. The primary disadvantage of the Roberts operator is its high sensitivity to noise, since fewer pixels are used in the calculation of the edge gradient. There is also a slight shift in the image when the Roberts edge detector is used. The Sobel edge detector does not produce such a shift.

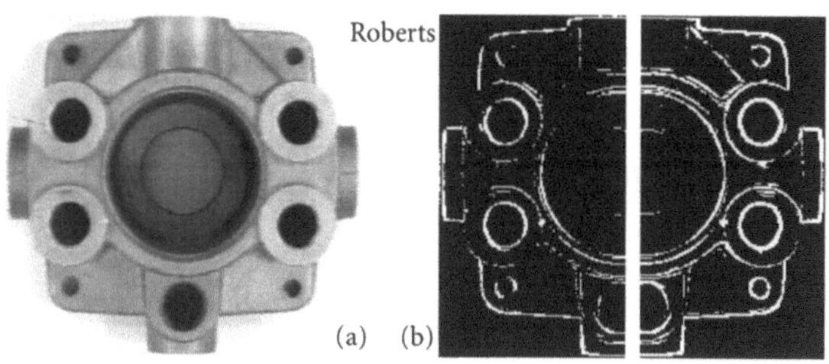

Fig. 2.5 Edge detection. (a) Original image. (b) To facilitate comparison, the left side of Roberts gradient image has been juxtaposed with the flipped left side of the Sobel gradient image (both after thresholding).

Prewitt edge detector. The Prewitt edge detector is similar to the Sobel operator, but is more sensitive to noise because it does not possess the same inherent smoothing. This operator uses the two 3×3 weighting matrices shown below to determine the edge gradient.

P_1:

−1	−1	−1
0	0	0
1	1	1

P_2:

−1	0	1
−1	0	1
−1	0	1

The Prewitt gradient magnitude is defined as $E \Leftarrow \sqrt{(P_1^2 + P_2^2)}$, where P_1 and P_2 are the values calculated from each mask, respectively.

Frei and Chen edge detector. This operator uses the two 3×3 masks shown below to determine the edge gradient.

F_1:

−1	$-\sqrt{2}$	−1
0	0	0
1	$\sqrt{2}$	1

F_2:

−1	0	1
$-\sqrt{2}$	0	$\sqrt{2}$
−1	0	1

The Frei and Chen gradient magnitude is defined as $E \Leftarrow \sqrt{(F_1^2 + F_2^2)}$, where F_1 and F_2 are the values calculated from each mask, respectively.

Rank filters [mdf, rid]. The generalized 3×3 rank filter is

$$c(i,j) \Leftarrow k1\ (A' \times W1 + B' \times W2 + C' \times W3 + D' \times W4 + E' \times W5 + F' \times W6 + G' \times W7 + H' \times W8 + I' \times W9) + k2$$

where
$A' = \text{LARGEST}\ (A, B, C, D, E, F, G, H, I)$
$B' = \text{SECOND\_LARGEST}\ (A, B, C, D, E, F, G, H, I)$
$C' = \text{THIRD\_LARGEST}\ (A, B, C, D, E, F, G, H, I)$
...
$I' = \text{NINTH\_LARGEST}\ (A, B, C, D, E, F, G, H, I)$

and $k1$ and $k2$ are the normalization constants defined previously. With the appropriate choice of weights $(W1, W2,..., W9)$, the rank filter can be used for a range of operations including edge detection, noise reduction, edge sharping, and image enhancement.

Direction codes [dbn] This function can be used to detect the *direction* of the intensity gradient. A direction code function DIR_CODE is defined thus:

$$\text{DIR\_CODE}(A,B,C,D,F,G,H,I) \Leftarrow \begin{cases} 1 & \text{if } A \geq \text{MAX}(B,C,D,F,G,H,I) \\ 2 & \text{if } B \geq \text{MAX}(A,C,D,F,G,H,I) \\ 3 & \text{if } C \geq \text{MAX}(A,B,D,F,G,H,I) \\ 4 & \text{if } D \geq \text{MAX}(A,B,C,F,G,H,I) \\ 5 & \text{if } F \geq \text{MAX}(A,B,C,D,G,H,I) \\ 6 & \text{if } G \geq \text{MAX}(A,B,C,D,F,H,I) \\ 7 & \text{if } H \geq \text{MAX}(A,B,C,D,F,G,I) \\ 8 & \text{if } I \geq \text{MAX}(A,B,C,D,F,G,H) \end{cases}$$

Using this definition, the operator *dbn* may be defined as

$$E \Leftarrow \text{DIR\_CODE}(A,B,C,D,F,G,H,I)$$

2.2.6 N-tuple operators

The N-tuple operators are closely related to the local operators and have a large number of linear and nonlinear variations. N-tuple operators may be regarded as generalized versions of local operators. In order to understand the N-tuple operators, let us first consider a *linear* local operator which uses a large processing window, say $r \times s$ pixels, with most of its weights equal to zero. Only N of the weights are nonzero, where $N << r \times s$. This is an N-tuple filter (see Figure 2.6). These filters are usually designed to detect specific patterns. In this role, they are able to locate a simple feature, such as a corner, an annulus, the numeral 2 in any position, etc. However, they are sensitive to changes of orientation and scale. The N-tuple can be regarded as a sloppy template which is convolved with the input image.

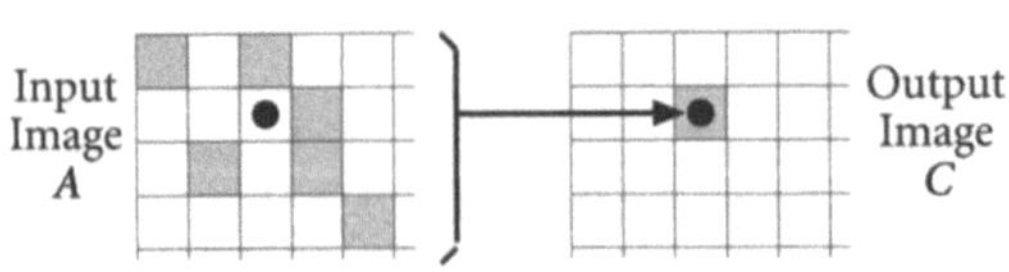

Fig. 2.6 An N-tuple filter operates much like a local operator. The only difference is that the pixels whose intensities are combined do not form a compact set. An N-tuple filter can be regarded as being equivalent to a local operator with a large window in which many of the weights are zero.

Nonlinear N-tuple operators may be defined in a fairly obvious way. For example, we may define operators which compute the average, maximum, minimum, or median values of the intensities of the N pixels covered by the N-tuple. An important class of such functions is the *morphological operators* (see Sections 2.4 and 2.5). Figure 2.7 illustrates the recognition of the numeral **2** using an N-tuple. Notice how the goodness of fit varies with the shift, tilt, size, and font. Another character (**Z**, for example) may give a score that is close to that obtained from a **2**, thus making these two characters difficult to distinguish reliably.

Fig. 2.7 Recognizing a numeral **2** using an N-tuple.

2.2.7 Edge effects

All local operators and N-tuple filters are susceptible to producing peculiar effects around the edges of an image. The reason is simply that, in order to calculate the intensity of a point near the edge of an image, we require information about pixels outside the image, which of course are simply not present. In order to make some attempt at calculating values for the edge pixels, it is necessary to make some assumptions, for example that all points outside the image are black, or have the same values as the border pixels. This strategy, or whatever one we adopt, is perfectly arbitrary and there will be occasions when the edge effects are so pronounced that there is nothing that we can do but to remove them by masking [*edg*]. Edge effects are important because they require us to make special provisions for them when we try to patch several low-resolution images together.

2.2.8 Intensity histogram *[hpi, hgi, hge, hgc]*

The intensity histogram is defined in the following way:

(a) Let

$$s(p,i,j) \leftarrow \begin{bmatrix} 1 & a(i,j) = p \\ 0 & \text{otherwise} \end{bmatrix}$$

(b) Let $h(p)$ be defined thus:

$$h(p) \leftarrow \sum_{i,j} s(p,i,j)$$

It is not, in fact, necessary to store each of the $s(p,i,j)$, since the calculation of the histogram can be performed as a serial process in which the estimate of $h(p)$ is updated iteratively as we scan through the input image. This *cumulative histogram*, $H(p)$, can be calculated using the following recursive relation:

$$H(p) = H(p-1) + h(p), \text{ where } H(0) = h(0).$$

Both the cumulative and the standard histograms have a great many uses, as will become apparent later. It is possible to calculate various intensity levels which indicate the occupancy of the intensity range [*pct*]. For example, it is a simple matter to determine that intensity level $p(k)$ which, when used as a threshold parameter, ensures that a proportion k of the output image is black. $p(k)$ can be calculate using the fact that $H(p(k)) = m \times n \times k$. The *mean intensity* [*avg*] is computed thus:

$$avg = \sum_p (p \times h(p))/(m \times n)$$

while the *maximum intensity* [*gli*] is equal to maximum$(p \mid h(p) > 0)$ and the *minimum intensity* is equal to minimum$(p \mid h(p) > 0)$.

One of the principal uses of the histogram is in the selection of threshold parameters. It is useful to plot $h(p)$ as a function of p. It is often found from this graph that a suitable position for the threshold can be related directly to the position of the "foot of the hill" or to a "valley" in the histogram.

An important operator for image enhancement is given by the transformation

$$c(i,j) \Leftarrow [\, W \times H(a(i,j))] \,/\, (m \times n)$$

This has the interesting property that the histogram of the output image $\{c(i,j)\}$ is flat, giving rise to the name *histogram equalization [heq]*. Notice that histogram equalization is a *data-dependent* monadic pixel-by-pixel operator.

An operation known as *local-area histogram equalization* relies upon the application of histogram equalization within a small window. The number of pixels in a small window that are darker than the central pixel is counted. This number defines the intensity at the equivalent point in the output image. This is a powerful filtering technique, which is particularly useful in texture analysis (see Section 2.7).

2.3 Binary images

For the purposes of this description of binary image processing, it will be convenient to assume that $a(i,j)$ and $b(i,j)$ can assume only two values: 0 (black) and 1(white). The operator "+" denotes the Boolean **OR** operation, "$\times$" represents the Boolian **AND** operation, "$\otimes$" denotes the Boolean **Exclusive OR** operation, and $\#(i,j)$ denotes the number of white points addressed by N(i,j), including (i,j) itself.

Inverse [not] $\quad c(i,j) \Leftarrow \text{NOT}(a(i,j))$

AND white regions [and, min] $\quad c(i,j) \Leftarrow a(i,j) \times b(i,j)$

OR [ior, max] $\quad c(i,j) \Leftarrow a(i,j) + b(i,j)$

Exclusive OR [xor] $\quad c(i,j) \Leftarrow a(i,j) \otimes b(i,j)$

This finds differences between white regions.

Expand white areas [exw]

$$c(i,j) \Leftarrow a(i-1,j-1) + a(i-1,j) + a(i-1,j+1) + a(i,j-1) + a(i,j) + a(i,j+1) + a(i+1,j-1) + a(i+1,j) + a(i+1,j+1)$$

Notice that this is closely related to the local operator *lnb* defined earlier. This equation may be expressed in the simplified notation: $E \Leftarrow A + B + C + D + E + F + G + H + I$.

Shrink white areas [skw]

$$c(i,j) \Leftarrow a(i-1,j-1) \times a(i-1,j) \times a(i-1,j+1) \times a(i,j-1) \times a(i,j) \times a(i,j+1) \times a(i+1,j-1) \times a(i+1,j) \times a(i+1,j+1)$$

or more simply $c(i,j) \Leftarrow A \times B \times C \times D \times E \times F \times G \times H \times I$

Edge detector [bed] $c(i,j) \Leftarrow E \times \mathrm{NOT}(A \times B \times C \times D \times F \times G \times H \times I)$

Remove isolated white points [wrm]

$$c(i,j) \Leftarrow \begin{cases} 1 & a(i,j) \cdot (\#(i,j) > 1) \\ 0 & \text{otherwise} \end{cases}$$

Count white neighbors [cnw] $c(i,j) \Leftarrow \#(a(i,j) = 1)$
where $\#(Z)$ is the number of times Z occurs. This results in a grey-scale image.

Connectivity detector [cny]. Consider the following pattern:

1	0	1
1	X	1
1	0	1

If X = 1, then all of the 1s are 8-connected to each other. Alternatively, if X = 0, then they are not connected. In this sense, the point marked X is *critical for connectivity.* This is also the case in the following examples:

1	0	0
0	X	1
0	0	0

1	1	0
0	X	0
0	0	1

0	0	1
1	X	0
1	0	1

However, those points marked X below are not critical for connectivity, since setting $X = 0$ rather than 1 has no effect on the connectivity of the remaining 1s.

1	1	1
1	X	1
0	0	1

0	1	1
1	X	0
1	1	1

0	1	1
1	X	0
0	1	1

A connectivity detector shades the output image with 1s to indicate the position of those points which are critical for connectivity and which were white in the input image. Black points and those which are not critical for connectivity are mapped to black in the output image.

Euler number [eul]. The Euler number is defined as the number of connected components (blobs) minus the number of holes in a binary image. The Euler number represents a simple method of counting blobs in a binary image, provided they have no holes in them. Alternatively, it can be used to count holes in a given object, providing they have no "islands" in them. The reason why this approach is used to count blobs, despite the fact that it may seem a little awkward to use, is that the Euler number is very easy and fast to calculate. It is also a useful means of classifying shapes in an image. The Euler number can be computed by using three local operators. Let us define three numbers *N1*, *N2*, and *N3*, where $N\alpha$, $\alpha = 1, 2, or\ 3$, indicates the number of times that one of the patterns in the pattern set α occurs in the input image.

Pattern Set 1 *(N1)*

0	0
0	1

0	0
1	0

0	1
0	0

1	0
0	0

Pattern Set 2 *(N2)*

0	1
1	0

1	0
0	1

Pattern Set 3 *(N3)*

1	1
1	0

1	1
0	1

1	0
1	1

0	1
1	1

The *8-connected* Euler number, where holes and blobs are defined in terms of 8-connected figures, is defined as $(N1 - 2 \times N2 - N3)/4$. It is possible to calculate the *4-connected* Euler number using a slightly different formula, but this parameter can give results which seem to be anomalous when we compare them with the observed number of holes and blobs.

Filling holes [blb]. Consider a white blob-like figure containing one or more holes (lakes), against a black background. The application of the hole-filling operator will cause all of the holes to be *filled in* by setting all pixels in the holes to white. This operator will not alter the outer edge of the figure.

Region labelling [ndo]. Consider an image containing a number of separate blob-like figures. A region-labelling operator will shade the output image so that each blob is given a separate intensity value. We could shade the blobs according to the order in which they are found during a conventional raster scan of the input image. Alternatively, the blobs could be shaded according to their areas, the biggest blobs becoming the brightest. This is a very useful operator, since it allows objects to be separated and analyzed individually (Figure 2.8). Small blobs can also be eliminated from an image using this operator. Region labelling can also be used to count

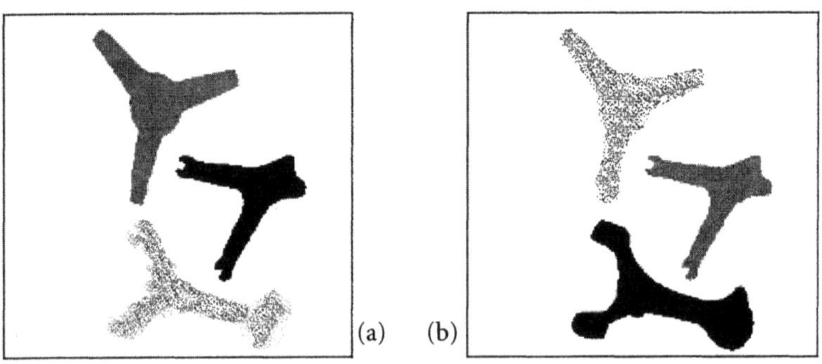

Fig. 2.8 Shading blobs in a binary image (a) according to their areas and (b) according to the order in which they are found during a raster scan (left to right, top to bottom).

the number of distinct binary blobs in an image. Unlike the Euler number, blob counting based on region labelling is not affected by the presence of holes.

Other methods of detecting/removing small spots. A binary image can be thought of as a grey-scale image in which only two grey levels, 0 and *W*, are allowed. The result of the application of a conventional lowpass (blurring) filter to such an image is a grey-scale image in which there is a larger number of possible intensity values. Pixels which were well inside large white areas in the input image are mapped to very bright pixels in the output image. Pixels which were well inside black areas are mapped to very dark pixels in the output image. However, pixels which were inside small white spots in the input image are mapped to mid-grey intensity levels (Figure 2.9). Pixels on the edge of large white areas are also mapped to mid-grey intensity levels. However, if there is a cluster of small spots, which are closely spaced together, some of them may also disappear.

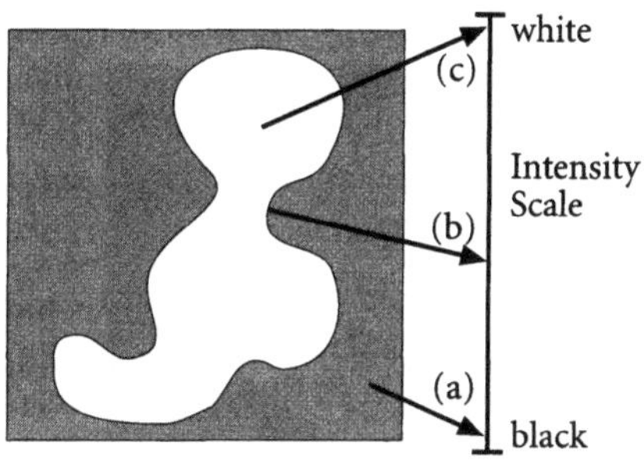

Fig. 2.9 Using a grey-scale blurring filter to remove noise from a binary image. (a) Background points are mapped to black. (b) Edge points are mapped to the central part of the intensity range. Thresholding at mid-grey has the effect of smoothing the edge of large blobs. (c) Central areas of large white blobs are mapped to white.

Based on these observations, the following procedure has been developed. It has been found to be effective in distinguishing between small spots and, at the same time, achieving a certain amount of edge smoothing of the large bright blobs which remain:

```
raf(11,11),        % Lowpass filter using a 11×11 local operator
thr(128),          % Threshold at mid-grey
```

This technique is generally easier and faster to implement than the blob-shading technique described previously. Although it may not achieve precisely the desired result, it can be performed at high speed.

An N-tuple filter having the weight matrix illustrated below can be combined with simple thresholding to distinguish between large and small spots. Assume that there are several small white spots within the input image and that they are spaced well apart. All pixels within a spot which can be contained within a circle of radius three pixels will be mapped to white by this particular filter. Pixels within a larger spot will become darker than this. The image is then thresholded at white to separate the large and small spots.

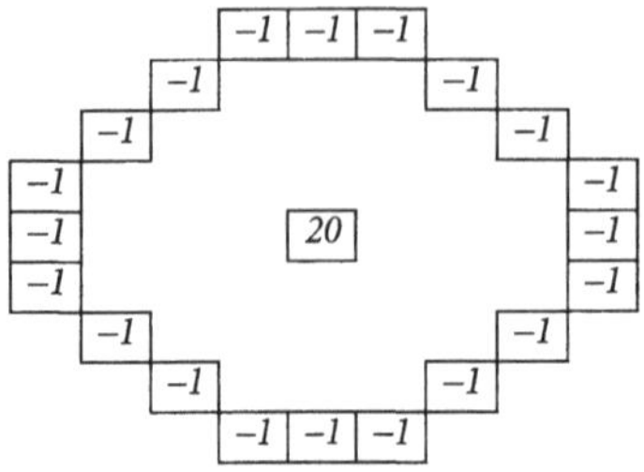

Grass-fire transform and skeleton [gfa, mdl, mid]. Consider a binary image containing a single white blob (Figure 2.10). Imagine that a fire is lit at all points around the blob's outer edge and the edges of any holes it may contain. The fire will burn inwards until at some instant, advancing fire lines meet. When this occurs, the fire becomes extinguished locally. An output image is generated and is shaded in proportion to the time it takes for the fire to reach each point. Background pixels are mapped to black.

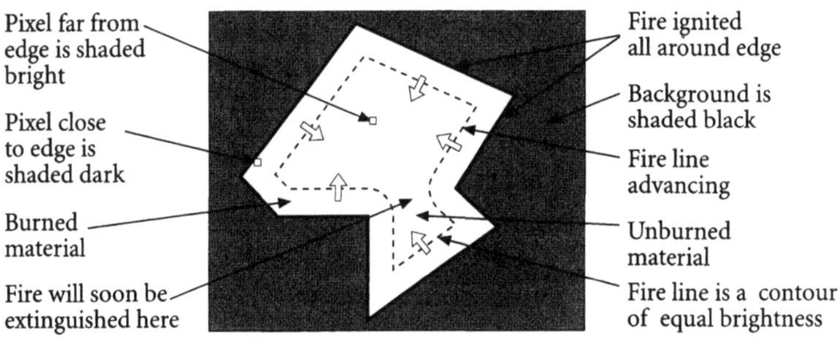

Fig. 2.10 Grass-fire transform.

The importance of this transform, referred to as the *grass-fire* transform, lies in the fact that it indicates distances to the nearest edge point in the image [BOR86]. It is therefore possible to distinguish thin and fat limbs of a white blob. Those points at which the fire lines meet are known as *quench* points. The set of quench points form a "matchstick" figure, usually referred to as a *skeleton* or *medial axis transform.* These figures can also be generated in a number of different ways [GON87] (Figure 2.11).

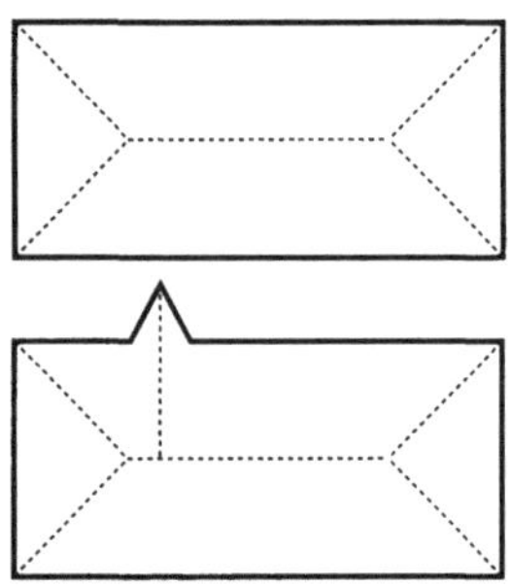

Fig. 2.11 Application of the medial axis transform.

One such approach is described as *onion peeling.* Consider a single white blob and a "bug" which walks around the blob's outer edge, removing one pixel at a time. No edge pixel is removed if by doing so we would break the blob into two disconnected parts. In addition, no white pixel is removed if there is only one white pixel among its 8-neighbors. This simple procedure leads to an undesirable effect in those instances when the input blob has holes in it; the skeleton which it produces has small loops in it which fit around the holes like a tightened noose. More sophisticated algorithms have been devised which avoid this problem.

Edge smoothing and corner detection. Consider three points *B1*, *B2*, and *B3* which are placed close together on the edge of a single blob in a binary image (see Figure 2.12). The perimeter distance between *B1* and *B2* is equal to that between *B2* and *B3*. The point *P* is at the center of the line joining *B1* and *B3*. As the three points now move around the edge of the blob, keeping the spacing between them constant, the locus of *P* traces a smoother path than that followed by *B2*. This forms the basis of a simple edge-smoothing procedure.

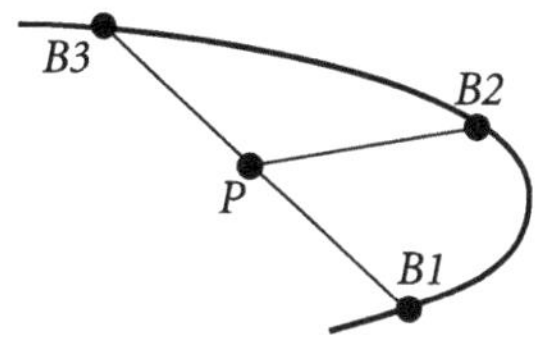

Fig. 2.12 Edge smoothing and corner detection.

A related algorithm, used for corner detection, shades the edge according to the distance between *P* and *B2*. This results in an image in which the corners are highlighted, while the smoother parts of the image are much darker.

Many other methods of edge smoothing are possible. For example, white pixels which have fewer than, say, three white 8-neighbors may be mapped to black. This has the effect of eliminating "hair" around the edge of a blob-like figure. One of the techniques described previously for eliminating small spots offers another possibility. A third option is to use the processing sequence: *[exw, skw, skw, exw]*, where *exw* represents expand white areas and *skw* denotes shrink white areas.

Convex hull [chu]. Consider a single blob in a binary image. The convex hull is that area enclosed within the smallest convex polygon which will enclose the shape (Figure 2.13). This can also be described as the region enclosed within an elastic string stretched around the blob. The area enclosed by the convex hull but not within the original blob is called the *convex deficiency*, which may consist of a number of disconnected parts and includes any holes and indentations. If we regard the blob as being like an *island*, we can understand the logic of referring to the former as *lakes* and the latter as *bays*.

Fig. 2.13 The convex hull of a "club" shape. The lightly-shaded region indicates the convex deficiency.

2.3.1 Measurements on binary images

To simplify the following explanation, we will confine ourselves to the analysis of a binary image containing a single blob. The area of the blob can be measured by the total number of object (white) pixels in the image. However, we must first define two different types of edge points, in order to measure an object's perimeter.

The *4-adjacency* convention (Figure 2.14) allows only the four main compass points to be used as direction indicators, while *8-adjacency* uses all eight possible directions. If the 4-adjacency convention is applied to the image segment given in Figure 2.14(c), then none of the four shaded segments (two horizontal and two vertical) will appear as touching, i.e., they are not connected. Using the 8-adjacency convention, these four segments are now connected, but we have the ambiguity that the inside of the shape is connected to the outside. Neither convention is satisfactory, but because 8-adjacency allows diagonally-adjacent pixels to be regarded as connected, it leads to a more faithful perimeter measurement.

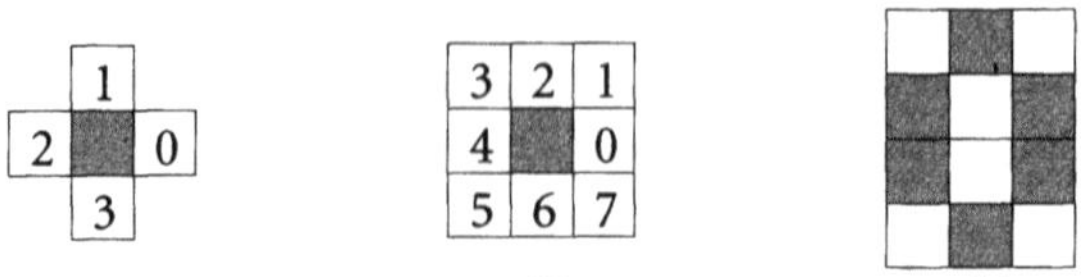

(a) (b) (c)

Fig. 2.14 Chain code. (a) 4-adjacency coding convention. (b) 8-adjacency coding convention. (c) Image segment.

Assuming that the 8-adjacency convention is used, we can generated a coded description of the blob's edge. This is referred to as the *chain code* or *Freeman code [fcc]*. As we trace around the edge of the blob, we generate a number, 0–7, to indicate which of the eight possible directions we have taken, relative to the center shaded pixel in Figure 2.14(b). Let N_o indicate how many *odd-numbered* code values are produced as we code the blob's edge, and N_e represent the number of *even-numbered* values found. The *perimeter* of the blob is given *approximately* by the formula $N_e + 2N_o$

This formula will normally suffice for use in those situations where the perimeter of a smooth object is to be measured. The *centroid* $(\bar{I}, \bar{J})$of a blob [*cgr*] determines its position within the image and can be calculated using these formulae:

$$\bar{I} \leftarrow \sum_j \sum_i (i(a(i, j)))/N_{i,j}$$

$$\bar{J} \leftarrow \sum_j \sum_i (j(a(i, j)))/N_{i,j}$$

$$\text{where } N_{i,j} \leftarrow \sum_j \sum_i a(i, j)$$

Although we are considering images in which the *a(i,j)* are equal to 0 (black) or 1 (white), it is convenient to use *a(i,j)* as an ordinary arithmetic variable as well.

2.3.2 Shape descriptors

The following are just a few of the shape descriptors that have been proposed:

a) The distance of the furthest point on the edge of the blob from the centroid.
b) The distance of the closest point on the edge of the blob from the centroid.
c) The number of protuberances, as defined by that circle whose radius is equal to the average of the parameters measured in a) and b).
d) Circularity = Area/Perimeter^2.
   This will tend to zero for irregular shapes with ragged boundaries, and has a maximum value of $1/4\pi$ for a circle.
e) The distances of points on the edge of the blob from the centroid, as a function of angular position. This describes the silhouette in terms of polar coordinates. (This is not a single-valued function.)

f) The number of holes. (Use *eul* and *ndo* to count them.)
g) The number of bays.
h) The Euler number.
i) The ratio of the area of the original blob to that of its convex hull.
j) The ratio of the area of the original blob to that of its circumcircle.
k) The ratio of the area of the blob to the square of the total limb length of its skeleton.
l) The distances between joints and limb ends of the skeleton.
m) The ratio of the projections onto the major and minor axes.

2.4 Binary mathematical morphology

The basic concept involved in mathematical morphology is simple: an image is probed with a template shape, called a structuring element, to find where the structuring element fits or does not fit within objects in the image [DOU92] (Figure 2.15). By marking the locations where the template shape fits, structural information can be gleaned about the image. The structuring elements used in practice are usually geometrically simpler than the objects they act on, although this is not always the case. Common structuring elements include points, point pairs, vectors, lines, squares, octagons, discs, rhombi, and rings. Since shape is one of the prime carriers of information in machine vision applications, mathematical morphology has an important role to play in industrial systems [HAR87].

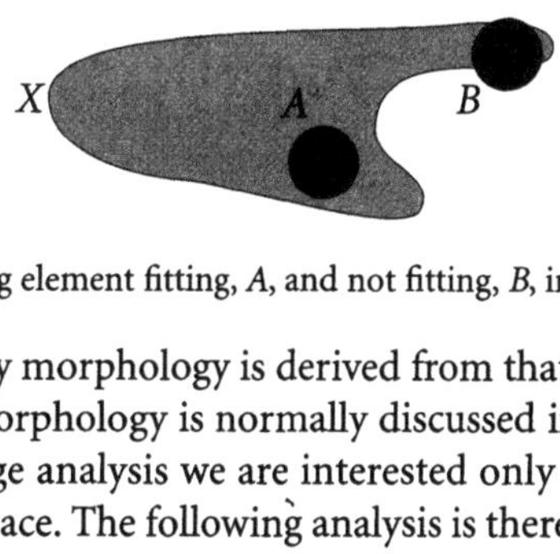

Fig. 2.15 A structuring element fitting, *A*, and not fitting, *B*, into a given object *X*.

The language of binary morphology is derived from that of set theory [HAR92]. General mathematical morphology is normally discussed in terms of Euclidean *N*-space, but in digital image analysis we are interested only in a discretized or digitized equivalent in two space. The following analysis is therefore restricted to binary images in a digital two-dimensional integer space, *Z2*. The image set (or object) under analysis will be denoted by *A*, with elements $a = (a1, a2)$. The shape parameter, or structuring element, that will be applied to object *A* will be denoted by *B*, with elements $b = (b1, b2)$. The primary morphological operations that we will examine are dilation, erosion, opening, and closing.

Dilation (also referred to as *filling* and *growing*) is the expansion of an image set *A* by a structuring element *B*. It is formally viewed as the combination of the two sets using vector addition of the set elements. The dilation of an image set *A* by a structuring element *B* will be denoted $A \oplus B$, and can be represented as the union of translates of the structuring element *B* [HAR92b]:

$$A \oplus B = \bigcup_{a \in B} B_a$$

where $\bigcup$ represents the union of a set of points and the translation of B by point a is given by $B_a = \{c \in Z^2 \mid c = (b + a)\}$ for some $b \in B$.

This is best explained by visualizing a structuring element B moving over an image A in a raster fashion. Whenever the origin of the structuring element coincides with one of the image pixels in A, then the entire structuring element is placed in the output image at that location. For example, in Figure 2.16 the image is dilated by a cross-shaped structuring element contained within a 3×3 pixel grid.

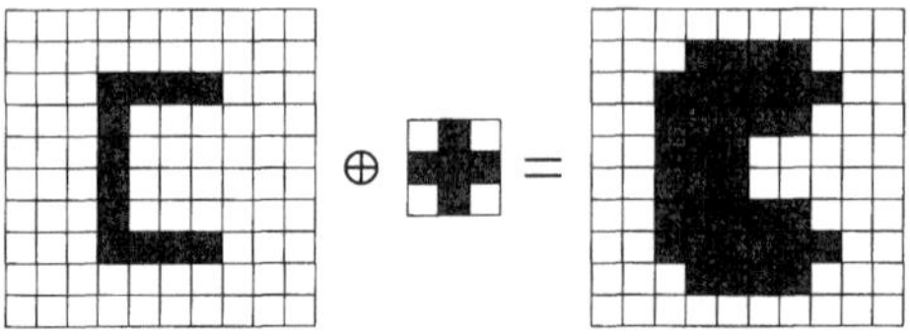

Fig. 2.16 Dilation of an image by a "cross" structuring element.

Erosion is the dual morphological operation of dilation and is equivalent to the shrinking (or reduction) of the image set A by a structuring element B. This is a morphological transformation which combines two sets using vector subtraction of set elements [HAR92b]. The erosion of an image set A by a structuring element B, denoted $A \ominus B$, can be represented as the intersection of the negative translates:

$$A \ominus B = \bigcap_{b \in B} A_{-b}$$

where $\bigcap$ represents the intersection of a set of points. Erosion of the image A by B is the set of all points for which B translated to a point x is contained in A. This consists of sliding the structuring element B across the image A, and where B is fully contained in A (by placing the origin of the structuring element at the point x) then x belongs to the eroded image $A \ominus B$. For example, in Figure 2.17 the image is eroded by a cross-shaped structuring element contained within a 3×3 pixel grid.

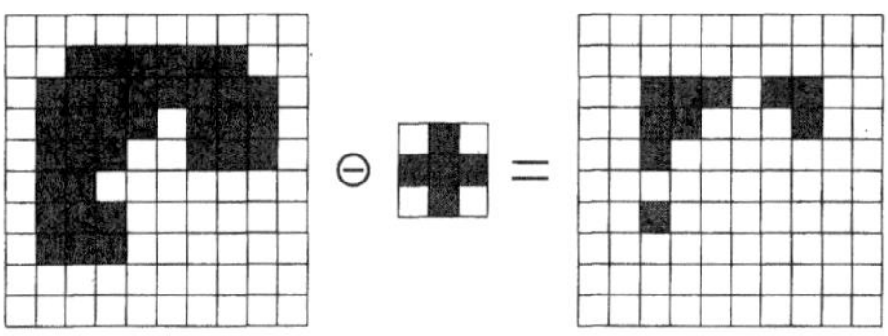

Fig. 2.17 Erosion of an image by a cross-shaped structuring element.

A duality relationship exists between certain morphological operators, such as erosion and dilation. This means that the equivalent of such an operation can be performed by its dual on the complement (negative) image and by taking the complement of the result [VOG89].

Although they are duals, the erosion and dilation operations are not *inverses* of each other. Rather, they are related by the following duality relationships:

$$(A \ominus B)^C = A^C \oplus \bar{B} \qquad (A \oplus B)^C = A^C \ominus \bar{B}$$

where A^c refers to the complement of the image set A and

$$\bar{B} = \{ x \mid \text{for some } b \in B, x = -b \}$$

refers to the reflection of B about the origin. (Serra [SER82, SER86] refers to this as the *transpose* of the structuring element.)

2.4.1 Opening and closing operations

Erosion and dilation tend to be used in pairs to extract or impose structure on an image. The most commonly found erosion–dilation pairings occur in the *opening* and *closing* transformations.

Opening is a combination of erosion and dilation operations that have the effect of removing isolated spots in the image set A that are smaller than the structuring element B and those sections of the image set A that are narrower than B. This is also viewed as a geometric *rounding* operation (Figure 2.18). The opening of the image set A by the structuring element B is denoted $A \circ B$, and is defined as $(A \ominus B) \oplus B$.

Closing is the dual morphological operation of opening. This transformation has the effect of filling in holes and blocking narrow valleys in the image set A, when a structuring element B (of similar size to the holes and valleys) is applied (Figure 2.18). The closing of the image set A by the structuring element B is denoted $A \bullet B$, and is defined as $(A \oplus B) \ominus B$.

Fig. 2.18 Application of a 3×3 square structuring element to a binary image of a small plant. (left) Original image. (center) Result of morphological opening. (right) Result of morphological closing.

One important property that is shared by both the opening and closing operations is *idempotency*. This means that successive reapplication of the operations will not change the previously transformed image [HAR87b]. Therefore,

$$A \circ B = (A \circ B) \circ B \quad \text{and} \quad A \bullet B = (A \bullet B) \bullet B.$$

Unfortunately, the application of morphological techniques to industrial tasks, which involves complex operations on "real-world" images, can be difficult to implement. Practical imaging applications tend to have structuring elements that are unpredictable in shape and size. In practice, the ability to manipulate arbitrary structuring elements usually relies on their decomposition into component parts.

2.4.2 Structuring element decomposition

Some vision systems [DUF73, STE78, WAL88] can perform basic morphological operations very quickly in a parallel and/or pipelined manner. Implementations that involve such special-purpose hardware tend to be expensive, although there are some notable exceptions [WAL94b]. Unfortunately, some of these systems impose restrictions on the shape and size of the structuring elements that can be handled. Therefore, one of the key problems involved in the application of morphological techniques to industrial image analysis is the generation and/or decomposition of large structuring elements. Two main strategies are used to tackle this problem.

The first technique is called *serial decomposition.* This decomposes certain large structuring elements into a *sequence* of successive erosion and dilation operations, each step operating on the preceding result. Unfortunately, the serial decomposition of a particular large structuring elements into smaller ones is not always possible. Furthermore, those decompositions that are possible are not always easy to identify and implement.

If a large structuring element B can be decomposed into a chain of dilation operations, $B = B_1 \oplus B_2 \oplus \ldots \oplus B_N$ (Figure 2.19), then the dilation of the image set A by B is given by

$$A \oplus B = A \oplus (B_1 \oplus B_2 \oplus \ldots \oplus B_N) = (\ldots((A \oplus B_1) \oplus B_2) \ldots) \oplus B_N$$

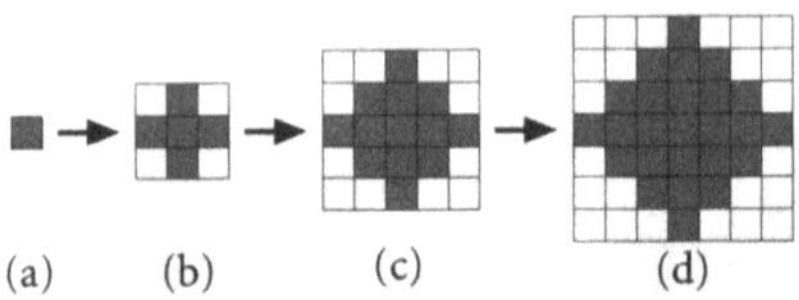

Fig. 2.19 Construction of a 7×7 structuring element by successive dilation of a 3x3 structuring element. (a) Initial pixel. (b) 3×3 structuring element and the result of the first dilation. (c) Result of the second dilation. (d) Result of the third dilation

Similarly, using the so-called chain rule [ZHU86], which states that $A \ominus (B \oplus C) = (A \ominus B) \ominus C$, the erosion of A by B, is given by

$$A \ominus B = A \ominus (B_1 \oplus B_2 \oplus \ldots \oplus B_N) = (\ldots((A \ominus B_1) \ominus B_2) \ldots) \ominus B_N$$

A second approach to the decomposition problem is based on "breaking up" the structuring element, B, into a union of smaller components, $B_1, \ldots, B_N$. We can think of this approach as "tiling" of the structuring element by subelements (Figure 2.20). Since the "tiles" do not need to be contiguous or aligned, any shape can be specified without the need for serial decomposition of the structuring element, although the computational cost of this approach is proportional to the area of the structuring element [WAL88]. This is referred to as *union* or *parallel decomposition.* Therefore, when B is decomposed into a union of smaller structuring elements, $B = B_1 \cup B_2 \cup \ldots \cup B_N$, then the dilation of an image A by the structuring element B can be rewritten as

$$A \oplus B = A \oplus (B_1 \cup B_2 \cup \ldots \cup B_N) = (A \oplus B_1) \cup (A \oplus B_2) \cup \ldots \cup (A \oplus B_N)$$

Likewise, the erosion of A by the structuring element B can be rewritten as

$$A \ominus B = A \ominus (B_1 \cup B_2 \cup \ldots \cup B_N) = (A \ominus B_1) \cap (A \ominus B_2) \cap \ldots \cap (A \ominus B_N)$$

This makes use of the fact that $A \ominus (B \cup C) = (A \ominus B) \cap (A \ominus C)$ [HAR87]. Due to the nature of this decomposition procedure, it is well suited to implementation on parallel computer architectures.

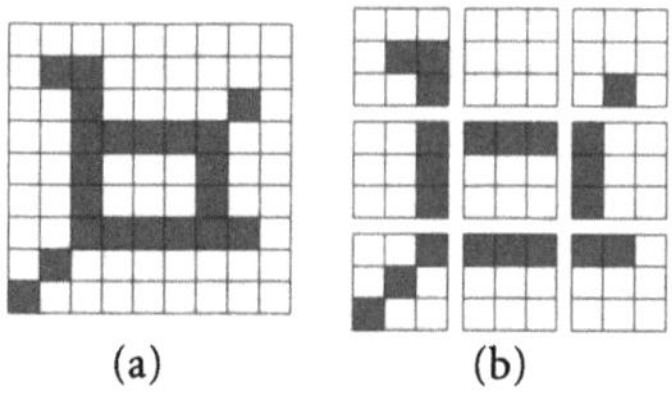

Fig. 2.20 Tiling of a 9×9 arbitrary structuring element. (a) The initial 9×9 structuring element. (b) Tiling with nine 3×3 subelements [WAL88].

Waltz [WAL88] compared these structural element decomposition techniques, and showed that the serial approach has a 9:4 speed advantage over its parallel equivalent. (This was based on an arbitrarily specified 9×9 pixel structuring element, when implemented on a commercially-available vision system.) However, the parallel approach has a 9:4 advantage in the number of degrees of freedom. (Every possible 9×9 structuring element can be achieved with the *parallel decomposition*, but only a small subset can be realized with the serial approach.) Although slower than the serial approach, it has the advantage that there is no need for serial decomposition of the structuring element.

Classical parallel and serial methods mainly involve repeated scanning of image pixels and are therefore inefficient when implemented on conventional computers. This is so because the number of scans depends on the total number of pixels (or edge pixels) in the shape to be processed by the morphological operator. Although the parallel approach is suited to some customized (parallel) architectures, the ability to implement such parallel approaches on *serial* machines is discussed by Vincent [VIN91].

2.5 Grey-scale morphology

Binary morphological operations can be extended naturally to process grey-scale imagery by the use of neighborhood minimum and maximum functions [HAR87]. Heijmans [HEI91] presents a detailed study of grey-scale morphological operators, in which he outlines how binary morphological operators and thresholding techniques can be used to build a large class of useful grey-scale morphological operators. Sternberg [STE86b] discusses the application of such morphological techniques to industrial inspection tasks.

In Figure 2.21, a one-dimensional morphological filter operates on an analog signal (equivalent to a grey-scale image). The input signal is represented by the "dashed" curve and the output by the solid curve. In this simple example, the structuring element has an approximately parabolic shape. In order to calculate a value for the output signal, the structuring element is pushed upwards from below the input curve. The height of the top of the structuring element is noted. This process is then repeated by sliding the structuring element sideways. Notice how this particular operator attenuates the intensity peak but follows the input signal quite accurately everywhere else. Subtracting the output signal from the input would produce a result in which the intensity peak is emphasized and all other variations would be reduced.

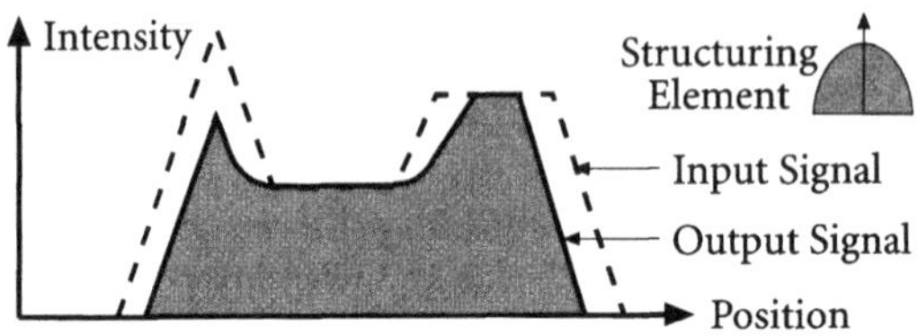

Fig. 2.21 A one-dimensional morphological filter eroding an analog signal.

The effect of the basic morphological operators on two-dimensional grey-scale images can also be explained in these terms. Imagine the grey-scale image as a landscape, in which each pixel can be viewed in 3-D. The extra height dimension represents the grey-scale value of a pixel. We generate new images by passing the structuring element above/below this landscape (See Figure 2.21).

Grey-scale dilation. This is computed as the maximum of translations of the grey surface. Grey-scale dilation of image *A* by the structuring element *B* produces an image *C* defined by

$$C(r,c) = \mathrm{MAX}_{(i,j)}\{A(r-i, c-j) + B(i,j)\} = (A \oplus B)(r,c)$$

where *A*, *B* and *C* are grey-scale images. Commonly used grey-scale structuring elements include rods, disks, cones and hemispheres. This operation is often used to smooth small negative-contrast grey-scale regions in an image. [*lnb*] is an example of a dilation operator.

Grey-scale erosion. The grey value of the erosion at any point is the *maximum* value for which the structuring element centered at that point still fits entirely within the foreground under the surface. This is computed by taking the *minimum* of the grey surface translated by all the points of the structuring element. (Figure 2.21). Grey-scale erosion of image *A* by the structuring element *B* produces an image *C* defined by

$$C(r,c) = \mathrm{MIN}_{(i,j)}\{A(r+i, c+j) - B(i,j)\} = (A \ominus B)(r,c)$$

This operation is commonly used to smooth small positive-contrast grey-scale regions in an image. [*snb*] is an example of an erosion operator.

Grey-scale opening. This operation is defined as the grey-scale erosion of the image followed by the grey-scale dilation of the eroded image. That is, it will cut down the peaks in the grey-scale topography to the highest level for which the elements fit under the surface.

Grey-scale closing. This operation is defined as the grey-scale dilation of the image followed by the grey-scale erosion of the dilated image. Closing fills in the valleys to the maximum level for which the element fails to fit above the surface. For a more detailed discussion on binary and grey-scale mathematical morphology, see Haralick and Shapiro [HAR92b] and Dougherty [DOU92]. The crack detector *[crk/1]* is an example of a closing operator.

2.6 Global image transforms

An important class of image-processing operators is characterized by an equation of the form $B \Leftarrow f(A)$, where $A = \{a(i,j)\}$ and $B = \{b(p,q)\}$. Each element in the output picture B is calculated using all or at least a large proportion of the pixels in A. The output image B may well look quite different from the input image A. Examples of this class of operators are *lateral shift, rotation, warping, Cartesian-to-polar coordinate conversion, Fourier transform,* and *Hough transform.*

Integrate intensities along image rows [rin]. This operator is rarely of great value when used on its own, but can be used with other operators to good effect, for example detecting horizontal streaks and edges. The operator is defined recursively:

$b(i,j) \Leftarrow b(i,j-1) + (a(i,j)/n)$ where $b(0,0) = 0$.

Row maximum [rox]. This function is often used to detect local intensity minima.

$c(i,j) \Leftarrow \mathrm{MAX}(a(i,j), c(i,j-1))$

Geometric transforms. Algorithms exist by which images can be shifted [*psh*], rotated [*tur*], magnified [*pex* and *psq*], and warped [*ctp* and *ptc*], and undergo axis conversion [*ctr, rtc*]. The reader should note that certain operations, such as rotating a digital image, can cause some difficulties because pixels in the input image are not mapped exactly to pixels in the output image. This can cause smooth edges to appear stepped. To avoid this effect, interpolation may be used, but this has the unfortunate effect of blurring edges. (See [BAT91a] for more details.)

The utility of axis transformations is evident when we are confronted with the examination of circular objects, or those displaying a series of concentric arcs, or streaks radiating from a fixed point. Inspecting such objects is often made very much easier if we first convert from *Cartesian* to *polar* coordinates. Warping is also useful in a variety of situations. For example, it is possible to compensate for *barrel* or *pin-cushion* distortion in a camera. Geometric distortions introduced by a wide-angle lens, or trapezoidal distortion due to viewing the scene from an oblique angle, can also be corrected. Another possibility is to convert simple curves of known shape into straight lines, in order to make subsequent analysis easier.

2.6.1 Hough transform

The Hough transform provides a powerful and robust technique for detecting lines, circles, ellipses, parabolas, and other curves of predefined shape in a binary image. Let us begin our discussion of this fascinating topic by describing the simplest version, the basic Hough transform, which is intended to detect straight lines. Actually, our objective is to locate *nearly linear* arrangements of disconnected white spots and *"broken" lines.* Consider that a straight line in the input image is defined by the equation $r = x \cos \phi + y \sin \phi$, where r and ϕ are two unknown parameters whose values are to be found. Clearly, if this line intersects the point (x_i, y_i), then $r = x_i \cos \phi + y \sin \phi$ can be solved for many different values of (r, ϕ). So, each white point (x_i, y_i) in the input image may be associated with a *set* of (r, ϕ) values. Actually, this set of points forms a *sinusoidal curve* in (r, ϕ) space. The latter is called the *Hough transform (HT)* image. Since each point in the input image generates such a sinusoidal curve, the whole of that image creates a multitude of overlapping sinu– soids in the resulting *HT* image. In many instances, a large number of sinusoidal curves are found to converge on the same spot in the HT image. The (r, ϕ) address of such a point indicates the slope ϕ and position r of a straight line that can be drawn through a large number of white spots in the input image.

The implementation of the Hough transform for line detection begins by using a two-dimensional accumulator array, $A(r, \phi)$, to represent quantized (r, ϕ) space. (Clearly, an important choice to be made is the step size for quantizing r and ϕ. However, we will not dwell on such details here.) Assuming that all the elements of $A(r, \phi)$ are initialized to zero, the Hough transform is found by computing a set $S(x_i, y_i)$ of (r, ϕ) pairs satisfying the equation $r = x_i \cos\phi + y_i \sin \phi$. Then, for all (r, ϕ) in $S(x_i, y_i)$, we increment $A(r, \phi)$ by one. This process is then repeated for all values of i such that the point (x_i, y_i) in the input image is white. We repeat that bright spots in the HT image indicate "linear" sets of spots in the input image. Thus, line detection is transformed to the rather simpler task of finding local maxima in the accumulator array $A(r, \phi)$. The coordinates (r, ϕ) of such a local maximum give the parameters of the equation of the corresponding line in the input image. The HT image can be displayed, processed, and analyzed just like any other image, using the operators that are now familiar to us.

The robustness of the HT techniques arises from the fact that, if part of the line is missing, the corresponding peak in the HT image is simply not quite so bright. This occurs because fewer sinusoidal curves converge on that spot and the corresponding accumulator cell is incremented less often. However, unless the line is almost completely obliterated, this new darker spot can also be detected. In practice, we find that "nearly straight lines" are transformed into a *cluster* of points, with a spreading of the intensity peaks in the HT image, due to noise and quantization effects. In this event, to calculate the parameters of the straight line in the input image we threshold the HT image and then find the centroid of the resulting spot. Pitas [PIT93] gives a more detailed description of this algorithm. Figure 2.22 illustrates how this approach can be used to find a line in a noisy binary image.

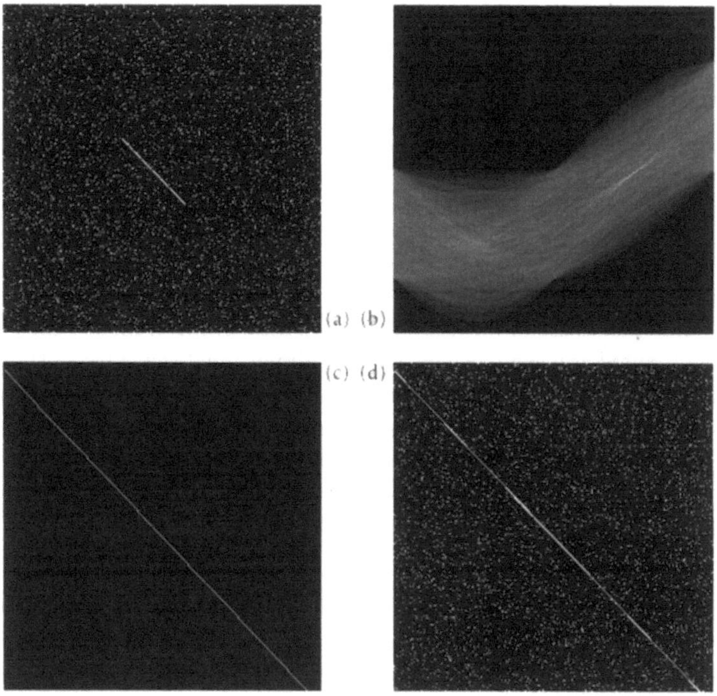

Fig. 2.22 Hough transform. (a) Original image. (b) Hough transform. (c) Inverse Hough transform applied to a single white pixel located at the point of maximum intensity in (b). (d) Inverse transform superimposed on original image. Notice how accurately this process locates the line in the input image, despite the presence of a high level of noise.

The Hough transform can also be generalized to detect groups of points lying on a curve. In practice, this may not be a trivial task, since the complexity increases very rapidly with the number of parameters needed to define the curve. For circle detection, we define a circle parametrically as: $r^2 = (x - a)^2 + (y - b)^2$ where (a, b) determines the coordinates of the center of the circle and r is its radius. This requires a *three-dimensional* parameter space, which, of course, cannot be represented and processed as a single image. For an arbitrary curve, with no simple equation to describe its boundary, a lookup table is used to define the relationship between the boundary coordinates, an orientation, and the Hough transform parameters. (See [SON93] for more details.)

2.6.2 Two-dimensional discrete Fourier transform

We have just seen how the transformation of an image into a different domain can sometimes make the analysis task easier. Another important operation to which this remark applies is the Fourier transform. Since we are discussing the processing of images, we will discuss the *two-dimensional discrete Fourier transform (DFT)*. This operation allows spatial periodicities in the intensity within an image to be investigated, in order to find, among other features, the dominant spatial frequencies. The two-dimensional DFT of an $N \times N$ image $f(x,y)$ is defined as follows [GON87]:

$$F(u,v) = \frac{1}{N} \cdot \sum_{x=0}^{N-1} \sum_{y=0}^{N-1} f(x,y) \exp[-i\, 2\pi (ux + vy) / N] \quad \text{where } 0 \le u,v \le N-1$$

The inverse transform of $F(u,v)$ is defined as

$$f(x,y) = \frac{1}{N} \cdot \sum_{x=0}^{N-1} \sum_{y=0}^{N-1} F(u,v) \exp[i\, 2\pi (ux + vy) / N] \quad \text{where } 0 \le x,y \le N-1$$

Several algorithms have been developed to calculate the two-dimensional DFT. The simplest makes use of the fact that this is a *separable* transform, which therefore can be computed as a sequence of two one-dimensional transforms. Thus, we can generate the two-dimensional transform by calculating the one-dimensional discrete Fourier transform along the image rows and then repeating this on the resulting image but, this time, operating on the columns [GON87]. This reduces the computational overhead when compared with direct two-dimensional implementations. The sequence of operations is as follows:

$f(x,y) \rightarrow$ Row Transform $\rightarrow F_1(x,v) \rightarrow$ Column Transform $\rightarrow F_2(u,v)$

Although this is still computationally slow compared with many other shape measurements, the Fourier transform is quite powerful. It allows the input to be represented in the frequency domain, which can be displayed as a *pair* of images. (It is not possible to represent both amplitude and phase using a single monochrome image.) Once the frequency domain processing is complete, the inverse transform can be used to generate a new image in the original, so-called, *spatial* domain.

The Fourier power spectrum, or amplitude spectrum, plays an important role in image processing and analysis. This spectrum can be displayed, processed, and analyzed as an intensity image. Since the Fourier transform of a real function produces a complex function: $F(u,v) = R(u,v) + i\, I(u,v)$, the frequency spectrum of the image is the magnitude function $|F(u,v)| = \sqrt{[R^2(u,v) + I^2(u,v)]}$ and the power spectrum (spectral density) is defined as $P(u,v) = |F(u,v)|^2$.

Figure 2.23 illustrates how certain textured features can be highlighted using the two-dimensional discrete Fourier transform. The image is transformed into the frequency domain and an ideal bandpass filter (with a circular symmetry) is applied. This has the effect of limiting the spatial frequency information in the image. When the inverse transform is calculated, the resultant textured image has a different spatial frequency content, which can then be analyzed. For more details on the Fourier transform and its implementations, see [PIT93] and [GON87].

2.7 Texture analysis

Texture is observed in the patterns of a wide variety of synthetic and natural surfaces (e.g., wood, metal, paint, and textiles). If an area of a textured image has a large intensity variation, then the dominant feature of that area would be *texture*. If

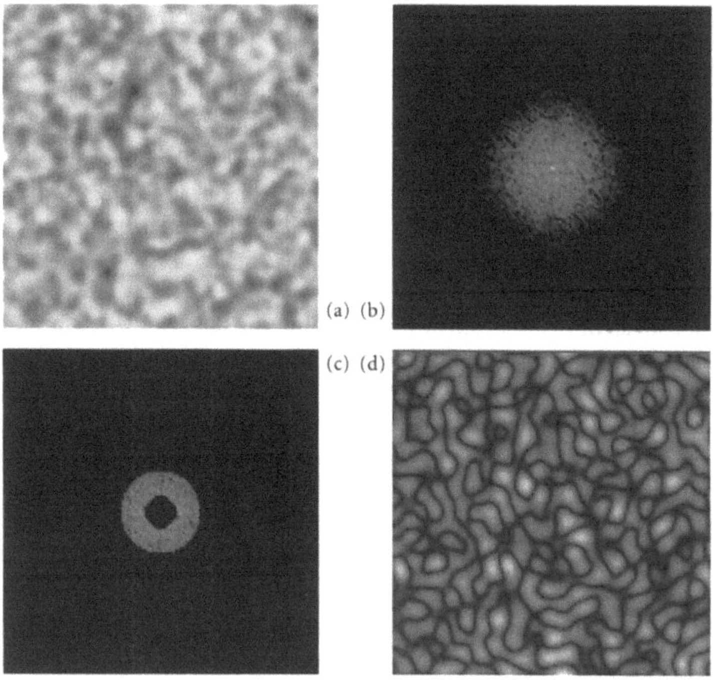

Fig. 2.23 Filtering a textured image in the frequency domain. (a) Original textured image. (b) Resultant transformed image after using the two-dimensional DFT. (The image is the frequency spectrum shown as an intensity function.) (c) Resultant frequency domain image after an ideal bandpass filter is applied to the image. (d) The resultant spatial domain image after the inverse two-dimensional discrete Fourier transform is applied to the bandpass filtered image in (c).

this area has little variation in intensity, then the dominant feature within the area is *tone*. This is known as the *tone–texture concept*. Although a precise formal definition of texture does not exist, it may be described subjectively using terms such as *coarse, fine, smooth, granulated, rippled, regular, irregular,* and *linear*. These features are used extensively in manual region segmentation. There are two main classification techniques for texture: *statistical* and *structural*.

2.7.1 Statistical approaches

The statistical approach is well suited to the analysis and classification of random or natural textures. A number of different techniques have been developed to describe and analyze such textures [HAR79], a few of which are outlined below.

Autocorrelation function (ACF)

Autocorrelation derives information about the basic 2-D tonal pattern that is repeated to yield a given periodic texture. The autocorrelation function $A(\delta x, \delta y)$ of an image matrix $[I(i,j)]$ is defined as follows:

$$A(\delta x, \delta y) = \left(\sum_{i,j}\left[(I(i,j))(I(i+\delta x, j+\delta y))\right]\right) \Big/ \left(\sum_{i,j}[I(i,j)]^2\right)$$

Although useful at times, the ACF has severe limitations. It cannot always distinguish between textures, since many subjectively-different textures have the same ACF. The variables *(i,j)* are restricted to lie within a specified window outside which the intensity is zero. Incremental shifts of the image are given by $(\delta x, \delta y)$. It is worth noting that the ACF and the power spectral density are Fourier transforms of each other.

Fourier spectral analysis

The Fourier spectrum is well suited to describing the directionality and period of repeated texture patterns, since they give rise to narrow high-energy peaks in the power spectrum. (See Section 2.6 and Figure 2.23.) Typical Fourier descriptors of the power spectrum include *the location of the highest peak, mean, variance,* and *the difference in frequency between the mean and the highest value of the spectrum.* This approach to texture analysis is often used in aerial/satellite and medical image analysis. The main disadvantage of this approach is that the procedures are not invariant, even under monotonic transforms of its intensity.

Edge density

This is a simple technique in which an edge detector or highpass filter is applied to the textured image. The result is then thresholded and the edge density is measured by the average number of edge pixels per unit area. Two-dimensional or directional filters/edge detectors may be used, as appropriate.

Histogram features

This useful approach to texture analysis is based on the intensity histogram of all or part of an image. Common histogram features include *moments, entropy dispersion, mean* (an estimate of the average intensity level), *variance* (this second moment is a measure of the dispersion of the region intensity), *mean square value* or *average energy, skewness* (the third moment which gives an indication of the histograms symmetry), and *kurtosis* (cluster prominence or "peakness"). For example, a narrow histogram indicates a low contrast region, while two peaks with a well-defined valley between them indicates a region that can readily be separated by simple thresholding.

Texture analysis based solely on the grey-scale histogram suffers from the limitation that it provides no information about the *relative position* of pixels to each other. Consider two binary images, where each image has 50% black and 50% white pixels. One of the images might be a checkerboard pattern, while the second one may consist of a salt and pepper noise pattern. These images generate exactly the same grey-scale histogram. Therefore, we cannot distinguish them using first order (histogram) statistics alone. This leads us naturally to the examination of the co-occurrence approach to texture measurement.

2.7.2 Co-occurrence matrix approach

The co-occurrence matrix technique is based on the study of *second-order grey-level spatial dependency statistics.* This involves the study of the grey-level spatial interdependence of pixels and their spatial distribution in a local area. Second-order statistics describe the way grey levels tend to occur together, in pairs, and therefore provide a description of the type of texture present. A *two-dimensional* histogram of the spatial dependency of the various grey-scale picture elements within a textured image is created. While this technique is quite powerful, it does not describe the shape of the primitive patterns making up the given texture.

The co-occurrence matrix is based on the estimation of the second-order joint conditional probability density function, $f(p,q,d,\alpha)$, for angular displacements, α, equal to *0, 45, 90* and *135* degrees. Let $f(p,q,d,\alpha)$ be the probability of going from one pixel with grey level p to another with grey level q, given that the distance between them is d and the direction of travel between them is given by the angle α. (For N_g grey levels, the size of the co-occurrence matrix will be $N_g \times N_g$.) For example, assuming the intensity distribution shown in the subimage given below, we can generate the co-occurrence matrix for $d = 1$ and α is 0 degrees.

2	3	3	3
1	1	0	0
1	1	0	0
0	0	2	2
2	2	3	3

Subimage with 4 grey levels.

Grey Scale	0	1	2	3
0	6	2	1	0
1	2	4	0	0
2	1	0	4	2
3	0	0	2	6

Co-occurrence matrix $\{f(p,q,1,0)\}$ for the subimage.

A co-occurrence distribution that changes rapidly with distance d indicates a fine texture. Since the co-occurrence matrix also depends on the image intensity range, it is common practice to normalize the grey scale of the textured image prior to generating the co-occurrence matrix. This ensures that first-order statistics have standard values and avoids confusing the effects of first- and second-order statistics of the image.

A number of texture measures (also referred to as *texture attributes*) have been developed to describe the co-occurrence matrix numerically and allow meaningful comparisons between various textures [HAR79] (see Figure 2.24). Although these attributes are computationally intensive, they are simple to implement. Some sample texture attributes for the co-occurrence matrix are given below.

Energy, or angular second moment, is a measure of the homogeneity of a texture. It is defined thus:

$$\text{Energy} = \Sigma_p \Sigma_q [f(p,q,d,\alpha)]^2$$

In a uniform image, the co-occurrence matrix will have few entries of large magnitude. In this case the *Energy* attribute will be large.

Entropy is a measure of the complexity of a texture and is defined thus:

$$\text{Entropy} = -\Sigma_p \Sigma_q [f(p,q,d,\alpha) \log(f(p,q,d,\alpha))]$$

It is commonly found that what a person judges to be a complex image tends to have a higher *Entropy* value than a simple image.

Inertia is the measurement of the moment of inertia of the co-occurrence matrix about its main diagonal. This is also referred as the *contrast* of the textured image. This attribute gives an indication of the amount of local variation of intensity present in an image.

$$\text{Inertia} = \Sigma_p \Sigma_q [(p-q)^2 f(p,q,d,\alpha)]$$

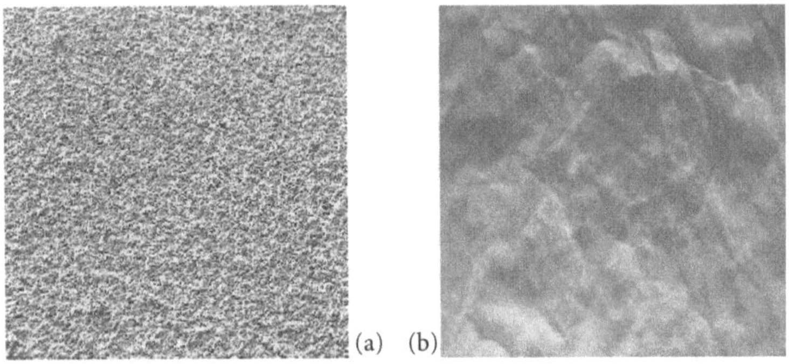
(a) (b)

(c)	Sand	Sand	Paper	Paper
	$f(p,q,1,0)$	$f(p,q,1,90)$	$f(p,q,1,0$	$f(p,q,1,90)$
Energy ($\times 10^6$)	1.63	1.7	3.49	3.42
Inertia ($\times 10^8$)	5.4	6.5	0.181	0.304

Fig. 2.24 Co-occurrence-based texture analysis.
(a) Sand texture. (b) Paper texture. (c) Texture attributes.

2.7.3 Structural approaches

Certain textures are deterministic in that they consist of identical *texels* (basic texture elements), which are placed in a repeating pattern according to some well-defined but unknown placement rules. To begin the analysis, a texel is isolated by identifying a group of pixels having certain invariant properties, which repeat in the given image. A texel may be defined by its *grey level*, *shape*, or *homogeneity of some local property*, such as size or orientation. Texel spatial relationships may be expressed in terms of *adjacency*, *closest distance*, and *periodicities*.

This approach has a similarity to language; with both image elements and grammar, we can generate a syntactic model. A texture is labelled *strong* if it is defined by deterministic placement rules, while a *weak* texture is one in which the texels are placed at random. Measures for placement rules include *edge density*, *run lengths of maximally-connected pixels*, and the *number of pixels per unit area* showing grey levels that are locally maxima or minima relative to their neighbors.

2.7.4 Morphological texture analysis

Textural properties can be obtained from the erosion process (Sections 2.4 and 2.5) by appropriately parameterizing the structuring element and determining the number of elements of the erosion as a function of the parameter's value [DOU92]. The number of white pixels of the morphological opening operation as a function of the size parameter of the structuring element **H** can determine the size distribution of the *grains* in an image. Granularity of the image **F** is defined as

$$G(d) = 1 - (\#[F \bigcirc \mathbf{H}_d] \,/\, \#F)$$

where H_d is a disk structuring element of diameter *d* or a line structuring element of length *d*, and *#F* is the number of elements in *F*. This measures the proportion of pixels participating in grains smaller than *d*.

2.8 Further remarks

The image-processing operators described in this chapter have all found widespread use in industrial vision systems. It should be understood that they are always used in combination with one another. Other areas of application for image processing may well use additional algorithms to good effect. Very often, these can be formulated in terms of the procedures described above, and implemented as a sequence of PIP commands. Two key features of industrial image-processing systems are the cost and speed of the target system (i.e., the one installed in a factory). It is common practice to use a more versatile and slower system for problem analysis and prototyping, whereas the target system must continue to operate in an extremely hostile environment. (It may be hot, greasy, wet, and/or dusty.) It must also be tolerant of abuse and neglect. As far as possible, the target system should be self-calibrating and able to verify that it is "seeing" appropriate images. It should provide enough information to ensure that the factory personnel are able to trust it; no machine system should be built that is a viewed by the workers as a mysterious black box. Consideration of these factors is as much a part of the design process as writing the image-processing software.

3 Algorithms, approximations, and heuristics

Murphy's Law was not formulated by Murphy, but by another person with the same name.
Unknown

3.1 Introduction

This chapter, indeed the whole of this book, is based on three axioms:

1. Machine vision is an engineering discipline, not a mathematical or philosophical exercise.
2. There is no unique way to perform any given image-processing function; every algorithm can be implemented in many different ways.
3. A given calculation may be very difficult using one implementation technology but very easy with another.

Given a certain method of implementation, simply reformulating an algorithm can often convert a "difficult" calculation into one that is straightforward, perhaps even trivial. It is important, therefore, that we gain some understanding of the way that algorithms can be reformulated before we consider the choice of implementation technology in detail in the subsequent chapters.

In every engineering design exercise, there are numerous factors specific to the task at hand. These must be carefully considered and properly balanced if we are to achieve a robust and effective system. In the case of machine vision, these include the shape, material, and optical properties of the object that is to be examined. Moreover, the *modus operandi* of the inspection system, its environment within the factory, and the skills and attitudes of the people who will work with it all have a critical role in system design. There are three critically important constraints, which are present in almost every engineering project. These relate to cost, performance, and operating speed. Another common constraint is the desire to use technology that is familiar, tried, and trusted. Together, these constraints may well preclude the use of certain implementation techniques and hence certain algorithms as well. We cannot simply rely on mathematical analysis alone to provide sufficient commendation for using a given algorithm. This point is of fundamental importance for our discussion in this chapter. Our experience has shown that operational, management, and performance constraints, such as those just listed, almost invariably impose far tighter control than "academic" discussions about whether we should use, say, one image filter rather than another.

We should never take it for granted that an algorithm must be performed in a given way. In particular, we should not expect that the initial explanation we gave in Chapter 2 necessarily defines how we are to implement a given algorithm. Indeed, we should always ask ourselves whether the algorithm is expressed in the best way possible and question whether reformulating it would lead to a better implementation. We should also investigate whether a similar algorithm, an approximate method, or an heuristic procedure is good enough for the given application. In the context of designing machine vision systems, sufficiency, rather than optimality, is the key concern. We must always be very wary of claims about optimality, because these are almost invariably based on some spurious assumption about what constitutes an "optimal solution." As in any engineering design process, all assumptions must be based on experience, judgement, and common sense. We must not rely blindly on mathematical rigor, since mathematics only very rarely provides a true model of the real physical world. An old adage states that *"If you can perform the mathematics, then you don't understand the engineering."* This is, of course, a gross simplification, but it paraphrases a vitally important truth. The reader should make no mistake about the authors' commitment to using rigorous mathematical analysis, wherever and whenever it is appropriate to do so. The most important point to note, however, is that there is a balance to be obtained: using algorithms, heuristics, and approximate methods in a sensible, reasoned way can lead to great improvements in the overall efficiency of a machine. The two latter categories of computational procedure may provide the basis for the design of a cost-effective machine vision system that is reliable and robust. On the other hand, mathematical analysis often yields deep insights that are not intuitively obvious. We must not forget, however, that as far as an engineer is concerned, one of the most foolish uses of mathematics is to prove what has been self-evident for years. More irritating still is the abuse of mathematics by using an incorrect or deficient physical model to "prove" that certain things are not true in the real world. To illustrate this point, we need only remind the reader that for many years the known laws of aerodynamics were not able to "prove" that bumblebees can fly [SIM00]. In other words, formal mathematical analysis based on erroneous/deficient physical or engineering models would lead us to believe that bumblebees cannot fly. The "proof" that this is not so is easily obtained: simply go into a garden on a summer's day. Sticking doggedly to only those algorithmic techniques with absolutely certain mathematical pedigrees is both severely limiting and likely to give us a false sense of confidence.

The choice of metric used to quantify algorithm performance almost inevitably involves our making an arbitrary and oversimplified choice. This can seriously affect the conclusions we reach about the performance of a vision system. There is an outstanding example of this in a paper that we encountered recently. An "objective" performance criterion was defined in terms of an *arbitrarily chosen* metric. A number of image filtering methods were then analyzed mathematically. As a result, one of them was found to be "superior" to all others. If a slightly different algorithm performance criterion had been selected, then a different filter might have been judged to be "best." In some situations, the ability to implement an image filter using an electronic circuit module with a power consumption of less than 100 μW might be a more appropriate criterion to apply, rather than some abstract mathematical performance metric. The latter might, for example, be based on the familiar concept of "least squares fit," but there cannot be any guarantee that this is in any way meaningful in any absolute sense. The reader should understand that we are not arguing that mathematical analysis has no place in the design of machine vision systems. It most certainly has a very important role, but mathematics must take its place alongside many other algorithm-selection criteria, in order to obtain a balanced engineering compromise. We are simply arguing that we must not judge the design of a machine vision system using any one-dimensional cost/performance criterion. (Even "cost" is a multi-dimensional entity.)

Our objective in this chapter is to explore the possibilities for expressing vision algorithms in a variety of ways. There is also the possibility of using approximate computational methods. Finally, we point out that techniques that are known to "fail" in some situations (i.e., heuristics) can sometimes provide a perfectly acceptable solution to the broader engineering design task.

3.1.1 One algorithm, many implementations

We will begin by discussing ordinary arithmetic multiplication. This will allow us to introduce the main points that we wish to emphasize in this chapter. By discussing an apparently straightforward task that is clearly relevant to many different areas of computation in addition to machine vision, we hope to explain why we felt it necessary to include this chapter in our book.

Consider the deceivingly simple task of multiplying two numbers together:

$$C = A\,B \tag{3.1}$$

Multiplication may be performed using a rather complicated digital hardware unit based on an adder, shifter, and recirculating data register. In some situations but by no means all, this may be the most convenient implementation technique. Special purpose integrated circuits, based on this idea or arrays of gates, are readily available but in certain applications may not provide the best solution.

Logarithms can, of course, be used to replace the process of multiplication with the simpler operations of table lookup and addition. We can use the well-known relationship:

$$C = \text{antilog}(\log(A) + \log(B)) \tag{3.2}$$

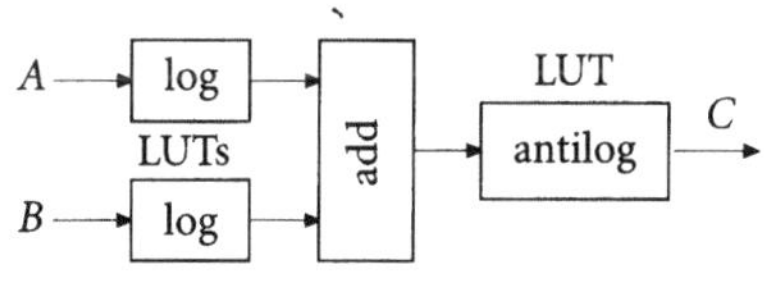

Fig. 3.1 Multiplying two numbers, *A* and *B* (typically 8-bit parallel) using logarithms. The *log* and *antilog* functions are implemented using lookup tables, possibly using ROMs.

If the variables A and B are short integers, the log and antilog functions can be performed conveniently using lookup tables, implemented in ROM or RAM (Figure 3.1). The reader will, no doubt, have already anticipated the complications that arise when the variables A and/or B are allowed to assume values close to zero. However, we will not dwell on this topic here.

Another possibility is to use the so-called *quarter-squares method*, which makes use of the following identity:

$$A\,B = [(A + B)^2 - (A - B)^2] / 4 \qquad (3.3)$$

Suppose that we are proposing to use an implementation technology which finds it easy to perform addition, subtraction, and squaring, perhaps by using a lookup table (Figure 3.2). In this case, the quarter-squares method can be very effective indeed. (Older readers may recall that this method was used in the heyday of analog computers to perform multiplication.) In certain circumstances, the quarter-squares or logarithm methods, implemented using lookup tables, might be more effective than the direct implementation of Eq. (3.1).

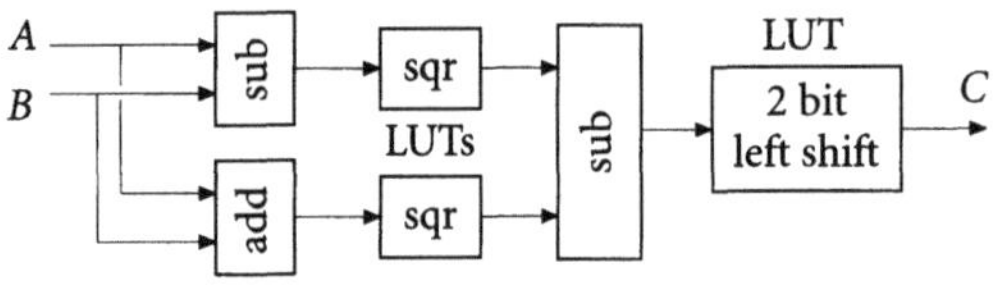

Fig. 3.2 Multiplying two numbers, A and B (typically 8-bit parallel), using the quarter-squares method. The *sqr* (square) functions are implemented using lookup tables, possibly using ROMs.

So far, we have tacitly assumed that both A and B are variables. However, if either of them is a constant, the implementation can be simplified quite significantly. If we wish to multiply a given number, represented in digital parallel form, by a constant, the use of a simple lookup table implemented using a read-only memory (ROM), is probably the simplest implementation technique (Figure 3.3).

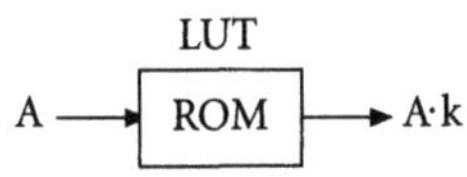

Fig. 3.3 Multiplying a number (typically 8-bit parallel) by a constant.

Suppose, however, that the constants are always integer powers of 2. In this case, "multiplication" reduces to a simple bit shift operation, which may be implemented using nothing more complicated than a piece of wire (Figure 3.4)!

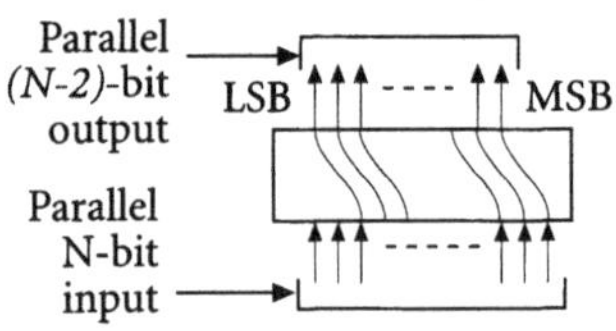

Fig. 3.4 Multiplying a number (typically N-bit parallel) by a constant of the form 2^M, where M is an integer. The figure shows the configuration for $M = -2$ (i.e., dividing by 4).

The reader might well question whether this discussion has any direct significance for image processing. To see one straightforward example where this relevance can be demonstrated, consider the arithmetic operations needed to implement the linear local operators, viz:

$$E \Leftarrow K_1(W_a A + W_b B + W_c C + W_d D + W_e E + W_f F + W_g G + W_h H + W_i I) + K_0 \quad (3.4)$$

At first sight this seems to indicate that we need to use 10 multiplications and 10 additions. Do we really need to use floating-point, or long-integer, arithmetic? Can we work with, say, an 8-bit representation of intensity (i.e., the variables, A, B, ..., I)? Can we obtain sufficiently accurate results using weights (W_a, W_b, ..., W_i) that are all integer powers of 2? If so, the problem reduces to one where we could use just one multiplication (i.e., K_0 (...)) and nine shift operations. Even if we must use general integer values for the weights, we can probably still manage to use a series of nine lookup tables, implemented using ROM/RAM, to compute the intermediate products $W_a A, \ldots, W_i I$ (Figure 3.5).

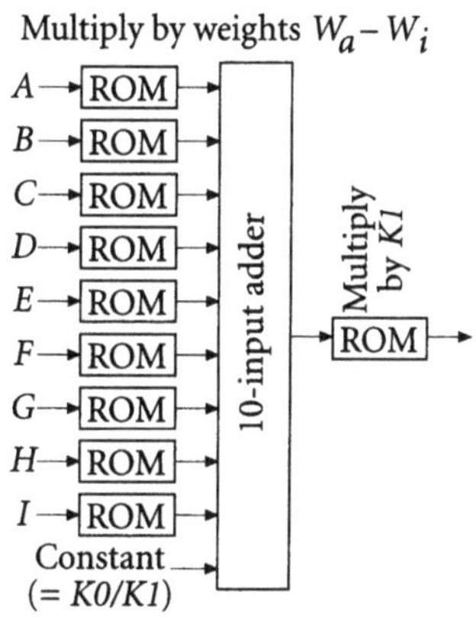

Fig. 3.5 A ROM-based implementation of the 3×3 linear local operator (Equation 3.4). This diagram does not explain how the inputs A, B, ..., I are generated.

3.1.2 Do not judge an algorithm by its accuracy alone

We cannot afford to make a decision about which algorithm to use on the basis of its performance alone. To do so would be nothing short of stupid! There are many different technologies available to us when we come to build a vision system. Indeed, in a single vision system there are inevitably several different image/information processing technologies coexisting and cooperating together. These include optics, optoelectronic signal conversion (camera/scanner), analog electronics, digital electronics, and software. They all have their own peculiar strengths and weaknesses. An image-processing algorithm is a mathematical prescription for a computational process which describes the operation of part of the data-processing chain. (When we use the term "image-processing algorithm" in connection with a vision system, we normally intend it to refer to the later stages in this chain, probably just the digital electronics and the software.)

Clearly, the speed of a machine is of great importance, but even this seemingly simple term has complications, because there are two components:

1. *Throughput rate.* (Number of images processed, or objects inspected, per second.)
2. *Latency.* (Time delay from digitization of an image to appearance of result.)

An on-line inspection system, working in conjunction with a continuously-moving conveyor, can usually tolerate a high latency, whereas a robot vision system cannot [BAT82b]. Hence, the term "real time" needs some qualification, in practice.

A popular way of representing algorithm complexity uses the $O(N)$ notation. Suppose that a certain algorithm requires an execution time of

$$A_0 + A_1 N + A_2 N^2 + A_3 N^3 + \ldots + A_m N^m \text{ seconds,}$$

where the A_i $(i = 1, \ldots, m)$ are constants and N is the number of data samples. (N is very likely to be related to the number of pixels in the image). When N is very large, the term $A_m N^m$ dominates all others and hence the algorithm is said to be of complexity $O(N^m)$. We do not always find that a simple polynomial function describes the execution time. For example, certain number-sorting algorithms require an execution time that is proportional to $N \log(N)$ (for very large values of N) and hence are said to have complexity $O(N \log(N))$. While some number-sorting algorithms have a higher complexity than this (typically $O(N^2)$, they may actually be easier and faster to implement in digital electronic hardware. For example, a simple *bubble-sort* algorithm (complexity $O(N^2)$) can be implemented in an iterative digital array. Figure 3.6 shows the organization of a hardware unit for imple-

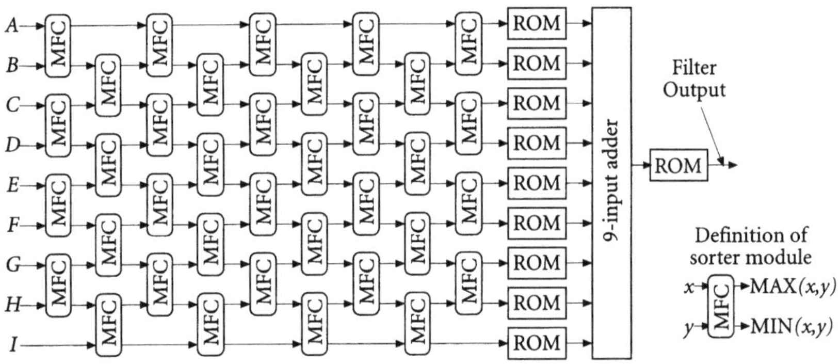

Fig. 3.6 Rank filter implemented using an iterative array performing a 9-element *bubble sort.*

menting *3×3 rank filters* (Section 2.2.5), which require the ranking of a small number of numeric items. This iterative array uses the ideas implicit in the bubble sort. On the other hand, the *merge sort* has complexity $O(N \log(N))$ and for this reason is often preferred in software implementations. However, this cannot be implemented in digital hardware in such a neat manner. The general point to note is that, although the $O(N)$ notation is very popular with mathematicians, it is not always a useful indicator of speed in a practical system.

3.1.3 Reformulating algorithms to make implementation easier

We have already made the point that there are many ways to implement a given algorithm and, in the latter parts of this chapter, we will give several examples to illustrate this. There are likely to be many similar image-processing algorithms resembling a given procedure. For example, there is a wide range of edge detection operators [e.g., PIP operators *sed, red, gra, edd*]. The point has been made elsewhere that, if the choice of edge detection operator is critical, then there is a problem somewhere else in the vision system – probably in the lighting! [PVB]. Bearing this in mind, we may be able to improve the quality of the input image by choosing the best possible lighting, optics, viewing position, and camera, and thereby allow ourselves the luxury of using a suboptimal image filter that is easy to implement.

An experienced vision engineer very often realizes that one algorithmic "atom" can be substituted for another in a processing sequence, producing very little effect on the overall system performance. To evaluate possible substitutions of this kind, it is essential to work with a convenient (i.e., *interactive*) image-processing "tool kit." An experienced system designer may well be able to anticipate which algorithms can safely be substituted in this way, but "blind" predictions like this are frequently mistaken. To be certain about the wisdom of employing the proposed revised algorithm, it must be assessed in a rigorous experimental manner. Much of the knowledge about what substitutions are sensible might, one day, be incorporated into an expert system but, to date, this has not been done except at the most trivial level. At the moment, an experienced engineer armed with appropriate tools (such as an interactive image processor) remains the best way to perform the high-level reasoning needed to design/redesign a vision algorithm.

In the previous chapter, we introduced several different types of image representation. As we will soon witness, these are particularly useful because certain algorithms can be reformulated very effectively using one or another form of image representation (e.g., chain code, run-length code, etc.), rather than the standard array representation. A prime example of this is to be found in the algorithms available for computing the convex hull of a blob-like figure. (Incidentally, this can be implemented even easier using the polar vector representation.)

3.1.4 Use heuristics in lieu of algorithms

A heuristic is a "rule of thumb" which is known to achieve a desired result in most cases, without necessarily being able to guarantee that it is always able to do so. On the other hand, an algorithm is always able to achieve the desired result. Why should we be prepared to accept a procedure which is known to fail in some situations? In our normal everyday lives, we use heuristics freely and without any serious misgivings. For example, the author of this chapter drives to work each morning along a road that experience has shown is able to get him there quickly. However, there are exceptions: one day last week, a car had broken down on the road and he was delayed for several minutes before he could complete his journey. That day, his heuristic for getting to work on time failed but it worked well this morning and it does so on most working days. There simply is no algorithm for getting to work on time. It is clearly impossible, in any practical way, to collect all of the information that would be needed to predict accidents or the breakdown of vehicles on the author's route to work.

3.1.5 Heuristics may actually be algorithmic

Heuristics can be very accurate and may actually be algorithmic without our knowing it. (We may not be clever enough to do the necessary mathematics to prove that a given procedure is algorithmic.) Very general heuristic rules can often cope with unexpected situations more comfortably and often more safely than "algorithms." Once we violate the conditions of use of an algorithm, we cannot guaranteed that it will generate valid results. Mathematical criteria defining conditions of use exist for good reason. If we violate them, we cannot properly anticipate the results. We might, for example, wish to relax the Dirichlet conditions for Fourier series expansion. It is highly unlikely that we could ever know for certain whether the result is mathematically valid, although by experimentation and/or computer simulation we may be able to satisfy ourselves that it is "good enough." The important points to note are that what we regard as an algorithm may not be so, and that a procedure that is not yet proven to achieve a "valid result" in every situation may in fact do so. The reader will observe a parallel here with Gödel's theorem. There are provable truths (algorithms that are know to be true), unprovable truths (procedures that are not known to be algorithmic but are true), unprovable untruths (e.g., procedures that fail in some as yet undiscovered way), and provable untruths (heuristics that are known to fail occasionally). However, we should not take this analogy with Gödel's theorem too far, since the analogy has no rigorous basis in logic.

3.1.6 Primary and secondary goals

Suppose that we are faced with the task of designing a machine that counts coffee beans, given an image such as that shown in Figure 3.7(a). Now, we know that we can count blobs in a binary image using the PIP operator *cbl*. This, combined with the prior use of a thresholding operator (*thr*), seems highly appropriate, since the latter converts the grey-scale input image into a binary image. A simple experiment (Figure 3.7(b)) soon shows that these two operators [*thr(_), cbl*] are unable to provide the needed precision. So, we soon conclude that we should have used some suitable image filtering (smoothing), to preprocess the images before thresholding. This seems a very reasonable approach and is certainly worth investigating further. However, we must be clear that the real (i.e., primary) objective is counting coffee beans, not blobs in a binary image. In order to make some progress, we were forced to make *some* assumptions but, as we will show, this one is not the best. In this instance, the assumption we have made is that the primary task (counting beans) can be replaced by two or more secondary tasks: filtering, thresholding, and blob counting. This is a reasonable working hypothesis but nothing more. After struggling for a long time to find a suitable filter, most people would probably conclude that they cannot generate a suitable binary image for this process to be reliable. (It is impossible, in practice, to obtain a "clean" binary image in which each bean is represented by just one blob.)

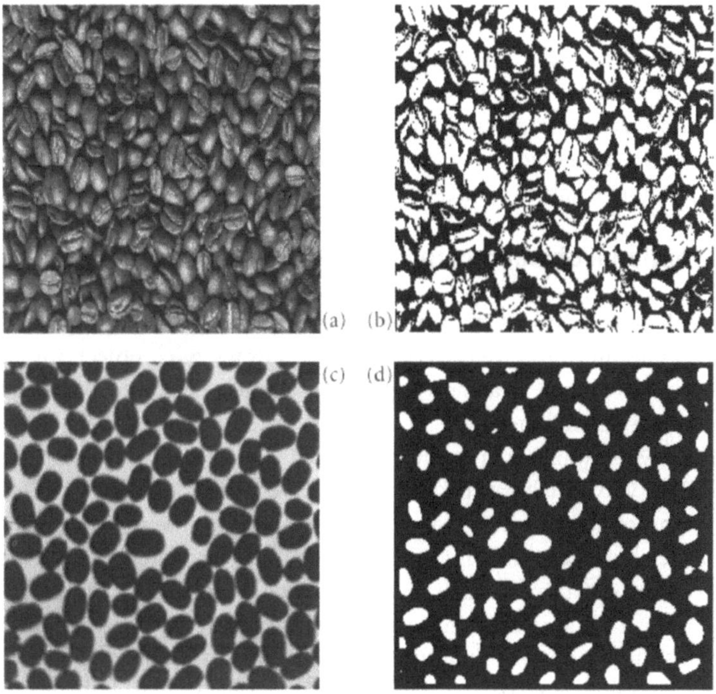

Fig. 3.7 Counting coffee beans. (a) Original image. Front lighting yields an image that is difficult to process. (b) Simple fixed parameter thresholding. Filtering can improve matters but not well enough to give a reliable count. (c) A different original image. Back lighting is superior. (d) Better separation, obtained using *[ecn, thr, neg, gft, enc, raf, thr(100)]*.

Eventually, we reassess our initial assumption and realize that we do not need to produce one blob for every bean. In this type of application, a statistically-valid answer is acceptable. We need to reassess our initial assumption that *cbl* is the way to count blobs. Suppose that a certain filtering and thresholding operation together produces on average 1.53 white blobs per bean, with a small standard deviation. Then, we can count blobs using *cbl*, the Euler number (*eul*), or any of a number of other procedures. In other words, there is nothing special about *cbl*. We might well find that we can substitute other techniques and still obtain a statistically-valid result. (We may find, of course, that the scaling factor relating the "blob count" to the "bean count" has to be revised. Recall that this was 1.53 for the operator *cbl*.) Our initial assumption that *cbl* is "the way" to count blobs and that we could obtain just one blob per bean was inappropriate, even though it seemed so reasonable when we began. Few image-processing operations have a better claim to being "algorithmic" than *cbl*, yet we are forced here to concede that this "guarantee of performance" means nothing; we are using an *algorithm* in a process which is being *judged statistically*. By substituting *eul* or another "blob count" procedure for *cbl*, we may be making little or no sacrifice in the overall performance of the bean counting system, or even conceivably improving it. As it was defined in Chapter 2, the PIP operator *cbl* counts 8-connected blobs, whereas we might have chosen to base the definition on 4-neighbors. Whichever we choose, the decision is quite arbitrary. For practical purposes when counting coffee beans, we might well find that *eul* produces results that are just as accurate in estimating the bean count.

In practice, we would probably find it more convenient to alter the method by which beans are presented to the camera. The image-processing task is then much simpler and more reliable (Figures 3.7(c) and (d)).

3.1.7 Emotional objections to heuristics

Many people have a serious emotional objection to using heuristics, because they cannot be guaranteed to give the correct result always. Such objections are more likely to arise when there is a known, often deliberately contrived, counterexample which makes a given procedure "fail." People often become confused at such a discovery, however unlikely or unimportant that "pathological" situation may be in practice. Even very intelligent people are often very poor at judging the statistical frequency and/or absolute importance of rare events. One has only to look at everyday life to see how frequently rational judgement is suspended, despite incontrovertible evidence to the contrary. (For example, the popularity of gambling and driving fast and the fear of flying and crime provide four such examples.) If an inspection system is known to ignore or give an inappropriate response, even on very rare or unimportant cosmetic faults, many managers will be unhappy to accept it. This caution will be "justified" on the grounds that senior management will not condone installing a machine with *any* known shortcomings. This thought process is not always rational, since there may be no alternative solution that is able to provide the "correct" results with 100% certainty. The important point to note is that, in reality, there is no sharp dichotomy between algorithms which *always* work and crude, approximate heuristics which do not. The concept of an algorithm-heuristic continuum is a more appropriate model for the situation in the real world. Another point is that the performance of a vision system is often judged *statistically*, not "*logically*." This is not a new idea, since we judge people on "statistical" evidence. All that we are asking is that machine vision systems and their component computational processes be judged in the same way.

3.1.8 Heuristics used to guide algorithms

Heuristics can be used to guide algorithms. For example, we may use a heuristic simply to speed up an algorithm. In this case, when the heuristic fails, the algorithm simply slows down but still yields a known result. A prime example of this is to be found in hill climbing. Heuristics can be very effective at finding local maxima, whereas an exhaustive search algorithm will always find the true maximum. Similarly, a heuristic may be used to speed up the process of finding "good" routes through a maze but an algorithm is able to guarantee that the final result is optimal. Heuristics are useful in those situations where knowledge of one good solution can lead to a rapid means of finding a better solution.

3.1.9 Heuristics may cope with a wider range of situations

A well-designed set of heuristic rules can often cope with a far wider range of situations than an algorithm is known to cover. This is possible because heuristics are able to encapsulate human experience, using the same language that people use to understand and express them. Such rules of thumb are usually not definable in strict mathematical form and are expressed in terms of a natural language, such as English. Normally such rules are then recoded into a computer language. (Prolog and hence PIP are very appropriate for use in this role.) It is entirely inappropriate to try to force mathematics where it does not fit naturally.

Lastly, we simply pose an important question: is an algorithmic solution really necessary? We often have to overcome ignorance, prejudice, and ill-informed opinion in order to transcend the emotional barrier that prevents many people from accepting "solutions that are known to fail." When a nontechnical manager is involved in making decisions about the use of heuristics rather than algorithms, there is always the danger that his natural inclination will be to reject heuristics in favor of algorithms, believing that the latter inevitably gives a better result. We hope to show in the pages that follow that this is often not the case.

3.2 Changing image representation

In Section 2.1 we discussed a number of ways of representing images in digital form. It will become clear later that some algorithms and heuristics can be expressed more naturally in one form rather than another. For this reason, we will explain first how images can be transformed from one representation to another.

3.2.1 Converting image format

The array representation arises naturally from standard area-scan cameras and line-scan cameras which are repeatedly scanned to build up a 2-dimensional image. It is natural to associate each photodetector site on the camera with one pixel in the digital image. The array representation therefore provides a reasonable starting point for specifying image-processing functions, and the other coding methods can be derived from it.

Array representation to run-code representation

The run code (or run-length code) is a more compact form of representation for binary images than the array code and can be derived from a video signal using fast

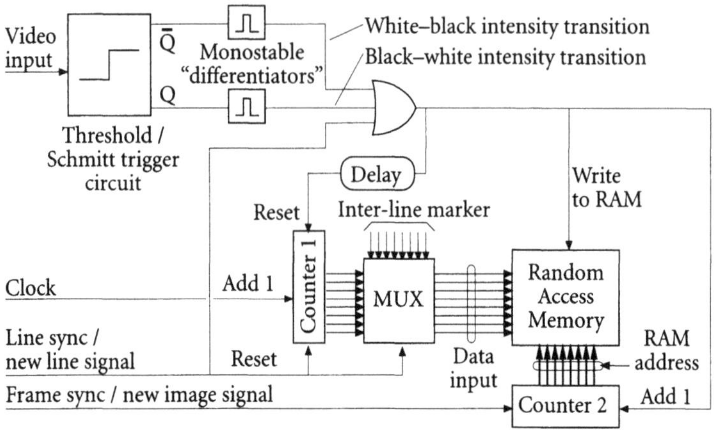

Fig. 3.8 The run code can be generated in simple digital or analog hardware. A simple low-cost circuit such as this can perform run-length encoding in real time on a video signal. This is important because the run code transforms many "difficult" calculations into trivial ones.

low-cost electronic hardware (See Figure 3.8). The operation of this circuit is quite straightforward: Counter 1 indicates how many pixels have been encountered in a raster scan through the image since the last black–white or white–black intensity transition. This value is stored in the RAM. (The box labelled "MUX" selects the input arising from Counter 1.) This counter is reset at the beginning of each new line, and whenever a black-to-white or white-to-black intensity transition occurs. Counter 2 indicates how many items have been stored in the RAM. Notice that a special symbol, called the *inter-line marker,* is stored in the RAM to indicate the end of each row of the image. This symbol is presented to the RAM's *data* input via, the multiplexor (MUX), when the *line sync/new line* signal is high. The format for data stored in the RAM is as follows:

$$\{R_{1,1}, R_{1,2}, \ldots, R_{1,n1}, {}^*, R_{2,1}, R_{2,2}, \ldots, R_{2,n2}, {}^*, R_{m,1}, R_{m,2}, \ldots, R_{m,nm}\}$$

where $R_{i,j}$ denotes the ith run in row j and '*' denotes the inter-line marker. Notice that $R_{i,1} + R_{i,2} + \ldots + R_{i,ni} \leq N$, where N is the number of pixels along each row. The same circuit can be used for both analog and digital video inputs with very little modification.

Array representation to chain-code representation

The chain code can be derived from the array representation of a binary image in at least two ways:

a) Perform a raster-scan search for an edge. Then, follow that edge until the starting point is revisited. This process is repeated until all edges have been traced (Figure 3.9). In this procedure, each edge point is visited not more than twice. If, as in some applications, we can be sure that there is only one object silhouette (i.e., only one blob) in the binary image, then the algorithm need visit only ($mn + N_{edge}$) pixels, where the image resolution is m n pixels and there are N_{edge} edge pixels.

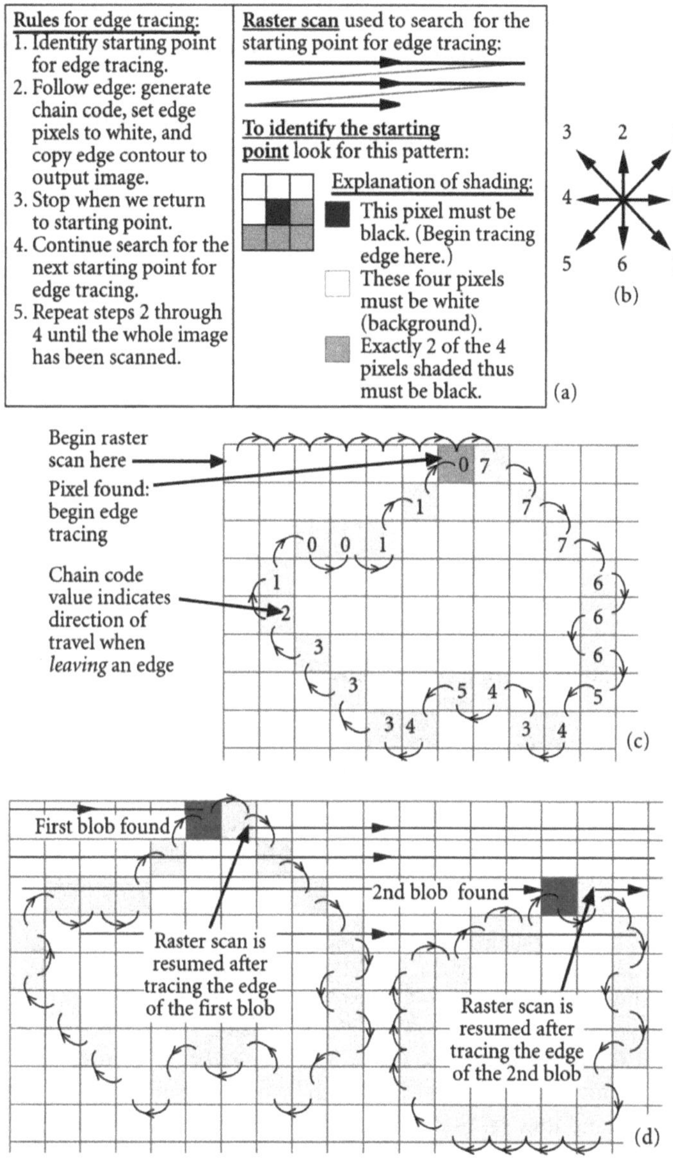

Fig. 3.9 Edge tracing. (a) Rules (b) Edge code directions (c) Locating the starting point and following the edge of an isolated blob. (d)Tracing the edges of multiple blobs

b) Use preprocessing to label each edge point in the image. The labels (Figure 3.10) indicate the chain code. (Some special symbol is used to indicate non-edge points.) Then, trace the edge to "thread" the chain code elements together into a string [BAT81].

While procedure (b) seems to be more complicated, it is well suited to implementation using dedicated hardware. Either a parallel processor or a pipeline processor can perform the initial edge labelling at high speed. The latter can, for example, perform the initial edge coding in a single video frame period. There then remains a serial process in which N_{edge} pixels are visited, as they are "threaded" together.

In algorithm (b), tracing the edges is performed by a serial process, which uses the binary neighborhood coding scheme explained in Figure 3.10. The preprocessing engine, whether it be a pipeline or parallel processor, assigns a number (a pseudo-intensity value) to each pixel to describe its 8-neighborhood. The serial processor then uses this value as an index to a lookup table, in order to locate the next edge point in the edge to be chain coded. The same binary neighborhood coding will be encountered in several places later in this chapter.

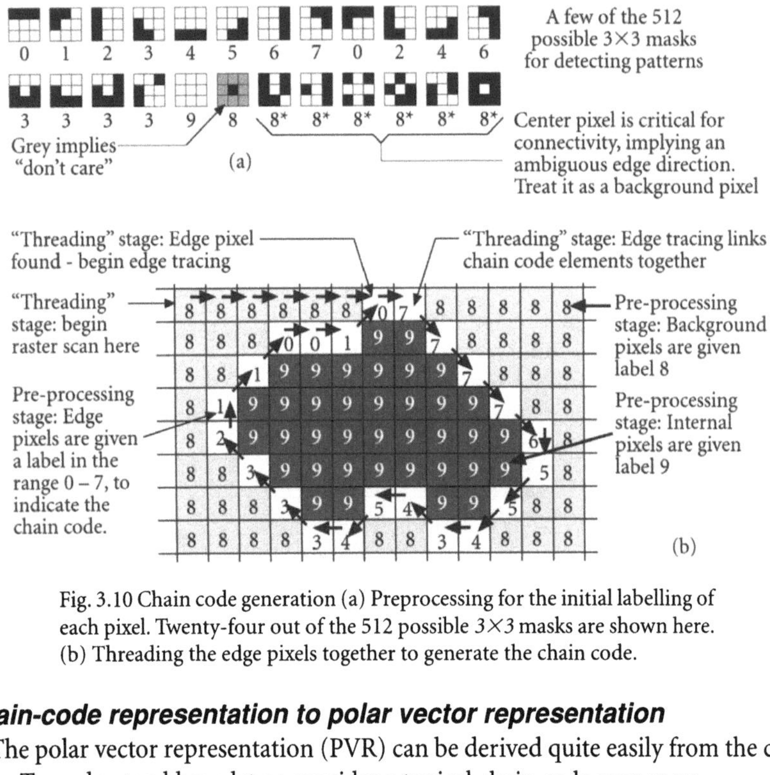

Fig. 3.10 Chain code generation (a) Preprocessing for the initial labelling of each pixel. Twenty-four out of the 512 possible *3×3* masks are shown here. (b) Threading the edge pixels together to generate the chain code.

Chain-code representation to polar vector representation

The polar vector representation (PVR) can be derived quite easily from the chain code. To understand how, let us consider a typical chain code sequence:

…,2,2,2,3,3,3,3,3,3,4,4,4,4,4,4,4,4,4,4,4,4,4,4,4,4,3,3,3,3,4,4,4,5,5,5,5,5,…

Consider Figure 3.11, from which it can be seen that a chain code value 2 is equivalent to a movement around the edge of one unit of travel at an angle of 90°. In polar vector form, this can be represented as $(1, 90°)$. However, a chain code value 3 is equivalent to a movement around the edge of $\sqrt{2}$ units of travel at an angle of 135°. $(\sqrt{2}, 135°)$. The chain code sequence just listed is therefore equivalent to the following sequence of polar vectors:

$$\ldots, (3, 90°), (6\sqrt{2}, 135°), (16, 180°), (4\sqrt{2}, 135°), (3, 180°), (5\sqrt{2}, 225°), \ldots$$

Thus, we have reduced a chain code sequence containing 37 elements to a polar vector representation continuing just 6 elements. It may, of course be possible to merge consecutive PVR elements to form longer vectors at angles that are non-integer multiples of 45°. This will make the PVR even more compact. However, this is venturing into edge smoothing, which will be discussed later, and the realms of high-level image understanding.

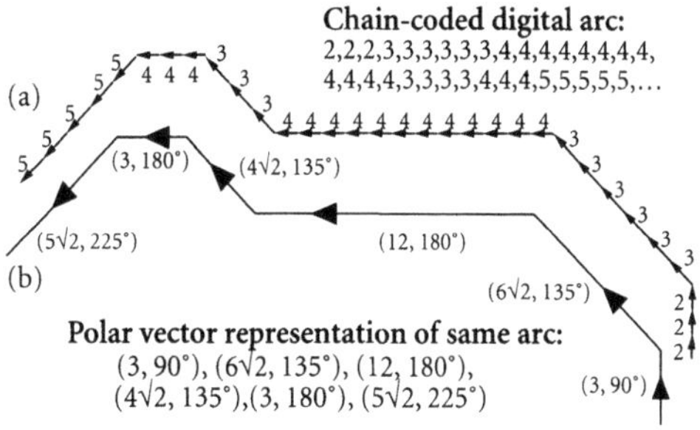

Fig. 3.11 (a) Chain-coded edge. (b) Equivalent polar vector representation (PVR).

Cartesian to polar coordinates

Many artifacts that are made in industry are circular, have circular markings, or possess a "wheel spoke" pattern. In order to simplify the processing of such images, it is possible to use a Cartesian-axis to polar-coordinate-axis transformation. If the axis transformation is centered properly, concentric circles are mapped into parallel lines. A series of radial lines is also mapped into a set of parallel lines that is orthogonal to the first set. Performing the axis transformation is not as straightforward as one might, at first, imagine. The reason is that pixels in the two coordinate systems do not have a one-to-one correspondence to one another (Figure 3.12). If we consider the pixels to be points, we can see how we can use interpolation in order to avoid sampling errors. Suppose that a point (i', j'), defined using in the polar-coordinate axes, maps to the point (i'', j'') located with respect to the Cartesian coordinate axes. Normally, (i', j') does not lie exactly on one of the grid points corresponding to the pixels in the Cartesian-axis image. To estimate the intensity at (i'', j''), we can use simple bilinear interpolation, just as we explained in connection with image rotation (Section 2.6).

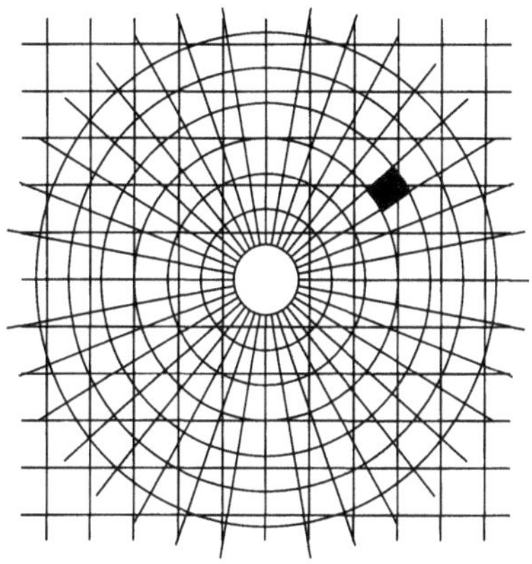

Fig. 3.12 Conversion from Cartesian to polar coordinates. Notice that there is no 1:1 correspondence between pixels in these two image formats.

Color axis conversion

Most color cameras used in machine vision generate the RGB color signals mentioned in Chapter 2. However, the HSI representation is usually regarded as being more useful for color recognition. The transformation is defined in terms of the following equations:

Hue: $H = \cos^{-1}((2R - G - B)/(2((R - G)^2 + (R - B)(G - B))))$

Saturation: $S = 1 - (3\,\mathrm{MIN}(R, G, B))/(R + G + B)$

Intensity: $I = (R + G + B)/3$

Given measured values of the signals *R*, *G*, and *B*, these equations can be evaluated using software. However, this process must be repeated for every pixel in the input image and hence is likely to be very slow. At first sight it seems unlikely that lookup tables could be used effectively, since there are three variables. However, in most applications, we can safely restrict each of the RGB channels to 6 bits without serious loss of color information. Even if we have to use 8 bits/channel, the situation is not too severe. The point is that, by using limited-precision color representation, we make lookup tables quite effective. Since there are three color channels, each with 6 (or 8) bits/channel, the total number of possible input conditions is $262\,144 = 2^{18}$ (or $16\,777\,216 = 2^{24}$). There are three signals (*H*, *S* and *I*) to be computed, each of which requires 6 (or 8 bits). Hence, we require a total of 0.75 MB (or 48 MB) of storage for the color lookup tables. The cost of ROM/RAM is quite modest, so color lookup tables provide a convenient option for color axis conversion. Notice that lookup tables based on RAM/ROM can perform the color transformation in real time on a digitized standard video signal.

3.2.2 Processing a binary image as a grey-scale image

Two separate series of operators appropriate for processing grey and binary images were defined in Chapter 2. At first sight, it may seem unlikely that grey-scale processing algorithms could be applied to good effect on binary images but, in fact, there is a lot of benefit to be obtained by this simple ploy. For example, we can smooth the edge of a blob in a binary image by applying a lowpass blurring filter, such as *lpf*, a few times and then thresholding at the mid-grey level. (We assume that the binary image intensity values are stored as 0 representing black and 255 denoting white.) This process will remove thin hair-like protuberances and "cracks" on the edge of a blob. The same grey-scale operator sequence [*M*•*lpf*, *thr*] can also eliminate so-called salt-and-pepper noise (consisting of small well-scattered black and white spots). The effect achieved is very similar to that of the binary operator sequence [*N*•*lnb*, *N*•*snb*, *N*•*snb*, *N*•*lnb*], where *N* is instantiated to some suitable (small) integer value.

The grass-fire transform is one obvious way in which a grey-scale image can be used to good effect to make it easier to analyze a binary image (Figure 2.10). This operator will be discussed in more detail later in this chapter. Recall that the result of the PIP operator *gft* is a grey-scale image in which intensity indicates the distance to the closest edge point. In addition, the operators *cny* and *cnw* also generate grey-scale output images from a binary image while still preserving the shapes of blobs. (The binary image can be reconstructed by taking black as 0 and all values in the range 1 to 255 as white).

The general linear convolution operator can be used to detect specified patterns in a binary image. For example, the diagonal edge

W	*W*	*W*
W	*X*	*B*
B	*B*	*B*

X = don't care

can be detected by [*con(1,1,1,1,0,-1,-1,-1,-1), thr(255)*].

This process can be taken further. The binary neighborhood coding scheme introduced earlier can be implemented with the general convolution operator using the following weights:

1	2	4
8	0	16
32	64	128

(The PIP command is [*con(1,2,4,8,0,16,32,64,128)*], which generates the value 15 [$= 255\times(1+2+4+8)/(1+2+4+8+16+32+64+128)$] on the diagonal edge pattern above.)

The PIP operator *rin* counts the number of white pixels in each row when it is applied to a binary image, while *rox* may be used to find the "shadow" of a blob.

The grey-scale operator *max* is very useful for superimposing white lettering (in a binary image) onto a grey-scale picture. On the other hand, a common ploy when processing grey-scale images containing several objects is to process the image first to produce a blob defining the edge of each object. We then select one blob and combine it with the original grey-scale image using the operator *min.*

Several binary operators can be implemented directly using their grey-scale counterparts. Here are some examples:

Binary	Grey Scale
exw	lnb
skw	snb
ior	max
xor	sub, thr(127,127), neg
bed	sed (approx.)
not	neg
cnw	con(1,1,1,1,1,1,1,1,1)
wmr	con(-1,-1,-1,-,1,16,-1,-1,-1,-1), thr(128, 254)

3.2.3 Using images as lookup tables

Images can be used as lookup tables. For example, a general image-warping transformation can be achieved with two images used as lookup tables. In the following equation, $\{u(i,j)\}$ and $\{v(i,j)\}$ are two images which, between them, define the warping function $c(i,j) = a(u(i,j), v(i,j)),\ 1 \leq u(i,j) \leq m,\ n \leq W.$

The output image is $\{c(i,j)\}$; the input image is $\{a(u,v)\}$. The pixel (i,j) in the output image $c(i,j)$ is derived from that pixel in the input image whose *x*-coordinate is $u(i,j)$ and whose *y*-coordinate is $v(i,j)$ (Figure 3.13). Notice that the number of grey levels must not be smaller than the number of rows and columns in the images. If there were no warping at all, $u(i,j) \equiv i$ and $v(i,j) \equiv j$. This technique will quite adequately perform geometric transformations where there is a small amount of warping; for example due to slight pin-cushion or barrel distortion, and is suitable for correcting for lens distortion, or the distortion caused by tube-camera scanning errors. It is also suitable for correcting trapezium (trapezoidal) distortion,

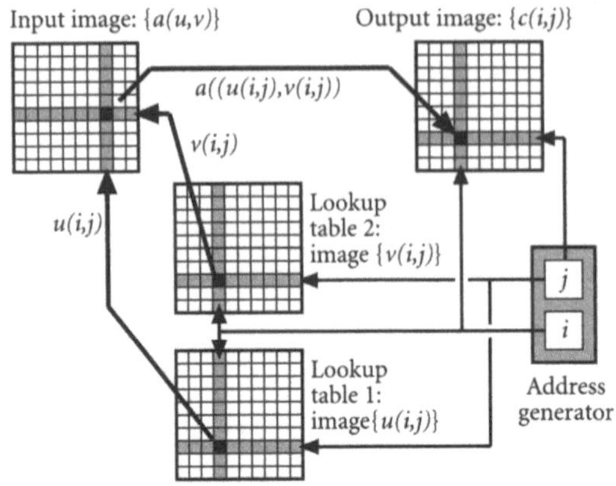

Fig. 3.13 Using images as lookup tables for image warping.

which occurs when a camera looks obliquely onto a surface. However, it does have problems if the warping is very severe. For example, an approximate Cartesian-to-polar coordinate mapping can be achieved in this way but there are severe difficulties near the center of the output image. Figure 3.12 shows why this is so. Since there is no interpolation, each output pixel is derived from just one input pixel. As a result, the mapping is inaccurate close to the center of the image. However, an annulus whose inner radius is not less than about half of its outer radius can be mapped quite accurately to a rectangle by this method.

3.3 Redefining algorithms

In this section, we consider how an algorithm may be reformulated while its performance remains unaltered. In later parts of this chapter, we will discuss approximate and heuristic methods.

3.3.1 Convolution operations

There are two ways to simplify the calculations required by the linear convolution operator [*con*]:

1. Decomposition into two 1-dimensional operators, acting separately along the vertical and horizontal axes of the image.
2. Iterative application of small-kernel filters, to produce the same effect as a large-kernel filter.

We will now consider these separately in turn and then combine them later.

Decomposition

Consider the 7×1 linear convolution operator with the following weight matrix:

1	1	1	1	1	1	1

This filter blurs along the horizontal axis only; vertical intensity gradients are smoothed, while horizontal ones are not. On the other hand, this 1×7 filter

1
1
1
1
1
1
1

blurs horizontal intensity gradients but not vertical ones. Applying them in turn produces exactly the same effect as the following 7×7 filter:

1	1	1	1	1	1	1
1	1	1	1	1	1	1
1	1	1	1	1	1	1
1	1	1	1	1	1	1
1	1	1	1	1	1	1
1	1	1	1	1	1	1
1	1	1	1	1	1	1

Of course, this filter blurs in all directions. If the normalization constants given in Section 2.2.4 are applied separately for the two 1-dimensional filters, the final result is also normalized correctly. Notice also that the simpler 1-dimensional filters can be applied in either order, without altering the final result.

Let us consider another simple example, where the weights are not all equal to unity and some are negative. Recall that the formula for the Sobel edge detector (Section 2.2.5) consists of two parts, each combining blurring in one direction and intensity differencing in the orthogonal direction. Here is the formula for the part which performs horizontal blurring and vertical differencing:

$$E \leftarrow (A + 2B + C) - (G + 2H + I)$$

This operation can be performed as a horizontal blurring process using the 1x3 weight matrix

1	2	1

followed by calculating the vertical difference using this 3×1 weight matrix:

1
0
-1

These filters can be applied in either order.

The other component of the Sobel edge detector is $E \leftarrow (A + 2D + G) - (C + 2F + I)$, which can be computed by applying the 1×3 differencing filter with weight matrix

1	0	-1

followed by vertical blurring with this weight matrix:

1
2
1

Again, these operations can be applied in either order. Hence, the Sobel edge detector can be performed using four 1-dimensional filters, which may be easier than implementing it directly with 2-dimensional filters.

Decomposition is important because it replaces a "difficult" 2-dimensional filtering operation by two simpler ones. Filtering along the rows or columns of a raster-scanned image is very straightforward using standard filtering techniques. It is possible to hold the interim results from the first 1-D (row scan) filter in a RAM-based store. This is then rescanned in the orthogonal direction (column scan), thereby generating another digital video signal which can be filtered (Figure 3.14).

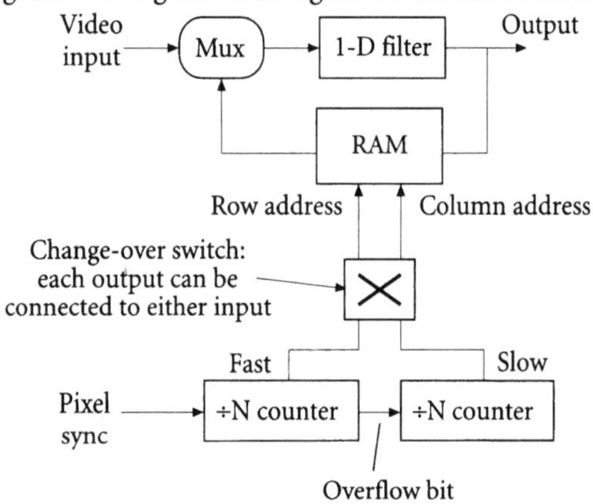

Fig. 3.14 Schematic circuit for implementing decomposable filters. In the first phase, the input image is scanned left-to-right/top-to bottom. (The signal on the line labelled "output" is ignored at this stage.) The image resulting from the row-filtering operation is stored in the RAM, filling it in the same manner. Then, in the second phase, the image is read out of the RAM, scanning the image in the order top-to-bottom/left-to-right. The output signal is now taken directly from the line leading out of the 1-D filter.

It is important to note that not all 2-D local operators can be decomposed into two orthogonal 1-D filters. To see that this is so, we only have to compare the number of degrees of freedom of a 2-D filter (49 for a 7×7 filter), with the equivalent figure for the two decomposed filters ($14 = 1 \times 7 + 7 \times 1$). However, most of the more important filters can be decomposed in this way, particularly if they possess both vertical and horizontal symmetry. (The number of degrees of freedom falls to 9 for 7×7 bi-symmetrical filters.)

Iterated filters

Consider the following filter

1	1	1
1	1	1
1	1	1

applied repeatedly to a very simple image consisting of zero (black) everywhere, except for a single central white pixel. For simplicity, we will ignore normalization and assume that the white pixel has an intensity equal to 1. The results of applying the filter once (left) and twice (right) are shown below:

...	...	...	...	...	...	...	...	...	...	...
...	0	0	0	0	0	0	0	0	0	...
...	0	0	0	0	0	0	0	0	0	...
...	0	0	0	0	0	0	0	0	0	...
...	0	0	0	1	1	1	0	0	0	...
...	0	0	0	1	1	1	0	0	0	...
...	0	0	0	1	1	1	0	0	0	...
...	0	0	0	0	0	0	0	0	0	...
...	0	0	0	0	0	0	0	0	0	...
...	0	0	0	0	0	0	0	0	0	...
...	...	...	...	...	...	...	...	...	...	...

...	...	...	...	...	...	...	...	...	...	...
...	0	0	0	0	0	0	0	0	0	...
...	0	0	0	0	0	0	0	0	0	...
...	0	0	1	2	3	2	1	0	0	...
...	0	0	2	4	6	4	2	0	0	...
...	0	0	3	6	9	6	3	0	0	...
...	0	0	2	4	6	4	2	0	0	...
...	0	0	1	2	3	1	1	0	0	...
...	0	0	0	0	0	0	0	0	0	...
...	0	0	0	0	0	0	0	0	0	...
...	...	...	...	...	...	...	...	...	...	...

Two more applications yield the following images:

...	...	...	...	...	...	...	...	...	...	...
...	0	0	0	0	0	0	0	0	0	...
...	0	1	3	6	7	6	3	1	0	...
...	0	3	9	18	21	18	9	3	0	...
...	0	6	18	36	42	36	18	6	0	...
...	0	7	21	42	49	42	21	7	0	...
...	0	6	18	36	42	36	18	6	0	...
...	0	3	9	18	21	18	9	3	0	...
...	0	1	3	6	7	6	3	1	0	...
...	0	0	0	0	0	0	0	0	0	...
...	...	...	...	...	...	...	...	...	...	...

...	...	...	...	...	...	...	...	...	...	...
...	1	4	10	16	19	16	10	4	1	...
...	4	16	40	64	76	64	40	16	4	...
...	10	40	100	160	190	160	100	40	10	...
...	16	64	160	256	304	256	160	64	16	...
...	19	76	190	304	361	304	190	76	19	...
...	16	64	160	256	304	256	160	64	16	...
...	10	40	100	160	190	160	100	40	10	...
...	4	16	40	64	76	64	40	16	4	...
...	1	4	10	16	19	16	10	4	1	...
...	...	...	...	...	...	...	...	...	...	...

Inspection of these results shows that we have progressively blurred the spot, which began as a 1-pixel entity and was transformed into an indistinct "fuzz" occupying 9×9 pixels. The same result could be obtained by applying the following 9×9 filter once:

1	4	10	16	19	16	10	4	1
4	16	40	64	76	64	40	16	4
10	40	100	160	190	160	100	40	10
16	64	160	256	304	256	160	64	16
19	76	190	304	361	304	190	76	19
16	64	160	256	304	256	160	64	16
10	40	100	160	190	160	100	40	10
4	16	40	64	76	64	40	16	4
1	4	10	16	19	16	10	4	1

It is, of course, no accident that the weights in this matrix are exactly the same as the values in the fourth image matrix above.

By combining the decomposition and iterated filter techniques, we can simplify this even further. We can obtain the same result as the 9×9 filter by applying the

1×3 1-D *row* filter with weight matrix [*1, 1, 1*] four times, and then applying the 3x1 *column* filter with weight matrix [*1, 1, 1*] four times.

3.3.2 More on decomposition and iterated operations

The decomposition technique just described can be applied to a number of other operators. For example, the PIP operator *lnb* can also be separated into row and column operations, thus:

$$E \Leftarrow \text{MAX}(D,E,F) \text{ followed by } E \Leftarrow \text{MAX}(B,E,H)$$

The operator *lnb* computes its output using all of the 8-neighbors. However, if we had defined *lnb* in a different way, using only the 4-neighbors, decomposition in this sense would not be possible.

So far, we have used the results of two filters in series, so that the result of one image-processing (row) operator is applied to another (column) operator. However, we could apply two 1-D filters in parallel and then combine the results. Suppose we perform the following operations

$$E \Leftarrow \text{MAX}(D,E,F) \text{ and } E \Leftarrow \text{MAX}(B,E,H)$$

in *parallel* and then combine the results using the dyadic operator, *max*. The result is exactly the same as the following operator, which combines only the 4-neighbors:

$$E \Leftarrow \text{MAX}(B,D,E,F,H)$$

We will refer to this process later, so we will use the mnemonic *lnb4*, to emphasis that it uses only the 4-neighbors. The equivalent decomposition process can be defined for the related operator *snb*.

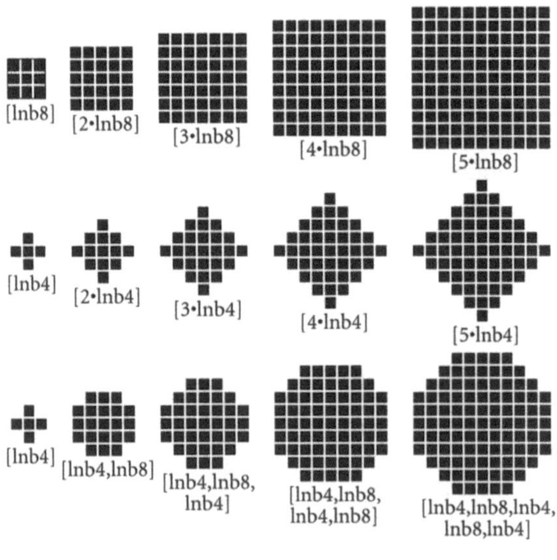

Fig. 3.15 Applying the dilation operators *lnb* and *exw* iteratively. (*lnb* operates on grey-scale pictures, while *exw* operates on binary images.)

By repeating the operator *lnb* or *snb*, we can extend the size of the processing window (Figure 3.15). For example, the PIP sequence *N•lnb* finds the maximum intensity within each window of size (*2N+1*) pixels. By repeating *lnb4* several times, we find the maximum intensity within a larger diamond-shaped region.

By alternating *lnb* and *lnb4*, we can find the maximum intensity within a regular octagonal region surrounding each pixel. Since this octagon is a better approximation to a circle than either *lnb* or *lb4* alone produces, we may find this combination attractive in certain applications. In the figure, [*5•lnb*] finds the largest intensity within an 11×11 window, while [*5•exw*] expands a single white pixel to cover an 11×11 square. Alternating *lnb4* and *lnb8* (or *exw4* and *exw8*) generates an octagonal covering. The erosion operators *snb* and *skw* operate similarly on black pixels.

Dilation [*dil*] and *erosion* [*ero*] are two important operators for processing binary images. Since they are so closely related to *lnb* and *snb* respectively, it will hardly come as any surprise that they can be decomposed in a similar way. Indeed all of the remarks made about *lnb* also relate directly to *dil* and those about *snb* to *ero*. This includes the comments made about the 4-neighbor versions.

3.3.3 Separated Kernel Image Processing using finite-State Machines

In a series of 23 papers beginning in 1994, Waltz (WAL94a...WAL99b, HUJ95a, HUJ95b, HAC97a, MIL97) described an extremely powerful and radically new paradigm for the implementation of many important neighborhood image-processing operations. These techniques, taken together, have been given the acronym SKIPSM (*Separated Kernel Image Processing using finite-State Machines*). The improvements in performance (speed or neighborhood size) provided by SKIPSM are so great in some cases as to demand a radical rethinking of what is or is not practical in industrial applications. The SKIPSM approach is mentioned briefly in this section on operator decomposition because it allows the decomposition of a broad range 2-D (and even 3-D or higher) linear and nonlinear operators into a sequence of 1-D operations. This includes many operators which were previously thought to be non-separable. An example of intermediate difficulty will be presented here. SKIPSM can do things much more powerful than these.

SKIPSM example:

We wish to apply *six 7×9* binary erosion structuring elements (SEs) *on a single raster-scan pass* through the input image. The six SEs are shown in Figure 3.16.

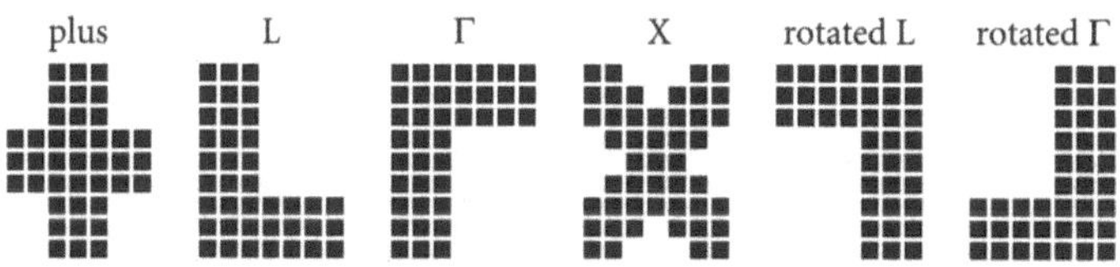

Fig. 3.16 The six *7×9* binary erosion structuring elements applied simultaneously. Among these, there are only seven distinct row patterns to be encoded by the row machine: 0011100, 1111111, 1110000, 1100011, 1110111, 0111110, and 0000111.

Using the automated SKIPSM code-generation programs, these six operations are individually separated into row operations and column operations. The row operations are combined into a single recursive FSM (finite-state machine), the *row machine.* The six column operations are combined into a second FSM, the *column machine.* These two FSMs are then typically (but not necessarily) implemented as lookup tables (LUTs), as shown in Figure 3.17. Each of the six separate binary outputs is assigned to a single bit plane of the 8-bit output image.

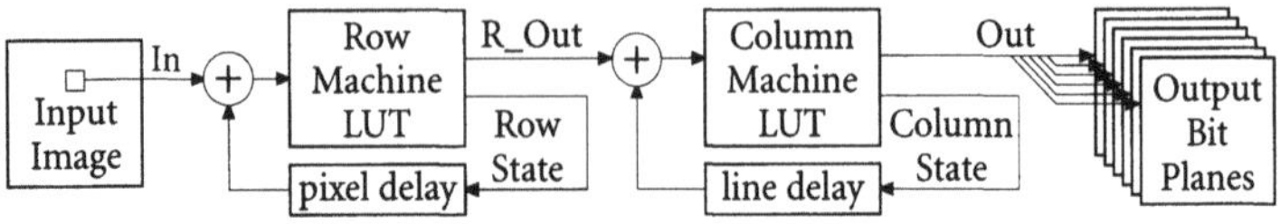

Fig. 3.17 SKIPSM flow chart (applicable to both hardware and software implementations) showing the row and column machines being implemented as lookup tables.

The pixel values from the input image are fed to this system in the usual raster-scan order. Each pixel is fetched *once and only once,* because all of the necessary "history" of the image neighborhood is contained in the *states* of the two FSMs.

The combined row machine has 216 states and, because the input has two levels, its LUT therefore has only 432 addresses. The combined column machine has 1,748 states and 16 input levels. Thus, its lookup table has 10,488 addresses. This is much larger than the row-machine LUT, but is still not excessively large, by today's standards. These LUTs are generated automatically by the SKIPSM software.

Please note a *very* important point here: The execution time for SKIPSM lookup-table-based implementations is *very* fast (requiring only four LUT accesses and two additions or bitwise-ORs), *independent of the size of the neighborhood and independent of the number of simultaneous operations performed.* Large operations and/or multiple operations require larger LUTs, but execute in the same very fast time as small ones. This fact has been demonstrated repeatedly on actual systems. The SKIPSM approach is described in more detail in Section 5.6.

3.3.4 Binary image coding

In Section 3.2.1 (Figures 3.8, 3.9, and 3.10), we introduced the idea of the binary image coding scheme and in Section 3.2.2 described how the linear convolution operator [*con*] can be used to generate it. A wide range of binary image-processing algorithms can make good use of this coding technique, which is incorporated into the implementation scheme outlined in Figure 3.18. The important point to note is

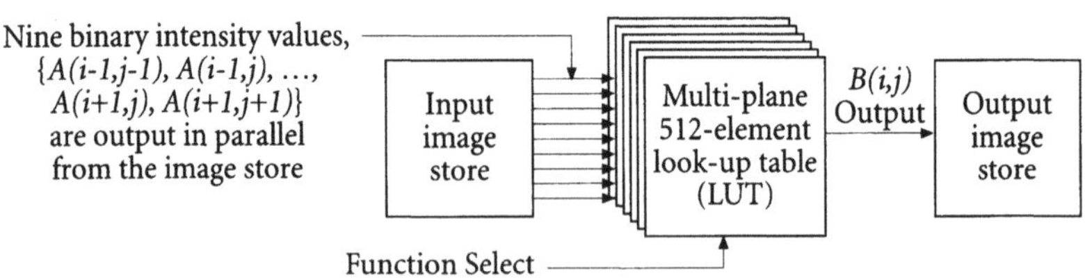

Fig. 3.18 Using a lookup table to perform a binary image-processing operation. The function implemented by this circuit can be altered very quickly by changing the "Function Select" input.

that any logic function of the 9 pixel intensities found within a 3×3 window can be "computed" by a lookup table. Table 3.1 presents a partial list of such functions.

Table 3.1 A partial list of binary functions which can be implemented using a lookup table.

Operation	PIP mnemonic	Comments
Dilate	dil	
Erode	ero	
Binary edge detection	bed	
Remove isolated white points	wpr	
Count white neighbors	cnw	
Critical points for connectivity	cny	
Skeleton joints	jnt	
Skeleton limb ends	lme	
Extend ends of arcs (skeleton limbs)	eel	
Euler number	eul	Extra processing needed
Chain code – preprocessing		See Section 3.2.2
Grass-fire transform (1 stage)	gft	Included in iterative loop
Skeletonization	ske	Included in iterative loop

Notice that the binary image coding and the implementation scheme of Figure 3.18 can be used to good effect in the grass-fire transform [*gft*] and the skeletonization operator [*ske*]. In both cases, a complex logical function is applied within an iterative loop, so high-speed implementation is of increased importance. Implementations using this configuration will be considerably slower than those for the corresponding SKIPSM implementations, as described in Section 3.3.3, but the required lookup tables would be smaller. Therefore, these might be preferred where large lookup tables are not feasible.

3.3.5 Blob connectivity analysis

The task to which we will give our attention in this section is that of labelling blobs in a binary image such as shown in Figure 3.19. All pixels in each 8- or 4-connected blob are assigned an intensity value to indicate the blob identity. Expressed in this way, the requirement seems very straightforward. However, when we consider general shapes about which nothing is known in advance, the real complexity of the algorithm becomes apparent. In particular, intertwined spirals or interlinked comb-like structures make the task far from straightforward.

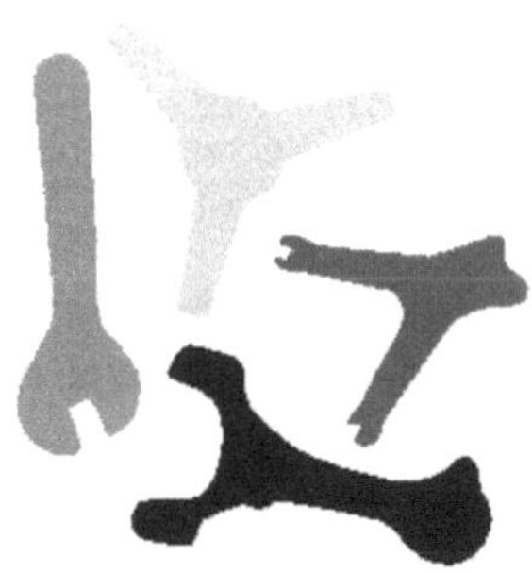

Fig. 3.19 All pixels within a given blob are given the same intensity value. Each blob is assigned a different intensity.

Several important image analysis tasks are made possible by blob labelling, including these:

- Identifying the blob with the largest area.
- Eliminating from an image all blobs that are smaller than a predetermined size.
- Associating the holes (or lakes) within a blob with the outer edge of that blob.

First blob-labelling scheme

We will describe three quite different algorithms for blob labelling. The first algorithm is explained in Figure 3.20 and is based on the array representation of a binary image. This algorithm requires two passes through the image. Thus, each pixel is visited twice. During the first pass, each pixel is assigned a provisional numeric value (label). A blob with a complicated shape may receive two or more different labels. On this first pass, each new "finger" that points upwards is assumed to be a new blob and hence is assigned a new label. During the second pass, these sub-regions are renumbered as required, so that all pixels within a given blob are given the same label.

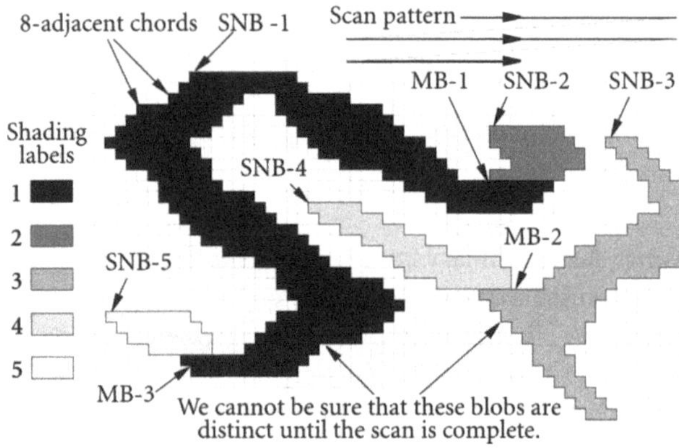

Fig. 3.20 First blob-labelling algorithm. This takes place in two stages, each comprising a raster scan of the image. During the first stage, each pixel is assigned an initial label. We suspect that a new blob has been found each time we encounter the binary pattern

W W W
W B X
X X X

where W = white (background), B = black (object), and X = "Don't care." Each pixel identified thus is given a new initial label (pixels SNB-1 to SNB-5). Other white pixels that are discovered to be 8-connected to SNB-i are given the same initial label (i.e., value *i*). The pixels marked MB-1 to MB-3 indicate where two regions that were originally thought to be separate are discovered to be 8-connected and hence form parts of the same blob. For example, at MB-1, we find that initial labels 1 and 2 represent regions within the same blob and should therefore be merged. Later in the first raster scan, at MB-3, we discover that the same blob also includes region 5. During the second raster-scan, regions 2 and 5 are reassigned with label 1. Of course, regions 4 and 3 are also merged, by mapping all pixels with initial label 4 to label 3.

Second blob-labelling scheme

The second blob-labelling scheme represents a fairly minor extension of the algorithm for generating the chain code (Section 3.2.1 and Figure 3.9(a)), and is shown in Figure 3.21. The modified algorithm requires that only the blob edge pixels in the input image be white. All other pixels, including the background and pixels "inside" the blobs, should be black. This can be achieved by applying a simple "binary edge" processing sequence before the following edge tracing process is started:

1. Set a counter (denoted here as *Counter*) equal to 1.
2. Initialize the output image to black (level 0).
3. Initialize parameters *U* and *V* to 1.

4. Start a raster scan of the input image at point [*U*,*V*]. When the first white pixel is encountered, record its position coordinates, [*X*Counter, *Y*Counter], then proceed to the next step.
5. Trace the edge contour just found. At each edge pixel [*X*edge, *Y*edge] perform the following operations:
   a) Set pixel [*X*edge, *Y*edge] in the input image to level 0 (black).
   b) Set pixel [*X*edge, *Y*edge] in the output image to level *Counter*.
   c) Stop edge tracing when we revisit pixel [*X*Counter, *Y*Counter] in the input image and proceed to the next step
6. Set [*U*,*V*] = [*X*Counter, *Y*Counter]
7. Increment *Counter*.
8. Repeat steps (4) to (7), using the new values for *Counter*, *U* and *V*.

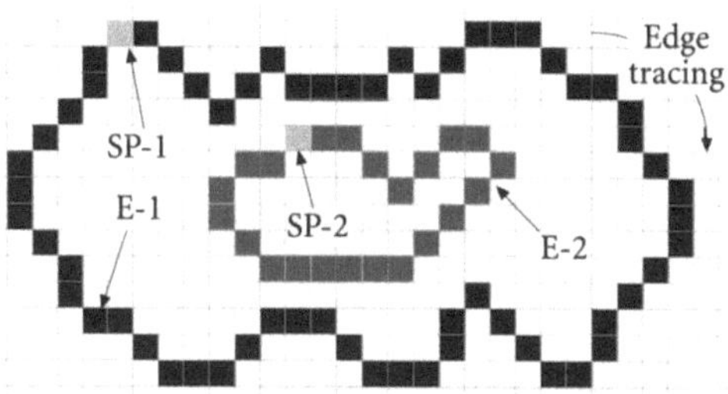

Fig. 3.21 Second blob-labelling algorithm. Here there are two blobs, with edges E-1 and E-2. Following the procedure (step 4), the first starting point (SP-1) is found, and edge E-1 is traced back around to SP-1 (step 5). Then the second starting point (SP-2) is found, and so on.

An important feature of this algorithm is that it cannot tolerate "hairy" edges, since they cause the edge tracer to travel along "blind alleys." ("Hair" is an open-ended digital arc that is only one pixel wide and protrudes outwards from the outer edge contour into the background or inwards from the inner edges of lakes (see Figure 3.22). "Hair" can be removed and a closed edge contour created by simple binary image filtering:

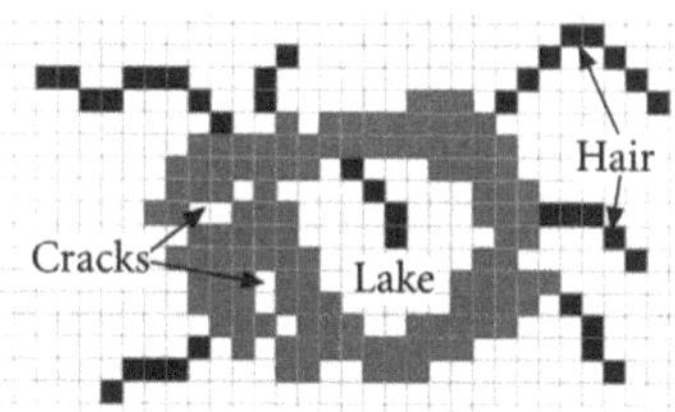

Fig. 3.22 "Hair", "cracks", and a "lake" in a binary object.

skw, exw, exw, wri(solid), xor % Creates an edge just outside object boundary

or

skw, exw, wri(solid), skw, xor % Creates an edge just inside object boundary

One small task remains: differentiating between outer and inner edges. We can use the following masks to identify outer or inner edges (edges surrounding lakes):

Mask for	B B X		*Mask for*	W W X
outer	B W X		*inner*	W W X
edges	X X X		*edges*	X X X

(X indicates don't care.) These masks are applied at step 4 to test the pixel located at [*X*Counter, *Y*Counter] in the preprocessed image. This is the image just saved in the file *solid*. (The masks are not applied to the edge contour image.)

Third blob-labelling scheme

The third method for blob labelling uses a technique borrowed from computer graphics, where it is used for object filling. We must first define our notation.

1. Let $[X_1, Y_1]$ be a point that is known to lie on the edge of or within the blob that we wish to isolate.
2. Let L denote a list of points to be analyzed.
3. Let S be a list. This will eventually hold the addresses of all pixels that are 8-connected to $[X_1,Y_1]$.
4. Let pixel intensities be represented by 0 (black), 1 (white and not yet analyzed), and 2 (white and already analyzed).
5. Let $N(X,Y)$ denote the list of addresses of pixels that are 8-connected to $[X,Y]$ and have intensity equal to 1. (Pixels of intensity 0 and 2 are ignored when computing $N(X,Y)$.)

Here is the algorithm, described as a set of simple rules:

1. a) Initialize L to $[[X_1, Y_1]]$.
   b) Set S to [ ], i.e., the empty list.
   c) Set pixel $[X_1, Y_1]$ to intensity value 2.
2. a) If L is empty, go to Step 4.
   b) Otherwise, find the first element of L and denote this by $[X,Y]$.
3. a) Derive the list $N(X,Y)$.
   b) Set the intensity value of pixel $[X,Y]$ to 2.
   c) Append $N(X,Y)$ to L.
   d) Append $[X,Y]$ to the list S.
4. Finish.

The result is a list (S) which contains the addresses of all of the pixels that are 8-connected to the starting point $[X_1, Y_1]$. However, the result (S) in an inconvenient format, since the image-scanning process is performed as we generate $N(X,Y)$.

In this section, we have seen that three completely different computational techniques can be used to perform blob labelling. We have also seen that sometimes it can be beneficial to store and employ more than one image representation.

3.3.6 Convex hull

The convex hull of a single connected object (S) in a binary image is easily imagined as the region contained within a rubber band stretched around S. The convex hull of a set of disconnected objects can also be envisioned in this way. There are several known algorithms for computing the convex hull of a single blob and rather fewer for the more general multiple-blob case. Algorithms for calculating the convex hull of a single blob can be based on the array, run-code, chain-code, or polar-vector representations. For the sake of simplicity, we will concentrate in this section on algorithms for computing the convex hull of a single connected blob. This situation is also of greatest interest for industrial applications. However, we must make it clear that, strictly speaking, the convex hull is defined for *any* set of points in a plane, whether or not they are connected to each other.

A very useful point to note is that for any given shape (S) there are four very easily identifiable points on its convex hull [WIL89] (see Figure 3.23):

a. a vertical line through the left-most point of S,
b. a vertical line through the right-most point of S,
c. a horizontal line through the top point of S, and
d. a horizontal line through the bottom point of S.

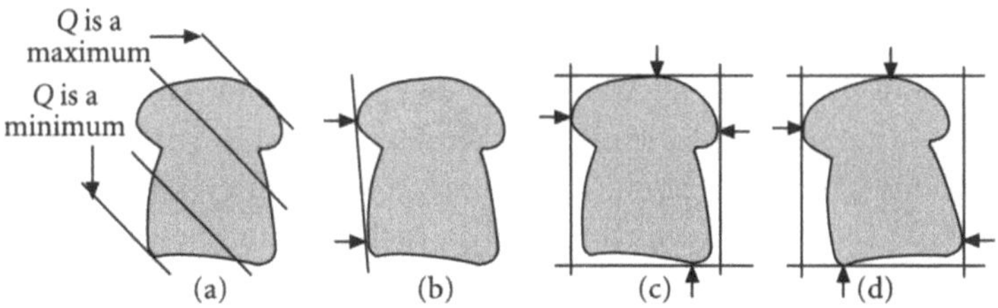

Fig. 3.23 Heuristics for finding a subset of the nodes of the convex hull of a binary object. (a) Finding the values for $[X,Y]$ which maximize/minimize $Q = (Y - mX - c)$ yields at least two nodes for each pair of values (m, c). (b) This heuristic may find more than two nodes. (c) The top-most, right-most, bottom-most, and left-most points of an object indicate (at least) four nodes. (d) Four more nodes can be found by rotating the object and repeating (c).

If we were to rotate S, we could obtain another four points on its convex hull with equal ease. Of course, to find all of the points on the convex hull requires that we rotate S many times and find these four extreme points each time.

This provides the basis for most of the known algorithms for computing the convex hull. Let us now discuss the details of some of these.

Array representation

Consider the set of parallel lines defined by the equation

$$y = x\tan(\theta) + c$$

where c is allowed to vary but θ is a constant. Some of these lines will intersect the given shape S, while others will not. The two lines that are tangential to S are of particular interest, since they define (at least) two points on the convex hull of S (Figure 3.23). We can find their coordinates in the following way:

1. Evaluate the expression $(Y - X\tan(\theta) - c)$ for all points $[X,Y]$ in the set of points which define S. (For convenience, we will also refer to this set as S.) Hence, we compute Q, where $Q = (Y - mX - c)$, for all $(X,Y) \in S$.
2. We record those values of (X,Y) which make Q a maximum and a minimum. Let us denote these values by $(X_{min,\theta}, Y_{min,\theta})$ and $(X_{max,\theta}, Y_{max,\theta})$.
3. Then, $(X_{min,\theta}, Y_{min,\theta})$ and $(X_{max,\theta}, Y_{max,\theta})$ are two points on the convex hull of S.
4. Additional points on the convex hull of S can be found by varying θ and repeating steps 1 through 4.

Notice that for a given value of θ there may be more than one convex hull point on each line (Figure 3.23(b)). This is a bonus and gives us more convex hull points for less computational effort. However, we cannot be sure that we will obtain more than one point per tangent line.

The above procedure can be used as the basis for a number of algorithms for deriving the convex hull. These will now be described.

Scanning the image array

The most obvious way to compute the convex hull of a shape *S* is to repeat steps 1 through 4 a large number of times for small increments of θ. Each pair of convex hull points ($X_{min,\theta}$, $Y_{min,\theta}$) and ($X_{max,\theta}$, $Y_{max,\theta}$) is computed by scanning the whole image array to find values which are members of *S*.

A second and slightly more efficient algorithm can be obtained by performing the operation *bed* on the image array before we apply the procedure just described. The savings in computational effort arise because we calculate only the value of Q for edge points – "internal" points are ignored. Notice, however, that each time we add two more convex hull points to the list, each pixel in the *whole image array* must be "visited" once.

Chain code

A third and even more efficient procedure for deriving the convex hull can be based on the chain code and is exactly as defined above, except that we "visit" only the edge pixels. (Background and "internal" points are simply ignored.) This algorithm requires that we compute ($X_{min,\theta}$, $Y_{min,\theta}$) and ($X_{max,\theta}$, $Y_{max,\theta}$) from the (X,Y)-coordinates of all of the edge points. This is a trivial calculation based on the following recursive procedure: $(X_{i+1}, Y_{i+1}) = ((X_i + H(C_i)), (Y_i + V(C_i))$.

Here, C_i is the chain code value which corresponds to a movement from the ith edge pixel (X_i, Y_i) to the adjacent $(i+1)^{st}$ edge pixel, (X_{i+1}, Y_{i+1}). $H(C)$ and $V(C)$ are derived from the following table:

C	0	1	2	3	4	5	6	7
H	1	1	0	-1	-1	-1	0	1
V	0	1	1	1	0	-1	-1	-1

Of course, we can modify the chain code generator instead, so that it also records the (X,Y) coordinate values of each of the edge pixels as the chain code is being computed. This will save some further computational effort.

Polar vector representation (PVR)

A fourth, radically different, algorithm for computing the convex hull can be defined in terms of the polar vector representation (PVR). Another similar procedure based on the chain code also exists but is less efficient [BAT80]. In the following description, we will make use of the fact that the PVR defines a polygonal representation of a shape. We will use N_v to indicate the number of vertices of this polygon, which we will represent by *P*. The algorithm finds a set of points, which we will call *nodes* and which are the vertices of that polygon which defines the convex hull of *P*. We will use N_n to indicate the number of nodes. The set of nodes is a subset of the set of vertices. Hence, $N_n \leq N_v$. Also, for many shapes, particularly after edge smoothing has been applied, we find that *Nv* is *very much less* than the number of edge pixels (*N*edge).

The first step in the algorithm is to find any one node and a reference axis which passes through it. A simple way to do this is to find the right-most vertex and then draw a vertical line through it (see Figure 3.24(a)). Any other line that is tangential to the polygon is acceptable as the initial reference axis. On subsequent steps in the algorithm, the reference axis is redefined by the line joining the two most recently discovered nodes.

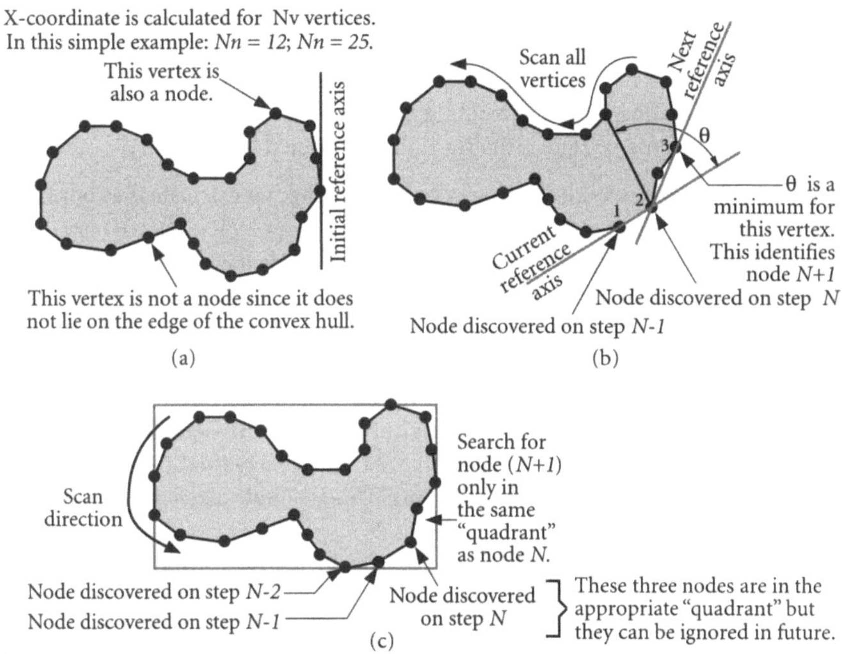

Fig. 3.24 Calculating the convex hull from the polar vector representation. (a) Initial reference line is the vertical line through the right-most vertex. (b) The $(N+1)^{st}$ node is derived from the $(N-1)^{st}$ and N^{th} nodes. (c) Segmenting the PVR at the top-most, right-most, bottom-most, and left-most vertices, creates four sub-tasks which can be performed in parallel.

Consider Figure 3.24(b), which shows the angle θ formed between the reference axis and the line which joins the most recently found node (labelled '2') to another vertex. The next node to be identified (3) is therefore identified as that vertex where the angle θ is a minimum. To minimize θ, we must analyze at most $(Nv - 2)$ vertices. (Vertices labelled '1' and '2' are already known to be nodes of the convex hull and therefore need not be considered again. However, we can improve the algorithm in two ways:

a) We do not need to calculate the angle θ for those vertices that lie between the node that was identified initially and the one discovered most recently.

b) We need analyze only vertices which lie in the same "quadrant" of the polygon *P*. See Figure 3.24(c). The justification for this follows directly from Figure 3.23.

As a crude approximation, rule a) halves the amount of computing effort needed, while rule b) reduces it to 25%. These two savings are cumulative. Hence, the total computational effort in finding all nodes involves roughly ($N_n N_v/8$) angle calculations. An adaptation of this algorithm but based on the chain code has also been devised [MAR80].

Parallel implementation of the PVR/chain code algorithm

It is possible to break the calculation into an arbitrary number of smaller tasks, which can be performed in parallel by a set of processors operating independently of each other. The number of parallel sub-tasks and hence processors is unlimited. The parallel version of the PVR algorithm relies on three phases: finding a subset of the nodes of the convex hull; segmenting the edge, and calculating the intermediate nodes for each segment. Here is the algorithm explained in more detail.

a) Find a subset of nodes of the convex hull. To do this, we calculate the coordinates $[X, Y]$ of each point which maximizes the quantity $Q = (Y - X\tan(\theta) - c)$ for $\theta = 0, 360/N, 720/N, 1080/N, \ldots, 360(N-1)/N$, where *N* is the number of processors used. Of course, all points must lie on the edge of the input shape. This is an *N*-way parallel process; each process can be performed independently by a (serial) processor.
b) Each processor passes the coordinates of the nodal point it has just identified to one of its neighbors. We are effectively sub-dividing the edge of the input blob into *N* segments. In Figures 3.23 and 3.24(c), we suggested that we break the edge of the input blob into four sections. This is simply a generalization of that process. See Figure 3.25.
c) Find the remaining nodal points, as we did in the fourth PVR method. Each processor has the coordinates of two nodes and therefore finds the other nodes which lie between them.

This algorithm is important because it has deliberately been designed to fit onto an *N*-way parallel processor. This means, of course, that there is the potential to achieve high operating speed.

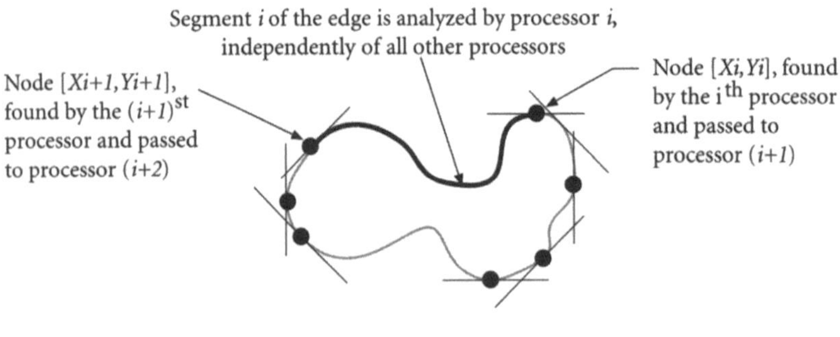

Fig. 3.25 The convex hull calculation lends itself to implementation in parallel processors.

Run code

Another ingenious algorithm for finding the convex hull uses the run code [RUT78]. However, this will not be explained in detail here, since it is not possible to do so quickly and easily. This algorithm is suitable only for a serial processor and hence is most appropriate for implementation in software.

3.3.7 Edge smoothing

As an abstract concept, edge smoothing needs very little explanation, even to the reader with little prior experience of machine vision. The algorithmic procedures which nominally perform edge smoothing are highly varied and are based on several different image representations. There are subtle differences among these competing techniques. These differences may be important in some applications but for most are insignificant. A procedure which embodies edge smoothing and relies upon these minor differences is likely to be over-sensitive for most industrial applications. In such situations, the system design engineer is more likely to find other more profitable ways of improving the machine. Adjusting the lighting/viewing subsystem is often far better! [PVB]

It should be understood that the objective for edge smoothing is not specified in absolute terms. Hence, we take it that we may make a free choice of whichever edge smoothing technique happens to fit in best with our broad objectives. We might, for example, find that for a particular system the chain code provides a suitable data structure for implementing a multi-step algorithm. In this event, we would naturally use one of the edge-smoothing operators that is based on the chain code. In certain other situations, however, the hardware might employ a different image representation. In this case, we find ourselves using an edge-smoothing procedure which behaves in a slightly different way. The point is that we simply should not put ourselves in a position where it matters that we select one edge-smoothing algorithm rather than another. This is yet another illustration of a very important general principle: *ease of implementation is as important as performance.*

Edge smoothing in PIP

Here is a PIP program which implements the morphological operator *closing*. This performs one type of edge smoothing, in which thin "cracks" are eliminated (see Figure 3.26).

```
closing(N):- N•exw, N•skw.
```

The related operator, *opening*, may also be defined thus:

```
opening(N):- N•skw, N•exw.
```

This eliminates "hair" (Figure 3.22). Both "hair" and "cracks" can be eliminated by concatenating these two operators:

```
smooth_edge1(N):- closing(N), opening(N).
```

In some circumstances, it may be advantageous to reverse the order of execution:

```
smooth_edge2(N):- opening(N), closing(N).
```

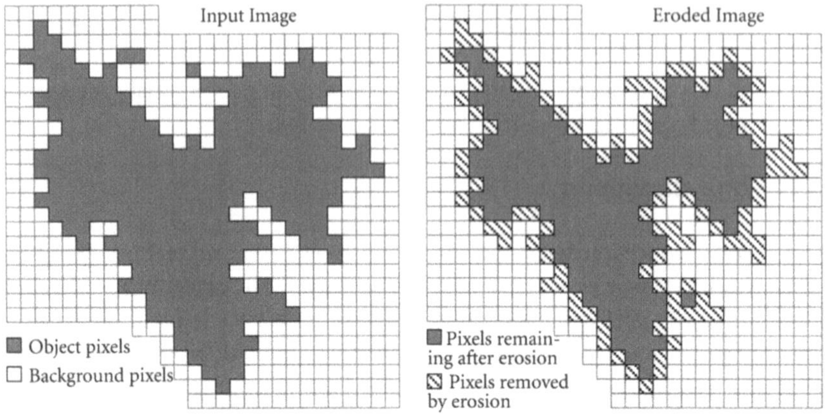

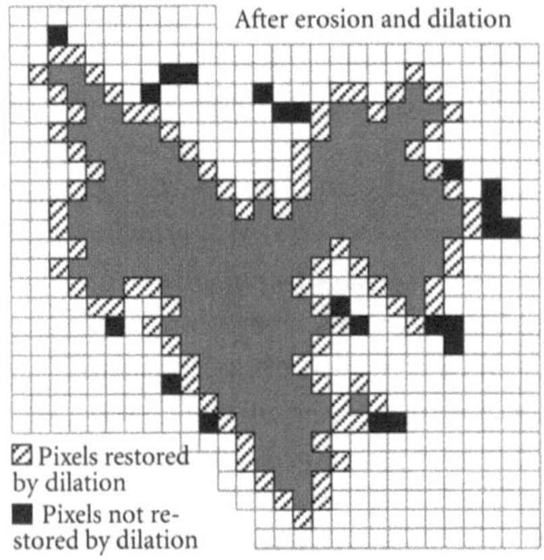

Fig. 3.26 Edge smoothing by erosion and dilation. (Upper left) Input image. (Upper right) After erosion [*skw*]. (Lower left) After erosion and dilation [*skw,exw*].

Smoothing the chain code

Consider the chain code sequence "..., 0, 0, 0, 1,7, 0, 0, 0,...." It is immediately obvious that the subsequence "1,7" is a small glitch. This might arise from camera noise and digitization effects, and should be replaced by "0, 0". This makes the edge represented by the sequence (...0, 0, 0, 0, 0, 0, 0, 0,....) smoother (Figure 3.27(a)).

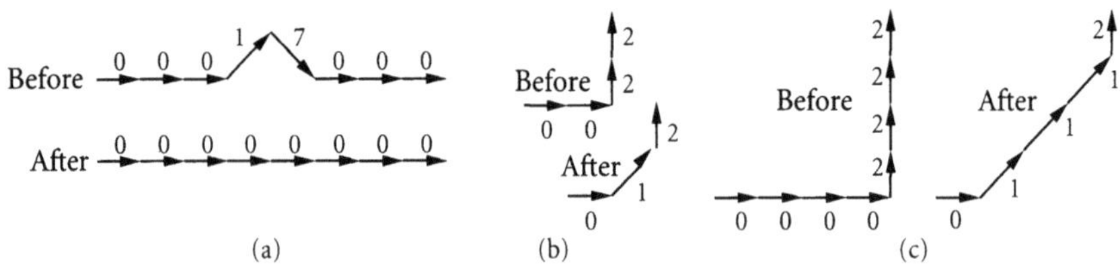

Fig. 3.27 Edge smoothing: filtering the chain code. (a) Replacing glitches. (b) Smoothing corners (mild). (c) Smoothing corners (more pronounced).

In a similar way, the sharp corner represented by the chain code sequence "..., 0, 0, 2, 2, ..." could be replaced by the smoother curve represented by "..., 0, 1, 2, ..." (Figure 3.27(b)). We can, of course, take this further and look for more complex subsequences which represent sharp corners and glitches. For example, the chain code sequence "..., 0, 0, 0, 0, 2, 2, 2, 2, ..." might be replaced by "..., 0, 1, 1, 1, 2, ..." (Figure 3.27(c)).

A fast low-cost edge-smoothing device based upon the chain code can be built using a shift register and lookup table (Figure 3.28). The same general principle could be used to good effect in software.

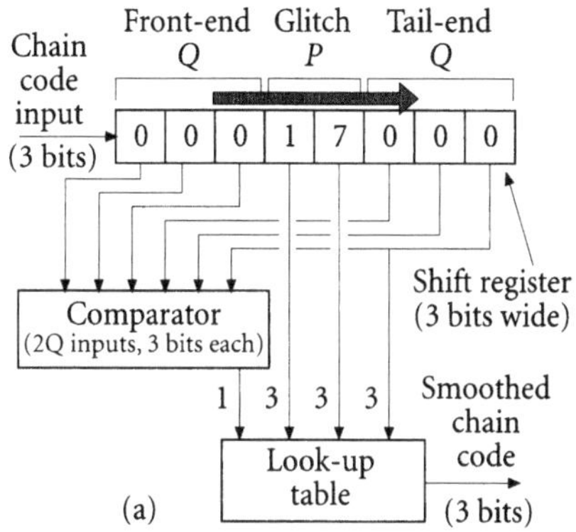

Glitch length P	Front-end/ tail-end length Q	Size of LUT (bits)
2	2	$3 \cdot 2^{10}$
3	3	$3 \cdot 2^{13}$
4	4	$3 \cdot 2^{16}$
5	5	$3 \cdot 2^{19}$
6	6	$3 \cdot 2^{22}$
P	$Q \geq P$	$3 \cdot 2^{(4+3 \cdot P)}$

(b)

Fig. 3.28 Filtering the chain code. (a) This circuit can detect and remove glitches of the form $A^Q,S(P),A^Q$, where P and Q are integers and $S(P)$ is a chain-code sequence of length P. The front-end and tail-end subsequences are both of length Q. (b) Size of the lookup table as a function of glitch length (P) and front-/tail-end length (Q).

Smoothing the polar vector representation

In one of the simplest edge-smoothing schemes based on the PVR, two adjacent elements are merged if they are nearly aligned. Thus, two consecutive PVR elements, $[r_1,\theta_1]$, $[r_2,\theta_2]$ are replaced by a single vector $[r_{1,2},\theta_{1,2}]$ if θ_1 and θ_2 are nearly equal (see Figure 3.29). The resultant vector is obtained using the standard vector addition formulae:

$$r_{1,2} = \mathrm{sqrt}\{[r_1 \cos(\theta_1) + r_2 \cos(\theta_2)]^2 + [r_1 \sin(\theta_1) + r_2 \sin(\theta_2)]^2\}$$

$$\theta_{1,2} = \tan^{-1}([r_1 \sin(\theta_1) + r_2 \sin(\theta_2)] \,/\, [r_1 \cos(\theta_1) + r_2 \cos(\theta_2)])$$

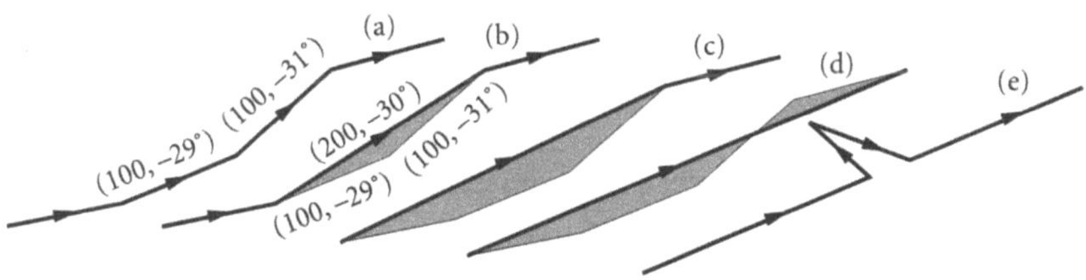

Fig. 3.29 Edge smoothing: using the PVR representation. (a) Original edge, consisting of 4 PVR elements. (b) Replacing two nearly collinear vectors by vector addition. (c) Replacing three vectors. Notice that the error, measured by the shaded area is small, whereas the length of the resultant vector is quite long. (d) Replacing four vectors. Notice that there are both small positive and small negative areas (i.e., shaded areas above and below the resultant vector), which tend to "cancel" one another. (e) This glitch can be detected quite easily since there are two long vectors that are nearly in line and two short ones between them.

The rule for deciding whether or not to merge two PVR elements may be expressed in a number of ways:

1. Merge all pairs of adjacent elements $[r_i,\theta_i]$ and $[r_{i+1},\theta_{i+1}]$, if $|\theta_i - \theta_{i+1}| \leq d_1$, where d_1 is some parameter we choose to suit the application requirements.
2. Merge all pairs of adjacent elements $[r_i,\theta_i]$ and $[r_{i+1},\theta_{i+1}]$, if $|r_{i,j} - r_i - r_{i+1}| \leq d_2$, where d_2 is some parameter we choose to suit the application requirements.
3. Merge all pairs of adjacent elements $[r_i,\theta_i]$ and $[r_{i+1},\theta_{i+1}]$, if the triangle enclosed by the three vectors $[r_1,\theta_1]$, $[r_2,\theta_2]$ and $[r_{1,2},\theta_{1,2}]$ has an area that is less than d_3, where d_3 is some parameter we choose to suit the application requirements.
4. Investigate all pairs of elements in the PVR coding of an edge but merge only one pair at a time. Obviously we could choose to merge a pair of PVR vectors if they minimize any of the criteria used in (1) – (3).

Defining suitable rules for detecting and removing glitches, such as

...,[100,37°],[1,127°],[1,-52°],[100, 37°],...

will be left as an exercise for the reader. (The rules become fairly obvious if we represent the PVR graphically.) Notice that a glitch can always be identified from the PVR as being either "hair" or "crack" (Figure 3.22), since the inside of the blob is always on the left (or right) of the vectors.

Figure 3.30 shows the effects of applying PVR edge smoothing to the outline of a model car. By applying the smoothing operator iteratively, long PVR elements are created, which in this case can be identified easily with the roof, windscreen, and hatchback of the vehicle. The edge-smoothing process is, in effect, performing a feature-identification operation. This very useful side effect is not shared by any other edge-smoothing procedure. This illustrates an important general principle: We must always be alert to such "side effects", since they can significantly alter our judgement of a procedure.

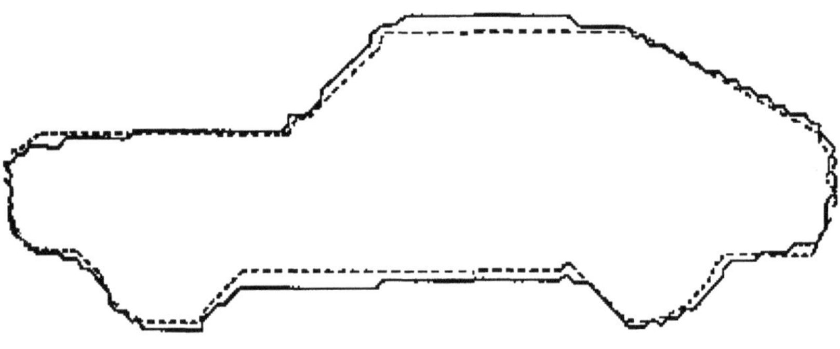

Fig. 3.30 PVR edge-smoothing example

Dog-on-a-leash algorithm

Another edge-smoothing operator can be defined in terms of the chain code, using a model based upon the idea of a person taking a lazy dog for a walk (see Figure 3.31). The point *C* represents the person and *A* the dog, which is always a given distance behind. On this occasion, distance is measured in terms of the number of chain code steps between dog the person. The dog is on a leash whose center point is *D*. (The model is somewhat unrealistic because the length of the leash varies, while the dog is a set number (*N*) of paces behind the man.) The locus of *D* as the person and dog traverse the edge of the blob forms the desired smoothed edge. This operation can be expressed conveniently in terms of the chain code and polar vector representations.

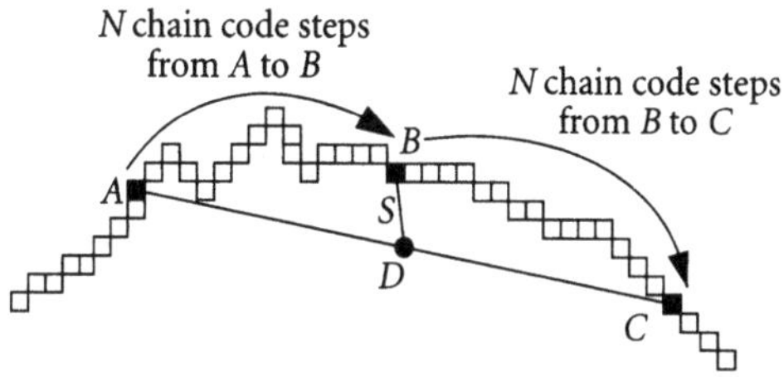

Fig. 3.31 Dog-on-a-leash algorithm for edge smoothing. The locus of *D* forms the smoothed edge. The distance *S* is a measure of the local curvature and provides a useful method for detecting corners.

An added benefit of this edge-smoothing algorithm is that it is very closely related to another which highlights corners and therefore acts as feature detector. Consider Figure 3.31 again. The point *B* is mid-way between *A* and *C*, again measured in terms of the number of chain-code steps. The Euclidean distance between *B* and *D* is measured for each edge point. (*B* is assigned an intensity value proportional to *S*.) The larger *S* is, for a given arc length *N*, the smaller the radius is of that circle that intersects points *A*, *B*, and *C*. Hence, bright points indicate corners (i.e., points of high curvature). Clearly, it is possible to combine both edge smoothing and corner detection into a single routine. As before, this "free benefit" can make one technique preferable over another.

Processing the binary image as a grey-scale array

Consider the following command sequence applied to a binary image (Figure 3.32):

smooth_edge(N):- N•lpf, thr.

for some suitable value of *N*. (Typically, $1 \leq N \leq 10$.) Using the operator *lpf* on binary images is perfectly acceptable, since they are simply special cases of grey-scale images. (We use the convention that white is represented by level 255 and black by zero.) By applying *N•lpf* to a binary image, we generate a multi-level ("grey scale") image in which the intensity at any given point indicates how many white pixels lie nearby. If we then threshold the resulting image, we create a new binary image. The overall effect is to smooth edges in a way that is at least intuitively acceptable. The parameter *N* can, of course, be adjusted to taste: larger values of *N* produce a greater degree of smoothing.

There is an additional point to be made here. We may wish to smooth only convex corners and not concave ones, or vice versa. This can be achieved by smoothing edges as above and then combining the processed image with the original (binary) image. Here is the PIP sequence which achieves this:

smooth_convex_corners (N):- wri, smooth_edge(N), rea, max.

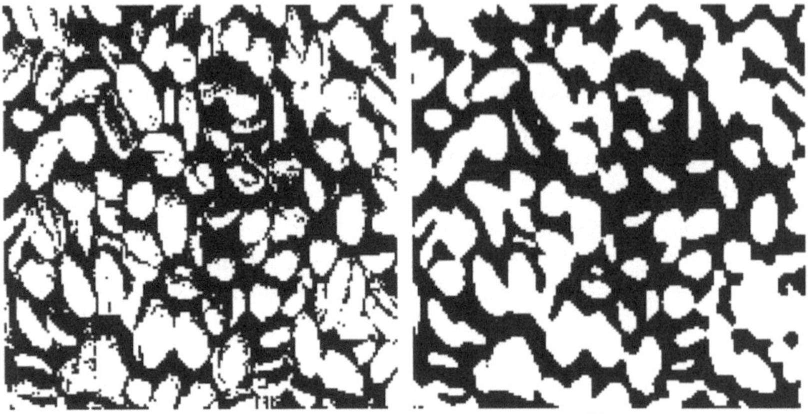

(a) (b)

Fig. 3.32 (a) Original (binary) image, derived from Figure 3.7 by applying [*heq, thr(100)*]. (b) Result of applying [*raf, thr*] to (a).

A typical result obtained by applying *smooth_convex_corners(3)* to a binary image is shown in Figure 3.33. Notice that the tips of the tines and the bottoms of the spaces between the tines can be detected and hence smoothed separately.

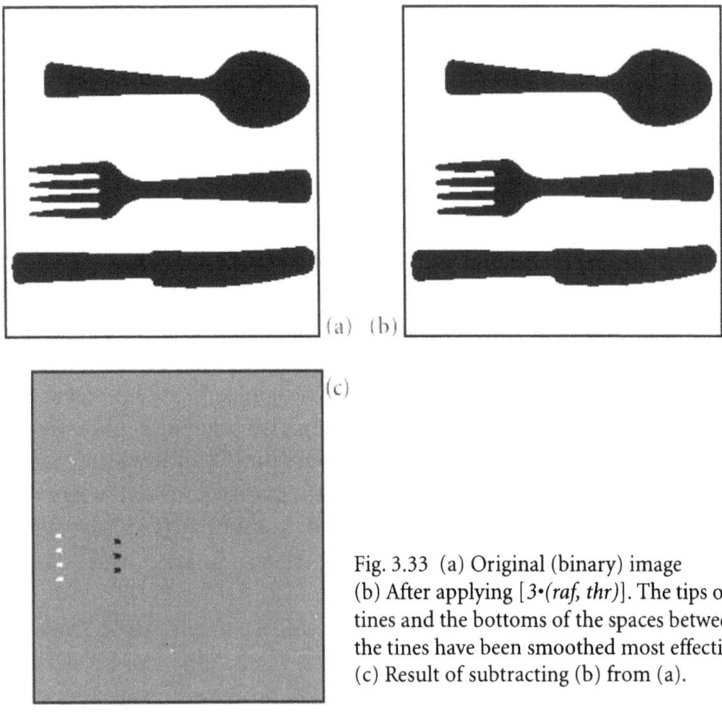

(a) (b) (c)

Fig. 3.33 (a) Original (binary) image (b) After applying [*3•(raf, thr)*]. The tips of the tines and the bottoms of the spaces between the tines have been smoothed most effectively. (c) Result of subtracting (b) from (a).

Edge coding

Several more algorithms for edge smoothing can be derived from a type of edge-tracing algorithm which has not been mentioned previously. This starts at some convenient edge point P which might, for example, be the right-most point lying along the horizontal line passing through the centroid (*CG* in Figure 3.34). Let D_i denote the distance around the edge from *P* to a given edge point Q_i. Furthermore, let R_i be the distance from the centroid of the blob to Q_i. Notice that D_i may be estimated from the chain code simply by counting odd-numbered values, each of which contributes an increment of $\sqrt{2}$, and even numbered values, which contribute an amount 1.0. Clearly, R_i is a single-valued periodic function of D_i. The period is equal to the crude perimeter estimate

$$\sqrt{2}\, N_o + N_e$$

where N_o is the number of odd-numbered chain-code elements and N_e is the number of even-numbered elements.

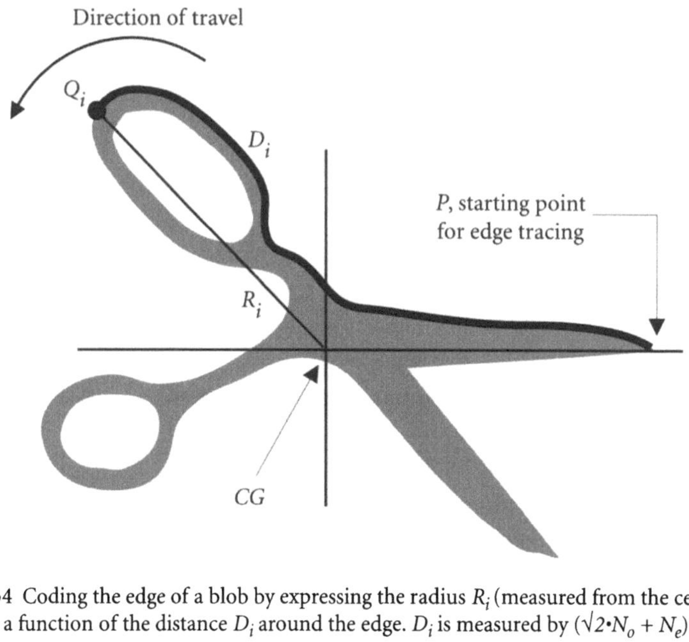

Fig. 3.34 Coding the edge of a blob by expressing the radius R_i (measured from the centroid *CG*) as a function of the distance D_i around the edge. D_i is measured by ($\sqrt{2}{\cdot}N_o + N_e$).

There are several ways to devise an edge-coding scheme based on this coding technique. All of the following suggestions are based on conventional low-pass filtering methods typically used by electronics engineers to attenuate the high-frequency components in a time series:

- Simple N-element averaging (typically $N = 3, 5, 7, \ldots$).
- Weighted-value smoothing, an extension of the previous idea which reduces the effects of distant events on the output signal.
- Fourier transform. This uses the fact that R_i is a periodic function of D_i.
- Walsh/Haar transform. Again, this makes use of the periodicity of R_i.

3.3.8 Techniques based on histograms

The importance of the histogram for machine vision can hardly be overstated, since parameters for many useful functions can be derived from it. Hence, the ability to find novel ways to compute the histogram opens up new opportunities for improving processing speed. Sometimes, as we will see, it is not necessary to calculate the complete histogram, since only one or two numeric values derived from it may be important for a specific application. We will touch on this point several times subsequently in this section.

The direct implementation of the histogram is straightforward. We scan the whole image and, at each pixel visited (i,j), we find the intensity $A(i,j)$ and then increment the $A(i,j)$-th element in a $(1+W)$-element vector. (The minimum intensity in the image is zero, and W denotes the maximum intensity.) Initially, every element in this vector is set to zero. The Uth element in the histogram indicates how many times the intensity value U occurs in the image.

Several approximate methods for computing the intensity histogram are known. First, we may sample the image to obtain an approximation to the histogram that is less accurate but can be calculated more quickly. The simplest way is to sub-sample the columns and/or rows and then calculate the intensity histogram on the lower-resolution image. It is often found that reducing the spatial resolution of an image makes little difference to the result, since calculating the histogram is an integrative process. Of course, pictures which contain a few large nearly-homogenous regions produce the most stable histograms when calculated in this way. The effects of changing the resolution can be investigated experimentally by running the following PIP program with different values of *N*:

```
reduce_resolution(N):-
    N•psq,    % Halve spatial resolution when we increment N by 1
    N•pex,    % Return image to the original size
    hpi.      % Calculate the histogram
```

Another reduced-precision method relies on the use of a Hilbert-curve sampling pattern (Figure 3.35). Notice that the Hilbert curve meanders through the image, providing a uniform sampling density everywhere. The sampling density can be controlled by a single parameter.

An option that we may use very effectively in some situations derives the histogram in the normal way but only on some important sub-region of the image (known as a *region of interest*, or *ROI*), ignoring other irrelevant parts. (This may achieve more useful results than the direct method, since non-relevant parts of the picture are ignored.)

It is possible to invent lots of other sampling schemes, which are combinations of or derivations from these basic ideas. One such hybrid scheme uses a low-precision scan (e.g., Hilbert-curve or raster) to locate a region of special interest and then employs higher-resolution sampling (perhaps of a different kind) to derive a high-precision histogram for a small part of the image.

Cumulative histogram

If the histogram is defined as the elements of the vector $\{H(i), i = 0,1,2, \ldots, W\}$, then the cumulative histogram is given by the recursive relationship

$$C(i) = C(i\text{-}1) + H(i),$$
$$\text{with } C(0) = 0.$$

Clearly, the cumulative histogram can be derived simply, by integrating the intensity histogram. However, there is another possibility which might, in some situations, be superior. Assume that we possess some very fast (hardware) device for performing thresholding and pixel counting. (A very simple Schmidt-trigger circuit and pixel counter will derive one new element for the cumulative histogram every frame period.) We can easily compute the cumulative histogram by progressively adjusting the threshold parameter. This technique lends itself very well to sampling the cumulative histogram. As before, we can exchange processing speed for precision. We have to realize that the cumulative histogram finds its main application in deriving the lookup table used for histogram equalization [*heq*] and that the cumulative histogram is monotonically increasing. With this in mind, we can see the advantage of sampling the cumulative histogram; we can use simple (e.g., linear) interpolation and just a few sample points on the cumulative histogram to obtain an approximation to the exact histogram-equalization intensity-mapping function (Figure 3.36).

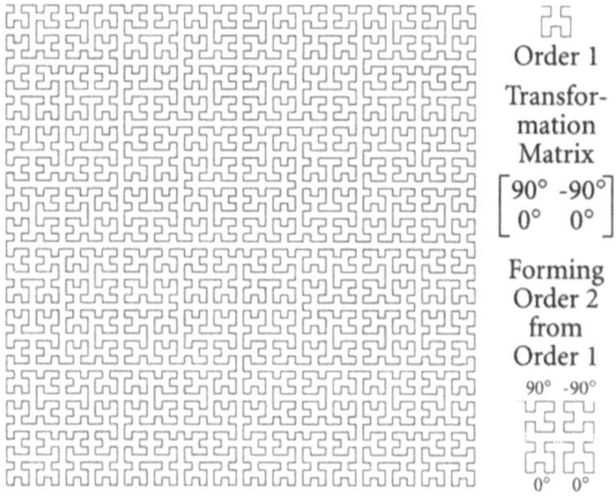

Fig. 3.35 Hilbert curve of order 5. Notice that the sampling density is uniform and can be varied at will by using Hilbert curves of different orders to sample the image. Higher-order Hilbert curves are derived by applying the transformation matrix shown above recursively.

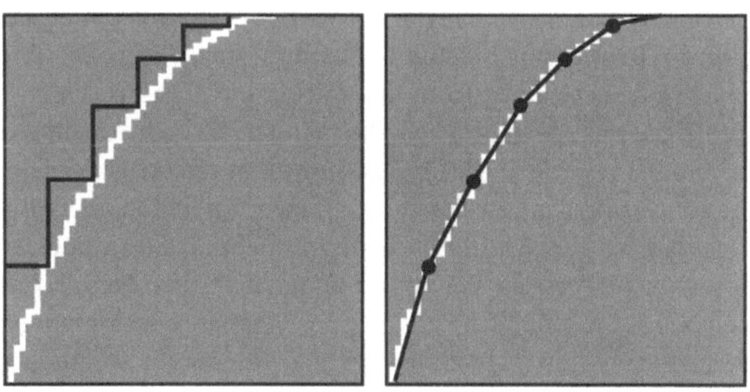

Fig. 3.36 (a) Approximating the cumulative histogram. *White*: cumulative histogram of the beans image. *Black*: cumulative histogram generated by applying *sca(3)* to the beans image. Notice that the staircase touches the true cumulative histogram exactly at 7 places. (b) Applying linear interpolation between these points produces a close fit to the true cumulative histogram.

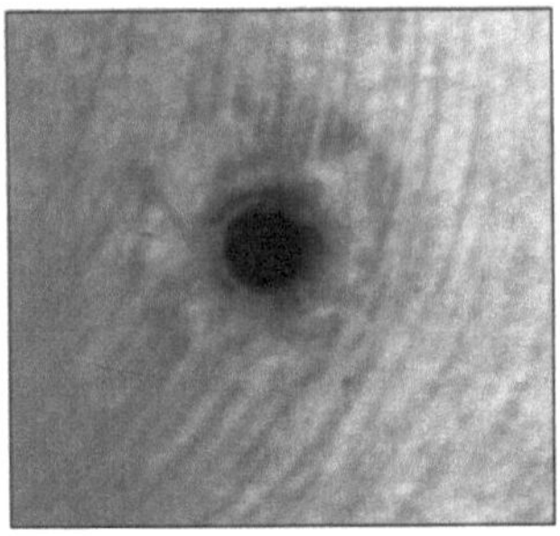

Fig. 3.37 The intensity histogram is very nearly constant, even if the major feature of the image wanders over a large area. Since the histogram of the machined background is nearly constant everywhere, it does not matter exactly where the camera is located in relation to the hole.

A completely different approach to approximating histogram equalization relies on the fact that, in many applications, the images obtained at nearby moments in time are statistically very similar. In this situation, successive frames in the signal obtained from a video camera will generate similar histograms. A small object moving in front of a featureless background will generate an almost identical histogram in successive video frames (Figure 3.37). A web-like product will similarly generate very similar histograms from images derived from areas that are close together. (Many changes in the web are due to slowly changing factors, such as roller wear, coating material viscosity, etc. Of course, defects in the web may cause significant and very sudden changes in the histogram.) When the histogram is changing slowly, the histogram-equalization mapping function derived from one image can be applied to another image, digitized soon afterwards. This provides the basis for an approximate method of performing histogram equalization in real time (i.e., full video frame rate) but with a 2-frame delay [MCC80].

1. Frame 1: Calculate the intensity histogram of image 1.
2. Frame 2:
   a) Derive the cumulative histogram. (Integrate the histogram.)
   b) Rescale the cumulative histogram to form the mapping function.
   c) Load this function into the lookup table.
3. Frame 3: Apply the mapping function to video frame 3.

Local-area contrast enhancement

This is a nonlinear filtering technique that is particularly useful for processing images derived from textured surfaces [BAT97]. Histogram equalization is performed on a relatively small region of the input image (e.g., 25×25 pixels) but only one pixel value is retained from this calculation. The processing window is scanned across the picture in a raster-like manner. At each step, the histogram-equalization process is performed. Let us assume that the image contains N^2 pixels and that the processing window contains P^2 pixels. Then, the direct implementation requires that we performing N^2 histogram equalization operations on P^2 pixels, keeping only N^2 pixels. We can do better than this! Let us consider what histogram equalization actually means. Suppose that after histogram equalization has been performed within a certain processing window $V_{i,j}$, we find that the new intensity value of a certain pixel (i,j) is $Q_{i,j}$. (As usual, the maximum value is W and the initial value of pixel $[i,j]$ is $A_{i,j}$.) Then, a proportion $Q_{i,j}/W$ of the pixels in the window $V_{i,j}$ were darker than $A_{i,j}$. This is useful because it allows us to perform histogram equalization for a single point, simply by counting how many pixels in the window $V_{i,j}$ are darker than $A_{i,j}$. This little trick has no real value for full-image histogram equalization but it can be useful for local-area histogram-equalization, where we have to perform the same operation many times. The revised algorithm for local area contrast enhancement is therefore as follows. (The processing window $V_{i,j}$ is centered on the point $[i,j]$. To ensure that $Q_{i,j}$ is a true count of the number of pixels in $V_{i,j}$ that are brighter than $A_{i,j}$, we assume that $V_{i,j}$ contains $P^2 \leq W$ pixels.)

1. Set the raster-scan counters $[i,j]$ to their initial values.
2. Count how many pixels in $V_{i,j}$ are darker than $A_{i,j}$. Denote this value by $Q_{i,j}$. (Notice that $0 \leq Q_{i,j} \leq P^2 \leq W$.) Set the intensity at point $[i,j]$ in the output image to $Q_{i,j}$.
3. Increment the raster-scan counters and repeat step 2 for all pixels in the input image.

Notice that we have managed to reduce the calculation to one in which we merely perform $(P^2 - 1)$ comparisons for each pixel. We have eliminated the need to calculate the histogram directly.

Percentage threshold

Let us turn our attention now to another histogram-based function, that of finding a threshold parameter which will divide the image in a given ratio. That is, given a parameter R (where $0 \leq R \leq 100$), what threshold parameter will segment the image so that R% of the pixels becomes black and $(100 - R)$% white? The cumulative histogram allows this parameter to be calculated directly. A particularly useful trick when using PIP is to perform histogram equalization followed by fixed-value theshholding. For example, [*heq, thr*] segments the image so that 50% of the image is black, while [*heq, thr(200)*] makes 78% black [= $(200/255) \times 100\%$]. In more general terms, the following program makes R% of the thresholded image black.

```
perecentage_threshold(R):-
    heq,              % Histogram equalization
    T is 255*R/100,   % Rescale percentage R to intensity scale, [0, W]
    thr(T).           % Theshold so that R% is black and (100 - R)% is white
```

(This is a possible way to implement the PIP operator *pct*.) If we compute the intensity histogram first, we can terminate the integration process early, when R% of the total number of pixels in the image has been taken into account.

Contrast enhancement

The PIP operator *enc* calculates the minimum and maximum intensities in an image and then rescales all of the intensity values, so that the darkest pixel becomes black and the brightest becomes white. This simple procedure is very effective in many applications but it is sensitive to noise. A more robust scheme for contrast enhancement is as follows: Let $T(R)$ be that threshold parameter which ensures that R% of the threshold image is black. Let R_1 and R_2 be two parameters defined by the programmer. (Typically $R_1 = 5\%$ and $R_2 = 95\%$.) Then, we modify all of the intensities in the image, so that pixels of intensity $\leq T(R_1)$ become black, pixels of intensity $\geq T(R_2)$ become white, and those in the intermediate range $[T(R_1), T(R_2)]$ are rescaled in a linear manner. This calculation is easily achieved using the cumulative histogram. However, note that we do not need all of the information in either the histogram or the cumulative histogram. We need only two values: R_1 and R_2. When using certain types of hardware (for example, the fast threshold and pixel-counting device mentioned above), we may not need to compute the histogram at all. Another point: once the values of R_1 and R_2 have been found, we can use a lookup table, rather than an arithmetic unit, to perform the intensity-scale mapping.

3.4 Approximate and heuristic methods

So far in this chapter, we have concentrated almost exclusively on algorithmic methods. Now, we turn our attention to approximate and heuristic techniques. There are several ways in which heuristics can be used in Machine Vision:

- to replace an algorithm
- to extend the range of application of an algorithm
- to make an algorithm run faster
- to do something useful when there is no known algorithm
- to do something useful when using an algorithm would be impractical

Machine Vision is a pragmatic subject. If we were to insist on using only algorithmic methods, our scope for building machines that are useful in practice would be severely curtailed. In this section, this point will be illustrated many times over.

3.4.1 Measuring distance

Distance is an important parameter that is required in a very wide range of inspection applications. If we are asked to monitor the size distribution of granular material, we do not require a precise measurement of each particle. We often need to know the general trend, so that we can determine whether the particles are getting bigger or smaller. When grading carrots, parsnips, etc., we do not need to know the length precisely, since the concept of length is itself ill-defined for such an object. Similarly, when measuring the eccentricity in the placement of a piece of cherry on top of a cake, we do not need to know the result precisely. Again, there are fundamental limits, since the position of an object of indeterminate shape such as a piece of cherry cannot be defined precisely. (The centroid is often used to define position but there is nothing absolute about this.)

As we will see in this section, there are many ways to measure distance and hence linear dimensions. These alternative measurements do not always coincide exactly with what we normally call "distance," which we can measure using a tape-measure, micrometer, or caliper gauge. However, they are often easier to calculate and will suffice for many applications. We use the term *Euclidean distance* to refer to the measurement that these instruments yield. In everyday life, we often use the phrase *as the crow flies* to emphasis that we are using the Euclidean distance. Astronomers and radar systems engineers relate distance to the time it takes for light or radio waves to travel between two points. These are simply ways of measuring Euclidean distance. However, vehicle drivers are well aware that a meaningful concept of "distance" must be related to the available roads. In this situation, the Euclidean distance metric is not the most meaningful.

The grass-fire transform was defined in Section 2.3 in terms of the Euclidean distance. As we will see later, it is often implemented using iterative techniques, which do not measure distance explicitly. Certain binary morphological operators can be regarded as performing a similar function to a *go/no-go* gauge and hence are size-sensitive. We will also describe a grey-scale morphological operator that implements the grass-fire transform and hence implicitly measures distances.

Let us now define the concept of Euclidean distance in formal terms. We will do so for a two-dimensional space since we are interested in pictures. However, we should point out that the concepts we are about to discuss can be extended easily to three- and higher-dimensional spaces. Let $[X_1, Y_1]$ and $[X_2, Y_2]$ be two points within an image. Then, the Euclidean distance between them is

$$D_e([X_1,Y_1],[X_2,Y_2]) = \sqrt{(X_2 - X_1)^2 + (Y_2 - Y_1)^2}$$

As we hinted earlier, there are numerous alternatives to the Euclidean distance, which may be rather easier/faster to calculate in practice and which are intellectually just as valid. Here are the definitions of just two of them:

Manhattan (city block) distance: $D_m([X_1,Y_1],[X_2,Y_2]) = |X_2 - X_1| + |Y_2 - Y_1|$

Square distance: $D_s([X_1,Y_1],[X_2,Y_2]) = \mathrm{MAX}(|X_2 - X_1|, |Y_2 - Y_1|)$

The Manhattan distance indicates how far a person would walk or drive in some North American cities, where the roads form a neat orthogonal grid. When travelling in such a city, it is perfectly natural to relate journey times to the Manhattan distance (i.e., number of city blocks to be traversed), rather than the Euclidean distance. The square distance is also more natural in some circumstances. For example, a *bang-bang* servomechanism operating a multi-axis robot will respond in a time that is proportional to the square distance to be travelled.

All three of these metrics have their uses for machine vision. In some situations, we may be able to use the *squared Euclidean distance*, D_{e2}, $(= D_e^2)$ and thereby avoid the square-root operation.

$$D_{e2}([X_1,Y_1],[X_2,Y_2]) = (X_2 - X_1)^2 + (Y_2 - Y_1)^2$$

In mathematical terms, D_{e2} is not, strictly speaking, a distance metric, since it violates one of the essential conditions, namely that

$$D(A,C) \leq D(A,C) + D(B,C)$$

In mathematics, two other conditions are imposed on a distance metric:

$$D(A,A) = 0 \quad \text{and} \quad D(A,B) = D(B,A)$$

Note that D_e, D_m, and D_s are all particular cases of the *Minkowski r-distance*, which is defined thus:

$$L_r([X_1,Y_1],[X_2,Y_2]) = \sqrt[r]{(X_2 - X_1)^r + (Y_2 - Y_1)^r}$$

In the limit as r tends to infinity, L_r approaches the square distance. Hence, we refer to D_s as the L_∞ metric.

Clearly, L_1 is the Manhattan distance and L_2 is the Euclidean distance. The equation

$$L_r([X_1,Y_1],[X_2,Y_2]) = K$$

where K and r are constants, defines a symmetrical closed figure. Just as a circle is defined by the Euclidean distance, D_m produces a diamond-shaped figure, while D_s generates a square. The linear weighted sum ($D_m + \sqrt{2}\, D_s$) generates an octagon, which is a more accurate approximation to a circle than either D_m or D_s alone.

The relevance of this discussion becomes clearer when we consider the effects of various operations in digital images. We have already seen in Figure 3.15 that applying the PIP operator *exw* more than once generates a 3×3 square from a single isolated white pixel. In other words, *exw* effectively draws contours of constant square distance. On the other hand, *exw4*, the 4-neighbor version of *exw*, generates a diamond and hence is linked to the Manhattan distance.

Note that *exw4* may be implemented using the PIP sequence

exw(4):- con(0,1,0,1,1,1,0,1,0), thr(1).

The sequence

N•(con(0,1,0,1,1,1,1,0,1,0), thr(1), con(1,1,1,1,1,1,1,1,1), thr(1))

generates an octagon.

Suppose now that we wish to find the farthest point in a blob, measured from some convenient reference point such as its centroid. (This forms a useful basis for measuring object orientation.) An obvious way to do this is to scan the image computing D_e, D_m, or D_s, as appropriate, for each pixel. However, this is wasteful of computing effort if we need to do this calculation often. We can, for example, employ the PIP operator *hic* to generate a reference image, which we then store for future use. *hic* generates an image in which the intensity at each point (i,j) is proportional to the Euclidean distance between (i,j) and the center of the image. The image to be analyzed is then shifted so that the reference point is coincident with the center of the stored image (i.e., the brightest point in the output of *hic*), and these two images are combined, using *min*. The pixel in the resulting image at which the intensity is a maximum indicates the pixel within the blob that is farthest from the reference point.

The grass-fire transform is based on the concept of distance from the nearest edge pixel (Figure 3.38). While we describe the grass-fire transform informally using the concept of (Euclidean) distance, a practical implementation is most unlikely to calculate this or any of the aforementioned distance metrics directly. One possible implementation of the grass-fire transform employs a 3×3 local operator, applied recursively until the image does not change. First, we have to decide whether to base the calculation on the 4- or 8-neighbors. (The latter effectively measures the square distance to the nearest edge pixel. The former measures the city block distance.) One possible implementation of the grass-fire transform is defined below. Initially, all pixels within an image are either white (level 255, inside a blob-like figure), or black (background). Consider pixel $[i,j]$, whose intensity will be denoted by $I(i,j)$. Those pixels in its (4- or 8-) neighborhood will be denoted by $N(i,j)$.

1. If $I(i,j) = 0$, $[i,j]$ is a background pixel, in which case, we do not alter it (Point P in Figure 3.38).

2. If $I(i,j) = 255$ and at least one pixel in $N(i,j)$ is black (level 0), we set $I(i,j)$ to 1. (These are the pixels just inside the edge of the blob, point Q in Figure 3.38.)
3. If a pixel $[i,j]$ already has a value other than black (0) or white (255), we do not alter it (point R in Figure 3.38).
4. If a pixel is white and all points in $N(i,j)$ are white, the fire has not reached point $[i,j]$ yet and we do not alter $I(i,j)$ (point S in Figure 3.38).
5. If at least one of the points in $N(i,j)$ has a value other than 0 or 255, we set $I(i,j)$ to $(Z+1)$, where Z is the minimum value for pixels in $N(i,j)$. We ignore level 0, i.e., black, when computing $I(i,j)$ (point T in Figure 3.38).

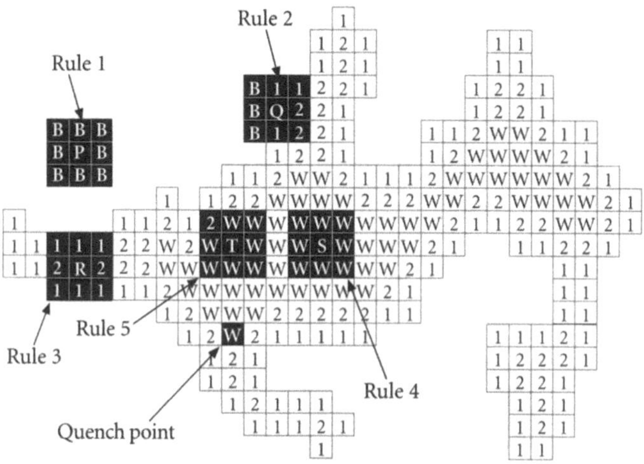

Fig. 3.38 Algorithm for the grass-fire transform. The black 3×3-pixel windows relate to cycle 1 (rules 1 and 2), cycle 2 (rule 3) and cycle 3 (rules 4 and 5)

Rules 1 to 5 are applied to all pixels in the image. The whole process is then repeated recursively, until no pixels are altered (i.e., rule 5 is not invoked during a complete cycle through the image). Notice that we have not specified the order in which the pixels are processed, which gives us the chance to employ either a raster scan or an "onion peeling" scan. (The latter is faster and is discussed in the following section.)

If $N(i,j)$ contains only the 4-neighbors of $[i,j]$, the grass-fire transform generates a shading which records the Manhattan distance to the nearest edge point. On the other hand, if $N(i,j)$ defines the 8-neighborhood, the procedure records the square distance. By alternating the procedure using 4- and 8-neighbors, we effectively combine the Manhattan and square distances, so that the grass-fire transform records the octagon distance mentioned above (Figure 3.15).

While other procedures for implementing the grass-fire transform exist, we will not describe them here, other than to say that the SKIPSM approach (Section 3.3.3 and Chapter 5) allows other distance measures to be defined by making use of arbitrarily-shaped structuring elements. The point to note is that there is no single "right way" to perform this calculation. We are accostomed to measuring the shortest route between two points in terms of the Euclidean distance between them.

However, we cannot say that the Euclidean distance is "right" and all other distance measures are "wrong." The implication for the grass-fire transform is that there is not just one procedure. Indeed, there are many and we must choose the one we want to use after considering its performance and its implementation.

Another point worth mentioning here is that the grass-fire transform can be regarded as providing a convenient alternative to an important subset of the binary morphological operators. The latter can be used to detect features that are within prescribed size limits (effectively measuring distance again). Notice, however, that the binary morphological operators do not measure distance or size explicitly; they are more akin to a *go/no-go* gauge, since they determine where an object that is the same size and shape as the structuring element will fit. The shape of the structuring element can, of course, be selected at will to be a circle, square, diamond or octagon, or any other figure of our choosing. Grey-scale morphological operators can be used to measure distance directly, using any distance metric. Figure 3.39 explains how a grey-scale morphological operator can be used to implement the grass-fire transform based on the Euclidean distance.

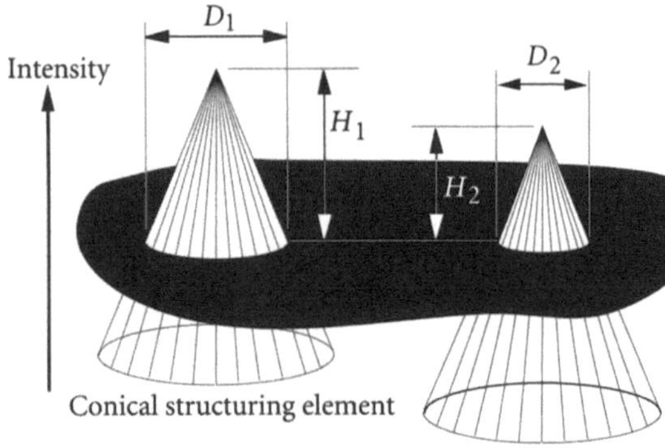

Fig. 3.39 Using grey-scale morphology to measure Euclidean distance. The structuring element is a right circular cone. The intensity surface (black region) contains two holes. The height of the tip of the cone (H_1, H_2) is directly proportional to the hole diameter (D_1, D_2).

Similarity and distance

Suppose that we have derived an ordered set (vector) of numeric measurements from an image. These measurements might, for example, define the components of the color vector at a given point, in which case they would simply be the RGB (or HSI) signals derived by a video camera. Another possibility is that these measurements summarize the overall texture within an image. Alternatively, the measurement vector might describe some feature such as a spot-like defect on a web. (Our inspection task then would be to identify the type of defect.) Whatever it represents, let us denote this set of measurements by $X = (X_1, X_2, \ldots, X_q)$, where q is a positive integer, typically in the range 2 – 20. In order to use X to identify a given image feature, it would be reasonable to compare it with a number of stored reference vectors, which we will represent by $Y_i = (Y_{i,1}, Y_{i,2}, \ldots, Y_{i,q})$, where i is a positive integer in the range $[1,N]$. A possible measure of *dissimilarity* between the features described by X and Y_i is given by the Euclidean distance:

$$D(X, Y_i) = \sqrt{((X_1 - Y_{i,1})^2 + (X_2 - Y_{i,2})^2 + (X_3 - Y_{i,3})^2 + \ldots + (X_q - Y_{i,q})^2)}$$

There are many other ways to perform classification using the concept of distance as a measure of dissimilarity [BAT74]. It is not our intention to review all of these in detail here. Instead, we will merely mention one important technique, *nearest neighbor classification.* In this scheme, we find the Y_i which minimizes $D(X, Y_i)$ in order to recognize the feature or pattern associated with the vector X. Each of the Y_i is carefully chosen to represent or typify some cluster or group within the data set and is therefore called an *exemplar.* Y_i is associated with an appropriate label, C_i, which indicates the type of pattern, feature, or event which gave rise to it. More than one Y_i may be given the same label, indicating that they correspond to the same pattern class. Several learning rules are known for designing the nearest neighbor classifier, some of which are liable to store a large number of exemplars. This can cause a problem not only for storage but also for the speed of computation. It is possible to improve the speed of the nearest neighbor calculation by subdividing the Y_i into groups, each of which is assigned a *super-exemplar.* The set of super–exemplars will be represented by $\{Z_j\}$. Notice that $\{Z_j\}$ is a proper subset of $\{Y_i\}$ (i.e., $\{Z_j\} \subseteq \{Y_i\}$). During a preprocessing phase, each of the super–exemplars Z_j is linked to another subset of the exemplars (Q_j), which are all known to be close to Z_j. (Again, $Q_j \subseteq \{Y_i\}$.) Then, classification of X proceeds as follows:

1. Find a near neighbor to X among the set $\{Z_j\}$. This will be denoted by Z_j.
2. Find the nearest neighbor to X among the set Q_j. Denote this by W_k. (Assuming that the preprocessing phase has been properly designed, step 2 always finds the *nearest* neighbor to X among the set $\{Y_i\}$ and the number of computational steps is very much reduced. More precise details are given elsewhere. They are tedious to explain and would distract us from our main theme.)
3. Associate the pattern which gave rise to the measurement vector X with class C_k. (Recall that W_k is the nearest neighbor to X among the set $\{Y_i\}$.)

This is an example of using a heuristic to improve processing speed. The procedure is not algorithmic because we cannot guarantee that it will achieve a speed increase, although we can be certain that it will always find the nearest neighbor. The improvement in processing speed is achieved at a certain cost, by increasing the computational load in the preprocessing phase. This is akin to using an optimizing compiler to reduce the computational effort needed at run-time.

Skeleton and the medial axis transform

The skeleton of a blob-like object can be defined in terms of the grass-fire transform as the set of *quench points.* A quench point is one where two or more advancing fire lines converge at the same time and derives its name from the fact that there is no more combustible material there left to burn (Figure 3.38). Another definition of the skeleton relies on the fact that every point on it lies equidistant between two or more edge points. Neither of these definitions is very convenient as the basis for use in a practical application where high speed is needed. A more pragmatic approach is represented by the so-called *onion-peeling* procedure. For the sake of simplicity in the following explanation, we ignore the possibility of there being lakes present, since they cause complications and obscure our understanding of the heuristic. The basic onion-peeling procedure may be described as follows (Figure 3.40). We trace the edge of the blob, as if we were chain coding it. As we do so, we delete (set to black) each pixel visited *unless it meets any one of these three conditions*:

1. It has only one white neighbor. Such a point is part of the skeleton and is therefore retained.
2. If it is critical for the connectivity among its remaining 8-neighbors (cf the PIP operator *cny*, Section 2.3). Again, such a point is part of the skeleton and is retained.
3. It is a skeleton point. (No skeleton point is ever erased.)

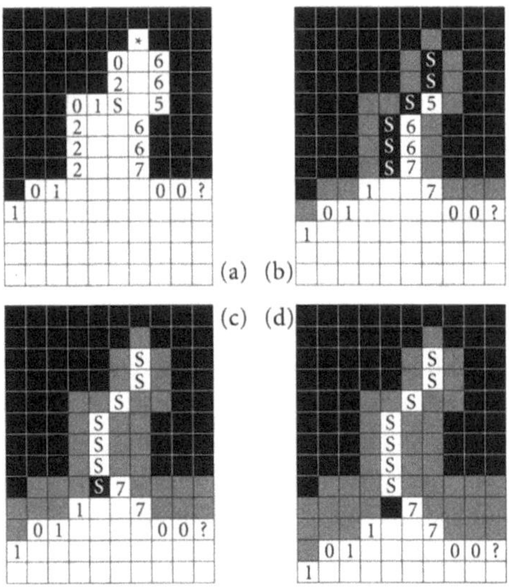

Fig. 3.40 Generating the skeleton of a blob by the onion-peeling procedure. For clarity, only part of the blob is shown. When the edge follower reaches the right-hand-side of the sub-image shown here, it continues tracing the outer edge of the blob in a anticlockwise direction until it eventually reaches the left-hand side again. Numerals 0 – 7 indicate the direction of travel of the edge follower (same as chain code). 'S' is a skeleton point. (a) First cycle: '*' indicates the starting point. When the edge follower reaches the top-most '0', it tries to revisit the starting point. This signals the beginning of the second cycle. (b) The second cycle consists of these moves: S,S,5,6,6,7,7,0,0,... ,1,0,1,1,S,S,S,S. (c) The third cycle consists of these moves: 7,7,0,0,...,1,0,1,1,S. (d) The fourth cycle consists of these moves: 7,7,0,0,...,1,0,1,1,S.

We repeat the process until there are no more changes possible. The resulting figure is the skeleton. However, the matchstick figure generated thus is not quite the same as that formed by the quench points generated by the grass-fire transform. (The latter is called the *medial axis transform*.) The onion-peeling procedure has been found to be perfectly acceptable in all practical applications that the authors have studied. Moreover, it somewhat faster and generates a "clean" skeleton, with very fewer short side branches than the quench-point method.

The onion-peeling procedure could be modified to generate the grass-fire transform, also. How this could be achieved is fairly obvious from Figure 3.40.

The problem mentioned earlier relating to lakes is illustrated in Figure 3.41. Notice how the skeleton is "tightened like a noose" around the lakes. To avoid this, we have to modify the procedure just described, so that we alternately trace once around each outer edge, then once around each inner edge. This involves additional rules, which inevitably slow down the procedure. For many purposes, however, the simpler heuristic will suffice, despite its known limitations on objects containing lakes.

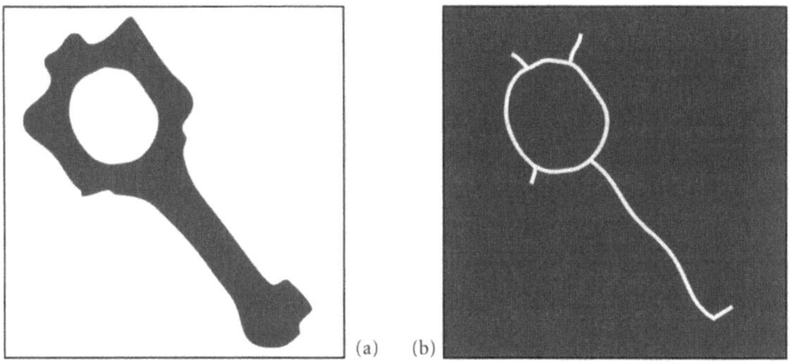
(a) (b)

Fig. 3.41 An undesirable feature of the simple onion-peeling procedure. The skeleton is "tightened like a noose" around the lakes.

Remarks

In this section, we have discussed several varied procedures, all of which are based on the concept of distance. First, we showed that it is possible, in many situations, to avoid the rather complicated calculation needed for the Euclidean distance. This can be achieved by using the square or Manhattan distance measures instead. These can often be substituted directly for the Euclidean distance. We have shown that it is often possible to avoid the distance calculation altogether, by using iterative filtering techniques. We can also use iterative processing to implement the grass-fire transform, while the skeleton can be derived by a process that is effectively a modified form of edge tracing. Morphological operators can measure object size in a binary image without the use of distance measurement and, as we have seen in Section 3.3.3, we can implement these using lookup tables. Although the nearest neighbor classifier is implicitly based on the concept of distance, the use of *near* neighbors allows us to improve the processing speed, by the simple expedient of eliminating calculations that are obviously unnecessary.

3.4.2 Fitting circles and polygons to a binary object

In this section, we will consider various ways of representing a blob-like object with another figure (i.e., a circle or polygon). We have already considered this topic in some detail in Section 3.3.6, where we discussed the derivation of the convex hull. Intuitively, we may feel that this figure has a unique relationship to the shape that it represents, since it satisfies these three criteria:

a) It is convex.
b) It totally encloses the given shape.
c) It is of minimal area.

While the figures about to be discussed do not possess all of these properties, they do, nevertheless, provide the basis for a range of useful analytical tools and in some circumstances can be used *in lieu* of the convex hull. Of course, there is no unique circle or polygon, including the convex hull, that properly represents *all* important features of a given blob. For this reason, we have to be careful to use the most appropriate one for a given application. In the following discussion, we will relax conditions b) and/or c), and impose other, equally valid criteria. Some of the figures that result are larger than the convex hull, while others do not contain all of the pixels within the given shape. We must not presuppose, however, that this invalidates them or makes them any less useful. Indeed, most of these alternative procedures are easier/cheaper to compute than the convex hull and, in many cases, possess features that are just as useful in practice. Since there is no "optimal" way of representing a blob by a circle or a polygon, the choice of which one to use in a given application depends upon its specific requirements.

Coin-in-a-slot procedures

Consider the task of inserting a coin into a slot-machine. We will generalize this to consider shapes other than circles being pushed through a slot. The size of the slot is, of course, deliberately chosen to match that of the object to be passed through it. An important parameter of a shape is the size of the smallest slot that it can pass through. However, this statement does not specify the problem fully In effect, there are several possible solutions, depending on the answers to the following questions:

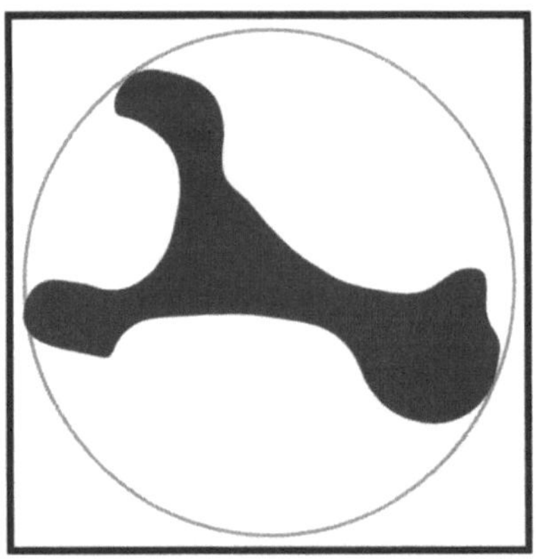

Fig. 3.42 The result of applying *mbc.* It usually touches the edge at three points.

a) Are we allowed to maneuver the shape as it goes through the slot, or must it pass "straight through"?
b) Are we allowed to reorient the shape before we pass it "straight through" the slot?

A complex iterative procedure is required to draw the minimum-size circle that encloses a given blob, touching it at three points (Figure 3.42). One important circle that can be drawn rather more easily around a given blob is the one that intersects the two edge points that are farthest apart. This can be obtained by the edge-following process, and generates the bounding circle of minimum area.

Now, suppose that we are allowed to rotate the shape before passing it through the smallest possible slot. In this case, to find the minimum size of the slot we can use another edge-tracing procedure. Let E be the set of edge points, which contains N_E pixels. The Euclidean distance between two edge points A and B will be denoted by $D(A,B)$, $A \in E$, $B \in E$. Given an edge point P, we simply find another edge point Q that maximizes $D(P,Q)$. Let us denote this value by D_P. We repeat this process for all edge points ($P \in E$), and then calculate the minimum value of D_P A simple procedure of this kind requires $O(N_E^{\,2})$ time. A modest reduction in computational effort (about 50%) can be achieved by tracing only those edge pixels that have not already been considered. (Recall that $D(A,B) = D(B,A)$. See Figure 3.43.)

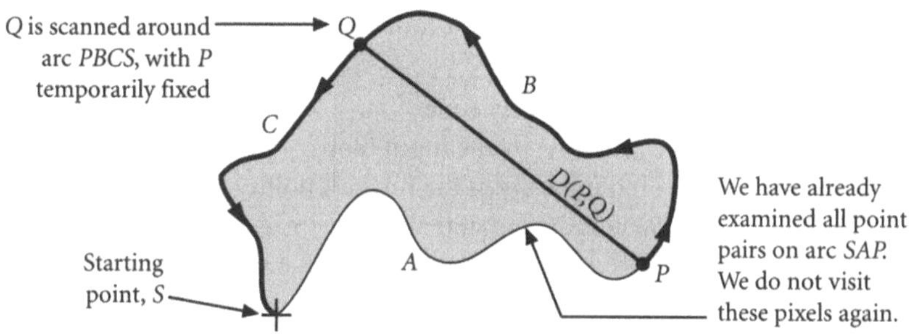

Fig. 3.43 Scanning the edge to find the two points which are farthest apart.

We will now describe a faster but approximate procedure for estimating the same parameter (Figure 3.44). Let us take two edge points A and B that are spaced $\lceil N_E/2 \rceil$ chain-code steps apart. (We are forced to round $(N_E/2)$ up or down quite arbitrarily, since we do not know whether N_E is odd or even. The chain code must be regarded as forming a circular list structure, since the starting point is quite arbitrary and must be treated in the same way as all other points.) We then calculate the Euclidean distance between them, $D(A,B)$. We then move A and B simultaneously

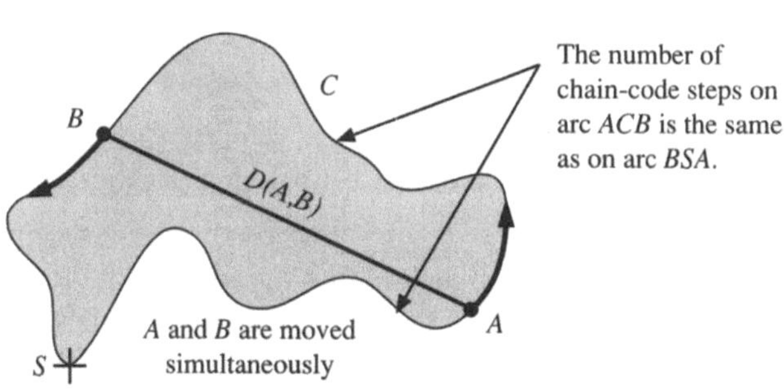

Fig. 3.44 Scanning the edge using two points (A,B) which are separated by $\lceil N_E/2 \rceil$ chain-code steps.

around the edge of the blob, and find the position which maximizes the value of *D(A,B)*. At this position, we find the mid-point of the line *AB*, which is then taken to be the center of the circle to be fitted to the given blob. This simple procedure does not always find the same circle as the previous method but it is likely to be much faster, since the computation time rises as $O(N_E)$.

Circumcircle based on the centroid

Another circle whose parameters are easy to compute is concentric with the centroid of the blob and has a radius equal to the distance from the centroid to the farthest edge point (Figure 3.45). This circle is usually larger than that produced by the three previous methods but is usually faster in software. Estimating the centroid requires $O(N^2)$ time, where N defines the image resolution, and locating the most distant edge point from the centroid takes $O(N_E)$ time. The following PIP command sequence also draws this circle:

```
ccc(X,Y,Z):-
    psk,          % Place image on the stack
    cbl(1),       % Check that there is only one blob
    cgr(X,Y),     % Centroid is at [X,Y]
    hic(X,Y),     % Draw intensity cone
    min,          % Mask cone by the input blob
    gli(_,Z),     % Find intensity at the furthest point from centroid
    swi,          % Switch current and alternate images
    thr(Z),       % Threshold the cone - produces a solid white disc
    pop,          % Pop image from the stack
    swi.          % Switch current and alternate images
```

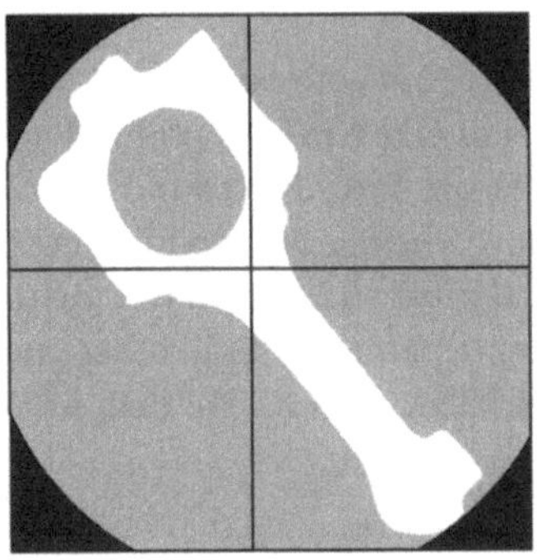

Fig. 3.45 The cross-lines indicate the centroid and the circle is tangential to the edge at its farthest point from the centroid.

Circle of equivalent area

Our next method for fitting a circle is to find the area (A) of the given blob. Then, take the radius to be $\sqrt{(A/\pi)}$. Again, the center of the circle is set to coincide with the centroid of the blob. Of course, this circle does not contain all of the points within the blob unless the blob happens to be a circle. This procedure is particularly useful for display for a human operator, since both size (area) and position are preserved (Figure 3.46). Here is the PIP program for generating this circle.

```
cav(X,Y,A):-
    psk,                     % Place image on the stack
    cgr(X,Y),                % Centroid is at [X,Y]
    cwp(A),                  % Count white points
    B is int(sqrt(A/pi)), X1 is X - B, Y1 is Y - B, X2 is X + B, Y2 is Y + B,
    zer,                     % Black image
    cir(X1,Y1,X2,Y2,255),    % Draw disc - specify corners of enclosing rectangle
    pop,                     % Pop image from top of stack
    swi.                     % Switch current and alternate images
```

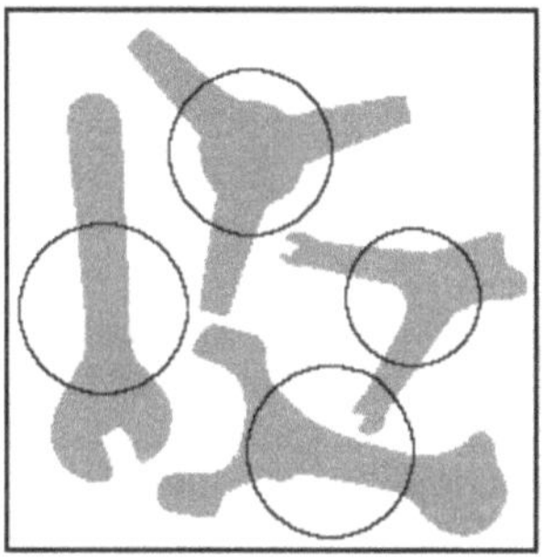

Fig. 3.46 Circles drawn by the goal *dab(cav)*. Each circle has the same area as the shape it represents.

Fitting a circle to a nearly-circular object

The next method for generating a circle to represent a given blob is most appropriately used when that blob is itself nearly circular, or it contains a well-defined region where the edge is a circular arc. It is particularly useful for estimating the radius and center of curvature and is used, for example in Section 8.5 (Plate 14), to compute the radius of curvature of the shoulder of a glass vial. Given three edge points, $[A_x,A_y]$, $[B_x,B_y]$ and $[C_x,C_y]$, we can use the following formulae to estimate the position of the center $[P_x,P_y]$ and radius, R, of that circle which intersects all three points:

$$D = 2(A_yC_x + B_yA_x - B_yC_x - A_yB_x - C_yA_x + C_yB_x)$$

$$P_x = (B_yA_x^2 - C_yA_x^2 - A_yB_y^2 + A_yC_y^2 + C_yB_x^2 + B_yA_y^2 + A_yC_x^2 - B_yC_y^2 - B_yC_x^2 - A_yB_x^2 + C_yB_y^2 - C_yA_y^2)/ D$$

$$P_y = (C_xA_x^2 + C_xA_y^2 + A_xB_x^2 - C_xB_x^2 + A_xB_y^2 - C_xB_y^2 - B_xA_x^2 - BxA_y^2 - A_xC_x^2 + B_xC_x^2 - A_xC_y^2 + B_xC_y^2)/D$$

$$R = \sqrt{((A_x - P_x)^2 + (A_y - P_y)^2)}$$

(This is the basis of the PIP command *fcd.*) To estimate the radius of curvature of a supposedly circular arc, three well-spaced edge points are chosen. The position coordinates (P_x,P_y) and radius (R) are then calculated. Notice that, if $[A_x,A_y]$,

$[B_x, B_y]$, and $[C_x, C_y]$ lie too close together, wild fluctuations in the estimated values for P_x, P_y, and R can occur in response to quantization effects and camera noise. In order to obtain an accurate fit, we can perform this process repeatedly, as $[A_x, A_y]$, $[B_x, B_y]$, and $[C_x, C_y]$ are moved progressively around the arc. At each position, we place a spot at the center of the estimated circle. Gradually, as more and more edge points are analyzed, a pattern emerges, often in the form of several bright fuzzy spots, each one indicating the centre of curvature of a separate circular arc (Figure 3.47). This provides a useful method for detecting circular arcs arising from

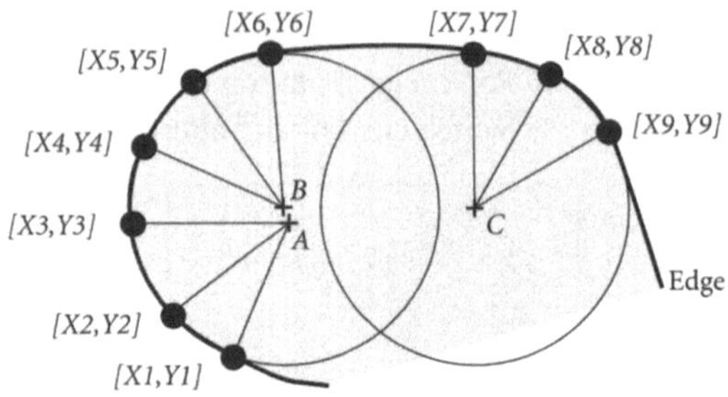

Fig. 3.47 Three edge points, such as $[X1, Y1]$, $[X2, Y2]$, $[X3, Y3]$, can be used to fit a circle (center A) to a small portion of the edge. This process can be repeated many times to yield a succession of circles (centers B and C). Notice the tendency of the centers of these circles to cluster together (points A and B) if the edge contains a long circular arc. This clustering is evident in Figure 3.48.

partially occluded circular discs. The other methods described here are unable to perform this function. If the arc being analyzed is a nearly circular ellipse, or in the shape of a egg or rugby ball, the set of points plotted in this way may trace a curve, rather than form a compact cluster (Figure 3.48).

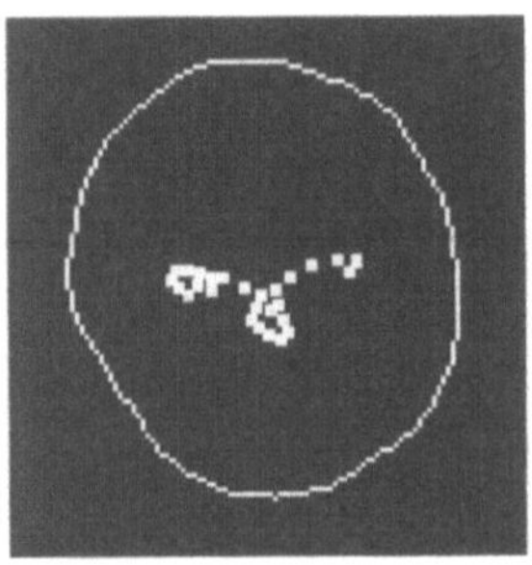

Fig. 3.48 The centers of the circles fitted to the edge of a satsuma (a member of the citrus family of fruit) form a diffuse cluster, since the edge is not a perfect circle.

Polygonal representations of a blob

Several algorithms for calculating the convex hull of a blob-like figure have already been described in Section 3.3.6. In contrast to this, our objective here is to demonstrate that, for many purposes, there exist perfectly acceptable alternatives (heuristics) that are likely to be easier to calculate but which generate approximations to the convex hull. It is possible, for example, to define a tree-like structure (akin to a concavity tree) to represent a shape in terms of its minimum bounding circle [*mbc*] rather than its convex hull. Sometimes the approximation to the convex hull can be quite crude. Apart from rectangles, we might choose to use triangles, pentagons, hexagons, octagons, hexadecagons, or one of the circles defined earlier in this section. In many application, these are all valid approximations. We need to adopt a pragmatic attitude, so that any technique that produces acceptable results at a tolerable computational cost can be used. Whether or not a given approximation will suffice depends, of course, on the application.

Minimum-area rectangle

One of the simplest of these approximate methods uses the minimum-area rectangle (PIP operator *mar*, Figure 3.49). Notice that the sides of this rectangle are parallel with those of the image. The minimum and maximum *X*- and *Y*-limits of a blob can be found very easily, whether we base our calculations on an array, the run-code, chain code, or the polar vector representation. The procedure for finding the maximum/minimum of a sequence of numbers is straightforward, and we will not dwell further on this point.

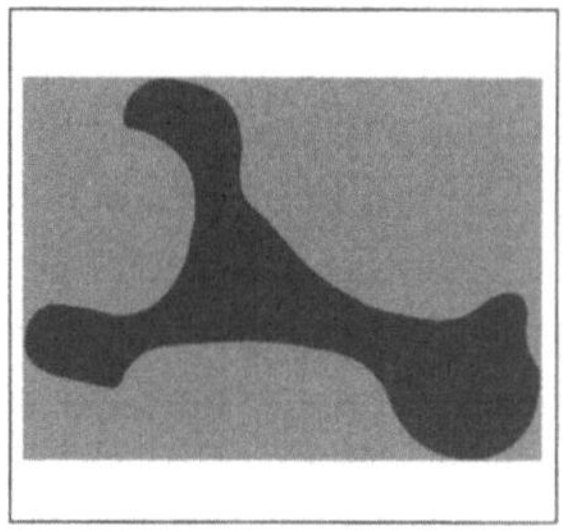

Fig. 3.49 The minimum-area rectangle whose sides are parallel to the image border.

Minimum-area octagon and higher-order polygons

A minimum-area octagon whose sides are all inclined at an integer multiple of 45° relative to the sides of the image can also be generated easily. One way to derive this figure is explained in Figure 3.50 and relies on the use of various intensity wedges.

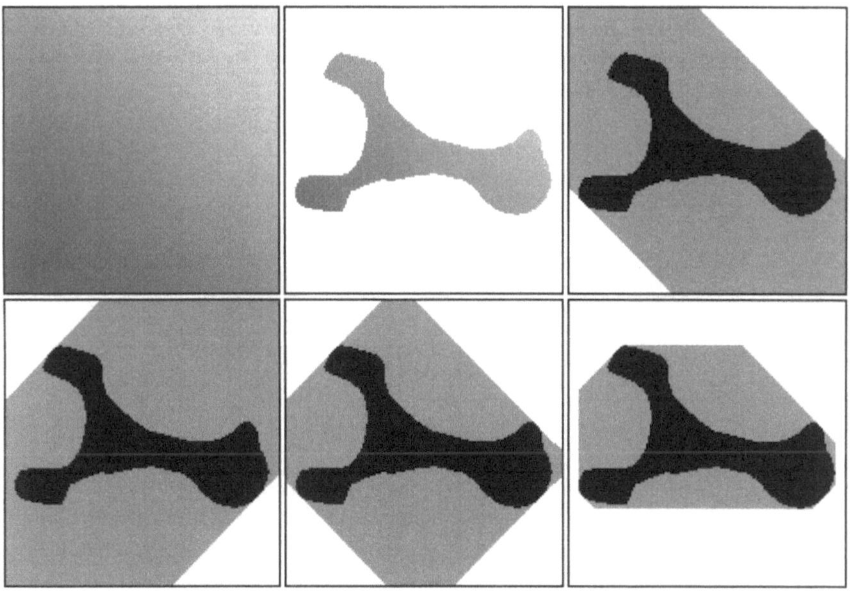

Fig. 3.50 Generating a minimal octagon to fit a given blob (i.e., automobile connecting rod).
(a) Intensity wedge, generated using [*wdg(2)*].
(b) Input shape applied as a mask to intensity wedge. (Grey-scale range expanded for clarity.)
(c) Max. and min. intensities found in (b) are used as parameters for thresholding (a).
(d) As in c) but beginning with [*wdg(4)*].
(e) The stripes in (c) and (d) are combined using [*min*].
(f) The octagon was generated in the same way as (e) but *four* stripe images, based on [*wdg(1)*], [*wdg(2)*], [*wdg(3)*] and [*wdg(4)*], were combined using [*min*].

(These have intensity contours (isophotes) inclined at different angles.) The wedge images can be stored, rather than recomputed for each new image that is to be processed. This ploy saves computation time but may require a little more hardware, i.e., extra memory. (Digital storage is so inexpensive nowadays that this is a small price to pay for increasing the processing speed.)

The minimum-area octagon is a closer approximation to the convex hull than is the minimum-area rectangle, simply because the former has twice as many sides. A multi-sided polygon can make the approximation even more accurate and can be generated by a simple extension of the same procedure. For example, we might choose to approximate the convex hull by a hexadecagon (16 sides), in which case, we generate and store 8 wedge images (Figure 3.51(a)). However, there is no reason why we should restrict ourselves to generating a *regular* polygon. Using a simple PIP program, we can generate a polygon with as many sides, inclined at any angles, as we wish. The program is not listed here because it relies on features of PIP that we have not yet explained.

Minimum-area rectangle

It is quite easy to construct a minimum-area rectangle whose sides are inclined at some arbitrary angle to the image axes. This can be achieved simply by rotating the input blob, then constructing the minimum-area rectangle as before, and finally reversing the rotation. As an illustration of this, the following PIP program derives the smallest rectangle whose sides are parallel/normal to the principal axis derived from the original image (Figure 3.51(b)).

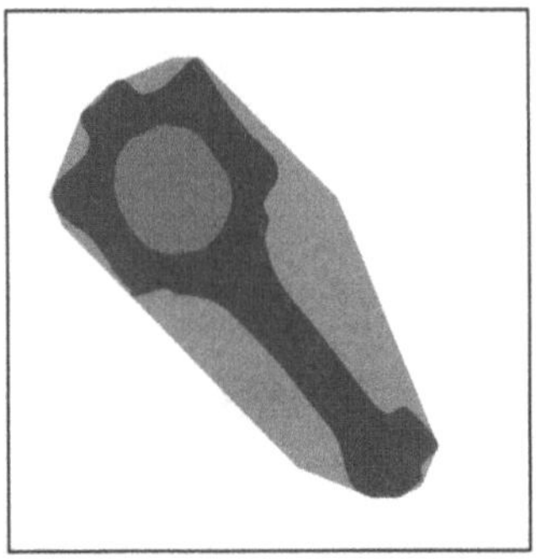

(a) (b)

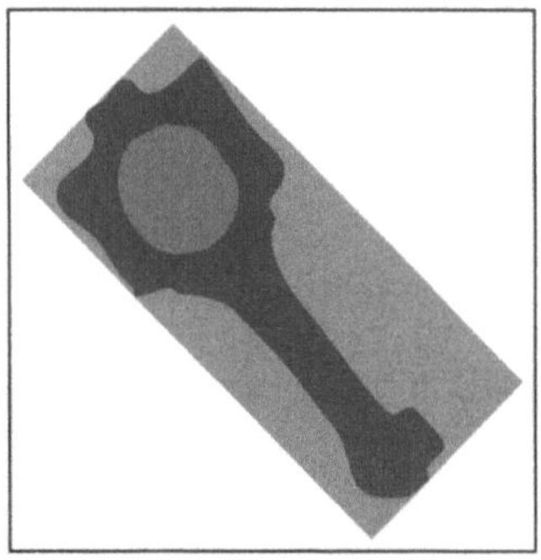

Fig. 3.51 (a) This hexadecagon was generated by the predicate *polygon*. The sides are oriented at angles of 0°, 22.5°, 45°, ..., 337.5°. (b) The minimum area rectangle whose orientation is determined by the principal axis. This rectangle was generated by the program *mar_princ_axis*.

```
mar_princ_axis:-
    psk,                          % Push image onto stack
    lmi(X,Y,A),                   % Centroid (X,Y) & orientation of principal axis
    dgw(Xw1,Yw1,Xw2,Yw2),         % Image size
    Xw is (Xw2-Xw1)//2,           % Width of the image
    Yw is (Yw2-Yw1)//2,           % Height of the image
    X1 is Xw - X,                 % How much to shift image in X direction
```

```
Y1 is Yw - Y,        % How much to shift image in Y direction
X2 is -X1,           % How much to perform reverse shift (X direction)
Y2 is -Y1,           % How much to perform reverse shift (Y direction)
psh(X1,Y1),          % Shift so blob is at center of the image
B is - A,            % Negative of angle of orientation
tur(A),              % Normalize orientation
mar,                 % Minimum area rectangle (hollow)
blb,                 % Fill the minimum area rectangle
tur(B),              % Reverse rotation
psh(X2,Y2),          % Reverse shift
pop,                 % Recover input image
swi.                 % Switch current and alternate images
```

Image rotation is not normally regarded as being very satisfactory because it frequently creates ragged edges. Notice, however, that *mar_princ_axis* always rotates a rectangle, which makes the effect less pronounced than it would be if we rotated the input blob a second time. The second image rotation can be avoided altogether, since we need only compute the coordinates of the *corners* of the rectangle generated by *mar*. We then calculate their new positions after rotation and construct a (hollow) rectangle, simply by drawing its sides as four digital straight lines.

The astute reader will have noticed by now that we do not need to rotate the input image either, since we could derive the orientation of the axis of minimum second moment (using *lmi*), then trace the blob edge to find the extreme points along that axis and normal to it. There are many ways to implement nearly every algorithm!

Remarks

In this section, we have discussed a variety of techniques for generating figures such as rectangles, polygons, and circles to represent a given shape. By the word "represent" we mean that we replace the given shape by another which in some way simplifies the subsequent calculation or display of the image. By choosing the most appropriate data-extraction procedure, we can remove unwanted information while retaining other data that is able to summarize some significant property of the given shape. We may wish to preserve position and orientation, but ignore the complexity of a winding edge contour. On the other hand, we may wish to compute the "density" of a shape by comparing its area with that of its convex hull, or the minimum bounding circle. There are numerous options; the biggest mistake of all is to believe that just one of them (e.g., the convex hull) is inherently "better" than all of the others.

3.4.3 Determining object/feature orientation

Perusal of the Computer Vision literature quickly reveals that the Hough transform is a very popular method for fitting a straight line to a set of disconnected colinear spots in a digital image. On the other hand, it has not gained anything like the same level of support with Machine Vision system designers. In this section, we will discuss why this has happened and we will look at alternatives to the Hough transform (Section 2.6.1) that can perform a similar function.

First, let us explain what the Hough transform (HT) does. Given a digital image containing a set of disconnected spots, the HT can detect a subset of those spots that lie along a straight line (Figure 3.52). That line can be at any orientation and be in any position. The HT can also be used to join together fragments of a "broken" straight line that might arise, for example, from applying an edge detector (e.g., *sed*) and thresholding to a low-contrast image. The fact that the HT can detect sets of collinear spots aligned at any orientation is both its strength and its weakness. Obviously, it is necessary sometimes to work with images whose structure is completely unpredictable. This is the domain of Computer Vision research. On the other hand, Machine Vision system designers nearly always know, in general terms, what format the images will have. This knowledge means that very general procedures such as the HT are often not needed, since less expensive techniques will suffice and may give better results, since they are more specific.

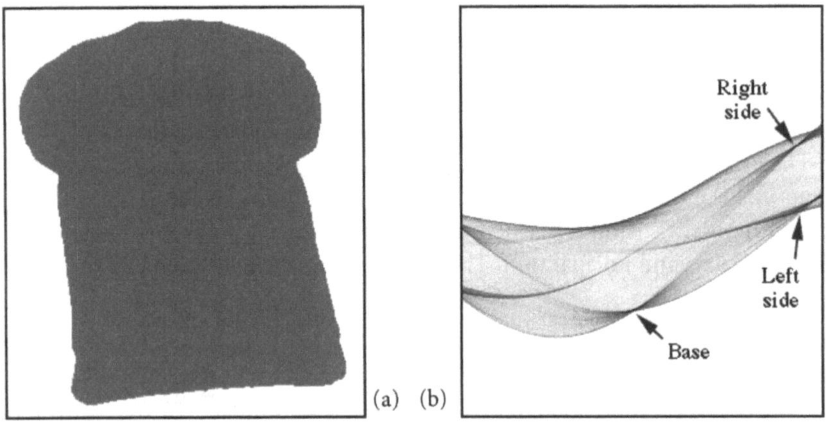

Fig. 3.52 Hough transform applied to the silhouette of a slice of bread. (a) Original image. (b) [*bed, huf, neg, 2•sqr*] applied to (a). (The subsequence [*neg, 2•sqr*] simply improves the contrast for human viewing.) Notice the three dark points, which indicate the base and sides of the loaf, and that the sides are nearly parallel (i.e., nearly same position along horizontal axis).

Consider the task of determining the orientation of an object such as a machine component. In many cases, we can use the outer edge to determine the object orientation. We might instead use holes, 3-D shapes, or vividly colored regions, etc., to perform the same function. We might sometimes be forced to use "internal" edges but we can, in some instances, make them more pronounced by using specialized lighting techniques. We must always be careful to distinguish between the primary task (i.e., finding the orientation of the object) and the secondary task (e.g., finding the orientation of a given edge in an image). Consider the image shown in Figure 3.53). This has several "strong" features, which create a high local contrast (e.g., holes, outer edge) and "weaker" ones, principally the "internal" edges. In order to determine the orientation and position of this particular object prior to directing a robot to pick it up, we would be foolish to rely on the weak features, whose structure is, in any case, complicated compared with that of some of the stronger ones. As a general principle we should, wherever possible, extract and measure features that are both robust and easy to analyze. In most case, internal edges are often not so satisfactory in one or both of these respects.

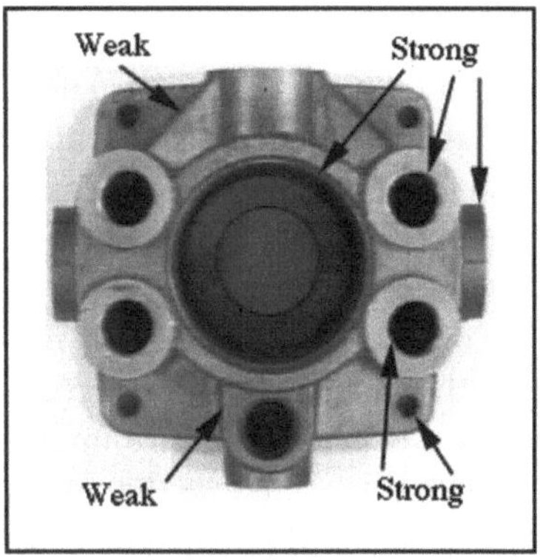

Fig. 3.53 The Hough transform could, in theory, be used to determine the orientation of this object (a hydraulics manifold) by using it to detect one of the four linear features, two of which are labelled "weak." However, this would not provide as reliable a result as a technique based on higher contrast features, such as the holes and the outer edge contour (labelled "strong").

Among the alternative procedures for determining object orientation that are open to us are these:

- Principal axis (axis of minimum second moment) (PIP command *lmi*).
- The line joining the centroid to the edge point that is farthest from the centroid.
- The line joining the centroid of the object silhouette to that of the largest bay.
- The line joining the centroid of the object silhouette to that of the largest lake.
- The line joining the centroid of the largest bay to that of the second largest bay.
- The line joining the centroid of the largest lake to that of the second largest lake.
- The line joining the centroid of the largest bay to the centroid of the largest lake.

Clearly, there are many other options. To be accepted for use in a vision system, the designer must believe that the HT is demonstrably better than all of these competitors. Moreover, to avoid using a computationally expensive technique such as the HT, the designer has many optical tricks at his disposal to make the task simpler for the image processor.

Now, let us consider analyzing sheet material made in the form of a web that is formed, treated, coated, and printed using rollers. All predictable features, such as printed patterns, are precisely aligned to the edge of the web, assuming that it is well made. Faults on the web are, by definition, not predictable and could, in theory, occur in any orientation. However, there are two points to note:

1. We do not normally need to know the orientation of a single blob-like fault, such as a spot caused by localized chemical contamination, material fault, coating blemish, or a squashed insect.
2. Certain faults which are caused by the manufacturing process are always aligned along the up–down axis of the web. For example, a very common type of fault is

caused by a piece of material being dragged along the surface of the web, leading to a scratch in the direction of the web motion. Another commonly occurring fault is manifest as a point-like fault repeated periodically. It is not difficult to see that this is caused by a foreign body adhering to one of the rollers used to manufacture the web.

In neither case do we need to measure the orientation of anything on the web, except perhaps along accurately defined axes. We will return to this later, since it enables far less computational work to be used. It is a common requirement for web products to have well-defined surface-texture properties but, once again, these can almost always be related to the web axes.

We are led then to the conclusion that the HT is not very useful for machine vision, where verification rather than recognition is required. There are, of course, many other application tasks, such as robotic trimming of vegetables, decorating cakes, guiding a sewing machine, etc., that we should consider to make our argument comprehensive in its scope. However, despite their many years of experience in this field, the authors have never felt the temptation to resort to using the Hough transform, because other techniques have proved equally or more effective and certainly more attractive in terms of the computational effort they require. To be fair in this discussion, we should point out that certain authors do use the HT extensively and have demonstrated good results [DAV00]. While not criticizing our colleagues, we simply point out that there are many instances where other techniques can achieve the same primary purpose as the HT without incurring its high computational cost. For this reason, we will discuss other procedures below:

1. Radon transform
2. "Island hopping"
3. Polar-to-Cartesian coordinate-axis transformation

Radon transform

The Radon transform (RT) is effectively a reformulation of the HT but does lend itself to greater versatility, as we will see later. To compute the Radon transform of a given binary image, we consider a thin strip inclined at an angle ϕ to the coordinate axes and located at a distance d from the center of the image (Figure 3.54) This strip will be denoted by $S(\phi,d)$. Its precise width need not concern us for the moment. Let us count the number of white pixels in $S(\phi,d)$ and denote this number by $\mu(\phi,d)$. Then, we compute $\mu(\phi,d)$ for all possible values of ϕ and d. As we do so, we construct an image whose coordinate axes are ϕ and d, and whose intensity is given by $\mu(\phi,d)$. This is the Radon transform of the input image. The Radon transform should, of course, be defined in formal mathematical terms as an integral involving two continuous

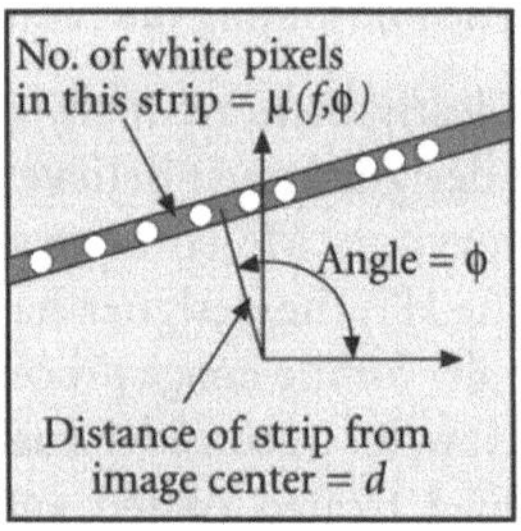

Fig. 3.54 The intensity is integrated along the dark gray strip $S(\phi,d)$. (In a binary image, we simply count white pixels within $S(\phi,d)$.) This yields the value $\mu(\phi,d)$.

variables, ϕ and d. In fact, the discrete nature of a digital image means that we can compute only an approximation to the Radon transform.

One of the most obvious ways to implement the Radon transform is to rotate the image by an amount ϕ and then integrate the intensities along the picture rows. (The PIP sequence [*tur(A), rin*] expresses the rotation-integration operation succinctly.) The right-most column of the resulting image will exhibit a strong peak, if we happen to have rotated the input image so that a set of collinear points is aligned horizontally (see Figure 3.55). We do not of course, always need to

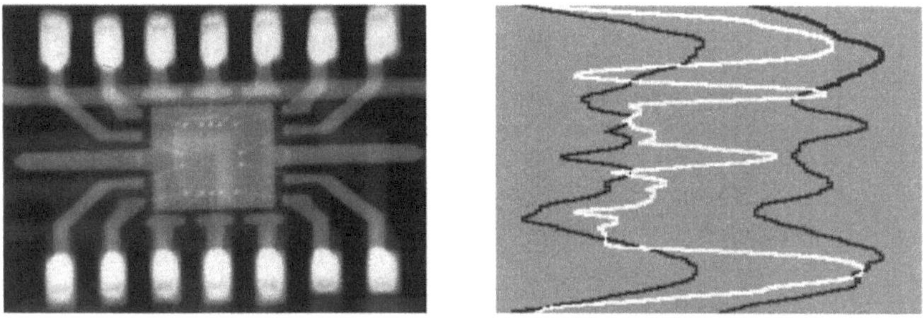

Fig. 3.55 The Radon transform detects linear features in grey-scale images. (a) Original image: microfocal x-ray of a small integrated circuit (b) Three intensity profiles, obtained after applying [*tur(Angle), rin*]. (White curve, angle = 0°. Black curves, angle = ±10°.) Notice the strong narrow peaks that are visible only in the white curve. These identify the linear feature just below the upper-most row of pads (i.e., bright rectangles) and across the center of the image.

construct the whole RT image. For example, we may know that the spots are likely to lie along a line whose orientation is within tightly defined limits. There are several special cases which make this ploy effective:

1. The horizontal and/or vertical axes of the input image may have some special significance in physical space. (For example, the horizontal axis of the image may represent the vertical (*Z*) axis in [*X,Y,Z*] space, or the vertical axis of the image may correspond to the direction of travel of a moving web or conveyor belt. In the latter case, the photo-detector array on a line-scan sensor would correspond to the horizontal axis of the image.
2. We can obtain a crude measure of orientation quickly and easily from some other features before applying a more refined procedure based on the RT.
3. We know beforehand what angles are critical. (For example, we may have learned that diagonal streaks are indicators of some undesirable "fault" condition, whereas horizontal ones are benign.)

The point to note is that in many instances we do not have to compute and analyze the full Radon transform image.

The PIP operator sequence [*tur(A), rin*] may be used as the basis for simply detecting the presence of a collection of "colinear" spots, without actually constructing the Radon transform image. This leads us to a simplified version of the Radon transform which is perfectly adequate for many purposes. Of course, we need to repeat the procedure many times in order to detect a set of spots, because these could be aligned at any angle.

In some cases, we might usefully split the analysis in two stages: a low-resolution initial RT scan, leading to and a second, more refined process based on the first. This can be expressed in PIP, but again the program cannot be included here because we have not yet described all of PIP's relevant features in sufficient detail.

"Island hopping"

In order to appreciate how straight lines can be reconstructed from a set of disconnected spots by the so-called "island hopping" procedure, we need to describe another idea first. We will therefore describe the fan image-processing operator built into PIP. Consider two points [*X1,Y1*] and [*X2, Y2*]. The fan operator searches a wedge or fan-shaped region of the image for a white pixel (See Figure 3.56). The

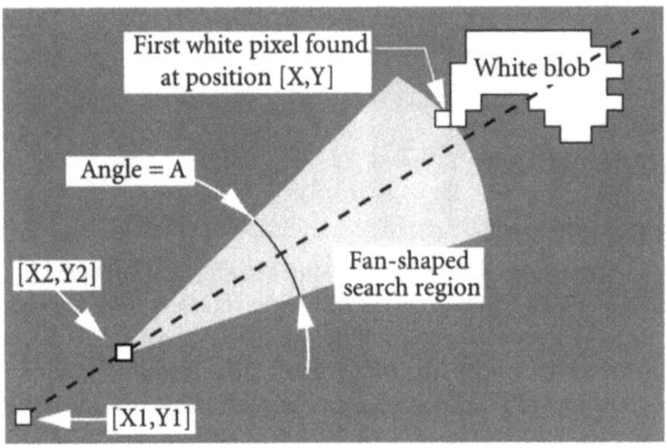

Fig. 3.56 The fan predicate used as the basis for "island hopping." The points [*X1,Y1*], [*X2,Y2*] and the angle *A* are specified when the fan is called. The point [*X,Y*] and its distance from [*X2,Y2*] are then computed.

search executed by fan(*X1,Y1,X2,Y2,A,X,Y*) begins at the point [*X2,Y2*] and is concentrated within a circular segment of angular width defined by the parameter A. The fan-shaped search area is centered on the line defined by [*X1,Y1*] and [*X2,Y2*]. The address of the first white pixel found is returned by instantiating the variables *X* and *Y*. It is easiest to think of the fan program being applied to an image consisting of single isolated white pixels. Then, given two starting points, a third point that is approximately aligned with them is found. A fourth point can be found by applying the fan operator recursively. A simple PIP program which reconstructs lines and gentle curves by the "island hopping" process and which uses the fan predicate:

```
island_hopping(A,B,C,D,E,F,G):-
    fan(A,B,C,D,E,X,Y),                        % New point found is [X,Y]
    writeseqnl(['New point found:', [X,Y]]),   % Tell the user the story
    island_hopping(C,D,X,Y,E,[[X,Y]|F],G).     % Jump to next island & repeat
island_hopping(_,_,_,_,_,A,A).% Force this predicate to succeed. Ends recursion.
```

Figure 3.57 shows the result of applying *island_hopping* to a picture containing isolated white pixels, some of which are aligned approximately. In this example, the angular width of the fan-shaped scan area is quite large.

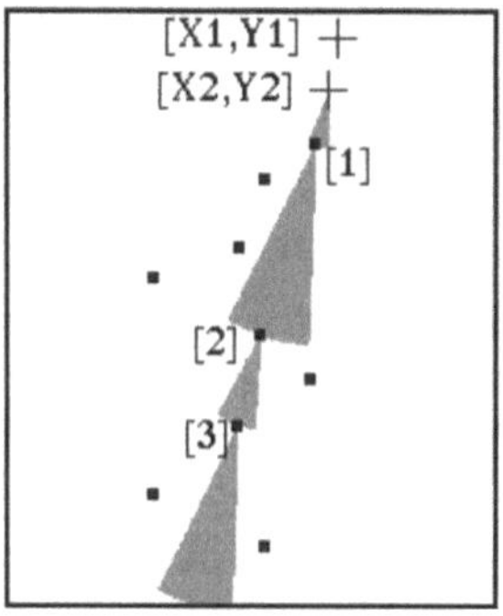

Fig. 3.57 Applying *island_hopping* to a simple test image. [*X1,Y1*], [*X2,Y2*] are the starting points. Numerals in brackets indicate the order in which the other points lying along the same line are detected.

General remarks

The Radon transform can be used to detect thin streaks and aligned spots in a grey-scale image, whereas the Hough transform, as it was defined in Chapter 2, cannot. Thus, the Radon transform provides not just a convenient alternative to the Hough transform but also offers interesting possibilities for analyzing grey-scale images, which the HT cannot match. On the other hand, the Hough transform can be modified to detect other curves of known form. For example, it can be used to detect circles, even when they are overlapping. The generalized Hough transform [DAV00] can even detect partially occluded shapes, such as touching/overlapping automobile connecting rods. While the Hough transform is able to accommodate broken contours, this property is shared with the Radon transform. Although this chapter is not concerned with implementation hardware, it is interesting to note that a cylindrical lens and a line-scan camera provide a very convenient means of performing the calculations necessary to implement the RT.

3.5 Additional remarks

We are near the end of this chapter, which has touched on a wide variety of algorithms and heuristics. It is appropriate therefore to review the lessons learned. The discussion above does not constitute a comprehensive review of these computational techniques; it is merely a discourse on the folly of taking too a narrow view of which algorithm to use. In the following chapters, we should bear in mind that virtually no algorithm is so important as to be regarded as "essential." Of course, it would be difficult to manage without using operators such as *threshold* [*thr*] and *negate* [*neg*] but bear in mind that even these can be implemented in several different ways.

3.5.1 Solving the right problem

Although this book concentrates on algorithmic/heuristic techniques suitable for use in machine vision systems and their implementation, we have repeatedly emphasized throughout that they must operate in harmony with the other parts of the system. Indeed, they should be designed/selected specifically with this in mind. Thus, we should not quibble about whether, for example, the Sobel edge detector is theoretically "better" than the Roberts operator, since they are similar in their performance. If it matters which one we use, then something else is wrong, probably the lighting-viewing subsystem [PVB]. There are numerous other examples of this general principle. We have at our disposal many different computational tools and often have several options for each algorithmic step in a processing sequence. We must be careful, therefore, that we address the most important questions. Very often, we find that whether we choose algorithm A or B is nearly irrelevant in terms of the results we finally achieve, compared with the far great differences that exist in the speed and cost of implementing them. We have constantly asserted that Machine Vision is a specialized area within Systems Engineering and, as a result, we must optimize all parts of a system separately and ensure that they will work together harmoniously. The real challenge is to ensure good system integration, since without this no system will ever be truly successful.

We must be clear about our primary purpose in building a vision system. It is not to prove how clever we are by using sophisticated techniques. Nor is it to analyze images in a preconceived way; we must be flexible in our thinking. Consider, for example, the task of finding the orientation of a machine component. We might be tempted to employ the Hough or Radon transforms, since they are so general. When demonstrating our academic prowess becomes a factor in our thinking, the design will inevitably be a poor one. We have at our disposal many different techniques for determining object orientation and these can be based on a variety of features, such as corners, bays, holes, colored regions (e.g., flashes, logos, etc.). We must bear in mind that our primary objective is to determine the orientation of an object, not find the orientation of a straight line in an image. We might decide that the latter forms a useful secondary goal. However, we might find instead that, by altering the image acquisition subsystem a little, we can achieve our primary goal very much more cheaply, and more reliably.

3.5.2 Democracy: no small subset of operators dominates

One thing has emerged clearly but has, so far, not yet been stated with sufficient force, namely that image-processing procedures for machine vision are highly varied. A machine vision system can and, at the moment, almost certainly should be designed for a specific task. Another important lesson is that the image-processing procedure needed for a real-world Machine Vision application is complex. In PIP terminology, it is not a single operator call. Instead, it requires a large set of operators, controlled in a sophisticated way. We cannot therefore pick on any one step and say that implementing this is crucial to solving the problem. Thus, we must never concentrate on implementing only one operator, since they are all vitally important. Hence, we must not allow ourselves to be persuaded to use a particular implementation technology simply because one of the elements in a processing sequence could make good use of it. A little while ago we mentioned the fact that the Radon transform can be implemented using a cylindrical lens and a line-scan camera. This combination is of no use whatsoever for such tasks as implementing the convex hull or Sobel edge detector, thresholding an image, or computing its histogram, etc. In the same way, a parallel processor that is appropriate for implementing a convolution operator is virtually useless for processing the chain code. A symbolic processor (e.g., Prolog) that can make inferences about canary yellow being "similar" to sulphur yellow is similarly limited when we need to perform histogram equalization. Thus, we must be prepared for the inevitable fact that we will have to use an implementation technology that is nearly optimal on part of the algorithmic sequence and less efficient on others. This is why we felt it desirable to include in this book a chapter where we highlight the potential for reformulating the algorithms. In this way, we can circumvent at least some of the difficulties that would ensue when we fix the implementation technology.

Let us end this section with a question for the reader to ponder. Suppose that we had an implementation technology that was nearly ideal for all operators except for one important sub-class, such as measuring blobs. Our hypothetical machine might be able to perform all of the "difficult" calculations, such as convex hull, blob shading, Hough transform, "island hopping," color recognition, etc. Would not such a machine be virtually useless in practice? Surely, we must consider the implementation of the command repertoire in its entirety.

3.5.3 Lessons of this chapter

- Every algorithm can be implemented in many different ways (Sections 3.1.1, 3.3.5, 3.3.6, 3.3.7). Even standard arithmetic operations, such as multiplication, can be implemented in more than one way.
- We should not judge an algorithm by its accuracy alone (Section 3.1.2). Many other factors are at least as important. These include speed, cost of implementation, compatibility with existing equipment, interfacing to people and machines, conforming with company policy, etc.
- Reformulating an algorithm may make its implementation very much easier.
- We must be careful to address the important primary questions and must not fall into the trap of believing that secondary issues are all-important (Section 3.5.1).
- No single algorithmic/heuristic procedure should be allowed to dominate our design, since all elements within an algorithmic sequence are of equal importance (Section 3.5.2).
- We can often use heuristics *in lieu* of algorithms to good effect (Section 3.1.4).
- Changing the image representation can make certain algorithms/heuristics very much easier to implement.

- We can sometimes employ grey-scale operators on binary images to good effect (Section 3.2.2).
- We can use images as lookup tables for operations such as image warping (Section 3.2.3).
- We can represent many of the more useful convolution operators as separate row and column operations (Section 3.3.1).
- We can approximate many of the more useful large-window convolution operators by iteratively applying a sequence of small-window operators (Section 3.3.1).
- Several nonlinear neighborhood operators (both binary and grey-scale) can be implemented by separating row and column operations.
- Separated-Kernel Image Processing using finite State Machines (SKIPSM, Section 3.3.3) allows a wide variety of morphological and other binary and grey-scale image-processing operators to be implemented (exactly) at high speed in both software and hardware.
- There is no unique algorithmic definition for certain operations that are specified in natural language (e.g., edge smoothing). Hence many operators exist.
- Histograms calculated on one image can be used as the basis for processing another image, if the images are derived in quick succession from a slowly changing scene.
- Image statistics derived using a low-resolution sampling of an image may be accurate enough for certain purposes.
- We can save a lot of processing effort by concentrating on those regions of a picture where the processing is relevant to achieving the desired result.
- Heuristics may be faster than algorithmic methods.
- Algorithms do not produce a reliable result if we violate their conditions of use.
- A given algorithm may be theoretically attractive but it may not solve our primary application task. The latter is not image processing *per se* but making decisions about real objects in the physical world.
- Heuristics can fail but this may be insignificant in practice, if they are designed and tested properly.
- Since heuristics are rules of thumb, we can add broad general rules which are safe, even if they are not always efficient.
- A heuristic may, in fact, be algorithmic, even though we do not know it to be so, because we have not been able to derive the necessary mathematical proof.
- Heuristics may be used to replace an algorithm, or to augment it.
- Heuristics may be used to make an algorithm run faster, without necessarily making its performance any less satisfactory.
- Heuristics may be used to extend the range of application of an algorithm.
- Heuristics can often "save the day" when there is no known algorithmic alternative.
- Heuristics can often achieve useful results when the use of an algorithm is impractical.

- We can often make heuristics as accurate as we wish, simply by doing more computational work.
- Elements in a processing sequence can often be combined to produce a simpler, more direct result. For example, the PIP sequence [*neg, thr(100,123), bed*] could be implemented in one new single-pass operator.
- lookup tables are very effective for implementing a wide range of image-processing operators and are potentially very fast, when implemented in both software and hardware.
- Techniques that are attractive for Computer Vision and other (i.e., non-industrial) application areas for Machine Vision are not necessarily of much use in our subject area. (For example, they may be too general, and therefore unable to exploit application constraints.)
- Heuristics may be very different in concept from the algorithms they displace.
- We must judge heuristics objectively in the light of statistically rigorous testing.

4 Systems engineering

When there is no vision, the people perish.
Book of Proverbs, ch 29, v 18

In Chapter 2, we concentrated on the mathematical formulation of digital image-processing operators, while in Chapter 3 we discussed some of the basic algorithmic variations that we can employ to facilitate their implementation in either software or dedicated electronic hardware. In this chapter, we move on to consider other much broader *Systems Engineering* issues relating to industrial Machine Vision. In particular, we will concentrate on the important topics of human-to-machine and machine-to-machine interfacing. We will base much of our discussion in this chapter on the premise that without trust and confidence, the workers in a factory or laboratory will rapidly become hostile to a vision system; they will look for every opportunity to disrupt its smooth operation and will surely prevent it from working effectively. If there is any perceived conflict between people and machines in a factory, human beings will always win! The lesson is clear: sound engineering and, in particular, good interfacing and attention to detail in system design are of crucial importance.

4.1 Interactive and target vision systems

First, we need to distinguish between two kinds of vision systems that have quite different rôles in industry:

a) *Interactive vision systems (IVS)* are used primarily in the laboratory or design office for the ill-defined tasks of analyzing new and previously unseen applications, and thereby choosing appropriate image-processing algorithms. They are therefore suitable for prototyping, algorithm development, and testing. In addition, they have a major rôle to play in demonstrations to would-be purchasers of vision systems, as well as for education and training. Interactive systems are likely to be implemented using a standard computer and a single plug-in card to digitize the incoming video signal.
b) *Target vision systems (TVS)* operate in the factory, often with minimal human supervision, performing a well-defined repetitive task, probably of limited scope, that makes a direct contribution to the quality of the product or the efficiency or safety of the manufacturing processes. Typical target systems perform routine tasks, such as inspecting, monitoring, controlling, guiding, measuring, sorting

grading, calibrating, etc. They normally observe the raw material, partially-made or finished product, the manufacturing machinery (e.g., tooling), machining processes, parts assembly, packaging materials, or the packing/wrapping processes. Target systems are often required to operate at high speed, thereby making the use of specialized electronic hardware quite likely.

Among their other functions, interactive vision systems are used primarily to design target systems. It would, of course, be very helpful if they both used the same control language. Failing that, an algorithm discovered using an IVS might be "compiled" to generate the instruction sequence for the TVS. At the time of writing, this ideal situation has not been achieved in full; a considerable amount of skilled engineering effort is still needed to design an efficient TVS. The reason for this is very simple: interactive and target vision systems are likely to use completely different architectures. It is commonly necessary to use a round-about method which runs inefficiently on the multi-purpose IVS and which must therefore be recoded in order to make it run faster on the target system.[1]

4.2 Interactive vision systems, general principles

We will consider the basic requirements of an IVS in general terms before going on to discuss the specific details of three different implementations: PIP (running on a Macintosh computer), WIP (Windows), and CIP (Internet).

4.2.1 Speed of operation

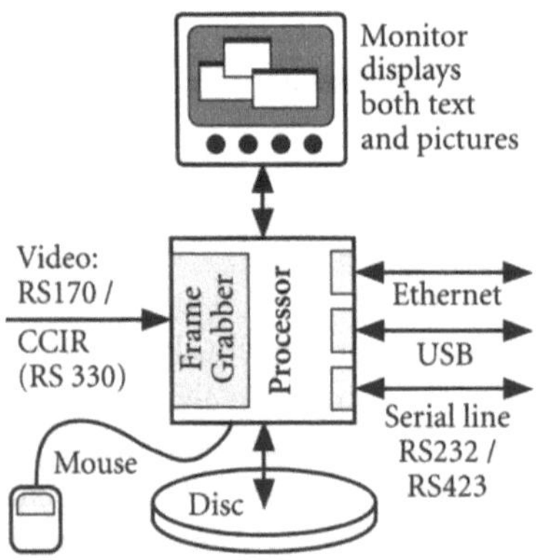

Fig. 4.1 Architecture for an interactive vision system.

Compared with a TVS, an IVS does not normally need to be very fast. It is much more important that the latter be versatile and easy to use. For this reason, interactive vision systems are often implemented using a conventional computer (Figure 4.1). Quite acceptable execution times can be obtained for many operations in this way (Table 4.1). From their experience gained with IVSs stretching over more than two decades, the authors are able to offer the following advice:

1. A good example of this is a program to plot intensity profiles along a given column of the image. The program to do this in PIP [*plt/1*] requires seven passes through the image, plus three image-to-image transfers. Virtually any programmer could write a few lines of highly efficient code to draw the intensity profile.

- The processing delay of an IVS should not normally exceed a few seconds. A delay on the order of two seconds is about the maximum that can be tolerated for frequently-used operators, such as *neg*, *thr*, *heq*, *enc*, *sed*, etc.
- Slower operation for infrequently-used commands is normally acceptable, although 30 seconds should be regarded as the absolute maximum delay for any operation that produces no intermediate pictorial results.
- Rather longer delays will be accepted by most users if the image display is dynamic. (That is, it shows a rapidly changing sequence of pictures.) Although the display of intermediate results in a long processing sequence may seem to be largely cosmetic, it reassures the user that the processing is being performed properly.

Table 4.1 Processing speed for PIP, a typical interactive vision system. These are measured execution times for PIP running on a Macintosh Powerbook G3 computer with a 250 MHz clock. Image resolution is 256×256 pixels. Such a computer was regarded as providing a good performance at the time of writing (September 1998), but this is probably no longer so when you are reading this.

Operation (PIP operator)	Execution time	Frequency of use	Comments Image size: $N \times N$ pixels
Add 2 images *(add/0)*	31 ms	Medium	Acceptable. Execution time is $O(N^2)$
Cartesian-to-polar image warping *(ctp/2)*	1210 ms	Low	Not used frequently
Expand white areas *(exw/0)*	33 ms	High	Acceptable. Execution time is $O(N^2)$
Histogram equalization *(heq/0)*	102 ms	High	Acceptable. Execution time is $O(N^2)$
Hough transform *(huf/0)*	2445 ms	Low	Just acceptable. Execution time is dependent upon number of white pixels in the image.
3×3 blurring filter *(lpf/0)*	86 ms	High	Acceptable. Execution time is $O(N^2)$
Interchange working images *(swi/0)*	1.3 ms	Very high	Acceptable
Negate (neg/0)	27 ms	Very high	Acceptable. Execution time is $O(N^2)$
Pop image onto stack *(pop/0)*	33 ms	High	High speed achieved despite 6-level hierarchical definition.
Push image onto stack *(pis/0)*	16 ms	High	High speed achieved despite 6-level hierarchical definition.
Recover image (from RAM) *(rea/1)*	31 ms	High	Acceptable
Rotate image *(tur/1)*	79 ms	Medium	Acceptable. Execution time is $O(N^2)$
Save image (to RAM storage) *(wri/1)*	12 ms	High	Acceptable
Shift image (psh/2)	25 ms	High	Acceptable. Execution time is $O(N^2)$
Skeleton of a binary image *ske/0)*	150 ms	Low	Acceptable. Not often used.
Sobel edge detector *(sed/0)*	78 ms	High	Acceptable. Execution time is $O(N^2)$
Threshold *(thr/2)*	78 ms	Very high	Acceptable. Execution time is $O(N^2)$
Interchange X & Y image axes *(yxt/0)*	30 ms	High	Acceptable. Execution time is $O(N^2)$

4.2.2 Communication between an IVS and its user

It is good practice to begin the design of any engineering system by reviewing the requirements and constraints imposed by the application. With this in mind, we have tabulated the major items of information that must be transferred between an

IVS and its user (Table 4.2). The *human–machine interface (HMI)* of an IVS is clearly of great importance, since the user must be able to concentrate on the application, not on the minutiae of an ill-considered dialog that was "designed" for the benefit of the machine or the convenience of the programmer, rather than the end user. We will discuss the (high-level) interface requirements of an interactive vision system in detail later in this chapter. For the moment, let it suffice to say that the user must be able to examine both "raw" and processed images visually. He must also be able to select and execute image-processing operators quickly and easily. It should be borne in mind that an interactive vision system will certainly possess a large and very rich repertoire of commands. (PIP, for example, has over 270 basic commands.) The user will also need to derive a variety of computed measurement values from the images. An IVS must also be programmable, although it will not be expected to operate in real time.

Table 4.2 Data transfer requirements between an interactive vision system and its user.

Data	Direction	Remarks
Command strings	User → IVS	Examples: *thr(123,145), program(54, largest).*
Cursor position & intensity/color information	User → IVS IVS → user	Mouse, light-pen, and/or touch-sensitive screen can be used. Visual display of text. Possibly speech synthesis.
Error messages	IVS → user	Examples: *File not found, Parameter out of range*
Images	IVS → user	Possibly with overlaid graphics
Menu selection	User → IVS	Pull-down and pop-up menus
Numeric data and text	IVS → user	Examples: image measurements data, Help information
Program code	User → IVS	PIP programs

Various modes of interaction have been devised to enable people to work effectively with computers and other machines with embedded processors (Table 4.3). These interfaces have different uses, exploiting their various strengths and weaknesses. A well-designed vision system should be able to provide a variety of control mechanisms, in order to accommodate a wide range of user abilities and operating procedures. For example, a great deal of attention has been paid to the human–machine interface of PIP, which provides all of the control mechanisms mentioned in Table 4.2, except for visual programming and selecting operators represented in iconic form. *JVision*, which is described in Chapter 8, relies on visual programming.

In order to develop effective demonstration prototypes of factory-floor (target) systems, IVSs are also required to operate a broad range of electromechanical devices,. More will be said later in this chapter about the requirements for the *machine–machine interface (MMI)* for both interactive and target systems. However, we will concentrate initially on the human–machine interface.

Table 4.3 Modes of interaction between user and vision system

Task Type and Example	Comments
Command keys (e.g., *Comand G* to digitize an image in PIP [BAT97])	Fast. Limited repertoire available. Ideal for very common operations.
Command line: (SUSIE [BAT79b] , AUTOVIEW [BAT82a], VCS [VCS], PIP)	Interactive and programmed modes are fully compatible. Consistent format when specifying command names and parameters Versatile but easy to make mistakes when typing.
Iconic VDL	Easy to represent some operations graphically but many others are difficult to distinguish in this way.
Key-activated menu (SUPERVISION [SUP])	A hybrid combining command-line and menu modes. The Help facility plays an integral part in the machine-user interaction. Operations are divided into classes. Each class has a unique initial letter. Another letter identifies the command within its class. Menu adjusts to show choices available at any given moment.
Pull-down/pop-up menus (Photoshop [PHO], Image [IMA], ...)	Well-suited to naive user. Optimal for commands which do not require parameters. Parameters require mouse and/or key-board entry of data.
Screen-based editing (Use mouse or light pen, plus image display and graphics, possibly overlaid.)	Limited in scope but extremely useful in appropriate places. Cursor may be used to measure distances between image features, find intensities (at a single point or in small window), define processing windows, morphological structuring elements, and convolution kernels, draw polygons and intensity/color mapping functions, move blobs around, move overlaid figures (e.g., caliper measuring tools), control camera pan, tilt, zoom, and focus.
Speech input (Available in PIP [BAT91a])	Overcomes some of the objections to using explicit commands for interactive image processing. Can be combined with natural language input.
Visual programming (OPTILAB, an extension to LabView [OPL], Chapter 8)	Closely allied to iconic mode. Well suited to programmed mode and for novice users.

4.2.3 Image and text displays

An essential element for both IVSs and TVSs is the image display. Experience has shown that it is impossible to design/choose an *effective* vision algorithm by the usual methods used for program development. Step-by-step interactive selection of image-processing primitives, based upon a good view of the images before and after each processing step, is absolutely essential. The image display must provide an appropriate spatial resolution and an adequate rendering of both the color and intensity scales. A spatial resolution of only 128×128 pixels is sufficient to provide a good rendering of a human face, provided that it nearly fills the screen (Plate 5). On the other hand, a complex industrial artifact may require an image resolution that is rather higher than this. (PIP is normally used with 256×256 pixels, although this is adjustable.) Our experience has shown that for most experimental purposes an interactive vision system with 512×512 pixels is perfectly adequate. (This is approximately the same resolution as that of a standard broadcast television picture.) It should normally be possible to see individual pixels. (If they are too small, the presence of isolated pixels is not evident to the user. In this case, he will not take them into account when designing an algorithm. This is one case where the usual maxim that *"more pixels is better"* does not necessarily hold.) On the other hand, the pixels should not be so large that the picture has an obvious block-like appearance.

The intensity and color performance of a visual display is also of considerable importance for interactive vision systems. While most people will tolerate working for a very short time with a very small number of grey/color levels (≤ 16), such an image is unnatural in appearance and the user quickly becomes irritated by its unrealistic appearance (Plate 5). Most importantly, an IVS that displays only a small number of grey levels could not possibly be effective, since a lot of subtle detail, such as low-contrast texture, is lost. In the 1970s, one of the authors (BGB) regularly used an IVS display providing 6-bit resolution (64 grey levels). For most purposes, this was perfectly adequate. After that experience, the early desk-top computers, which displayed only 16 grey levels, appeared woefully inadequate. For most purposes, the 256 grey levels displayed on a modern video/computer monitor is perfectly acceptable. There is rarely any need to exceed this particular aspect of the specification, even though the processing may require higher resolution. Ultra-high precision of rendering of color is not normally needed for interactive systems.

The ability to transform monochrome images into pseudocolor (also called *false color*) provides a distinct advantage for an IVS. Subtle changes in intensity can be detected much more effectively by a person if the intensity has been transformed using quite simple color-mapping functions (see Fig. 4.2 and Plate 19). Pseudocolor is especially helpful if the user is able to adjust the color-mapping table at will. The reader may have noticed that the x-ray baggage-inspection systems used in airports have pseudocolor displays, which can often be flicked on/off. This is a valuable facility for an IVS too.

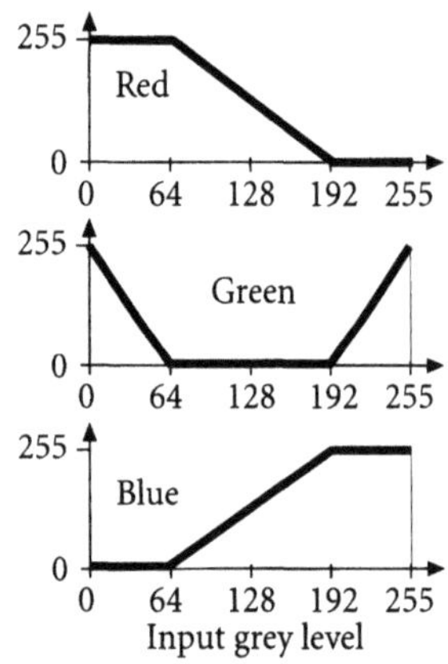

Fig. 4.2 Lookup tables for a pseudocolor display. This is only one of many possible sets of mapping functions. An ideal system allows a range of pseudocolor mappings to be tried in quick succession.

Overlaid graphics can be useful to provide a means of "annotating" an image, for example by showing where certain critical features, such as corners, centers of holes, etc., have been found, or where the intensity edges/contours lie (Plate 21(TL)). However, there can be no compromise whatsoever on the maxim that the text and image displays must be kept quite separate, probably by using separate windows. Under no circumstances should text, such as command strings and measurement data, ever be superimposed on the images (Plate 21(TR)). Graphs, such as intensity plots, spectra, and histograms, should normally be displayed separately from the images in a separate window. Where graphical symbols are overlaid onto the images (e.g., by the vision system superimposing image dimensions), it should be possible for the user to flick the graphics on/off with a simple manually-operated (software) switch.

4.2.4 Command-line interfaces

The earliest interactive image-processing systems, developed in the late 1970s and early 1980s, used a conventional *command-line* operating protocol. For example, SUSIE, the original progenitor of PIP, used 2-letter mnemonic commands of the form outlined in Table 4.4 Superficially, PIP looks very similar, except that 3-letter mnemonics are used instead (Table 4.5).

Table 4.4 Command-line operation and defining macros in SUSIE (1976)

Command	Function
NE	Negate
TH 123,145	Threshold
AV @12	Calculate the average intensity and store it in register number 12
AV@ 6; TH @6	Compound command; threshold at the average intensity in reg. #6
DM CONTOUR LP; LP; LP; SC 3; GR; TH 1; EM	Define a macro called *CONTOUR*. This macro blurs the image, reduces the number of grey levels to 3, finds the edges, and thresholds the resulting contours to form a binary image. The result is a set of *intensity contours*, or *isophotes*.
DM HPF WR TEMP; {1}; RE TEMP; SU; EM	Define a general high-pass filter. The image is blurred using the low-pass filter, defined by the (first) argument of *HPF*. N. B. *{1}* represents the first argument of the macro, which is specified at run time. Typical macro call: *HPF (LPF;LPF;LPF)*

Table 4.5 Command-line operation and defining higher-level predicates in PIP (1996)

Command	Function
neg	Negate (equivalent to *NEG* in SUSIE)
thr(123,145)	Threshold between levels 123 and 145 (equiv. to *TH 123,145* in SUSIE)
avg(X)	X is the average intensity. (*avg* is equivalent to *AV @12* in SUSIE)
avg(X),thr(X)	Compound command; threshold at average intensity, stored by variable X.
contour:- lpf, lpf, lpf, sca(3), gra, thr(1).	Define a new higher-level predicate, called *contour*. This is conventional Prolog syntax. This program blurs the image, then reduces the number of grey levels to 3, finds the edges, and thresholds the resulting contours to form a binary image. *contour* always succeeds, since all image-processing sub-goals (i.e., *lpf*, *sca*, *gra* and *thr*) succeed. The result is a set of *intensity contours*, or *isophotes*.
contour:- 3•lpf, sca(3), gra, thr(1).	Alternative definition of *contour* using the operator '•'. N.B. *N•G* satisfies goal *G* a total of *N* times.
hpf(Lpf):- wri(temp),call(Lpf), rea(temp), sub.	Define a general high-pass filter, based on an unspecified low-pass filter *(Lpf)*. Examples: *hpf((lpf,lpf,lpf))* or *hpf(3•lpf)*

4.2.5 How many images do we need to store in RAM?

The answer is *a minimum of two.* The reasons for this are as set out below:

1) Monadic operations could be performed in a serial (von Neumann) computer using just one stored image, which is overwritten pixel by pixel as the processing takes place. (As we will see later, storing two images allows us to display both "input" and "output" pictures simultaneously.)

2) Dyadic operators require two images, provided that we allow one of them to be overwritten during processing.
3) Local operators require just one stored image, although some additional memory is required. For a local operator based on an $N \times N$ processing window, this additional store must be able to hold a minimum of $(N-1)$ rows, plus $(N-1)$ pixels. However, this solution is not wholly satisfactory, since the output image is shifted along both X and Y axes. A more straightforward implementation uses a second image store to avoid introducing any such shift.
4) Additional images can be read from or written to backing store, as needed. Only one of the RAM images need be designated for transfers to/from backing store.

Current and alternate images

Remembering the arguments just given, we can now introduce an operating paradigm for interactive image processing that was developed in the late 1970s [BAT79b] and which still forms the basis of the latest software packages, such as PIP, WIP, and CIP. The *2-image model* has two stored images, called the *current* and *alternate images.* A monadic operator (e.g., *neg, log, exp, thr, sqr, enc*) or local operator (e.g., *con, lpf, sed, lnb, bed, xor*) operates as follows:

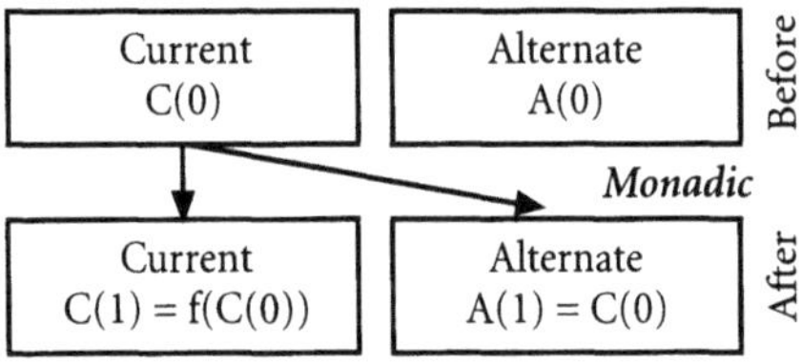

a) The current image is preserved by the operation but is transferred to the alternate image buffer.
b) The (original) alternate image is destroyed by the image-processing operation.
c) The result of the processing is placed in the current image buffer.

Dyadic operators combine the current and alternate images and therefore function as follows:

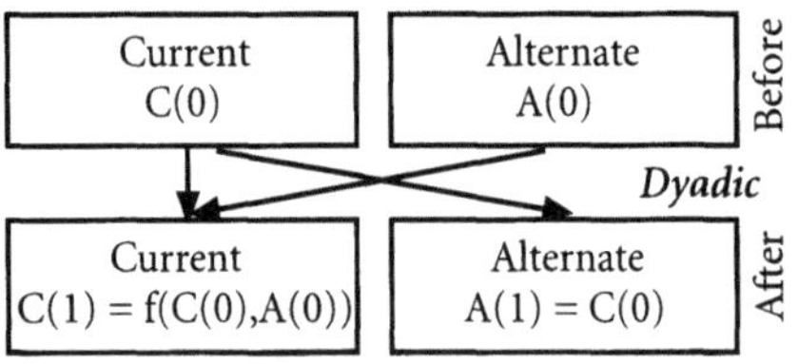

Images may be read from the backing store (disc file).

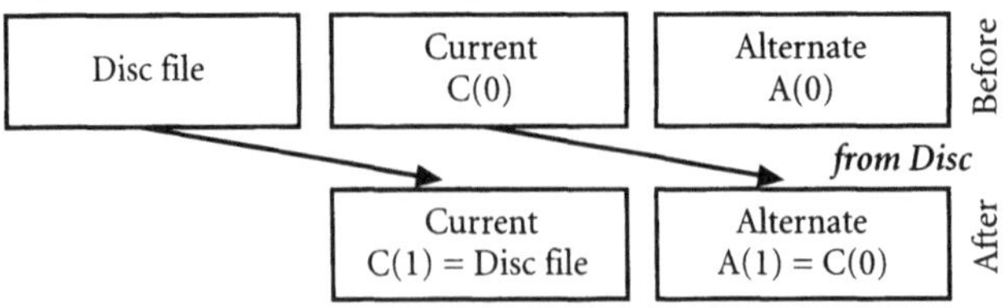

A similar model holds for *image-generation* operators, such as *wgx* and *hic.*

Writing images to disc is even more straightforward.

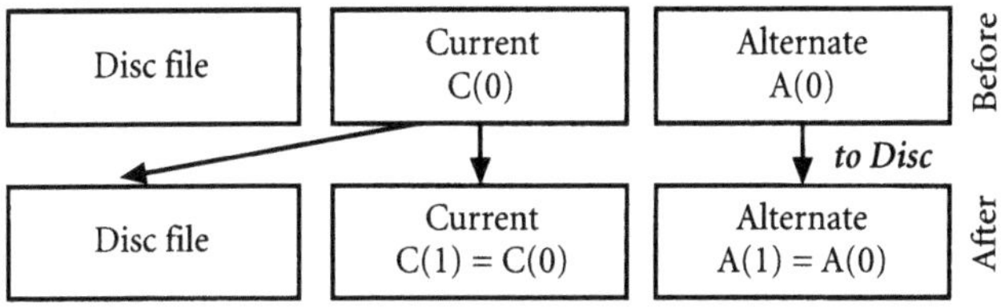

Lastly, image measurement operators (e.g., *gli*, *cwp*, *cbl*, *per*, *eul*, *dim*, *lmi*) involve no image model, since neither the current nor the alternate image is modified.

4.2.6 How many images do we need to display?

So far, we have discussed the number of images needed to implement various vision algorithms. Now, we need to consider the same question, taking into account the needs of the user, rather than the image-processing algorithms.

A usable but sub-optimal IVS can be based on a device which displays only one image. This will be most effective if the hardware allows the user to flip quickly between the current and alternate images. The first generation system (SUSIE) did just that. Switching between images was achieved by a single key-stroke. The hardware switched to display the new image at the beginning of the next video frame (i.e., not more than 40 ms after the user pressed the appropriate key). Switching between two images in this way allows a person to compare them quite effectively, since they are registered exactly in space. However, many later IVSs have employed a 2-image display, which allows the current and alternate images to be viewed simultaneously (Figure 4.3). This display format has been very effective and still forms the basis of PIP, WIP and CIP, as well as various commercial products.

A convenient extension of the basic 2-image interface displays a third image in which the "raw" (i.e., unprocessed) image is permanently visible to the user, as well as the current and alternate images. This additional picture display will be called the *Scratch-pad* or *Scratch* image. During interaction, the user may well wish to recover the original image, in order to combine it with some processed image that he has just derived from it. In order to recover pictures from the scratch image, a signal flow path is needed from the additional image display to the current image. Thus, two commands (*dpi* and *rpi*) are sufficient to support this extra image display.

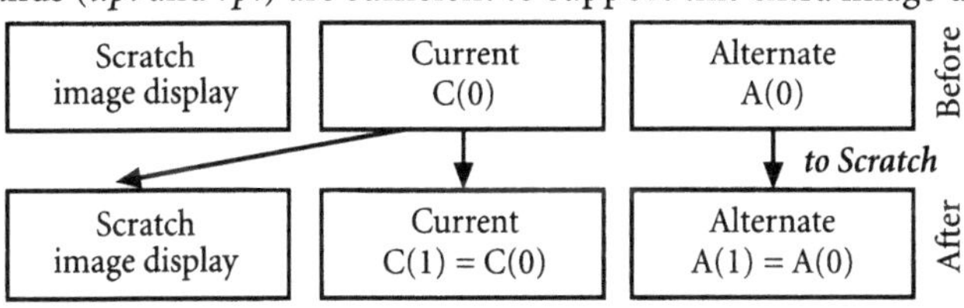

It is appropriate to highlight two commercial systems that have used these ideas: *Autoview* [BAT82a] displayed the current and alternate images as well as two scratch-pad images. All of these images were displayed in full resolution. The *Intelligent Camera* [ICM] displayed the current image in full resolution, while the alternate image and one scratch-pad image were shown in quarter resolution.

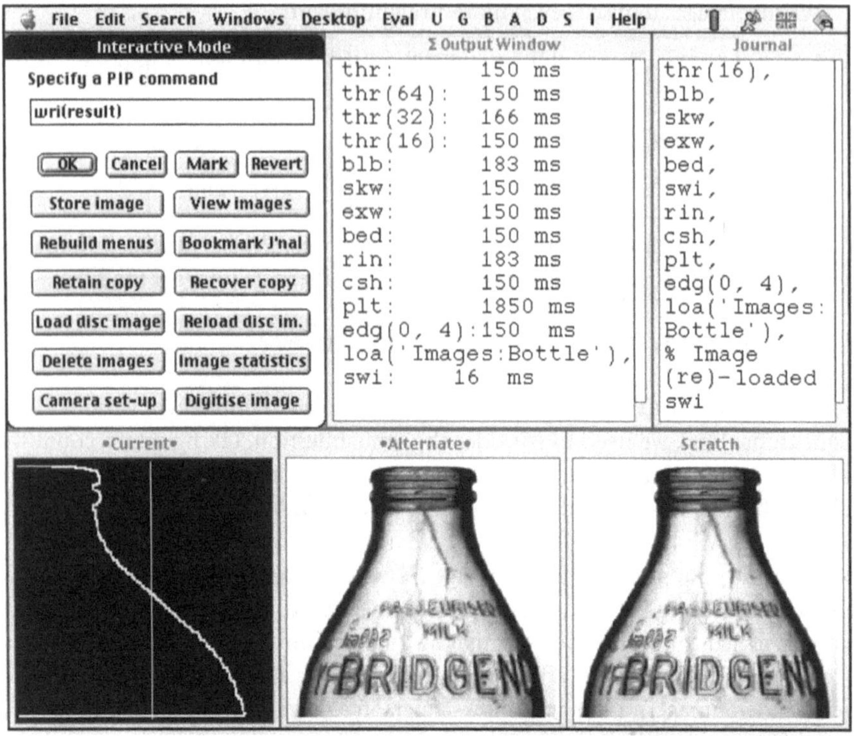

Fig. 4.3 Screen layout for the PIP vision system. The image on the right *(Scratch)* does not change during the execution of PIP commands. It is used to display an undisturbed view of the original image, enabling the user to study the effects of the processing. (Also see Figures 4.6 and 4.8.) The pull-down menus U, G, B, A, D, S, and I are specific to PIP, while the others are part of the MacProlog environment.

A screen generated by the PIP system is shown in Figure 4.3. The current, alternate and scratch images are all displayed in full size, if they do not exceed a given resolution. (Higher resolution images are rescaled automatically, so that they fit within the same physical size limits.) PIP allows a much larger number of pictures to be displayed on the screen. The goal *sim(Name)* creates a picture called *Name*, which can be retained on the computer screen indefinitely until the goal *him(Name)* is satisfied. Hence, it is possible for PIP to create a new image for each image-processing goal that is satisfied. However, this is not particularly helpful, since the computer screen soon becomes seriously cluttered. Another approach has been adopted, namely that of the *image stack*, which is discussed in Chapter 7.

4.3 Prolog image processing (PIP)

It is helpful to turn our attention now to one specific implementation of an IVS called PIP. The authors' intention has been to devise an IVS that incorporates the general principles outlined above and which embodies as many useful features as possible, in order to make the system easy to use and improve its effectiveness.

PIP is an extension of the standard AI language *Prolog*. Readers who are familiar with Prolog merely need to note that PIP incorporates its image-processing functions as *Built-in Predicates* within the standard language. Readers who have not encountered Prolog before should consult the brief introduction to this fascinating language given in Appendix B or one of the many excellent textbooks on the subject [BRA86]. The PIP command repertoire is listed in Table 4.7.

4.3.1 Basic command structure

The reader has already been briefly introduced to the PIP command structure in Table 4.5 and on earlier pages, although some important features have yet to be explained.

Arity In the discussion to follow, we will use the following notation to define the *arity* (i.e., number of arguments) required by a given operator:

qwer/N	Operator *qwer* has arity N. Typical command: *neg*.
qwer/[M,N]	Operator *qwer* has arity M or N. Typical commands: *plt*, *plt(100)*.
qwer([M-N]	Operator *qwer* has any arity in the range M, M+1, M+2, ..., N.

Thus, the threshold operator will be represented by *thr[0-2]*.

Passing parameters It is possible to pass parameters from one PIP operator to another by *instantiating* ("giving an instance of") a variable, which is then used later. Here is a simple example:

```
avr(X),  % The variable X is instantiated² to a value equal to the average intensity.
thr(X)   % The value X calculated above is used as the threshold parameter.
```

Compound goals The simple command sequence just given should properly be regarded as a compound goal in Prolog. The variable X is instantiated by satisfying the goal *avr(X)*. This value is then used when trying to satisfy the goal *thr(X)*. The image-processing "commands" in PIP should be regarded as built-in Prolog predicates that always succeed. They fail on backtracking. This is the same mode of operation as the standard built-in Prolog predicates *write/1, nl/0, tab/1* all use [CLO81].

Terminology We will often use the term *command* when discussing PIP's built-in image-processing predicates. The reason is that the user thinks of them as commands when using PIP interactively. Moreover, they behave as if they were commands, since they always succeed, assuming that any parameter values passed to them are within the allowed ranges. However, they fail if any of their parameters is out of range.

Defining higher-level predicates Consider the following command sequence:

```
gli(A,B),            % A is the minimum intensity and B is the maximum intensity.
C is int((A+B)/2),   % Standard Prolog arithmetic. C is integer part of average of A and B.
thr(C).              % Threshold midway between brightest and darkest intensity values
```

The user of the PIP system might find it convenient to define a new operator, *threshold_at_mid_intensity/1*, which performs the same function. This can be achieved in the conventional manner for programming in Prolog:

---

2. Instantiation is a term borrowed from Prolog programming. It means that X is temporarily given a value, although this might be reassessed later. However, in this particular program X cannot be reinstantiated due to the fact that *avr/1* cannot be resatisfied on backtracking.

```
threshold_at_mid_intensity(C):-
    gli(A,B),           % A is the minimum intensity and B is the maximum intensity.
    C is int((A+B)/2),  % Standard Prolog arithmetic.
    thr(C).             % Threshold midway between brightest and darkest intensity values
```

Synonyms Although the name *threshold_at_mid_intensity* is descriptive of the function that it performs, this name of this operator is rather long and inconvenient to use in practice. Hence, the user might prefer to use the synonym *tmi/1*, defined thus:

```
tmi(A):- threshold_at_mid_intensity(A).
```

Synonyms can be useful for several reasons:

- Accommodating synonyms in English. For example, the following families of synonymous terms might be useful for certain categories of users:
  {blur, low_pass_filter, raf}
  {chu, rubber_band, smallest_enclosing_convex_polygon}
  {rotate, tur}
- Providing a more descriptive form of mnemonic for novice users
- Providing a version of PIP that is better suited for specific applications areas, such as microscopy, metallurgy, web inspection, etc.
- Defining foreign-language versions of PIP

Although synonyms may seem to be mere cosmetic "froth," they can make a big difference to a user's understanding and willingness to accept a language such as PIP. It is easy for the user to define his own terms and hence to tailor the language for his specific needs.

Varying the arity The user might also find it convenient to define *tmi/0*, which prints the threshold value being used:

```
tmi:-                  % This defines a clause of arity 0.
    tmi(A),            % This has arity 1.
    writeseqnl(['The average value of the minimum and maximum intensities is ',A]).
```

The same technique can be used to define default values. For example, the definition of the standard threshold operator (*thr*) includes two auxiliary clauses:

```
thr:- thr(128,255).      % Defines "thr" with arity 0 and default values 128 and 255.
thr(X):- thr(X,255).     % Defines "thr" with arity 1. 2nd argument default = 255.
```

Using parameters to control the program Of course, there is no reason why we cannot use the parameter calculated by one command, such as *tmi(A)*, to control the operation of the program. In the following example, the computed parameter *A* is used to improve the lighting if the image is too dark.

```
automatic_lighting_adjustment:-
    grb,                 % Digitize an image from the camera
    tmi(A),              % Threshold midway between min. and max. intensity (level A)
    A≤ 100,              % Is mid-range value (A) ≤ 100?
    increase_lamp_brightness, % It is, so increase lighting levels. (Predicate not defined here.
    automatic_lighting_adjustment.
    automatic_lighting_adjustment.  % "automatic_lighting_adjustment" always succeeds
```

Mnemonic or explicit commands? Some users like short, cryptic (mnemonic) command names, while others prefer long (explicit) ones. The relative merits of these two forms are summarized in Table 4.6.

Table 4.6 Mnemonic and explicit commands

Type and Examples	Comments
2-letter mnemonics (e.g., SUSIE): NE, TH 123,145, AV @1	Quick – enables rapid interaction with the user. Mnemonics must be learned but it is relatively easy to teach a new user. Limited number of functions available ($676 = 26^2$). Mnemonic codes are often obscure.
3-letter mnemonics (e.g., basic-level PIP): neg, thr(123,145), avg(X)	Quick – enables rapid interaction with the user. Scope for many functions and synonyms ($17\,576 = 26^3$). Mnemonics must be learned but it is relatively easy to teach a new user.
Explicit: negate, threshold(123,145), convex_hull difference_of_low_pass_filters(F1, F2)	Slow to type – impedes interaction. Prone to typographical and spelling errors. Often difficult to specify function in a few words. Hence there is very little advantage over 3-letter mnemonics.
Mnemonic with named source and destination images: neg(Im1,Im2) thr(Im1,I2, 123,145)	Very slow to type. Prone to typographical and spelling errors. Impedes interaction. Requires high skill level. Satisfactory for programmed systems. Some command names are obvious, but many must be learned explicitly.
Mnemonic/explicit command with named source and destination images, plus specified processing window: negate([L,B,T, R], Im1, Im2), threshold([L,B,T, R], Im1, Im2, P1, P2)	Very slow to type – seriously impedes interaction. Requires high skill level. Satisfactory for programming. Very prone to typographical and spelling errors.

The choice between short and long command names is not a significant issue for PIP, because it can easily accommodate both forms. Since PIP is based on Prolog, it is a very simple matter to convert from one form to another. For example, suppose that we wish to extend the standard PIP software for a novice user who has a particular liking for long explicit command names. Then, we can use a one-line definition, as illustrated below:

```
a_new_threshold_function(A,B):- thr(A,B).
```

Of course, if *a_new_threshold_function* is to be defined with a range of arities, then we need to include one such definition for every possible value for the arity. A useful predicate *new_command_name/3* may be defined which does this automatically. For example, the program sequence

```
New_name = a_new_threshold_function,
Old_name = thr,
List_of_arities = [0,1,2],
new_command_name(New_name, Old_name, List_of_arities)
```

will generate and assert three clauses for *a_new_threshold_function*, with arities of 0, 1, and 2. Writing a program to implement *new_command_name/3* is straightforward but requires a deeper knowledge of Prolog than we have explained here.

The other forms of command names mentioned in Table 4.6 can also be accommodated within PIP. For example, the fourth one listed there uses named source and destination images and can be implemented easily. However, we will defer discussion of this until we consider the multi-image operating paradigm explained in Section 4.4.3.

Sample PIP program Here is a program which computes the value (V) and suit (S) of a non-picture playing card. This illustrates some of the points made so far.

```
non_picture_playing_card(V,S):-
    loa,                     % User selects image
    wri(temp1),              % Save image in file "temp2"
    tmi,                     % Adaptive threshold – works very well with playing cards
    neg,                     % Negate
    big,                     % Select biggest blob
    wri(temp2),              % Save image in file "temp2"
    cwp(A),                  % Count white points
    B is int(A/2),           % Divide by 2, round to integer value
    swi,                     % Interchange current and alternate images
    kgr(B),                  % Keep only those blobs with area ≥ B
    rea(temp2),              % Read image in file "temp2"
    cvd,                     % Convex deficiency (regions inside convex hull
                             % but not in original figure)
    cbl(F),                  % Count number of blobs=bays in suit symbol
    suit(F,S),               % Consult database for meaning of this result
    rea(temp1),              % Restore original image.
    writeseqnl(['The card is the ',V,' of ',S]).    % Message for the user.
    % Database relating number of bays for each suit-symbol to the suit name:
    suit(4,'clubs').         % See Figure 4.4.
    suit(2,'spades').
    suit(1,'hearts').
    suit(0,'diamonds').
    suit(_, 'unknown suit').   % Any other result is an error.
```

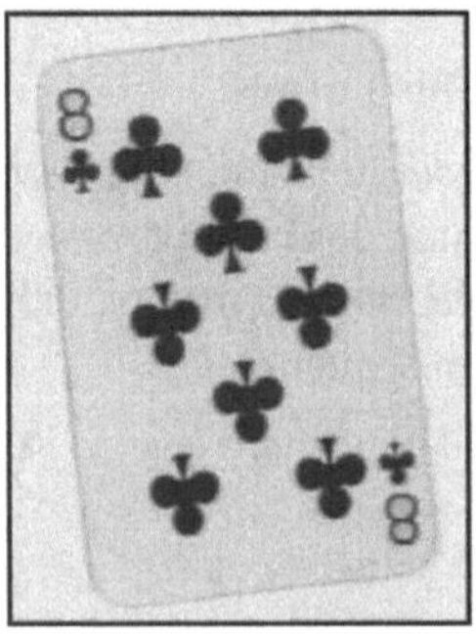

Fig. 4.4 Non-picture playing card (8 of clubs). Also see Figure 4.19.

Notice that *tmi/0* is used in just the same way as PIP's basic functions. At this level of programming, we are not concerned at all with how *tmi/0* works, only with what it does. PIP programs are usually hierarchical in nature, sometimes with 20 or more different levels of nesting. (No effective limit is imposed on the degree of nesting allowed in PIP programs.) *non_picture_playing_card/2* happens to be a very simple program and does not have a high degree of nesting. We will encounter much more complicated programs later, having much deeper nesting structures.

Extending PIP Consider the following program, which plots the intensity profile along a given column of the current image.

```
plt(A):-
    dgw(_,_,C,B),          % Find dimensions of the image
    pis,                   % Push an image onto the stack
    D is C - A,            % Simple Prolog-style arithmetic
    psh(D,0),              % Shift the image horizontally by an amount D
    csh,                   % Make all columns the same as the R.H.S. of the image
    wgx,                   % Generate an intensity wedge
    sub,                   % Subtract this image and the wedge
    thr,                   % Threshold the result
    sed,                   % Sobel edge detector
    vpl(A,1,A,B,255),      % Draw a digital straight line between [A,1] and [A,B]
    thr(1),                % Threshold at intensity level 1 (very dark grey)
    pop,                   % Pop image from the stack
    swi.                   % Interchange the current and alternate images
```

While *plt/1* is a simple "linear" sequence of elementary operations, this is not a limitation inherent in PIP. Although *plt/1* might seem to provide a complicated and slightly devious way to perform a very simple function, it is very convenient to be able to generate a new PIP command without needing to resort to "pixel level" programming. A large proportion of the commands currently available in PIP are implemented in this way, although the user is normally quite unaware of this fact. It is important to understand that many PIP commands are built on top of other operators that have been defined earlier, since this affects the processing speed. Here again we have encountered the hierarchical construction technique for PIP programs, which is an important feature of the language and results in a tree-like program structure. Although it is not unique in this respect, Prolog makes it very easy to define new operators without carrying the burden of complicated and distracting sub-routine headers and data-type declarations. The need for fast implementation of the "core" PIP operators is very obvious when we realize that the tree-like programs generated in this way may be quite deep. (The operators *pis/1* and *pop/1* used within *plt/1* are both 6 levels deep.)

Readers familiar with Prolog can now program in PIP! Of course, for real effectiveness, experience and detailed knowledge of PIP's large command repertoire (Table 4.7) are essential. However, we have not yet described all of the operating features of PIP, which were provided to enhance its rôle in problem analysis and algorithm selection/design.

Table 4.7 PIP command list.

Mnemonic/Arity	Function
#	Backtracking operator for image-processing commands
&	Equivalent to Prolog's goal concatenation operator ','
if	Equivalent to Prolog operator ':-'
\	\ X performs operation X and then saves result in file name X
§	§ A is equivalent to [dab(A)]; operation A is performed on all blobs
•	N•G repeats goal G a total of *N* times
æ	Issue AppleEvent. Interfacing MacProlog to other program modules
aad/[0,3]	Aspect adjust
abs/0	Fold intensities about mid-grey ("absolute value" function)
acn/1	Add a given constant to all intensities
add/0	Add intensities at corresponding points in two pictures
and/0	Logical AND of two binary images
ang/6	Angle formed by line joining 2 given points and horizontal axis
avr/[0,1]	Average intensity value (synonymous with avg[0,1])
bay/0	Isolate bays (edge concavity)
bbb/[0,3]	Coordinates of centroid of biggest blob and its intensity.
bbt/1	Is biggest concavity (bay or hole) above second largest?
bcl/[0,1]	Binary closing (mathematical morphology)
bed/[0,1]	Edge detector (binary image)
bic/1	Set given bit of each intensity value to 0
bif/1	Flip given bit of each intensity value
big/[0,1]	Select the *i*-th biggest blob
bis/1	Set given bit of each intensity value to 1
blb/0	Fill all holes ("blob fill")
blo/1	Increase contrast in center of intensity range
blp/6	Calculate six parameters for all blobs in image
bop/[0,1]	Binary opening (binary morphology)
box/5	Draw a rectangle with given corner points and intensity
bpt/[0,2]	Bottom-most white point
bsk/0	View bottom image on stack
bug/0	Corner detection function and edge smoothing (bug) function
bve/8	Find coordinates of extrema along defined vector
cal/1	Copy all pixels above given grey-level into another image
cbl/[0,1]	Count blobs
ccmb/[0,1]	Combine red, green, and blue images to form a color image
ccp/0	Cursor controlled crop
cct/[0,1]	Concavity tree of object in a binary image
cgr/0[0,2]	Centroid coordinates
cgrb/[0,2]	Grab color image and separate into [r,g,b] component images
cgsg/3	Set up a LUT (used with clve, cpsu)
chf/[0,1]	Flip horizontal axis if longest vertical chord is left of image center
chu/0	Convex hull
cin/0	Column integration, top to bottom
cir/5	Draw circle inside given bounding box
circ/9	Parameters of circle intersecting 3 given points
clc/0	Column run-length coding (travelling from top to bottom)

cloa/[0,1,2]	Load color image into 3 separate pixel planes
clve/[0,1,2,3]	Show live image in pseudocolors
cnw/0	Count neighbors
cob/0	Corners of blobs
com/[0,1]	Compare two images
con/[0,9]	General linear convolution operator, 3×3 pixels
cox/0	Scan columns top to bottom to right to find column maximum
cpsu/[0,2]	Set pseudo color
cpt/2	Set pixels along vector to values defined by a given list
cpy/0	Copy the current image into alternate image
crk/[0,1]	Crack detector (morphological grey-scale filter)
crp/4	Crop current image to rectangle specified by user
csca/[0,1]	Color scattergram
cscm/[0,1]	Mask color scattergram
csh/0	Copy right-most column of current image to all columns
csk/0	Clear the image stack
ctm/0	Camera to monitor (set-up camera)
ctp/2	Cartesian to polar transformation
cua/4	Use cursor to find the position of a rectangular area defined by user
cur/[0.3]	Cursor (point defined by user with mouse, synonymous with cup/0,3)
cvd/0	Convex deficiency
cvr/[0,1]	Find set of non-black points in rectangular area defined using cursor
cwd/[1,2]	Set the current image to white
cwp/[0,1]	Count white points
dab/1	Do given task for all blobs in image (synonymous with §)
dbn/0	Direction of brightest neighbor
dcg/[0,1]	Draw centroid and print coordinates
dci/0	Draw center of the image
dcl/[0,2,3]	Draw cross lines
dcn/1	Divide current image intensities by constant
dgw/[0,4]	Get image size
dif/0	Absolute value of difference between current and alternate images
dil/1	Dilate white areas of binary in direction specified by user
dil4/0	4-neighbor dilation (takes place in 4 directions simultaneously)
dil8/0	8-neighbor dilation (takes place in 8 directions simultaneously)
dim/[0,4]	Bounding box of the white pixels in binary image
din/0	Double all intensity values
disc/4	Draw a white disc given its center and radius
div/0	Divide intensity in current image by intensity in alternate image
dlp/2	Difference of low pass filters
dpa/[0,2]	Draw principal axis
dpi/[0,1]	Copy current image into passive image display
dsl/3	Draw straight line given one point on it and its slope
eab/1	Evaluate given goal for each blob
ect/0	Threshold mid-way between min. and max. intensity
edd/0	Edge detector (grey-scale morphology)
edg/[0 - 2]	Set the picture boundary of given width to defined grey-level
egr/[0,4]	Grow ends of limbs of skeleton-like objects in binary image

enc/0	Enhance contrast; set darkest pixel to black, brightest to white
ero/1	Erode white regions in binary image in given direction
ero4/0	4-neighbor erosion
ero8/0	8-neighbor erosion
eul/[0 - 3]	Euler number, area and perimeter
exp/0	Exponential of all intensities in current image
exw/0	Expand white regions
fac/0	Flip about vertical line through the centroid
fbr/0	Find blobs touching border and remove
fcb/4	Fit circle to blob
fcd/9	Fit circle to data: coordinates of 3 points
fil/5	Fill given rectangle shape with intensity value defined by user
fld/4	Fit line to data: coordinates of 2 points
frz/0	Digitize an image
fsr/1	Remove blobs below given size (synonymous with kgr)
gcl/[0,1]	Grey-scale closing (mathematical morphology)
gdh/0	Extract measurements from intensity histogram
gft/0	Grass-fire transform (also called "Prairie fire" transform)
gis/[0,1]	Grab (digitize) image sequence
gli/[0,2]	Minimum and maximum intensities
gob/0	Get one blob
gop/[0,1]	Grey-scale opening (mathematical morphology)
gpx/1	List of all non-black pixels and their intensities
gra/0	Intensity gradient (edge detector)
grb/[0 - 3]	Grab (digitize) an image from the camera
gri/2	Extract regions of grey level above given value
gry/[0,1]	Set all intensities in current image to given grey-level
hfl/0	Fill holes (synonymous with blb)
hgc/	Cumulative histogram, output as a list
hgi/[0,1]	Intensity histogram, output as a list
hgr/0	Horizontal gradient function
hid/0	Horizontal intensity difference
hil(/[0,3]	Highlight given range of intensity values
hin/0	Halve all intensity values
hlp/0	Help - PIP system documentation
hmx/[0,2]	Find peak in the intensity histogram
hpf/0	High pass filter
hpi/0	Plot histogram
hsm/0	Horizontal smoothing
huf/0	Hough transform
hyp/0	Hyperbolic intensity mapping
ict/[0,1]	Intensity contours (smoothed or unsmoothed. Also called isophotes)
iht/0	Inverse Hough transform (principal peak only)
imx/[0 - 3]	Center point and intensity of largest region with max intensity
inv/0	Invert each pixel value in a binary image
ior/0	Logical OR of two binary images
isd/1	Display sequence of images in Source or Destination folder
isg/1	Digitize image sequence from camera and store them on disc

isi/0	Display interactive control window for image sequence processing.
isp/1	Apply the given operator on the image sequence in the Source folder.
isr/1	Read image from Source Folder, given its relative address
isu/2	Read image, given its absolute address (numeric index)
isv/2	Write image, given its absolute address (numeric index)
isw	Write image into Destination Folder (relative address mode)
isx(A)	Delete all files from named folder
itv/0	Interactive mode
jnt/[0,1]	Isolate and count joints (of skeleton-like figure)
kgr/1	Keep blobs with area greater than defined limit
ksm/1	Keep blobs with area smaller than defined limit
kwi/1	Kill all windows and images
lak/0	Isolate lakes
lat/1	Local averaging with thresholding
lav/[0,1]	Local average (blurring filter)
lgr/0	Largest gradient
lgt(/[0,2]	Get pixel intensities lying along given vector
lhq/1	Local-area histogram equalization
lin/0	User defines point which is then used to invert Hough Transform
ljt/1	List of all skeleton joints
lle/1	List of all skeleton limb ends
lme/[0,1]	Isolate and count limb ends of skeleton-like figure
lmi/[0,3,4]	Principal-axis parameters
lnb/[0,1]	Largest neighbor in 3×3 neighborhood (local operator)
lni/[0,1]	Load image and associate with a new name
loa/[0,1]	Load image from disc (path is defined by user)
log/0	Logarithm of all intensities
lpc/[0,1]	Laplacian filter (spot detector)
lpf/0	Low pass filter
lpt[0,2]	Left-most white point
lrt/0	Left-to-right (flip horizontal axis)
mar/[0,4]	Minimum area rectangle enclosing white points
max/0	Maximum intensity in two images (dyadic pixel-by-pixel operator)
mbc/[0,3]	Minimum bounding circle
mcn/1	Multiply all intensities by given constant
mdf/1	Median filter (also implements rank filters)
mim/[0,1]	Coordinates of the center of the image
min/0	Minimum intensity in two images (dyadic pixel-by-pixel operator)
mul/0	Multiply current and alternate images
ndo/0	Numerate distinct objects (blob labelling)
neg/0	Negate image
neg_log/0	Negative logarithm intensity mapping
nmr/[0,2]	Normalize [*X*,*Y*] position of minimum–area rectangle
not/0	Logical inversion (negation) of binary image
npo/[0,3]	Normalize position and orientation
nxy/[0,2]	Normalize [*X*,*Y*] position of centroid
pcc/[0,2,3]	Find white point closest to the center of image and its distance
per/[0,1]	Perimeter

pex/[0,2]	Picture expand
pfx/3	Draw one given pixel at defined grey-level
pgn/[0,1]	Draw polygon or arc using cursor, construct list of its nodes
pii/3	Test whether given point is inside the image
pis/0	Push image onto the stack (synonym for psk/0)
plt/1	Plot intensity
pop/0	Pop the image stack
ppi/2	Print the intensity of a given point
psh/2	Shift image, no wrap-around
psi/1	Process a sequence of images
psk/0	Push image onto the stack (synonym for pis/0)
psq/[0,2]	Picture squeeze
psw/[0,2]	Shift image, with wrap-around
ptc/0	Polar to Cartesian transformation
pth/2	Percentage threshold
put/3	Draw one given pixel at defined grey-level (synonymous with pfx)
raf/[1]	Blurring filter (repeat lpf many times)
rbi/0	Recover current and alternate images from the stack
rea/[0-2]	Read named image from RAM
red/0	Roberts edge detector
ria/1	Read image from "Archive Images" folder
rim/[0,1]	Read image from the "Temporary Images" folder
rin/0	Row integration, left to right
rip/[0,1]	Remove white isolated points
ris/0	Read frame of an image sequence from disc store
rlc/0	Row run-length coding (scanning from left to right)
rnd/0	Pseudo-random number generator (uniform distribution)
rni/1	Read image, name assigned by lni
roa/0	Rotate anti-clockwise by 90 degrees
roc/0	Rotate clockwise by 90 degrees
rox	Scan rows left to right to find row maximum
rpi/[0,1]	Read image from passive image display
rpt[0,2]	Right-most white point
rsh/0	Copy bottom row to all rows
sbi/0	Save both current and alternate images on stack
sca/1	Remove least significant bits of each intensity value
sco/0	Circular wedge. (Intensity increases with angle rel. to vertical axis)
set/0	White image
sgb/0	Initialize image-grabber
shf[0,1]	Shape factor (=area/(perimeter*perimeter))
shp/[0,1]	Sharpen image
sim/[1,2]	Open window with a copy of current image
sio/0	Show image sequence (original)
sip/0	Show image sequence after processing
ske/0	Skeleton
skw/0	Shrink white regions
slt/1	Select LUT for image mapping
snb/[0,1]	Smallest neighbor in 3×3 neighborhood (local operator)

sqr/0	Square all intensity values
sqt/0	Square-root of all intensity values
ssi/0	View temporary stored images
sto/[0 - 2]	Write current image picture to disc file
sub/0	Subtract images
swi[0,2]	Switch images (Current and alternate images are switched by default)
tbt/0	Top-to-bottom (flip vertical axis)
thp/[1,2]	Threshold at given centile values
thr[0,1,2]	Threshold between two given intensity values
thx([0,1]	Threshold at maximum intensity (Isolate points with max. intensity)
tia/0	List files in "Temporary images" folder
tia/0	List of temporary images available
tpt/[0,2]	Top-most white point
tsk/0	View top image on stack
tur/1	Turn (rotate) image about its center
usm/[0,1]	Unsharp masking
vgr/0	Vertical gradient function
vgt/2	Instantiate list to values in right-hand column of image
vid/0	Vertical intensity difference
vpl/5	Draw a line with between two given points
vpt/2	Set right hand column of image to values defined in given list
vsk/0	View image stack
vsm/0	Vertical smoothing
vsw/0	Variable frequency sine-wave generator
wdg/[0,1]	Intensity wedge
wgx/[0,1]	Intensity wedge (synonymous with wdg)
wia/1	Writing to the "Archive Images" folder
wim/[0,1]	Write image to the "Temporary Images" folder
wis/0	Write frame of an image sequence to disc store
wri/[0,1]	Write image into named RAM file
wrm/[0,1]	Remove isolated white or black pixels in binary image
xor/0	Logical exclusive OR of two binary images
yxt/0	Interchange X and Y axes (transpose image)
zer/0	Black image

4.3.2 Dialog box

PIP provides a custom dialog box enabling the user to enter commands individually, as well as perform a number of other operations (Figs.4.3 and 4.8; also Table 4.8). The primary function of an IVS is to explore a variety of paths through the algorithm search tree (Figure 4.5). During interaction, the user frequently discovers that the path that he has been exploring is unproductive (i.e., it leads to a series of dead ends). In this situation, it is natural for him to want to revert to an image generated earlier and from there explore another promising branch of the tree. The dialog box that is used to control PIP has been designed to facilitate this kind of exploratory process. In Figures 4.3 and 4.8, notice that there are two buttons labeled *Mark* and *Revert*, which have the following functions:

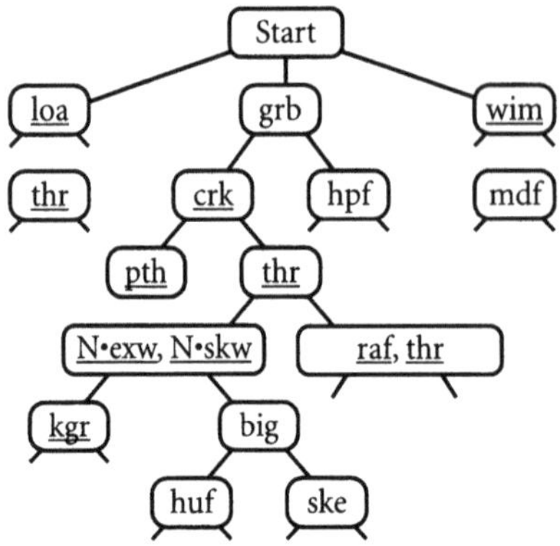

Fig. 4.5 Algorithm search tree. (This tree has been simplified for purposes of illustration and includes only a sub-set of the "sensible" options. It may be necessary to experiment in order to find suitable values for some of the commands/command sequences. These are indicated by underlining.

Mark This button is used to record the fact that an interesting picture has been obtained. Images that are "marked" in this way are pushed onto the top of an image stack maintained by PIP.

Revert This button is used to pop images from the stack. Hence, the user can quickly return to an intermediate result (image) generated and marked earlier.

PIP maintains a record (in a window called the *Journal*) of all goals that are specified by this dialog box (Figs. 4.3 and 4.6). Each time an image is marked, a comment line is inserted into the Journal, so that the user can find out how he generated the marked images. We will discuss the Journal in Section 4.3.6.

Table 4.8 Buttons provided by the interactive dialog box. When the user clicks on any one of these, the *Journal* window is annotated appropriately.

Button	**Function**
OK	User enters the command in the edit field, then clicks on "OK" to execute it.
Cancel	End of interactive session. Dialog box disappears.
Mark	Put (push) current image onto the image stack. Image is "marked" as being significant, so that the user can re-visit this point in the algorithm search tree.
Revert	Pop image from the image stack. Recover the most recently "marked" image, following the discovery of a dead end in the algorithm search tree.
Store image	Save image in a file named by the user.
View images	Display images saved by "Store image" or *wim/1* one at a time.
Rebuild menus	Kill all image windows. Hide all program windows. All menus are rebuilt.
Bookmark	User enters comment in the edit field. This is then placed in the Journal window.
Retain copy	Store current image in 'Scratch' image file and display it on the screen as a "passive" image, called 'Scratch'.
Recover copy	Read the 'Scratch' image file (effectively transfers the "passive' on-screen image called 'Scratch' into the current image)
Load disc image	Scroll menu is displayed, inviting user to choose a file to be loaded from disc.
Reload disc image	Reload the last image previously loaded using the **I** menu.
Delete images	Delete all images displayed on screen and all those saved by *wri/1*.
Image statistics	Useful image statistics: image size, minimum intensity, maximum intensity, average intensity and most densely populated intensity level.
Camera set-up	Allows user to see "live" image on computer screen. Same operation as *ctm/0*.
Digitise image	Digitize an image from the camera. Same operation as *grb/0*.

Notice that we have deliberately avoided using the term *backtracking* when discussing the use of the *Mark* and *Revert* buttons, since this is a technical term applied to Prolog programs. While the user mentally backtracks when he clicks on the *Revert* button, Prolog does not do so.

Another point to note here is that an image stack has other uses in programmed-mode operation, and for this reason will be discussed again later.

```
Journal
loa('Brake pads')  % Image loaded from the 'I' menu
enc,  % Picture in file Image0
thr,  % Picture in file Image1
neg,  % Picture in file Image2
exw,  % Picture in file Image3
skw,  % Picture in file Image4
big(2),  % Picture in file Image5
npo,  % Picture in file Image6
% USER COMMENT: Normalised position & orientation
cwp,  % Picture in file Image7
% USER COMMENT: Area
mar,  % Picture in file Image8
blb,  % Picture in file Image9
cwp,  % Picture in file Image10
% USER COMMENT: Area of minimum area rectangle
% End of interactive session
true,
```

Fig. 4.6 *Journal* window. These commands were entered individually by the user, via the dialog box shown in Figure 4.3. Notice that the activity recorded in this window is syntactically-acceptable Prolog.

4.3.3 Pull-down menus

PIP provides access to all of the standard MacProlog menus, plus several others, designed specifically for developing Machine Vision algorithms. Their functions are summarized in Table 4.9. Notice that the **G**, **B**, **A**, and **D** menus are initially empty but that they and the **U** menu can all be extended easily. (This process will be described in the following section.) The **I** and **S** menus require some explanation.

Table 4.9 Menus available in PIP.

Name	Menu Item	Comments
File	File	Standard MacProlog menu
Edit	Edit	Standard MacProlog menu
Search	Search	Standard MacProlog menu
Windows	Windows	Standard MacProlog menu
Desktop	Desktop	Standard MacProlog menu
Eval	Evaluate Prolog goal	Standard MacProlog menu
U	Utility	Contains several "fixed" items that are normally operated by command keys (Table 4.10) as well as extensions added using **U** → *Extend menus*. These are typically demonstrations
G	Grey-scale image processing	Initially empty. Extended by user selecting menu **U** → *Extend menus*
B	Binary image processing	Initially empty. Extended by user selecting menu **U** → *Extend menus*
A	Image Analysis	Initially empty. Extended by user selecting menu **U** → *Extend menus*
D	Device control	Initially empty. Extended by user selecting menu **U** → *Extend menus*
S	Speech	Not normally activated directly by user. Speech recognition system selects items from this menu
I	Images	Offers user choice of images to load

The **I** menu allows the user to gain easy access to a library of images held in a defined folder on the host computer's hard disc. To extend this library and the associated **I** menu, the user simply inserts the new image file (PICT format) into that folder and then rebuilds the menus by pressing the key combination *Command N*. (Notice that thereafter the user can use the key combination *Command 9* to reload this image into PIP's Current image.)

The user is not expected to operate the **S** menu with the mouse, since it has nothing at all to do with the normal operation of the PIP system. It is reserved for use with a speech recognition system and will be discussed in Section 4.3.7. (It is possible to hide the **S** menu from the user by renaming it using a single "white" (nonprinting) character, such as SPACE (ASCII 32).)

4.3.4 Extending the pull-down menus

The **U**, **G**, **B**, **A**, and **D** pull-down menus can be extended using a simple dialog, illustrated in Figure 4.7. The user is invited first to specify which menu is to be extended, then to indicate the name of the new menu item, and finally to define what action is to be taken when that item is selected. The new menu item can be added permanently to PIP simply by re-compiling the Prolog code (key combination *Command K*) and then saving for future use (*Command S*). The user can thereby build up a personalized selection of utilities and demonstrations.

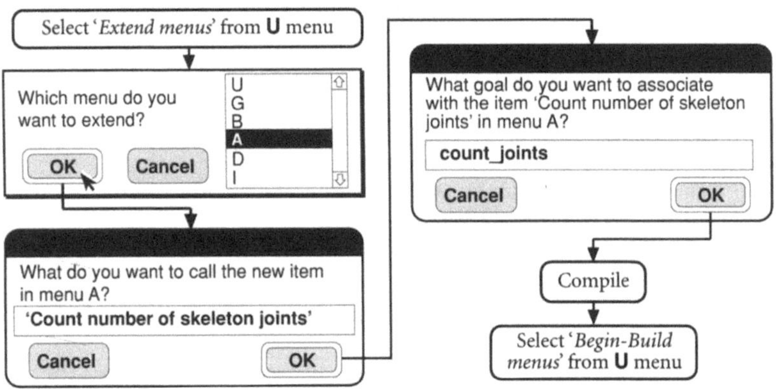

Fig. 4.7 Extending one of PIP's pull-down menus.

4.3.5 Command keys

Experienced users of PIP find the command keys particularly useful, since they speed up the selection of commonly used functions. A list of the command keys currently in use in PIP is given in Table 4.10.

Table 4.10 Command key short-cuts in PIP.

Key	Menu Item	Comments
*	Convolution	Display window contains weight matrix for 3×3 linear convolution operator
+	Extend menus	Extend any of the menus. Enters dialog with user
,	Mask edge of image to black	Set image border to black. Equivalent to *edg(0,4)*
-	Modify pull-down menus	User is invited to edit Prolog database holding details of menu extensions
/	Morphology	Display window inviting user to select type of morphology operator (i.e., erosion/dilation), size and shape of the structuring element. See Section 4.4.2
0	(zero) Delete archive images	Delete all images in the *Archive Images* folder
1	Clear image stack	Ensure that image stack has no images stored in it
2	Cycle through image stack	View images held on the stack, one at a time
3	Push image onto stack	Put another image on top of the stack
4	Pop image from stack	Recover and delete image from top of image stack
5	Image size	Find size of the current image
6	Write image	Save image in disc file *menu_image*
7	Read image	Recover image from disc file *menu_image*
8	Enlarge image	Option to show image in full resolution if larger than standard size
9	Get stored image	Reload most recent image found using **I** menu
;	Mask edge of image to grey	Set image border to mid-grey. Equivalent to *edg(127,4)*
=	Speech synthesizer	Test speech synthesizer
G	Grab image	Digitize an image from the camera
H	Help – system documentation	System documentation is held in HyperCard stacks (Section 4.3.10)
J	Clear Journal	Clear the *Journal* window
L	Set up camera	Live image on screen allows user to set up the camera
N	Begin – Build menus	Delete visible and hidden images held in RAM. Rebuild menus after extension and reconstruct list of images that are accessible via the **I** menu
O	(letter) Switch images	Interchange current and alternate images
T	Interactive mode	Dialog box (Figure 4.8) becomes visible
W	Kill images and windows	Delete visible and hidden images held in RAM
[	Automatic image saver	Save every image generated in interactive session, in the *Archive Images* folder
]	See temporary files available	View images held in *Temporary Images* folder
\	Backtrack to saved archive image	Reload an image generated earlier during the interactive session. (Click anywhere on a line in the *Journal*.)

4.3.6 Journal window

The Journal window provides facilities for recording and replaying command sequences generated during the interactive session. It provides several modes of operation (Table 4.11).

Table 4.11 Using the Journal window.

Operation	Function
Enter the appropriate text into the edit field of the interactive dialog box and click on the button *Bookmark J'nal*	During interaction, the user is able to place any convenient remark in the *Journal* window. The compiler ignores these comments.
Press *Command 0*, or select menu **U** → *Delete archive images*	Delete all images in the *Archive Images* folder.
Press *Command J*, or select menu **U** → *Clear Journal*	Clear Journal Delete all text in the *Journal* window.
Press *Command T*, or select menu **U** → *Interactive mode*	Whenever the user asks for the interactive mode dialog box to be brought to the front, the system enters a comment in the *Journal* window.
Press *Command [*, or select menu **U** → *Automatic image saver*	Switch automatic image archiving *on/off*. The automatic image saver keeps a record of all images PIP generates during the interactive session; after each interactive operation, the current image is saved in the *Archive Images* folder and the file name is recorded in the *Journal* window.
Press *Command* \ (back-slash), or select menu **U** → *'Backtrack' to saved archive image'*	Reload an image stored by the automatic image saver. (The image to be reloaded is defined by the position of the cursor in the *Journal* window, just before "*Command* \ "is pressed.)
User selects a block of text in the *Journal* window, then presses the *Enter* key	Rerun PIP operator sequence recorded in *Journal* window.

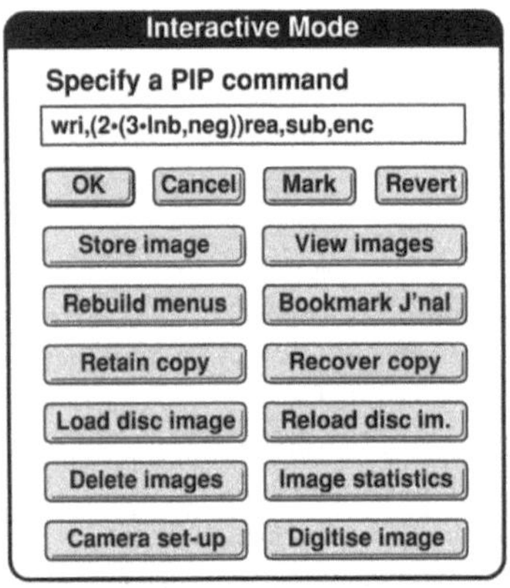

Fig. 4.8 Function of the buttons in PIP's user dialog box (also see Fig. 4.3).

In its simplest most basic mode, the Journal records each image-processing command as it is entered via the dialog box (Figure 4.8). In fact, individual command lines are separated from one another by commas, thereby generating syntactically correct compound goals in Prolog. Hence, a multi-command sequence can be replayed or copied into a program without any editing. One easy way for the user to replay a command sequence is to select a block of text in the Journal window and then press the *ENTER* key. MacProlog then takes that block of text to define a compound goal.

Comments can be inserted in the Journal window by the user typing an appropriate message into the edit field of PIP's dialog box (Figure 4.8) and then clicking on the *Bookmark J'nal* button. Comments written in this way are ignored by the MacProlog compiler.

Each time the user loads an image (using the **I** menu, by pressing the key combination *Command 9*, using [*loa*], or digitizing an image from the camera [*grb*]), an appropriate comment is inserted in the Journal window. Comments created in this way are also ignored by the compiler.

There also exists a facility for automatically recording every image generated by PIP as it executes commands issued by the user. To initiate image archiving, as this is called, the user presses the key combination *Command [*. Thereafter, until the same key combination is pressed again, every command entered via the standard dialog box is recorded in the *Journal* window, together with a comment indicating the name of the file where the Current Image has been stored. Here is a sample of the contents of the *Journal* window after a short interactive session in which automatic image archiving was switched on:

```
% New interactive session
loa('Macintosh HD:PIP:Images:Bottle'),
% Image (re)-loaded and saved in file 'Scratch'
% New interactive session
wri,                    % Picture in file Image2
3•lnb,                  % Picture in file Image3
neg,                    % Picture in file Image4
                        % User comment line ignored by the compiler
3•lnb,                  % Picture in file Image5
neg,                    % Picture in file Image6
rea,                    % Picture in file Image7
sub,                    % Picture in file Image8
```

Now, if the user places the cursor anywhere in a line of text in the Journal window and selects menu item **U** → *'Backtrack to saved archive image'*, (or presses the corresponding command-key combination *Command* \), the file named on that line will be reloaded. This restores the current image to the state it was in just after that command was executed.

Other facilities supported by the Journal window are summarized in Table 4.11.

4.3.7 Natural language input via speech

One of Prolog's strengths is its ability to handle Natural Language *(NL)*, albeit in a limited form. While there are several ways to approach this task [GAZ89], *Definite Clause Grammars (DCG)* provide a convenient mechanism to define grammars for well-defined tasks, such as operating moderately-complex machines (e.g., domestic video recorder), controlling a robot, switching lights on/off, etc. Although Prolog cannot possibly cope with the full richness of unconstrained human language, it can perform a useful function analyzing sentences relating to a limited domain of discourse.

The syntax of DCGs closely resembles Backus–Naur notation; the basic principles may be summarized thus:

1) $a \rightarrow b, c$ states that phrase *a* can be formed by taking phrase *b* and following it by phrase *c*.
2) $a \rightarrow b; c$ states that phrase *a* can be formed by taking it to be phrase *b* *or* phrase *c*.
3) Taken together, the two following rules are equivalent to the formula in rule 2.

$$a \rightarrow b. \qquad\qquad a \rightarrow c.$$

4) Recursion is permitted when defining DCGs. Hence, the rule $a \rightarrow a,b$ indicates that *a* consists of an indefinite number of occurrences of *a*, followed by a single occurrence of *b*.
5) Testing a given sentence *(B)* to determine whether or not it conforms to a given set of DCG rules *(A)* is achieved using the MacProlog goal *phrase(A,B,[])*. (The third argument for *phrase/3* need not concern us here.)
6) The MacProlog parser *phrase/3* allows the programmer to insert regular (i.e., Prolog) goals into a DCG, as a simple example will show. Consider the DCG rule

$$a \rightarrow b, \{goal1, goal2, ..., goal(N\}, c.$$

While it is trying to satisfy the goal *phrase(a, [b,c], [ ])*, the parser will accept that phrase *a* can begin with the symbol *b*. Next it will try to satisfy the sequence of sub-goals: *goal1, goal2, ..., goalN*. If they all succeed, it will finally accept that *a* ends with the symbol *c*.

While there is insufficient space here to describe in detail how DCGs might be employed in practice, we are able to explain in outline terms how Natural Language might be used to control a vision system and its attendant opto-electromechanical devices [BAT91a]. A grammar has been defined for controlling an (X,Y,θ)-table (Figure 4.9) and can be extended easily to allow it to accommodate a wide range of sentence types. The "meaning" can be extracted from a sentence that conforms to the given grammar by removing all terms other than the following:

a) right/left
b) numeral
c) noun type n2.

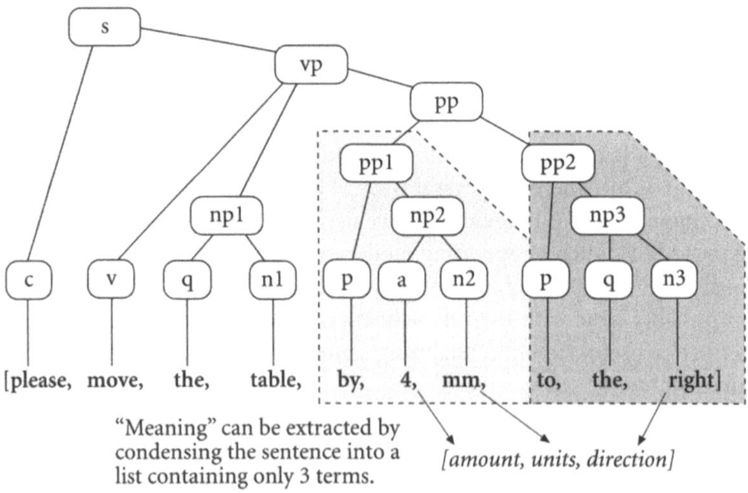

Fig. 4.9 Definite Clause Grammar for controlling an (X,Y,θ)-table: parsing a simple sentence. It is possible to extract the "meaning" of each sentence of this grammatical form by simply deleting non-critical elements. Table 4.12 shows the parse tree expressed as a Definite Clause Grammar. Notice that another acceptable sentence form can be obtained by simply reversing the positions of the sub-trees headed by nodes *pp1* and *pp2* (shown shaded).This new sentence form can be accommodated by simply adding the second clause in the definition of *pp*.

Table 4.12. Explanation of terms and structures shown in Figure 4.9.

Symbol	Represents	Sample DCG Rule/Remarks
a	amount	a → [0]; [1]; [2]; ... [9]. (single digit) a → a, a. (multiple digits)
c	courtesy	typical phrases: [please], [i, X, you, to], [will, you, please]
n1 – 3	nouns (3 separate sets of nouns are useful)	a) n1 → [table]; [platform]; [stage]. b) n2 → [mm]; [inches]; [pixels]; [steps]. c) n3 → [left]; [right].
np1 – 3	noun phrase (3 separate forms are useful)	a) np1 → q, n1. b) np2 → q, n2. c) np3 → q, n3.
p	preposition	p → [ ]; [by]; [to].
pp1 – 2	preposition phrase (2 separate items)	a) pp1 → p, q, n2. b) pp2 → p, q, n3.
pp	preposition phrase list	pp1 → pp1, pp2. (pp is pp1 followed by pp2.) pp2 → pp2, pp1. (Allows pp1 & pp2 to be in reverse order.)
q	qualifier (article, or numeral)	q → [ ]; [a]; [an]; [the].
s	sentence (top level)	s → c,vp (s is c followed by vp)
v	verb	v → [move]; [shift], [translate].

This extraction of "meaning" is possible using the mechanism just described. Another very simple program can then be used to translate the resulting 3-element list into a command for the robot mechanism. In another example (Figure 4.10), a set of DCG rules is used to control a set of lamps. Both the(X,Y,θ)-table and the lamps form part of the *Flexible Inspection Cell (FIC,* Figure 4.11 and Section 4.4.11).

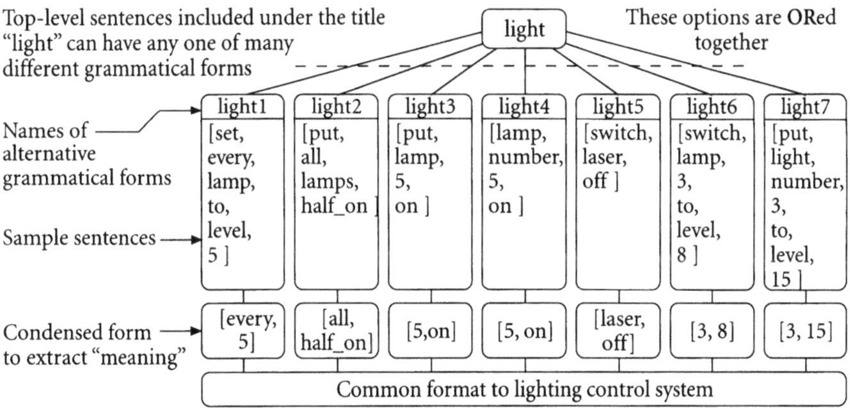

Fig. 4.10 A task requiring the ability to handle a wide variety of sentence forms can be simplified by analyzing each type separately. Each phrase type *(light1, light2, ...)* is treated in the same general way, as was outlined in the previous diagram.

The speech recognition device that has been used with PIP is able to select any item within any pull-down menu created by any application [BAT91a]. This unit is therefore used in conjunction with PIP's **S** menu, so that the user can select any

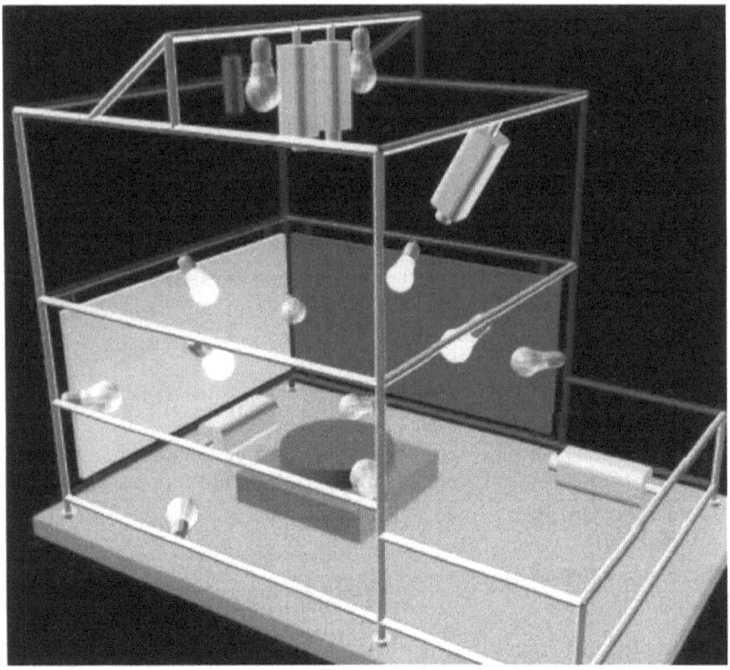

Fig. 4.11 Flexible inspection cell (FIC). (Drawing by R. J. C. Hitchell, used by permission.)

item in that menu by uttering the appropriate vocal command. (The speech recognizer has to be taught to recognize these utterances before it can be used, but this does not involve PIP.) The **S** menu also contains a small number of key-word items:

Begin Clear the input buffer, ready for the next sentence.

End Parse the sentence just entered and, if successful, try to perform the operation it defines.

Cancel Delete the last word entered.

Abort Stop receiving spoken commands. Then, wait for the next "Begin" command."

The speech recognition facility has also been used to operate PIP in its normal rôle as an interactive image-processing system. The user can therefore employ spoken commands for operations such as *negate, blur* (low-pass filter), *histogram equalize*, etc. which do not require any arguments. However, it is difficult (i.e., unnatural) for a person to accept the limitation that he has to utter spoken commands in a highly constrained format, such as

[threshold, number, twenty, nine, number, one, hundred, and, thirty, four]

even though he is happy to type *thr(29, 134)*. There seems to be little point in trying to use speech recognition for intensive tasks like this, when it is subject to this constraint.

4.3.8 Speech output

At first sight, the provision of a speech synthesizer within a Machine Vision system may seem odd. However, experience has shown that speech output, whether synthesized or recorded, provides a very effective way to follow backtracking and recursion, and hence to monitor program "flow." Experienced Prolog programmers are very well aware that the "flow" within a program is often difficult to predict. Speech synthesis provides an excellent way for the program to indicate what a program is doing. This can make debugging programs easier and therefore more effective than conventional printed output. Speech synthesis holds particular attractions for a TVS, where "eyes free" operation can be important while watching a vision system in operation. A major rôle for speech synthesis in PIP is therefore in providing a facility for demonstrating feasibility and the usefulness of this technology. In short, since target systems can make good use of speech synthesis technology, interactive systems should have it too.

4.3.9 Cursor

The primitive PIP cursor command (*cur/[0-3]*) provided in PIP simply allows the user to investigate an image interactively, enabling him to find the locations and intensities of points of particular interest. This command may be used on its own or in combination with other operators, as the following examples show.

First, we define a simple PIP predicate which draws a pair of crossed lines at a point [*X*,*Y*], defined by the user operating the computer mouse or track-pad.

```
dcr(X,Y) :-
    cur(X,Y,Z),                      % Point address [X,Y]. Intensity = Z
    writeseqnl(['Intensity at point ', [X,Y], ' is ', Z]),   % Message for the user
    dcl(X,Y).                        % Draw crossed lines
```

Similarly, the user can plot the intensity profile along any column.

```
cpt(X) :-
    cur(X,_,_),                      % Interactive cursor
    plt(X)                           % Plot intensity profile for column X
```

The following program invites the user to define a rectangular region in the image. The coordinates of two of its corner points are then passed to the command *crp/2*, which crops the image.

```
ccp :-
    cur(A, B, _),                    % Corner of rectangle
    cur([C,D,_]).                    % Another corner of the rectangle
```

The following program allows the user to draw a polygonal arc, which may be either open or closed.

```
pgn :-
    pgn(A),
    writeseqnl(['/* List of points selected using the cursor: ',A,' */']).
```

```
pgn(A) :-
   pis,                          % Push current image onto the stack
   cur(U,V),                     % Cursor. User defines [U,V] using mouse or track-pad
   dcl(U,V,1),                   % Draw crossed lines at [XY]
   remember(cursor_starting_point,[U,V]),   % Save starting point coordinates
   remember(cursor_nodes,[U,V]), % Initialize the node-coordinate list
   tsk,                          % Copy image from top of the stack
   max,                          % Superimpose the crossed-lines onto the input image
   pgn(U,V),                     % Draw rest of the polygon
   recall(cursor_nodes,A),       % Recall list of node coordinates
   pop,                          % Pop the input image from top of the stack
   swi.                          % Interchange current and alternate images

pgn(U,V) :-
   cur(X,Y),                     % Cursor. User defines [X,Y] using mouse or track-pad
   remember(cursor_finishing_point,[X,Y]),  % Save latest point found
   recall(cursor_nodes,P),       % Get list of nodes found previously
   remember(cursor_nodes,[[X,Y]|P]),        % Add new node to the list and save it
   swi,                          % Interchange current and alternate images
   vpl(U,V,X,Y,255),             % Draw white line between [U,V] and [X,Y]
   dcl(X,Y,1),                   % Draw crossed-lines at [X,Y]
   swi,                          % Interchange current and alternate images
   max,                          % Cross and arc are both superimposed onto input image
   yesno(['More points?']),      % Draw any more points?
   !,                            % We are not coming back again so throw everything away
   pgn(X,Y).                     % Add some more points

pgn(_,_) :-
   yesno(['Do you want to draw a closed figure?']),   % User decides
   recall(cursor_starting_point,[U,V]),     % Recover first point found
   recall(cursor_finishing_point,[X,Y]),    % Recover last point found
   swi,                          % Interchange current and alternate images
   vpl(U,V,X,Y,255).             % Draw white line between [U,V] and [X,Y]

pgn(_,_) :- swi.                 % Finished, so interchange current and alternate images
```

Of course, the cursor has a much wider range of uses (Table 4.13). All of the functions defined there can be implemented by writing simple PIP programs based around *cur/3*, as we have demonstrated above.

4.3.10 On-line documentation

PIP's large command repertoire makes the provision of good Help facilities particularly important. (Even PIP's authors cannot remember the precise details about all of its commands!) Three approaches have been considered for displaying Help messages about PIP:

1) MacProlog
2) HyperCard, linked to MacProlog.
3) WWW browser, with Help files written in HTML.

Table 4.13 Typical operations that can be performed in PIP using the cursor command *cur/3*. See Figure 4.18 for further details of the gauge predicates.

Function	Comments
Draw polygonal arcs	The user draws a polygonal arc, which may be either open or closed.
Draw intensity-mapping functions	The user draws an open polygonal arc. This defines an intensity-mapping function, implemented as a LUT, which is then applied to the input image.
Balloon gauge	The user locates the center of the "balloon", which then expands uniformly in all directions, starting from zero radius, until it encounters a white point. Hence, the program finds the address of the closest white pixel to the starting point (i.e., the balloon center).
Caliper gauge (binary)	The user places an icon resembling a caliper gauge onto the image. The jaw orientation and separation can also be set using the cursor. The caliper jaws then close automatically, to measure the width of the white blob between the caliper jaws.
Caliper gauge (grey scale)	This is a software version of a tool-maker's caliper gauge. Initially, the user places the gauge onto the image. The program then closes the jaws, to measure the diameter of a white blob.
Caliper gauge (internal)	Similar to the caliper gauges just described except that the jaws open wider to measure internal dimensions (e.g., neck of a bottle).
Fan gauge	Similar to the balloon gauge, except that the search angle is restricted. The user defines a fan-shaped search region within the current image, stopping when it discovers a white point.
Compass gauge	The search for a white point is performed by a line rotating about a pivotal point defined by the user.
Edge gauge	The user defines two points with the cursor. The point of highest intensity gradient along the line joining them is then found.
Morphological operators	The user employs the cursor to determine the size of the structuring element size needed for morphological operators.
Inverse Hough transform	The user selects one of the bright points in the image generated by the Hough transform. The program then performs the inverse Hough transform on that point, to draw the corresponding digital straight line. This line intersects the set of points which contributed to that point in the HT image.

The ability of MacProlog to provide a basis for building a hypertext display is evident in the facilities built around the Journal window and particularly the MaViES Expert System. This approach was not considered initially because MacProlog then lacked some of the important facilities needed for the task. However, the language is now sufficiently well developed to make the integration of the image-processing engine and its associated documentation much easier than it would have been when the second option was adopted several years ago.

Option 2 was the approach favored until recently [BAT97]. HyperCard is a hypertext software package. Although it is limited to running on a Local Area Network (LAN) of Macintosh computers, HyperCard behaves very much like a WWW browser in its ability to navigate through a series of linked pages (HyperCard calls them cards. It is interesting to note that HyperCard pre-dates the WWW by about nine years.) An important feature of the PIP documentation facility based on HyperCard is the ability it provides for controlling the image-processing engine from the Help display cards (Figure 4.12). This enables the user to read about a command and then conduct experiments without ever leaving the Help display. This is particularly helpful for students and other people learning to use PIP.

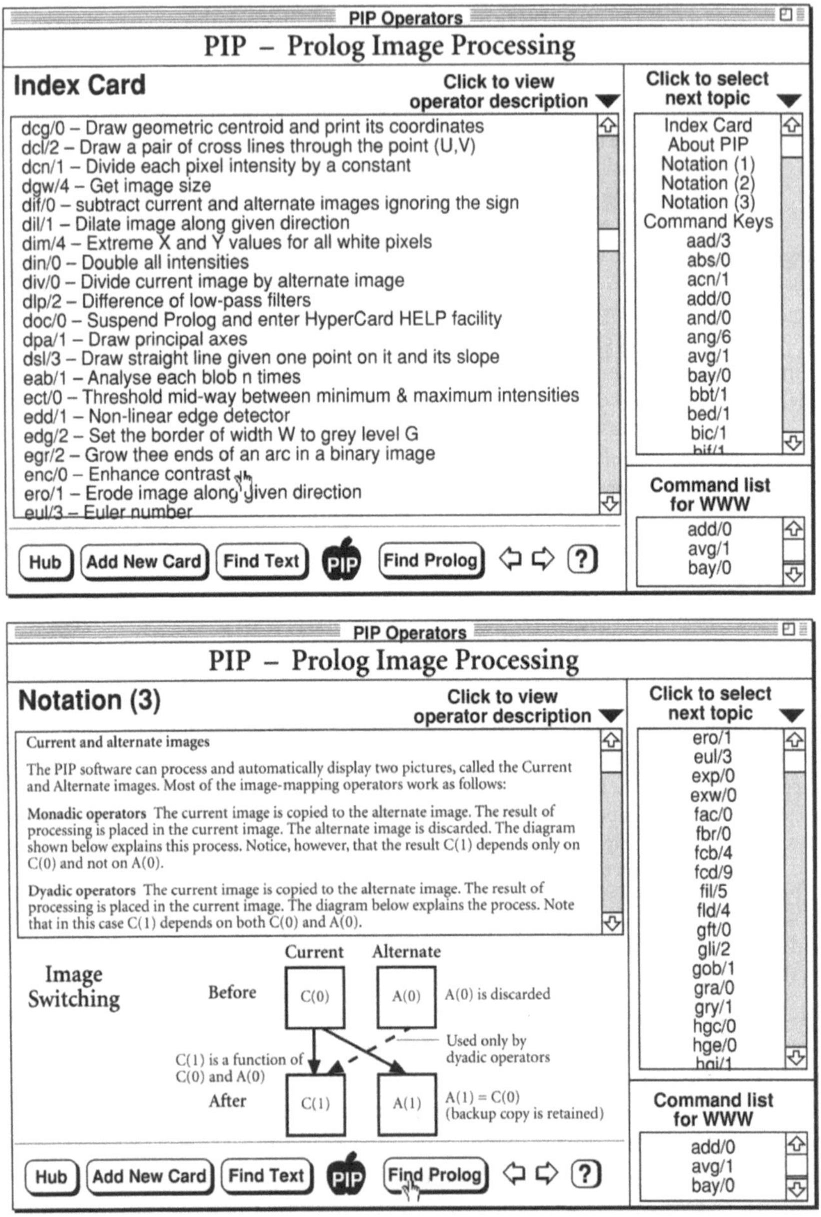

Fig. 4.12 (a) (top) Index card of PIP's HELP facility (implemented in HyperCard). To view the details of a given PIP operator, the user clicks on the appropriate entry in one of the large scroll menus.
(b) (bottom) Typical card showing operating details and navigation menu (right-hand side).

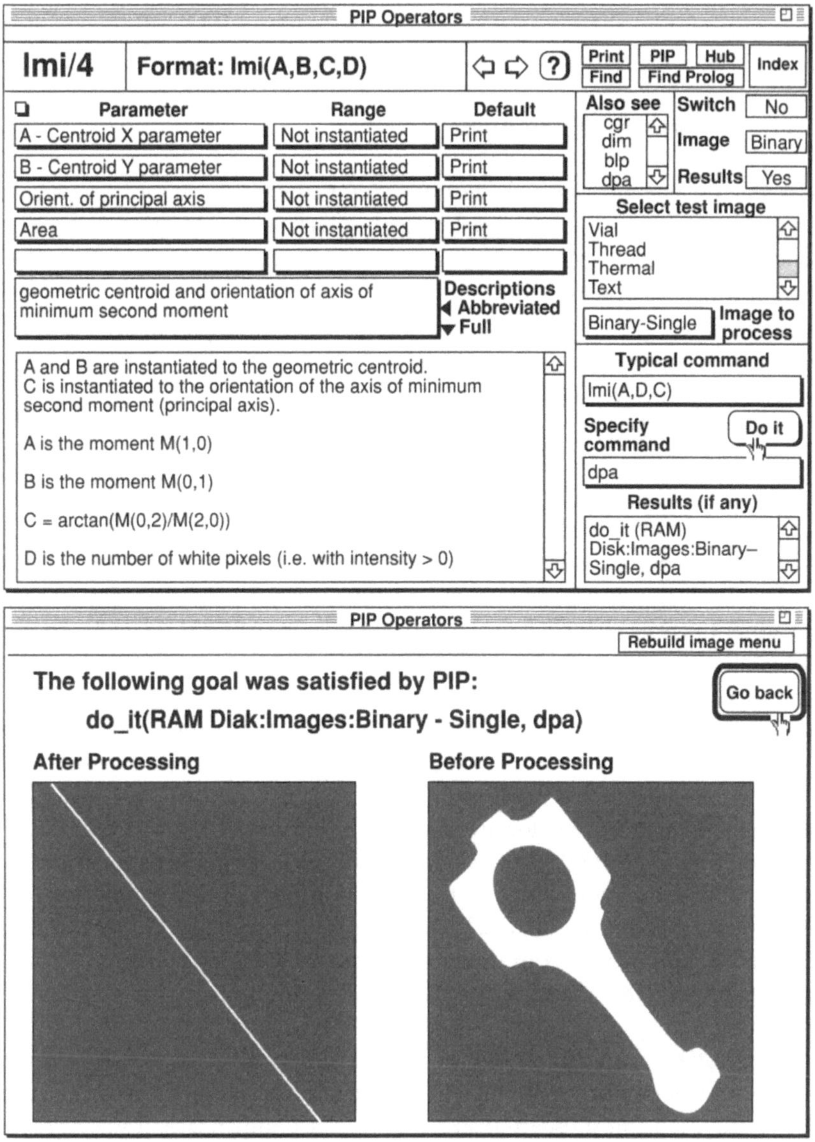

Fig 4.12 (c) (top)Typical card describing one of PIP's image-processing operators. Notice that it is possible to select a test image and then enter and execute any PIP command from within the HELP library. This is valuable for training purposes and especially for understanding the functions of the arguments. (d) (bottom) Window showing the result of an image-processing operation ordered from HELP card (c).

In view of the universal access it provides, Option 3 is preferred now. However, it takes a long time to translate the large amount of Help material about PIP from one medium to another. At the time of writing, a rather unsatisfactory set of WWW pages is available describing PIP. Option 3 has the great advantage that only one set of HELP files has to be maintained and can be read anywhere in the world on a range of computers. At least some of the same Help files could (in theory) be shared with the WIP and CIP systems, although this economy has not been achieved.

4.3.11 Generating test images

PIP is able to generate a range of test images for itself. This can be very useful for fault diagnosis during software development, as well as for education and training. The *wedge generator, wdg/1* is perhaps most useful of all. *wdg(0)* simply generates an image in which the intensity varies in direct proportion to the horizontal position coordinate (Plate 1(BL)). Image processing applied to the wedge image can create a variety of other useful test patterns, including

- staircase [*wdg,sca(3)*]
- chess board [*wdg, sca(3), sub, enc, thr, yxt, xor*]
- parallel lines [*wdg, sca(3), sed, thr(1)*]
- grid [*wdg, sca(3), sed, thr(1), yxt, max*]
- composite patterns, such as that shown in Plate 1(BL).

A circular wedge, in which the intensity varies in proportion to angular position, is also available as a standard function [*wdg(9)*]. This can be transformed in a similar manner to create circular staircase and wheel-spoke patterns (Plate 8). An image in which the intensity varies in proportion to the distance from its center can also be created by using [*wdg, ptc*] Concentric circles and spirals can be derived by combining these patterns in various ways. PIP can also generate a set of sine waves of varying frequency [*vsw*]. This is useful for testing 1-dimensional filters [BAT93b]. Pictures with a random intensity variation can also be created [*rnd*]. These are particularly useful for investigating the effects of noise on image-processing algorithms. Plate 1(BR) shows an image with a uniform intensity distribution. A reasonably good approximation to a Gaussian distribution can be created by [*rnd, sqr*]. More exotic test images, based for example on Hilbert curves and Walsh/Haar/Hadamard functions, can also be generated using Prolog's power of recursion. Many other useful test images can be created using PIP, the user's imagination being the limiting factor.

A range of standard digitized images is also very useful, particularly for training and educational purposes. Although PIP is intended primarily for industrial applications, one of the most useful of all test images is a human face. It was discovered long ago that a face is the best picture to illustrate the effects of many grey-scale image–image mapping functions; no other picture can portray the subtle effects of so many different image-processing operators.[3] For this reason, many of the PIP operators are illustrated using the digitized picture of a child (Plates 1 – 5).

3. There are good reasons for this. Very young children can identify certain faces long before they display any sign of recognizing any other scene. For this and other reasons, many psychologists believe that there are distinct mechanisms in the human brain for recognizing faces and analyzing other scenes.

4.3.12 PIP is not just an image-processing system

By now, the reader will be aware that PIP possesses several features which extend its capabilities beyond the limits normally associated with image-processing systems. These additional facilities include natural language processing (albeit in a rudimentary form), speech recognition, and speech synthesis. In the next section, we will develop this theme further by adding facilities for controlling a range of external devices and interfacing PIP to a Wide Area Network (WAN). Two very important differences exist between a standard image-processing package and PIP:

1) The output of the former is an *image*, or possibly an *algorithm*, whereas the primary output of PIP is a *program*, or a more generally, a system design concept.
2) An industrial vision system must be able to interact with people and/or other machines around it. Hence, the prototyping system (i.e., PIP) should also have this capability.

These are essential differences, because they place the emphasis on different aspects of a system: to an image processor, its image-manipulation capabilities are of prime importance, whereas a Machine Vision system must also possess the ability to interact with both people and other machines. To equate PIP to an image-processing system is like expecting a workman's tool box to contain only carpenter's tools. Of course, a well-fitted "general purpose" toolbox will also contain utensils for plumbing, electrical work, metal work, and building work. Admittedly, PIP's *primary* function is to provide a tool for prototyping image-processing algorithms, but it also contains a variety of other tools as well. The important point to note is that these are not optional extras, since they form an essential part of PIP.

Many of the standard image-processing packages lack PIP's "intelligence" and hence cannot be expected to perform complex tasks such as cutting oddly-shaped pieces of leather to make shoes from an animal hide, calibrating a moving-needle instrument, checking the quality of cakes, reasoning about the stability of a pile of boxes, etc. (These are all problems that have been studied successfully, because PIP is based on Prolog.) To summarize, PIP provides far more varied functionality and a higher level of "intelligence" than conventional image processors do. In the next section, we will see that the story is not yet complete and that PIP is even more powerful than we have so far described!

4.4 Advanced aspects of PIP

In this section, we will describe some of PIP's more advanced features, which demonstrate that it is capable of being developed in many additional directions. The lesson to be conveyed here is clear: Prolog was a good choice for the top-level control of PIP, and we have not yet begun to approach the limits of its capabilities.

4.4.1 Programmable color filter

Color perception and color image processing are very complex issues that could fill many books. Rather than try to condense such a large amount of scholarship into one short section, we will concentrate on a single specific topic: learning to recognize colors symbolically (i.e., by name). We will therefore ignore such issues as theories of human color perception, the technology of photography, printing, scanning, television/video displays, as well as the broader aspects of color image processing. Our discussion will be restricted to the design of a device that learns the names of colors from a person in a process of "teaching by showing."

A *Programmable Color Filter (PCF)* is a versatile tool for analyzing color images. Strictly speaking, a PCF is one of a range of a mathematical techniques for recognizing colors, but we will also use the term to refer to software (i.e., PIP programs). Very fast low-cost hardware implementations are also possible. As a result, this approach to color recognition is very convenient for both software and hardware-based TVSs. Unlike an optical filter, a PCF can be controlled by software. In addition, it is well suited to the declarative style of programming embodied in PIP, and permits teaching by showing. The latter is popular because it is very easy to use. Conceptually, a PCF may be regarded as a "black box", whose inputs are the digitized RGB signals arising from a video camera or scanner or from a processed 3-channel digital image (Figure 4.13). Notice that its output consists of an 8-bit number, i.e., in the range [0,255]. However, this may be regarded as being simply a coded version of a symbolic name for a set of 255 different colors, such as green, green-turquoise, blue-turquoise, sky-blue, royal-blue, etc., with output 0 (zero) normally reserved to indicate black/low-light level. It is important to realize that consecutive integers in the range of output values from the PCF do not necessarily represent colors that are perceptually similar. Since the color of each pixel is identified individually, the PCF cannot take into account macroscopic (i.e., inter-pixel) effects, such as texture on color perception.

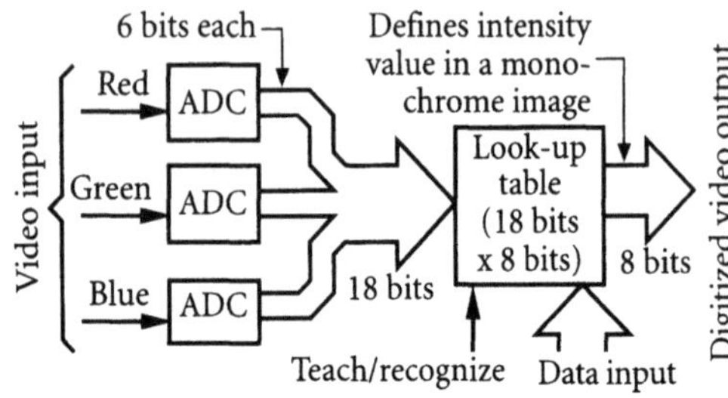

Fig. 4.13 Programmable Color Filter block diagram.

Table 4.14 PIP commands to control the PFC.

PIP operator	Function
psc/1	Plot the color scattergram, with/without the outline of the color triangle
dct/1	Draw/delete the outline of the color triangle
ppf/0	Program the PCF from the current image
pcf/0	Apply the PCF to the current image
spf/1	Save this programmable filter LUT in the file nominated by the user
rpf/1	Read the LUT in the named file to set up a PFC
npf/1	Define a new PFC LUT from set of values specified in a Prolog list

The major functions of the PCF and their PIP mnemonics are listed in Table 4.14. The PIP command *psc(0)* generates a grey-scale image containing the *color scattergram* (Appendix A). This image normally consists of a series of diffuse clusters, and indicates what colors are present in the input image (Figure 4.14 and Plate 22). The color scattergram can be processed just like any other grey-scale image.

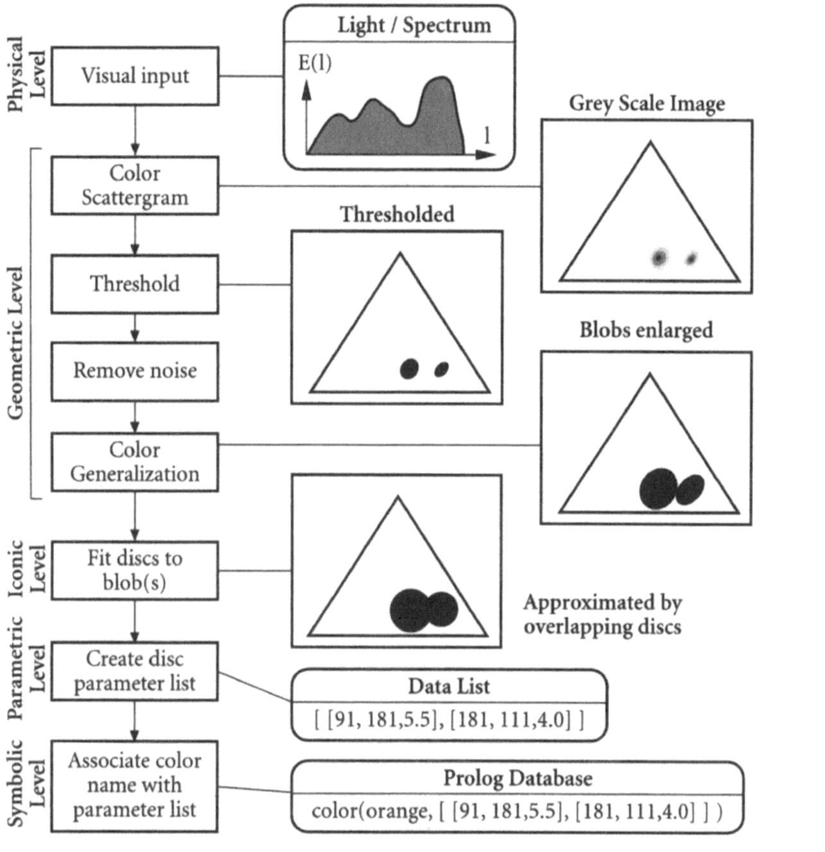

Fig. 4.14 Using a Programmable Color Filter: (a) Designing the PCF by processing the color scattergram. Notice that color can be represented in several different ways: the spectrum of light emitted by a surface, the color scattergam (i.e., a grey-scale image), the processed color scattergam (binary image), a set of overlapping circles, a list of numeric values (i.e., parameters of those circles), and abstract symbols (i.e., the names of colors that a person might use).

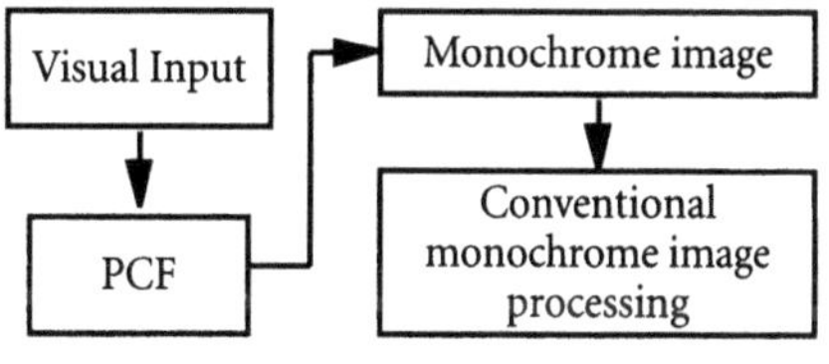

Fig 4.14(b). Using the PCF.

Using PIP in the usual way, it is possible to perform various functions on the color scattergram, such as thresholding, blob-edge smoothing, removing outliers, and blob labelling [*ndo*]. Once a series of "clean" and labelled blob-like figures has been created from the color scattergram, the current image can be used to program the PCF [*ppf*]. If the blobs are first enlarged (using *lnb*), the resulting PCF will recognize a wider range of colors corresponding to each symbolic color label. (This is the basis of so-called *color generalization.* See Figure 4.15.) It is possible to use this fact to adjust the blob sizes to make the PCF more robust [*lnb*], or more specific [*snb*] in the range of colors that it recognizes.

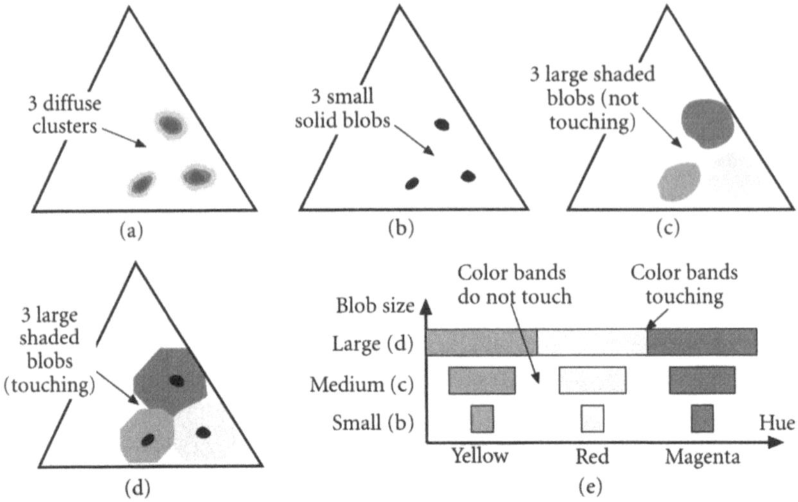

Fig. 4.15 Color generalization (a) Color scattergram is a grey-scale image contains several diffuse clusters. (b) Initial thresholding applied to the color scattergram produces a set of small blobs. (c) Expanding the blobs by applying the PIP operator [*lnb*] several times. but not allowing them to touch. (d) Expanding the blobs until they touch. (This uses a so-called watershed operator.) (e) Effects of expanding the blobs on the PCF's ability to recognize colors. Notice that enlarging the blobs produces a filter which associates a wider range of colors with each output value.

In order to illustrate how the PCF is used in practice, we will describe how we can use it to recognize members of a small set of national flags, each consisting of red, white, and blue elements and viewed against a clearly-distinguishable background, such as a clear mid-day sky (cyan). For the sake of simplicity, we will simply count these elements and will not attempt to identify them as stars, triangles, rectangles, L-shapes, etc. Nor will we consider their relationships to one another in space. Let us assume that a color filter has already been trained to recognize four colors: *red, white* (i.e., neutral shade, corresponding to the central region of the color scattergram), *blue,* and *cyan* (clear mid-day sky) and that these four color labels are associated with PCF output values 49, 156, 214 and 237 respectively. (This is a perfectly arbitrary assignment and reflects the blob intensities in the processed color scattergram just before *ppf* was satisfied.) It would clearly be useful to include several clauses of the following form in PIP's database:

```
isolate(red) :- rea(image), pfc, thr(49,49).
isolate(white) :- rea(image), pfc, thr(156,156).
isolate(blue) :- rea(image), pfc, thr(214,214).   % Assume that blue is separate from cyan
isolate(cyan,) :- rea(image), pfc, thr(237,237).
```

This avoids the necessity of the programmer having to remember the arbitrary association of integers with the names of colors. It is possible to recognize several national flags using the following program:

```
flag(C) :-
   grab_image,
   isolate(red,R),                      % Count red parts
   isolate(white,W),                    % Count white parts
   isolate(blue,B),                     % Count blue parts
   identify_flag(R,W,B,C),              % Consult the database
   writeseqnl(['The flag has been identified as being that of ', C]).
identify_flag(1, 1, 1, 'France').       % French and Dutch flags are not ...
identify_flag(1, 1, 1, 'Netherlands').  % ... distinguishable by this simple program
identify_flag(1, 4, 4, 'Iceland').
identify_flag(1, 4, 8, 'United Kingdom').
identify_flag(4, 1, 0, 'Denmark').
identify_flag(4, 4, 1, ' Norway').
identify_flag(7, 56, 1, 'United States').
......
   identify_flag(_, _, _, 'No known country').  % Many flags are not recognized
```

In order to identify more than these few national flags, the program must be extended to recognize other colors and take into account spatial relationships, such as *above/2, left/2, inside/2,* etc.

A summary of the theoretical basis for this topic is presented in Appendix A.

4.4.2 Mathematical morphology

The traditional approach to the study of the morphology operators has been through the use of mathematical analysis [SER82, SER86]. This has played a major rôle in building our present understanding of the importance of these very useful techniques. However, mathematics does not help us to decide which morphology operators, if any, are most appropriate for a given application.

As we saw in Chapter 3, many morphology operators can be implemented using the SKIPSM approach, which offers significant advantages for both software and hardware implementations. At the time of writing, PIP has over 250 different SKIPSM lookup tables, representing *structuring elements (SEs)* in the form of circles (both hollow and solid), squares, diamonds; upright/diagonal crosses, vertical, horizontal and diagonal bars. In addition, SEs in the form of the letters U, T and L are all available, each in four different orientations. PIP also possesses SEs in many other shapes that cannot be described succinctly. The SEs currently included in PIP have been classified into more than twenty categories. These various shapes are all available in a wide range of sizes. Some operators (e.g., discs and squares) exist in sizes up to 51×51 pixels, although not all shapes are available in all possible sizes.

Since there are so many morphology operators available in PIP, and bearing mind that they are often used in combination with one another, there is an urgent need for software to assist a vision engineer in choosing the most suitable SE for a given situation. When making such a choice, the user's skill, knowledge, and experience are, of course, of paramount importance, since there is no algorithmic method for choosing an appropriate SE. Hence, we perceived a need for a facility that enables a vision engineer to explore a range of morphology operators quickly and easily. To achieve this, PIP has been provided with a custom dialog box (Figure 4.16), which allows the user to select the size and shape of the SE, as well as deciding

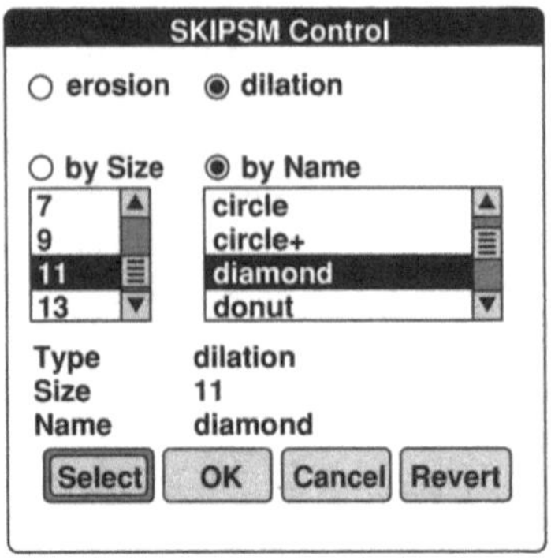

Fig. 4.16 Dialog box for selecting SKIPSM operators. In view of the large number of options available via SKIPSM, it is important that the user of PIP can experiment easily with different sizes and shapes of SEs.

whether to erode or dilate the image. When the selected morphology operation has been completed and the resulting image displayed, this dialog box is reinstated on the computer screen, so that the user can iteratively explore other choices for the size or shape of the SE. (He can also elect to return to PIP's normal interactive mode, exemplified in Figures 4.3 and 4.8.) As a direct result of using the SKIPSM implementation technique, every one of the available morphology operators can be performed in the same (short) period of time. This enables effective interaction to be achieved for even the largest SEs. It is important to note that this ability to experiment with different SE shapes and sizes would be completely impossible using conventional implementation methods. The traditional ways of implementing SEs are far too slow to achieve effective interaction for any but the very smallest SEs. In fact, the ability to interact effectively by applying a series of morphology operators in PIP is a direct result of the SKIPSM implementation technique.

One important point of detail about the implementation is worth noting here: the SE shape and size menus available within the custom dialog box are built automatically at start-up. Hence, the only user action needed to add a new SE to PIP's repertoire is to place a file containing the new lookup tables into the appropriate folder. (A simple file-naming convention has been adopted so that the PIP software can build the menus automatically.)

The predicate *imc([0,4]* can be used to select the appropriate SE. It uses the balloon predicate (Table 4.12), combined with the cursor, to isolate a blob selected by the user. The program then fits the minimum bounding circle (*mbc*) to this blob, thereby allowing its (outer) diameter to be estimated.

```
% User-selected point: [X,Y]. Blob area: C. Radius of MBC: D
imc(X,Y,C,D) :-
    pis,                        % Push image onto the stack
    cur(X,Y,_),                 % Cursor. User selects blob he wishes to measure
```

```
ndo,                      % Shade blobs
pgt(X,Y,B),               % Find blob identity i.e., intensity at point selected by user
thr(B,B),                 % Isolate this blob
pis,                      % Push another image onto the stack
cwp(C),                   % Measure area of biggest blob
mbc,                      % Minimum bounding circle (MBC)
blb,                      % Fill it
cwp(E),                   % Find its area
pop,                      % Pop image from the stack
add,                      % Combine the blob that has been isolated and its MBC
D is sqrt(E/pi),          % D is radius of the MBC
pop,                      % Recover input image from the stack
swi.                      % Switch current and alternate images.
```

The value of D returned by *imc/4* may be taken to be an upper limit on the size of SEs that will detect similar blobs to the one selected by the user (Figure 4.17). An estimate of the inner diameter of the blobs in an image may be calculated using the grass-fire transform. In fact, a local intensity peak in the output of the grass-fire transform (*gft*) provides a direct indication of the maximum size of square-shaped SE that will detect that corresponding blob (Figure 4.17).The caliper gauge predicates (Table 4.12, Figure 4.18) can also be used to estimate blob diameter, enabling both upper and lower bounds to be found. (Figure 4.18 also shows a range of other useful predicates for analyzing images. They are intended to model the familiar tool-maker and draftsman's tools: caliper gauges, ruler, compass and protractor.) Notice that these empirical techniques provide "soft" estimates of blob size but they can nevertheless be useful in setting lower and upper bounds on the sizes of SEs that are appropriate for analyzing a given scene.

Notice the following points:

1) These techniques for estimating blob size can be applied just as easily to the black background in a binary image as they can to a (white) figure. (Simply apply the negation operator [*neg*] first.)

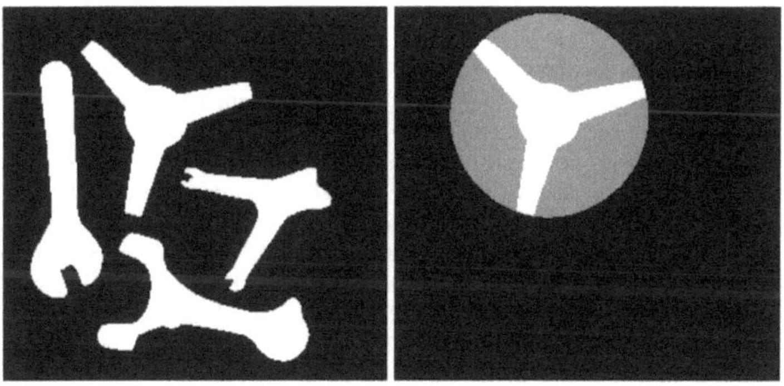

Fig. 4.17 Finding bounds for the sizes of morphology operators needed to detect a given blob-like object. (a) (left) Original image. (b) (right) *imc(X,Y,C,D)* applied to (a). The program asks the user to indicate the object of interest, in this case the 3-pointed "star". The diameter of its minimum bounding circle indicates the size of the largest possible SE of any shape that could fit within that object. (Figure continued on next page.)

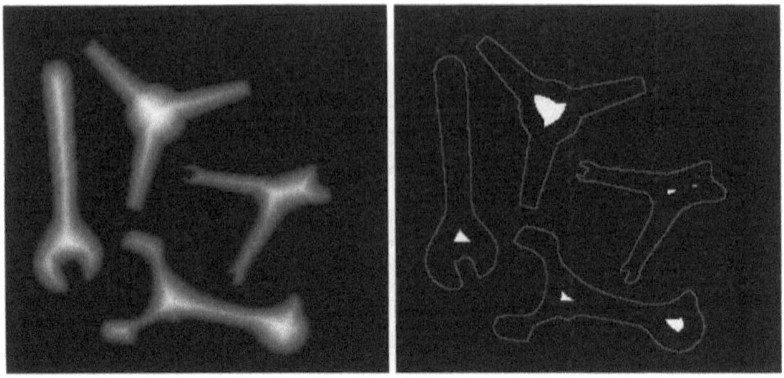

Figure 4.17 (continued) (c) The grass-fire transform allows us to investigate the effects of applying (square) SEs with different sizes. Finding the peak intensity in this image allows us to find the largest square SE that will fit within the white regions of the image. (d) By thresholding image (c) at various levels, it is possible to determine where other, smaller square SEs will fit into the image.

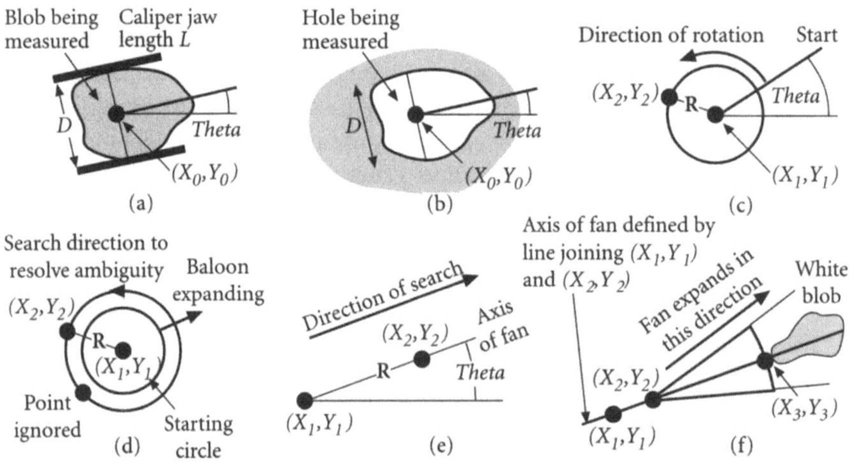

Fig. 4.18 Caliper gauge predicates. (a) *caliper/5* for measuring outer diameter. The action of this program is similar to that of a tool-maker's caliper/micrometer gauge. The user specifies the position (X_0, Y_0) of the instrument, its jaw length (*L*), and orientation (*Theta*). The jaws then close, beginning from an infinite separation. Each jaw stops moving when it encounters a white point. The final jaw separation, D, is then calculated. (b) *internal_caliper/4* for measuring inner diameter. The starting position, (X_0, Y_0), and orientation (*Theta*) are defined by the user. The program hunts along a straight line until it finds two white points. Their separation, D, is then calculated. (c) *compass/6* searches for a white point (X_2, Y_2) at a fixed distance R from a given point (X_1, Y_1). The scan begins at angle *Theta* and proceeds in anti-clockwise direction. (d) *balloon/5* inflates a circle, centered at (X_1, Y_1) and of initial radius R, looking for a white point (X_2, Y_2). If more than one white points exists at a given distance from (X_1, Y_1), the one that is found first during an anticlockwise scan defines the output. (e) *protractor/7* searches for a white point (X_2, Y_2) along a line inclined at an angle *Theta* through the point (X_1, Y_1). (f) *fan/7* searches for a white point (X_3, Y_3) by expanding a fan-shaped region. The orientation of the fan is defined by two points (X_1, Y_1) and (X_2, Y_2) and its angular spread is determined by the parameter *Theta*.

2) It is possible to write a program that automatically estimates both lower and upper limits on blob size. Hence, it is feasible to consider selecting a set of appropriate SEs for analyzing a given image without any human intervention whatsoever. We do not dwell on details here, because the main theme here is the user interface.
3) Since the SKIPSM technique can be used to implement a wide range of morphology operators, it is important that we provide an interface that allows the user to experiment with them easily and simply.

4.4.3 Multiple-image-processing paradigms

PIP and all of its predecessors have been based on a 2-image paradigm. This term refers to the current and alternate images, both of which are active in image-processing operations. The *Scratch* image in Figure 4.3 is not active during processing and, in effect, forms nothing more than a visible image-storage area. However, it is possible to conceive of operations which involve processing more than two images. There are several obvious situations in which this could be useful:
1) *Processing color and other multispectral images.* Recall that color images are usually represented by three, four, or more color-separation channels (e.g., CMYK, RGB, or HSI), each of which is a monochrome image. Clearly, there is a need to process each channel separately, or in pairs, triples, or even larger groups. We have not even begun to investigate the use of PIP-style interactive image processing in this area.
2) *Sensor fusion.* Images derived from a range of very different sensors are combined and/or inter-related in some way. For example, we may wish to correlate visible-light, UV, IR and x-ray images. (The images to be combined would normally be derived from sensors which effectively view the world from the same position.) Another situation which requires multi-image processing is likely to arise when analyzing range maps. (These measure surface height, rather than reflectance.) It may be helpful to relate external image features (e.g., corners, edges, etc.), that are visible to the eye or a standard camera, with surface-height data.
3) *Processing image sequences.* It is possible to define a wide range of filters that operate in both space (i.e., across an image) and time (i.e., between images). For example, we may wish to differentiate in space and smooth in time, or *vice versa.* It would clearly be of helpful, if the vision engineer could simultaneously view several images, representing samples taken at different moments in time.

As we will see in Chapter 5, the PIP software allows multiple images to be displayed on the computer screen, in order to facilitate these and other advanced types of processing. However, to date this facility has not been explored properly, even in the areas just mentioned. In order to illustrate the facilities that PIP offers, we will explain in Section 4.4.5 how the user-interface has been extended to program simple processing of image sequences.

4.4.4 Image stack and backtracking

The image stack has several fairly obvious uses, some of which we have already encountered:

1) To hold images during processing, particularly in nested image-processing operators. The reader's attention is drawn to the fact that many of the programs listed in this book have the following general form:

```
main_goal :-
pis,                      % Push image onto the stack
processing_sequence,      % Defined in the same general way as "main_goal"
pop,                      % Pop image from top of stack. Restore current image
swi.                      % Switch the current and alternate images
```

Notice that *main_goal* behaves in the same way as any standard PIP command, such as *neg*, *sed*, *thr*, or *lpf*. By following this programming convention, the language can be extended indefinitely, without the user ever becoming aware of there being a distinction between the "core" language and its Prolog-level extensions. The stack makes the hierarchical nesting of commands very much simpler than it would be if named files were used to hold intermediate results.

2) To permit the restricted form of backtracktracking needed during interaction. The *Mark/Revert* buttons in the standard interactive dialog box (Figures 4.3 and 4.8) operate the image stack. The function of these buttons was explained in Section 4.4.1 and so will not be repeated here.

3) To enable full backtracking to take place during programmed processing of images. To illustrate the use of backtracking, consider the following program sequence:

```
process_with_backtracking :-
   loa(image_file),            % Load an image from disc
   pre_processing,             % Preliminary processing of the image
   grey_scale_processing,      % Process the pre-processed (grey-scale) image
   choose(X),                  % This is not really a mysterious process - see notes below
   thr(X),                     % Threshold filtered image at level X
   binary_image_processing,    % Processes the binary image
   test_binary_image.          % Fails if image does not have properties defined by user
```

So far, we have insisted that image-processing operations always succeed but are never resatisfied on backtracking. (The "core" operators listed in Table 4.7 all behave in this way.) We will now remove this constraint to good effect. Indeed, the benefit can be quite startling.

Each of the predicates *pre_processing/0*, *grey_scale_processing/0*, *choose/1*, *more_processing/0* might all be defined in terms of several clauses. We might, for example, base the *pre_processing* predicate on several alternative edge detectors:

```
pre_processing :- pis, lnb, sub, acn(-127), din, pop,swi.
pre_processing :- pis, snb, swi, sub, acn(-127), din, pop, swi.
......
pre_processing :- sed.
```

The definition of *choose/1* may be very simple (e.g., a pseudo-random number generator), or very complicated indeed, using intelligent parameter-adjustment rules, to improve on values that have already been tried and found to produce good results. This extremely naive definition will suffice for our purposes:

```
choose(X) :-
    member(X,[0,16,32,48,64,80,96,112,128,144,160,176,192,208,224,240]).
```

We might define each of the other predicates using several clauses, in which case the number of possible solutions for *process_with_backtracking* rises exponentially. We can let Prolog investigate all of these possibilities. Of course to do so really effectively, we must provide a suitable definition for *test_binary_image.* Prolog performs backtracking naturally enough but the program *process_with_backtracking* does not restore the images as it does so. To make PIP do so, we insert the operator Δ at each point in the program where we might want to "undo" the processing. The operation Δ *thr(X)* thresholds the image at level X in the usual way, but it permits the input (grey-scale) image to be restored on backtracking. Here is the definition of the Δ operator:

```
:- op(900,fx,Δ).                % Declaring a prefix operator with precedence 900
    Δ A :- pis, call(A).        % Satisfied whan A is satisfied
    Δ _ :- pop, fail.
```

This is used in the following program to select a threshold parameter automatically.

```
automatic_threshold(A,B,C,D,E) :-
    member(Q,[true, neg]),      % We may need to negate image to get good result
    Δ call(Q),                  % Do "neg" or nothing at all
    member(E,[0,16,32,48,64,80,96,112,128,144,160,176,192,208,224,240]),
                                % Select a value to try as a threshold
    Δ thr(E),                   % Threshold at chosen value
    cbl(C),                     % Count blobs
    cwp(D),                     % Count white points
    C ≥ 1, C ≤ A, D ≥ 10*C, D ≤ B.% Test the result against values given by the user
```

Figure 4. 19 shows the result obtained by applying *automatic_threshold/5* to the image of a playing card.

4.4.5 Programming generic algorithms

In the previous section, we defined an abstract program (i.e., the predicate *process_with_backtracking*), simply by listing several auxiliary predicates (*pre_processing*, *grey_scale_processing*, *binary_image_processing*, and *test_binary_image*), each of which is itself expressed in terms of a range of possible algorithmic/heuristic procedures. Following this example, we are able to model the way that a vision engineer might formulate the notion of an abstract class of "similar programs". The process involves the construction of a tree-like representation of a group of conceptually similar programs (Figure 4. 20). Terminal nodes in this tree represent specific programs, while other nodes correspond to groups of algorithms that are, in some sense, similar to one another. Nodes near the root of such a tree, representing so-called *generic programs*, are characteristically associated with a high

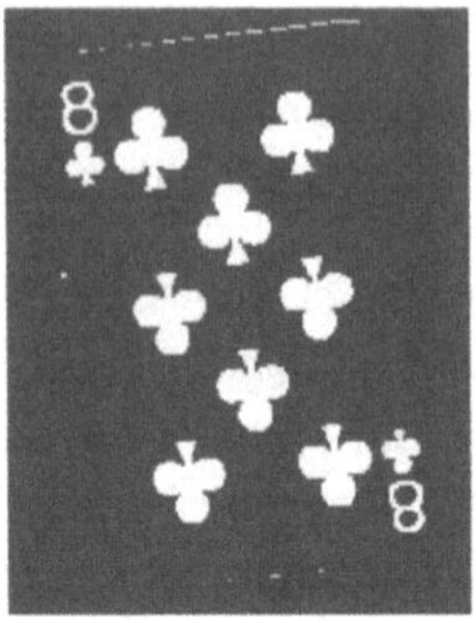

Fig. 4.19 Automatic thresholding . The original image is shown in Figure 4.4. "Input" parameters: A = 30; B = 10 000.

level of generalization and abstraction. The ability to represent a whole family of algorithms which all share a common conceptual basis is a major feature of PIP and arises from our using Prolog as the top-level control language.

In this section, the term *specific program* will be taken to mean an ordinary PIP program of the type that we have encountered many times before and which defines a single sequence of actions. On the other hand, *generic program* is the name given to a PIP program which identifies a set containing several or many conceptually similar specific programs. A typical generic program defines a set of edge detection operators (predicate *edge_detector*), whereas *[sed]*, *[red]* and *[lnb, sub, neg]* are all specific programs contained within this set. (More examples will be given later.) Generic programs make use of a library of pre-defined predicates, rather than primitive PIP commands. Once such a library has been prepared, it can be used in many design exercises.

We envision using generic programs to allow an experienced vision engineer to define a "skeleton solution" (a generic program) for a given vision problem. A less

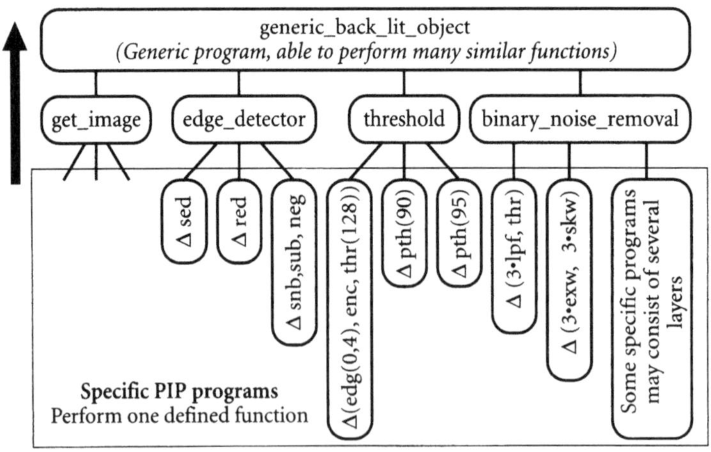

Fig. 4.20 Generic programming tree structure. Notice that some programs may be several layers deep but cannot be described as generic because they do not convey any abstract information.

experienced assistant will then be employed to assess the performance of each of the specific programs that this encompasses. We would prefer that the vision engineer be involved once and only very briefly, at the beginning of each design exercise and then delegate responsibility for completing it to his less experienced assistant (Figure 4. 21). In this way, the "skeleton solution" generates a set of specific algorithms semi-automatically. This is achieved by converting the generic program into a series of specific programs, some of which will inevitably be useless, while others will, we hope, be of much greater value in practice. All of the specific algorithms generated in this way will then be judged by the assistant and good ones selected,

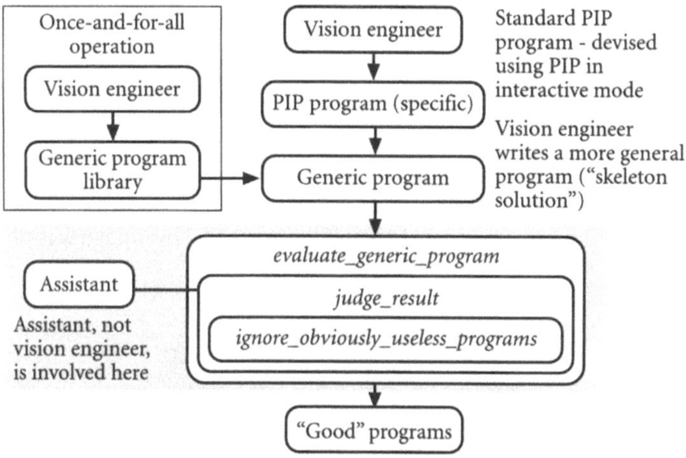

Fig. 4.21 Explaining how generic programming techniques might be employed in practice.

without any further intervention from the vision engineer. To minimize the work load on the assistant, another PIP program (*evaluate_generic_program*) is needed that can eliminate programs that are obviously impractical. This scheme has been devised to maximize the benefit obtained from the efforts of an experienced vision engineer. This is important in the present state of affairs where there are numerous unsolved applications of Machine Vision and too few properly experienced vision engineers to cope with the work load. This is entirely in line with earlier work to develop a range of design aids for Machine Vision [BAT90].

There are several ways in which we can use PIP to define generic algorithms but we will describe just one. They all rely on two key features of the language. (N is a small positive integer.)

a) Many individual PIP commands and command sequences can be grouped together, so that members of a given group perform roughly similar functions. Here are just a few examples:

- Edge detector: { [sed], [red], [gra], [lnb, sub], [snb, wri, swi, lnb, rea, sub] }
- Counting: blobs: { [cbl], [eul], [blb, eul], [N•exw, N•skw, eul] }
- Enclose in convex polygon: { [chu], [mar], [ccc], [mbc] }
- Noise removal (grey scale): { [raf], [N•lpf], [mdf], [N•lnb, 2•N•snb, N•lnb], [lnb,wri,swi,snb,rea,add] }

- Overall intensity: { [avr], [gli(A,B), C is (A+B)/2], hmx, [pth(50,A), swi], [rin,csh,hmx] }
- Enhance contrast: { [enc], [heq], [enc, sqr], [enc,log], [heq log], [enc, neg, log, neg] }

Of course, these groups are incomplete. Furthermore, many more groups like these can be defined. In this short section, we will explain briefly how we can represent and use generic programs written in PIP, leaving precise details until a later publication. The ability to define *groups* of operators which perform similar functions is possible only because PIP possesses a rich command repertoire.

b) It is possible to delay defining a sub-goal within a Prolog (i.e., PIP) program until the very last moment, during goal satisfaction. This can be used to pass control parameters and even specify sub-goals (i.e., short PIP command sequences) at "run-time".

In order to illustrate how generic programming might be used in practice, let us consider the deceivingly simple task of segmenting high-contrast images, such as the silhouette of a matte-grey metal gear (Figure 4.22). This image was obtained by

Fig. 4.22 Gear. This image was generated using back-lighting.

viewing the gear against a plain bright background. In order to make the segmentation algorithm as robust as possible, it must be designed to tolerate variable-intensity illumination of the background, as well as some light falling on the fore-ground object. The following program was designed in the usual way, using PIP in interactive mode, and is, of course, just one of many possible techniques for segmenting this image:

```
back_lit_object :-
    loa('Gear'),        % Load gear image from disc
```

```
    sed,            % Sobel edge detector
    edg(0,4),       % Mask off the image border, 4 pixels wide
    enc,            % Enhance contrast
    thr(128),       % Threshold at mid-grey (half-way between min. & max. intensities)
    3•exw,          % Expand white regions. Apply 3 times (part of noise filtering process).
    3•skw.          % Shrink white regions. Apply 3 times (part of noise filtering process).
```

We will now show how this can this specific procedure can be regarded as one particular example of a more general algorithmic structure. Here, then, is a generic algorithm for segmenting the image of a back-lit scene:

```
generic_back_lit_object:-
    get_image,          % Generic auxiliary command - obtain an image for processing
    edge_detector,      % Generic auxiliary command - highlight edges
    threshold,          % Generic auxiliary command - convert the image to binary form
    binary_noise_removal.
                        % Generic auxil. command: eliminate noise from binary image

% The following are sample definitions for the sub-goals used in this program.
% Two sample clauses for "get_image "
get_image :- grb.                  % Digitize an image
get_image :- loa.                  % Select and load an image from disc
% Three sample clauses for "edge_detector". Another one will be defined later.
edge_detector :- Δ sed.            % Sobel edge detector is one possibility
edge_detector :- Δ red             % Roberts edge detector is another possibility
edge_detector :- Δ snb, sub, neg.  % Morphology operator (grey-scale erosion)

% Three sample clauses for "threshold ". Recall that the Δ restores the image on backtracking.
threshold:-
    Δ (edg(0,4),            % Mask off the image border 4 pixels wide
    enc,                    % Enhance contrast
    thr(128)).              % Threshold at mid-grey
threshold :- Δ pth(90).     % Threshold so that 90% of the image is white
threshold :- Δ pth(95).     % Threshold so that 95% of the image is white
% Two sample clauses for "binary_noise_removal "
binary_noise_removal :-
    Δ (3•exw,   % Expand white regions 3 times (part of noise filtering process).
    3•skw).     % Shrink white regions. Apply 3 times (part of noise filtering process).

binary_noise_removal :-
    Δ (3•lpf,               % Apply low pass (grey-scale) blurring filter
    thr.)                   % Threshold
```

It is easy to see that *back_lit_object/0* is subsumed by this definition of *generic_back_lit_object/0*. In fact, this version of *generic_back_lit_object/0* is the exemplar for a total of 36 specific programs. These will all be executed in turn if we simply force backtracking by placing *generic_back_lit_object/0* in a loop:

```
evaluate_generic_program:-
    generic_back_lit_object,    % Substitute any generic program here
    judge_result,       % Program performs initial sifting . Assistant makes final decision
    fail.               % Force backtracking. This is why we included Δ in above definitions
    evaluate_generic_program .  % Goal always succeeds in the end
```

Certain functions, such as *threshold*, may require that some control parameters be specified at "run time" (i.e., when attempting to answer a query). Our present program does not make use of any parameters. These can be handled in the usual manner employed in Prolog. However, we must be careful to avoid spoiling the general nature of a program, since some clauses defining sub-goals (e.g., *threshold*) may require parameters, while others do not. Suppose, for example, that we defined *threshold* without regard for consistency in its arity:

```
% Two sample clauses with arity equal to 0 (zero)
threshold :- Δ thr.
threshold :- Δ enc, thr.

% Two sample clauses with arity equal to 1
threshold(A) :- Δ thr(A)
threshold(A) :- Δ pth(A).

% One sample clauses with arity equal to 2
threshold(A,B) :- Δ thr(A,B)
```

We cannot now include *threshold* in a generic program without causing some complications. (What arity do we use when invoking *threshold* ?) We can, of course, consult the user (i.e., assistant) to obtain suitable values for certain parameters, in order to reduce the arity of *threshold* to a constant value (zero) for all clauses. Alternatively, we can employ an automatic parameter-value generator for the same purpose, as we did earlier (see definition of *automatic_threshold/5*). The latter is preferable, since it removes some very tedious user intervention during program execution. It is wise, as far as it is possible, to reduce the responsibility placed on the assistant. Clearly, the problem of varying arity in low-level predicates is one that requires very careful attention during the construction of the library of generic programs. The subject is too complicated to be discussed in detail here.

The reader should understand that there is no fundamental difficulty in passing parameters between sub-goals in a generic program, as the following program explains:

```
inspect_bays :-
    cvd,                        % Convex deficiency
    generic_blob_count(N),      % See notes above for options to be included here
    for(J,1,N, (big(J), cwp(A),tell_user(A))).
        % For J = 1,...,N, isolate the J-th biggest blob, measure its area, tell user
```

Problems arise only when a low-level predicate is defined with variable arity, since the very essence of generic programming is that the user should not have to worry about tedious details, such as matching and generating suitable values for arguments.

There is a very real possibility of writing a generic program that is useless in practice, because it leads to a combinatorial explosion and therefore cannot be assessed properly. Some means should therefore be devised to enable the user (i.e., the assistant) to interact with the program during "run time," to eliminate/escape from obvious dead ends in the search tree. One possible way to do this is to run the generic program via a meta-interpreter [STE86a]. This would allow the user to interrupt the generic program, and thereby avoid wasting time by interactive editing during "run time". Again, this subject is beyond the scope of this discussion.

We anticipate that generic programs will ultimately rely on a combination of program-based and human judgment to select the most promising program for a given application task. Many of the specific programs that are generated by a generic program are likely to be nonsensical, in that they produce intermediate or final results that are obviously useless. (For example, the picture is all black or all white. Alternatively, it may contain far too many spots, or the contrast may be too low.) The program to implement these simple rules is naive but is nevertheless useful in practice as a means of reducing the demands placed on the assistant:

```
judge_result :- ignore_obviously_useless_programs.
judge_result :-
    display_results_to_user
    results_ok.      % Ask the user whether the results are acceptable. Fail if not.
ignore_obviously_useless_programs :-
    cwp(N), N is 0, !, fail.               % Picture is all black
ignore_obviously_useless_programs :-
    neg, cwp(N), N is 0, neg, !, fail.     % Picture is all white
ignore_obviously_useless_programs :-
    cbl(N), N ≥ 100, fail.                 % Picture contains too many blobs
ignore_obviously_useless_programs :-
    gli(A,B), 10 ≥ B - A, !, fail.         % Contrast is too low
```

This definition of *ignore_obviously_useless_programs* can be extended in several ways, although it will not be possible to write programs to detect some of the more subtle conditions that can arise in practice. Hence, we must rely ultimately on human judgment to make the final decision about which small sub-set of algorithm(s) should be kept for detailed study and final implementation in a TVS. We do not know yet whether we can write really effective "final judgment" programs that are selective enough to avoid overloading the engineer's assistant. Much more work is needed in this and other areas before we can be sure that generic algorithms offer any real prospect of improving the design process for Machine Vision systems. At the time of writing, we can only say that this approach seems to be promising enough to warrant serious investigation.

4.4.6 Batch processing of images

Batch processing of images is easy in PIP, as the following program shows.

```
bip(A) :-
    source_folder(B),      % Identify the folder (B) containing the "input" images
    files(B,C),            % List of images in this folder (C)
    forall((member(X,C), concat([B,X],D)), (loa(D), call(A), wim(X))).
                           % Perform operation A on all images
```

The last line ("forall((member...") requires a little more explanation. *X* is the name of one of the files in folder *B*, and *D* is its full path name. This file is then read [*loa(D)*] and operation *A* is performed on it [*call(A)*]. Finally, the result is saved [*wim(X)*]. This process is repeated for all files in folder *B*.

4.4.7 Simulating a line-scan camera

A line-scan camera samples the light distribution along a single line and is often used in conjunction with a linear or rotational transport mechanism [see Section 1.7.1]. In this way, it is possible to construct a 2-dimensional view of a moving object. If that object is moving along a straight line, a rectangular scanning pattern is generated. If it is rotating, the scanning pattern resembles a set of wheel spokes. The geometric transformation thus created is very similar to that performed by the PIP operator *ptc/0* and hence can be removed using *ctp/2* (Plate 17(CL)).

Although line-scan cameras are often used in industrial vision systems, they do not employ an appropriate video standard (i.e., CCIR, RS(343)/RS170, IEEE 1394, etc.) and hence cannot be interfaced directly to the prototyping system. For this or some other reason, a line-scan camera may not be at hand during a design exercise. The ability to simulate a line-scan camera may be useful, enabling the ideas that we have described above to be applied to realistic images, typical of those generated by an industrial scanning system that may not yet exist. PIP is able to simulate a line-scan camera. In fact, a simulation was used to generate Plate 17(TR). PIP achieves this by extracting one line from every frame of the digitized output from a regular (area-scan) camera. The output image is therefore built one line at a time by extracting just one row from each image derived from the area-scan camera. After each step, the object is shifted/rotated by a small amount. Although the PIP program to do this is straightforward, we do not have space to list it here.

4.4.8 Range maps

A range map, sometimes called a *depth map*, is an image in which the intensity indicates the 3-dimensional structure of an object surface. Plate17(CR) shows an example, obtained by scanning the object shown in Plate 17(TL) as it was rotating slowly.

A range map is constructed from a series of images like that shown in Plate 17(TL), except that the ambient light level is normally reduced to zero. Such a picture can be created by projecting onto the object surface a thin fan-shaped beam of light from a diode laser fitted with a cylindrical or diffraction lens. A camera placed at an oblique angle is used to view the scene. The camera therefore sees a curve like that shown in Plate 17(TL). The vertical coordinate of each point on this curve allows us to estimate the height of one point on the object surface. Hence, for each image like that shown in Plate 17(TL), we can determine the height of the surface along a single cross-section of the object. This cross-section lies in the plane defined by the fan-shaped laser beam.

The procedure needed to construct a full range map like that shown in Plate 17(CR) involves the following steps:

1) Set loop counter *I* to 1. Initialize the range map image, making every pixel black.

2) Project the fan-shaped beam onto the object surface. The laser should point vertically downwards.[4]
3) Obtain an image like that shown in Plate17(CR), using a camera whose optical axis is at about 45° to the vertical axis.
4) Process the image to generate a binary image consisting of a thin white line against a black background. Simple thresholding may suffice, or it may be necessary to employ a highpass filter [*wri,raf,sub*] or [*neg, crk*] first.
5) Estimate the vertical height of each point on the curve generated in step 4. This can be achieved using [cox, cin] and results in a set of height values stored as intensity values along the bottom row of the image.
6) Copy the bottom row of the image created in step 5 into the Ith row of the range map.
7) Move the object being examined a small distance and increment the loop counter *I* by 1.
8) Repeat steps 2 to 7 many times (i.e., once for each row in the range map).

Since there is insufficient space here to list the complete program, the reader should simply note that PIP is able to perform all of these tasks easily, including that of controlling the object motion.

4.4.9 Processing image sequences

The principal difference between processing image sequences and batch processing described in Section 4.4.6 is that they require different ways of indexing picture files. PIP uses *named* image files; the following are typical commands for accessing picture files:

- rea(picture12) % Read picture12 from RAM
- wim(result) % Write to disc file "result"
- loa('My computer:Pip:Images: Bottle'). % Load image from a defined file path

Of course, like most other operating systems, the Macintosh OS requires that a file be addressed using a symbolic rather than a numeric index. On the other hand, an image sequence is, by definition, an ordered collection of pictures and hence requires some indexing scheme that preserves its linear structure. Image sequence processing therefore requires the use of *numeric* indexing of image files. While there is no inherent difficulty in extending it, PIP has to be programmed to allow the programmer to specify images in this way. In addition, several higher-level predicates have been provided for processing image sequences (Table 4.15) and a customized dialog box has been designed for this purpose (Figure 4.23).

4. A common mistake is made in the technical literature, which often shows the camera placed vertically above the object, with the laser beam striking it an inclined angle. The range map is easier to interpret if the laser is above the object and the camera is offset and views it from an oblique angle.

Table 4.15 PIP predicates for batch processing, digitizing, and displaying image sequences.

Command	Function
isd(A)	Display sequence of images in A (i.e., *Source* or *Destination).*
isg(N)	Digitize sequence of N images from the camera and store them on disc. Note that *isg/1* is not necessarily able to operate in real time; it cannot guarantee to be able to grab successive video frames.
isi/0	Display interactive control window (see Figure 4.23).
isp(A)	Process the image sequence in the Source Folder. See text for code examples.
isr(A)	Read images from Source Folder. Relative address mode. *A* specifies the offset value.
isu(A,B)	Read image in folder *A*. *B* defines the absolute address.
isv(A,B)	Write image into folder *A*. *B* defines the absolute address.
isw	Write image into Destination Folder. Relative address mode. *A* specifies the offset value.
isx(A)	Delete all files from folder *A*.

We need to build in the ability to combine individual images (frames) within a sequence. Failure to do so would severely limit the types of operations that can be performed. We therefore need to be able to specify files using both *relative* and *absolute* numeric indexing. We will explain these in more detail. When we are simply digitizing or displaying an image sequence, it is sufficient to employ a single counter that is incremented after each step is completed. This parameter effectively constitutes an *absolute index.* On the other hand, to detect motion in a sequence, we might need to subtract consecutive frames, or close neighbors in a sequence. In order to define such an operation, we use two parameters, i and j, as the following formula shows.

$$\text{Result}(i) \leftarrow \text{Image}(i) - \text{Image}(i+j), \text{ for all } i \text{ in the range } [1,N]$$

(Normally, $j << N$.) The parameter j is a *relative index.* Such an operation will be performed for all values of i, which will be called the *base* index. Notice that, just as local operators (e.g., *lpf* and *sed*) create edge effects, processing image sequences inevitably produces undefined and undesirable effects at the beginning and/or end

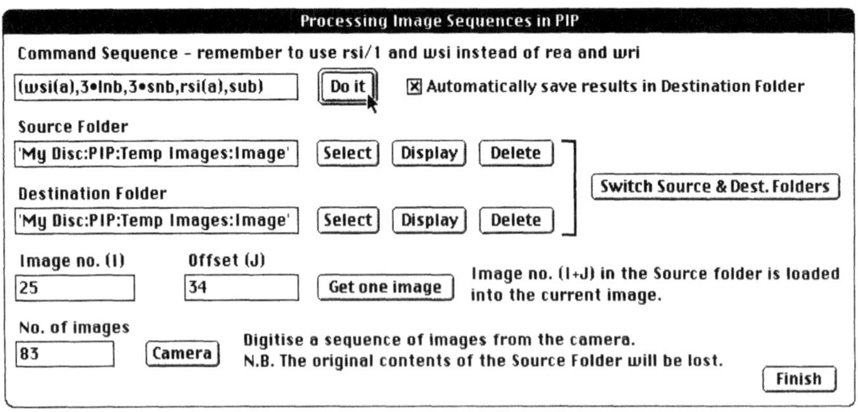

Fig. 4.23 Dialog box used to control the processing of image sequences.

of the sequence (i.e., when the value of *(i+j)* lies outside the range *[1,N]*). Of course, our programs must take this into account and should not, for example, crash when we attempt to load a file that does not exist.

Programs for storing and reading images in a sequence

The normal disc read/write operators (i.e., *wri*, *rea*, *wim*, *rim*, *sto* and *loa*) are unsuitable for image-sequence processing, since they cannot accept a numeric relative index. Higher level programs, based on numeric indices, have therefore been defined which perform similar functions. In the following program definitions, we will assume that there are two folders, called *Source* and *Destination*. Furthermore, two parameters are stored as Prolog properties.

is_files	2-element list: [Source folder, Destination folder]
is_file_number	Integer: Base address

```
% ________________Working with relative addresses________________
% Read an image from the Source Folder. A is the relative address
isr :- isr(0).                   % Default case, when no argument is specified
isr(A) :-
    recall(is_file_number,B),    % B is base address
    C is A+B,                    % C is absolute address
    recall(is_files,[D|_]),      % Source file name less numeric index
    concat([D,C],E),             % E is path-name of chosen file
    loa(E),                      % Load image file E
    !.                           % Allow only one solution
% If the clause immediately above failed (i.e., loa(E) failed), do the following.
isr(_) :- isr(0).                % Load first image in sequence. (This an arbitrary choice)
% Store image from the Source Folder. Relative address is always 0 (zero).

isw :-
    recall(is_file_number,A).    % Base address
    recall(is_files,[_,D]),      % Destination file identity, less numeric index
    concat([D,A],E),             % E is full path-name of chosen file
    eprint([D,E]),               % Reformat, to create a valid file name
    sto(E,1),                    % Store image. Over-write existing file, if necessary.
    !.                           % Allow only one solution

% ________________Working with absolute addresses________________
% Reload the image that was read most recently, using isu/1
isu :-
    recall(temp_image_sequence, B),
    % What was last picture loaded?
    loa(B).                      % Now reload it

% Read image from folder F (i.e., Source or Destination). A is absolute address
isu(A) :- isu('Source',A).       % Default is the Source Folder
isu(F,A) :-
    concat(['My Disc:PIP:Image Sequence:',F,':Image',A],B),   % Building the path
                                 % name from the absolute address
    remember(temp_image_sequence, B),
    loa(B), !.                   % Load the appropriate image
% isv(F,A) - write image A into folder F (Source or Destination).
```

```
isv(A) :- isv('Destination',A).   % Destination Folder is used as default
isv('Source',A) :-
    recall(is_files,[B,_]),       % Destination file identity, less numeric index
    concat([B,A],C),              % C is path-name of chosen file
    sto(C,1).                     % Store image. Over-write existing file if necessary.
    isv('Destination',A) :-
    recall(is_files,[_,B]),       % Destination file identity, less numeric index
concat([B,A],C),                  % C is path-name of chosen file
    sto(C,1).                     % Store image. Over-write existing file if necessary.
```

Programming image-sequence processing

The following program applies the command sequence defined by A on all images in the Source Folder, placing the results in the Destination Folder.

```
% Process all images in the Source Folder
isp(A) :-
    (remember(is_file_number,1),
    % Initialize the loop counter
    source_folder(B),             % Find the Source Folder
    files(B,C),                   % Now find the files in the Source Folder
    length(C,N),                  % How many file are there?
    isx('Destination'))           % Delete all files in the Destination Folder. Not defined here.
     ->                           % Do not backtrack within this clause
    apply_to_all(A,N).            % Perform operation A on all N files
    isp(_).                       % Goal isp(_) always succeeds
    apply_to_all(_,0).            % Terminal clause. Stop when counter gets to zero.
apply_to_all(A,N) :-
    isr,                          % Read one image in the sequence
    call(A),                      % Perform operation defined by A
    isw,                          % Save resulting image
    increment_file_counter,       % Increment loop counter by 1. Not defined here
    M is N - 1,                   % No. of images to be processed is reduced by 1
    !,
    apply_to_all(A,M).            % Repeat for all images in Source Folder
```

Sample programs

Example 1: This goal negates all images in the Source Folder. Each image is processed separately:

```
i sp(neg)
```

Example 2: The following slightly more complex goal performs a (grey-scale morphological) edge-detection operation on each image. Again, each image is processed separately:

```
isp((lnb,wri,swi,snb,rea,sub,edg(128,4),enc))
```

Example 3: In the following example, consecutive images are first blurred (*raf*) and then subtracted.

```
isp((raf,wri,isr(1),raf,rea, sub,enc)).
```

This performs the same operation as the following "conventional" program which is listed only in part:

```
......
sub, enc,                % End of 7th cycle
rea('Image8'),           % Begin 8th cycle
raf, wri,
rea('Image9'),           % Performs the equivalent operation to isr(1).
% More generally, on cycle M, this reads file Image(M+1)
raf, rea, sub, enc,      % End of 8th cycle
rea('Image9'),           % Begin 9th cycle
raf, wri,
......
```

If there are *N* images in the Source Folder, *isp((raf,wri,isr(1),raf,rea, sub,enc))* generates a sequence of (*N-1*) "useful" images in the Destination Folder; *ImageN* contains an anomalous result. More generally, if we specify the goal as

isp((raf,wri,isr(P),raf,rea, sub,enc))

where $0 < P << N$, the result is a sequence of (*N-P*) "useful" images in the Destination Folder. Table 4.16 explains.

Table 4.16 Processing a sequence of images using *isr*.

isr(0) reads	Result	Comment
Image1	Valid	*Image1* is combined with *Image(P+1)*
Image2	Valid	*Image2* is combined with *Image(P+2)*
...	Valid	
Image(*N-P*)	Valid	*Image(N-P)* is combined with *ImageN*
Image(*N-P+1*)	Invalid	*Image(N-P+1)* is combined with *Image1*
...	Invalid	
ImageN	Invalid	*ImageN* is combined with *Image1*

If P is negative, *Image1*, *Image2* ..., *ImageP* are invalid, while *Image(P+1)*, *Image(P+2)*, ..., *ImageN* are all valid. This is, of course, akin to the production of edge effects in local operators (*lpf, sed, lnb*, etc.).

Example 4: The program

isp((isr(-1),lpf,wri, isr(1),lpf,rea, sub, abs, thr(135), cwp))

compares two images *(isr(-1) and isr(1))* and measures the difference between them. In this case, both *Image1* and *ImageN* are anomalous and must be ignored.

Garbage collection spoils real-time operation

Notice that PIP cannot guarantee to perform any operation, including digitizing, processing, and displaying an image sequence in real time. The reason is that Prolog may "spontaneously" initiate garbage collection at any time, without the programmer being able to anticipate or control the event. Any image-sequence created using the rather naive operator currently implemented in *isg/1* may contain "glitches", corresponding to those moments when garbage collection was taking place. Of course, it is possible to overcome this problem by programming image-sequence digitization using a different language, or by synchronizing the camera to the PIP program by a handshake mechanism.

Modern computers are fast enough to be able to display a sequence of images in near real-time.[5] However, when displaying an image sequence, there are occasional short pauses, again while Prolog performs garbage collection. Once again, the user has no effective control over this, although it is not normally a serious problem when displaying short image sequences. In its intended rôle as a prototyping tool-kit, PIP is not expected to process images in real time, so garbage collection does not have any significant effect. On the other hand, for a real-time factory-floor inspection system, garbage collection could have disastrous consequences.

4.4.10 Interfacing PIP to the World Wide Web

MacProlog possesses a mechanism for interfacing to the World Wide Web (WWW), thereby enabling a user working on a remote computer to present a query and receive the results generated in response to it. This facility employs *AppleEvents*, which form a well-established tool within Macintosh OS for inter-process communication. Using this basic mechanism and a WWW browser on the remote computer, it is possible for the user to initiate a PIP program but he cannot view any of the images that it may generate. To construct a more effective PIP-WWW interface that allows the user to view the current and alternate images, this basic AppleEvents interface mechanism must be extended. This is straightforward, although tedious to describe in detail. Let it suffice to say that this process involves many layers of software and is slow in operation. In particular, it is currently too slow for effective interactive image processing and hence misses the whole point of using an IVS for problem analysis.

The PIP-WWW interface was developed initially to provide an on-line documentation facility and for training novice IVS users. There have been problems in obtaining reliable performance and sufficiently high speed on image-processing tasks. As a result, the PIP-WWW interface has been abandoned in favor of the Java-based system CIP, described later.

4.4.11 Controlling external devices

As we have taken care to explain, both interactive and target vision systems should be able to control a wide range of external hardware devices, including lamps, optics, cameras, and a variety of electromechanical manipulators. With this in mind, PIP has been provided with a low-cost multi-purpose interface module, which allows it to control a variety of low-speed devices of this type. This unit, referred to as *MMB*,[6] currently provides facilities for operating the following devices/data ports, via a single serial (RS423/RS232) port on PIP's host computer:

5. The authors have used the RAM disk in their work to optimize the speed of display. At the time of writing, a Macintosh G3 processor, with a 250 MHz clock can display about 11 - 13 images of resolution 320x240 pixels in a second.

a) ten ON/OFF mains devices (lamps, pattern projector, etc.)
b) four ON/OFF pneumatic valves. (These are mounted on board the hardware module.) Two push–pull pneumatic valves are also provided.
c) one 8-way video multiplexor
d) six programmable-speed serial (RS232) communication ports
e) six opto-isolated 8-way parallel I/O ports

Using MMB, PIP is able to select different video cameras and to control several lamps and pneumatic air lines, a laser light-stripe generator, and the position of an (X,Y,θ)-table (stepper-motor drives). These electromechanical devices form part of a so-called *Flexible Inspection Cell (FIC)*, which is illustrated diagrammatically in Figure 4.11. However, it should be understood that MMB is a general-purpose interfacing unit that has also been used with other computing systems and electromechanical devices (e.g., workstation, pan-and-tilt mechanism, laser-beam-steering optics).

Figure 4.24 shows the organization of the MMB hardware. At its heart is a standard microprocessor, connected to a series of serial and parallel interfacing chips, including

- one 8-channel UART
- three 3-port PIO chips.

The latter are, in turn, connected to

- one 8-way video multiplexor
- sixteen solid-state relays, capable of switching mains power. Six of these relays are dedicated to operating pneumatic control valves.

Programs for the MMB can be held in either battery-backed RAM or ROM. The latter can, of course, be programmed using a conventional "free-standing" ROM programmer. Alternatively, the MMB can be reprogrammed through its control port. (This is the port which, in normal operation, is connected to PIP's host computer.) It is not appropriate to program the MMB from PIP, since Prolog is not a suitable language for bit-level programming. Assembly-level programming is probably best suited to this task. MMB programs can be very simple or quite complicated. For example, the calibration of a multi-axis manipulator, such as the FIC's (X,Y,θ)-table, might require a sophisticated procedure that is best implemented using MMB, rather than PIP software. The important point to note is that, once it has been programmed, MMB can operate as a semi-autonomous slave that is able to act intelligently and independently on certain tasks defined by PIP. In certain specific applications, it might be appropriate to reprogram MMB so that PIP is required to do the minimal amount of work. However, MMB is provided with a range of standard software functions for performing "low-level" tasks: switching lamps on/off, switching specific pneumatic lines on/off, operating the video multiplexor, sending/receiving text data via a serial line, setting/sensing the

---

6. MMB is the acronym for *Mike's Magic Box*, after its designer Michael Daley, Cardiff University, Wales, UK.

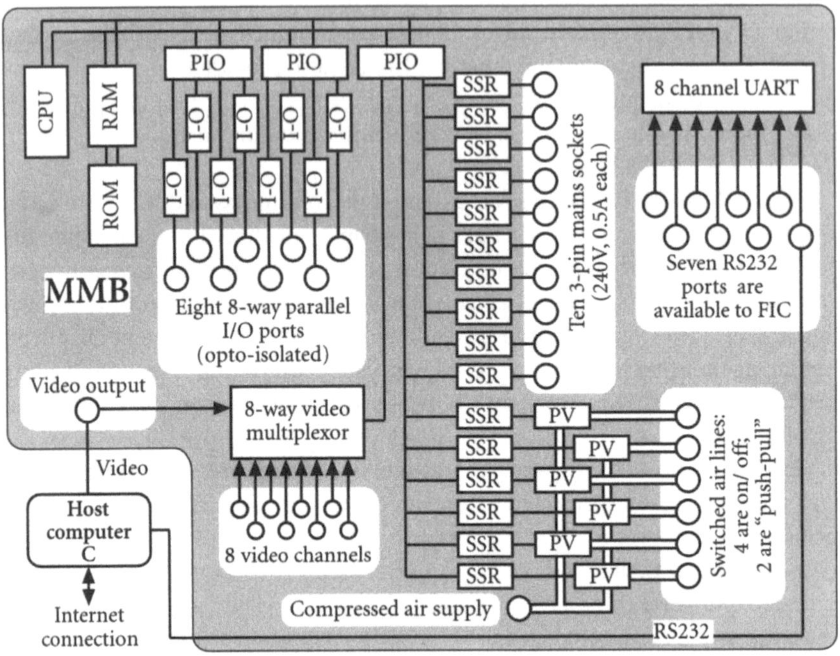

Fig. 4.24 Block diagram of the MMB hardware-control unit. Key: PIO, parallel input/output module; O-I, opto-isolator; SSR, solid-state relay; PV, pneumatic valve; UART, universal asynchronous receiver/transmitter.

parallel I/O lines, etc. The standard MMB software has been found to be very useful on a wide variety of applications.

As we indicated earlier, the PIP host computer is connected to one of MMB's serial ports. This serial line has to handle all I/O traffic. Hence, there is a potential data-flow bottleneck. Communication along the control line currently takes place at 9600 Baud, although MMB's other serial ports can operate at different speeds. (This facility is particularly useful for the FIC, since the controller for the (X,Y,θ)-table currently accepts data only at 2400 Baud.) In fact, this has not been found to present a serious bottleneck for such low-speed tasks as operating the FIC. Here, as in many other prototyping applications, the versatility of the PIP-MMB system, its low cost, and its ease of use are more important than high operating speed.

4.5 Windows image processing (WIP)

While PIP has been successful technically, it has not found popular support on a commercial basis for two reasons:

a) The *Macintosh OS* (operating system) is not nearly as popular as Microsoft *Windows*, even though the former possesses better memory-handling facilities and a superior user interface.

b) Prolog is not widely understood, compared with "conventional" languages such as C, Basic, Fortran, Java, etc. Of course, the latter are intended for quite different tasks from those that favor the use of Prolog. Regrettably, many supposedly computer-literate people have very little idea of what Prolog is or when it should be used in preference to these other languages.

On seeing PIP, visitors to the authors' research laboratories have many times made comments such as *"Very good! ... but I want similar facilities running on a PC."* and *"I don't understand Prolog. I want to write control programs in language X instead."* (Usually X = C.)

One possible response to such comments is to offer PIP running on a Macintosh OS emulator under Windows. (This is quite successful technically and system response times are adequate for most purposes.) Of course, this addresses only the first part of the criticism. The second point relating to the choice of Prolog can also be answered by writing a higher-level scripting facility in that language. (It is a straightforward matter to write looping, decision-making, and other conventional program-control mechanisms in Prolog. Many books on the language include code for such facilities.) However, there remains a significant body of opinion that stubbornly favors the *Windows/C* option for *all* programming activity, even though there is no technical, commercial, or aesthetic justification for this point of view in this situation.

With this in mind, development of new system called *WIP (Windows Image Processor)* was begun in the mid-1990s. WIP is based on the *Windows* operating systems and, at the time of writing, has a command repertoire that is rather smaller than that offered by PIP. WIP is an "unadorned" interactive image processor, lacking the important features that have been emphasized above. Unlike its cousin, WIP is not constrained to use Prolog, although this remains an option. Other "top level" control languages such as Visual Basic, C, JavaScript, and Delphi have been used so far. The choice of which one to employ is at the user's discretion. This ability to use other host languages *in lieu* of Prolog is WIP's main feature of note. As we saw earlier, many of PIP's most attractive features are made possible because it is based on MacProlog, which in turn relies on the Macintosh OS. At the time of writing, an interface to WinProlog is being added to WIP. In many respects WinProlog and MacProlog are identical. This is hardly surprising, since they are produced by the same company. It should be possible, therefore, to import many of the features of PIP directly into the Prolog-WIP system, although we should emphasize that this has not been done at the moment.

The Windows98® operating system permits inter-application communication using what are termed *Dynamic Link Libraries (DLLs)*. This has the advantage that any program that conforms to the defined standard can interact with any other program that also does so (Figure 4.25). WIP's main image-processing DLL is written in C, which provides sufficient speed for nearly all IVS requirements. It is a simple matter to interface other DLLs (user libraries) to the main DLL. In this way, several students at Cardiff have been able to write self-contained image-processing modules that extend WIP. The architecture makes this quite straightforward. The present command repertoire of WIP is not as large as that of PIP and the user interface has not been developed to the same level of sophistication.

The original motivation for developing WIP has been undermined, to a large extent, by the advent of the first widely available multi-platform computing language, Java. Nevertheless, there are several lessons to be drawn from our work with WIP:

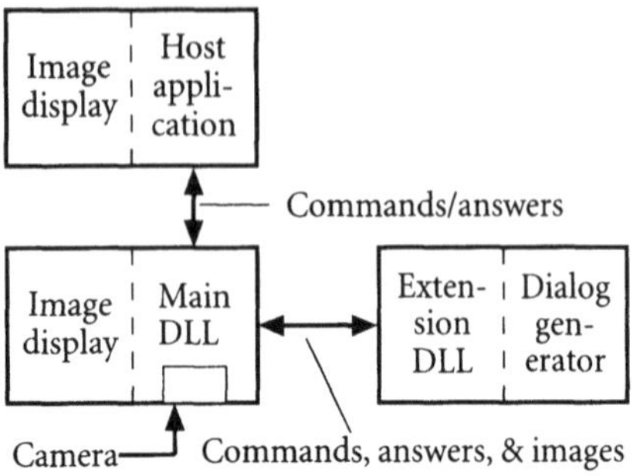

Fig. 4.25 WIP organization.

a) Dynamic Link Libraries provide a good general-purpose inter-application communication mechanism.
b) WIP allows almost any "top-level" control language to be used *in lieu* of Prolog.
c) So far, WIP has not significantly challenged the authors' fundamental belief in Prolog as a control language for an IVS. Nor has Windows mounted any serious challenge to Macintosh OS in its rôle as the host for an IVS. For our present needs, the Macintosh operating system remains unsurpassed, even compared with UNIX, OS/2, or any other of the more significant operating systems.
d) WIP provides an excellent platform for studying other computer languages in the rôle of host for an IVS. An interesting research program lies ahead in this area.

Both WIP and PIP are limited to running on a "stand-alone" computer. There are many advantages to be obtained from developing a PIP-like system for operating over the Internet/Intranet. This largely undermines the motivation for developing WIP further.

4.6 Web-based image processing (CIP)

Following the advent of Java in the mid-1990s, it is now possible to provide effective interactive machine vision software that is able to operate via the World Wide Web (WWW). At the time of writing (May 2000), a package called CIP is still under development at Cardiff and lacks some of the refinements of PIP [CIP]. Nevertheless, CIP already possesses a useful command repertoire, a scripting language, and the ability to control external electromechanical devices. Work is continuing to extend CIP so that it will have approximately the same specification as PIP, including a Prolog top-level controller. For a variety of technical reasons, it is not appropriate to try to match PIP and CIP exactly.

Figure 4.26 shows a system block diagram explaining the internal organization of CIP. Notice that the CIP software and an image library are held on a central server. It can be down-loaded and run in "stand-alone" mode as a Java application, or interactively as an applet. In both modes, CIP offers several important advan-

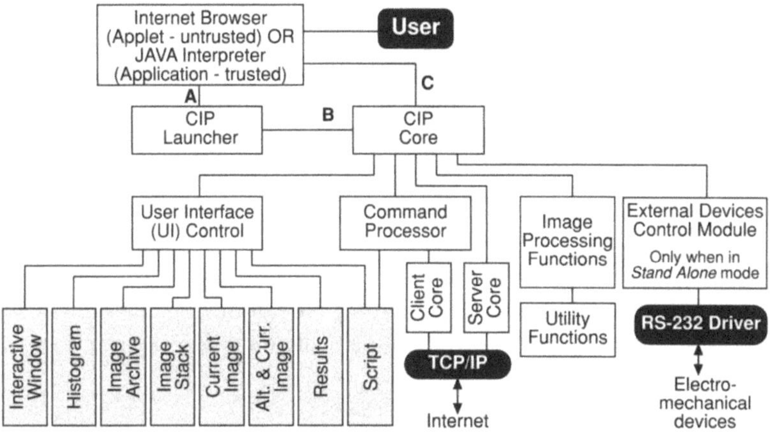

Fig. 4.26 CIP system architecture.

tages over a system such as PIP, which is based on a "stand-alone" computer and a fixed operating system:

- Exactly the same software can be used by anyone who has been granted the appropriate access rights. That person can be working anywhere in the world, provided there is an appropriate network connection available.
- Since the CIP software is written in Java, it is platform-independent and can run on any of the major operating systems, including Windows 98, Windows NT, UNIX, and Macintosh OS.
- It is easy for its authors to keep the CIP software and its associated documentation up to date, since there is only one copy.
- It is easy to extend CIP using a well-defined protocol for writing user libraries.
- Given the appropriate WWW-based Machine Vision tools, engineers should be able to avoid visiting customer sites altogether, or at least reduce the number and duration of such visits. CIP is just one of several tools of this type under development with this objective in mind [BAT99].
- A customer can be granted access rights to exactly the same software tools as the vision equipment supplier uses. This enables them to work closely together on such tasks as collecting data samples and running system evaluation experiments. There are also considerable benefits to be obtained in teaching the customer the concepts and language of Machine Vision technology. Such developments can do much to build customer confidence. Of course, it is not necessary for the customer to be granted access rights to all parts of the CIP software; a specially prepared training package, based on a "cut down" version of CIP, might be appropriate for distribution to customers.
- The CIP software is very well suited to education and training, since the tutor and his students can run exactly the same software, using whatever computers they have available. Students can be given a Java application, allowing them to continue to run CIP after they disconnect their computers from the WWW

when they go home. Even students of moderate ability can write software modules to extend CIP.

We will return to some of these points later. Before that, we will describe what a person experiences when he uses CIP and how it can be used in practice.

CIP is still under active development and thus any detailed description will soon become out of date. Let it suffice for the present to say that CIP has been integrated with a program called *JIFIC* (Java Interface to the Flexible Inspection Cell) that is able to perform on-line control of the Flexible Inspection Cell via the World Wide Web (Figure 4.27).

Together, CIP and JIFIC form part of an integrated operating environment called ROPE (Remotely Operated Prototyping Environment) [BAT99]. The importance of this is that the ROPE software should allow a vision engineer to set up an

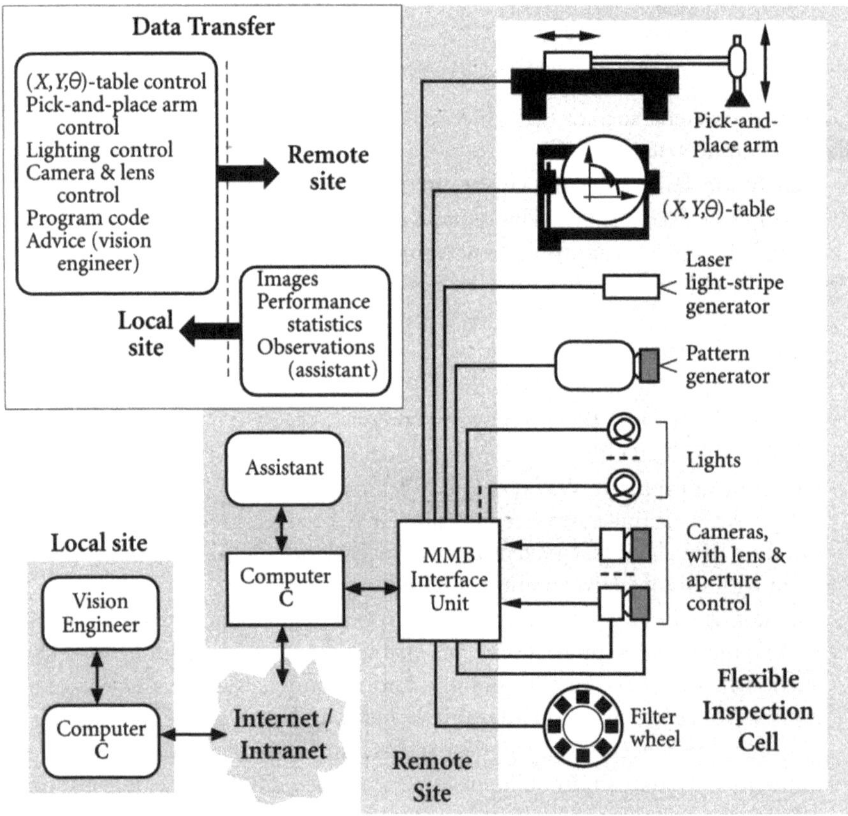

Fig. 4.27 Organization and data flow in the Remotely Operated Prototyping Environment (ROPE). The FIC currently incorporates 5 internal cameras, scene camera, 10 lights, laser light-stripe generator, pattern generator, (X,Y,θ)-table and pick-and-place arm. The present design of the MMB allows us to add numerous extra lights (TTL control, external power switches), multi-position filter wheel, control for the cameras (RS232), control of lenses (zoom, focus and aperture), pan-and-tilt positioning of the cameras, extra stepper motors, solenoids, relays, air/fluid valves, as well as numerous TTL signal inputs.

image-acquisition system at a remote site, without ever visiting it. There are many issues still to be resolved in this development, including inter-personal communication between the vision engineer and an intelligent assistant working at the factory site and who has no special knowledge of Machine Vision technology.

4.7 Target (factory floor) vision systems

It may seem odd at first that a system that is meant to operate autonomously should require that we give detailed consideration to its interface with human beings. However, even a factory-floor inspection system that is supposedly fully automatic must be able to communicate effectively, at some time or other, with the shop-floor personnel who work alongside it. A minimum requirement for a target vision system is that it must be able to reassure the shop-floor workers that it is operating properly. Its main function may simply be to operate an *accept/reject* mechanism. In this case, the human–machine interface(HMI) might consist of a simple visual display panel which summarizes the statistical history of the target machine's recent performance. However, it is *always* good practice to extend the scope of this interface, so that the user can monitor the system's health and recent performance in some detail. In addition, the vision system should be able to guide the user through the relevant set-up and system calibration procedures.

The reason that an image display is needed for a target system is less obvious but very compelling. Shop-floor personnel are likely to feel less threatened and hence be much more willing to accept a vision system if they feel that they "understand" it. A series of images showing the various steps in a processing sequence provides an excellent way to teach people what a vision system does. (For example, people do not need to know anything about the computational implementation of a filter in order to learn that its effect is to blur a picture or emphasizes edges.) An image display also provides an excellent diagnostic tool, as a means of detecting equipment malfunction. Since people are very sensitive to changes in pictures, they can quickly identify a picture that is in some sense anomalous, either among a set of supposedly similar images or in a time-sequence. An experienced vision engineer can quickly identify faults in inspection hardware or software if he can observe it operating in a step-by-step manner through a complex algorithm. The reason is that the human eye can very quickly detect even very subtle changes in a picture; an image display is to a vision engineer what an oscilloscope is to an electronics engineer!

A factory-floor inspection system must be synchronized to and be able to control some kind of electromechanical actuator. As we have already indicated, this may be quite simple, such as stopping/starting a conveyor belt and operating a binary *accept/reject* gate. However, it may require much more sophisticated control of, for example, a complex multi-axis robot. In addition, it is very likely that the target vision system will be required to control its camera(s), lens(es), and lamps within a closed feedback loop. This is very frequently necessary, because a target system is expected to operate autonomously for most of its working life, often in a hostile environment. In short, a wide variety of peripheral devices may be attached to a target vision system. Some of them may operate at high speed and require real-time control. A target system's prime requirement is that it should be fast enough to keep up with the production processes that it is meant to monitor or control. The use of standard machine–machine interfaces is clearly very important, to make connecting new devices to the vision system as simple and as straightforward as possible. It is not possible to describe these interfaces in detail here. It is simply important to note that vision systems cannot justifiably claim to require any special I/O standard. To ensure compatibility with other industrial devices, the available I/O standards must be made to suffice in almost all applications.

We have devoted many pages to the needs of the development system, often pointing out that this simply reflects the requirements for the TVS. Hence, we will not repeat the lessons again here.

4.8 Concluding remarks

Machine Vision is a multi-disciplinary subject, involving several different branches of engineering. To be successful, a vision system must be *well balanced*, by giving proper consideration to all of its component sub-systems. Failure to do this will almost inevitably result in a system that is much less reliable and effective than it should be. The central theme running through this chapter is the emphasis that is required on the systems aspects of Machine Vision. Indeed, it is this which most clearly distinguishes Machine Vision from Computer Vision. Of critical importance for the design of a good vision system are the communication links to both human beings and other machines. Obviously cost, high processing speed, reliable operation, and the ability to perform a wide range of image-processing functions are also very important. These remarks apply with equal force to both interactive and target systems. The ability to control external devices and act intelligently are also essential to both. To judge a vision system properly, we must take into account all of these features. There is no single parameter that can alone represent the overall effectiveness of an industrial vision system. On no account should cost alone *ever* be used in this way. In the same way, we must never measure "accuracy" by a single number. There is an unfortunate tendency in modern society to try to describe all systems, however complex, and even human institutions, by a single (scalar) quantity. When judging vision systems, this approach is utter folly. Unfortunately, all too many vision systems are justified (and sold) on the basis of cost alone. In this chapter, we have tried to convince the reader that there are many other factors to be considered, both in commissioning and designing a vision system.

5 Algorithms and architectures for fast execution

"Will you walk a little faster?" said a whiting to a snail. "There's a porpoise close behind us, and he's treading on my tail."
Alice's Adventures in Wonderland
Lewis Carroll

The primary purpose of this chapter is to discuss image-processing implementations on desktop computers (PCs) and special-purpose hardware. The preliminary discussions apply to both kinds of systems. Although most readers will be more familiar with the former, they are asked to keep both possibilities in mind.

Sometimes the speed of image-processing operations is not of primary importance; often, it is more important that the system be inexpensive. For interactive algorithm development, computing times of 1–2 seconds for each basic function are not objectionable. However, execution times in excess of ten seconds can be very annoying and hamper user concentration. A complete inspection algorithm will typically consist of a sequence of many operations, perhaps as many as 50–100. Clearly, when we are conducting exhaustive statistical performance tests of an algorithm (probably just before transferring it to the TVS), the overall execution speed of the IVS is especially important.

For many inspection tasks, much higher speeds are needed to keep pace with modern production processes. (For this reason, the speed requirements for a TVS are almost always considerably more demanding than for an IVS.) In these situations, the choice of computing platform and the method of implementation are critical. Repeated past experience has shown that there are always inspection applications for which the fastest economically-justifiable systems are not fast enough. As faster systems come on the market, new areas of application become economically feasible, but other more difficult applications remain tantalizingly out of reach. Therefore, software implementations or hardware architectures that facilitate faster image-processing computations are a worthwhile topic for study. In this chapter, a wide variety of options for achieving high execution speed are examined. As so often happens, software and hardware issues are interdependent. Furthermore, implementation details cannot always be considered separately from the underlying algorithmic processes that we are trying to perform. Thus, there is a degree of overlap between the material presented here and Chapter 3.

5.1 Classification of operations

Image-processing operations can be classified in various ways, based on size, type, and resolution of the images, as well as the nature of the algorithms. In this chapter, image-processing operations are classified according to two criteria:

(a)Monadic vs. dyadic images
(b)Pixel-by-pixel vs. neighborhood vs. global operators.

As may reasonably be expected, operations within a given category usually require broadly similar implementation approaches.

5.2 Implementation of monadic pixel-by-pixel operations

By definition, a monadic operation has one input image. A pixel-by-pixel (or point-by-point) operation is one in which the value of each output pixel depends only on the value of the correspondingly-positioned pixel of the input image, or a pixel at a fixed offset. Here are some typical examples (see Chapter 2). $A = \{a(i,j)\}$ and $B = \{b(i,j)\}$ are input images, while $C = \{c(i,j)\}$ is an output image. Appropriate normalization constants are included. These operations are performed on every pixel.

- Negate (*neg*):
  $c(i,j) \Leftarrow 255 - a(i,j)$
- Absolute value (*abs*):
  $c(i,j) \Leftarrow \text{maximum}(a(i,j), 255 - a(i,j))$
- Logarithm (*log*):
  if $a(i,j) = 0$ then $c(i,j) \Leftarrow 0$; else $c(i,j) \Leftarrow \text{integer}(255 \times (\log(a(i,j)/2.4065)))$
- Antilogarithm (*exp*):
  $c(i,j) \Leftarrow \text{integer}(\exp(a(i,j)/45.9)$

All such operations are very easily and quickly accomplished on virtually any kind of system. One option is to take each input pixel in turn, and carry out the computation embodied in the definition of the operation. Alternatively, one can fetch the output value from a lookup table (LUT) containing precomputed values for every possible input value. This approach is almost invariably faster than direct computation of the result. A LUT with m address bits (2^m addresses) and n output bits will be denoted as an $[m,n]$-LUT in the discussion to follow.

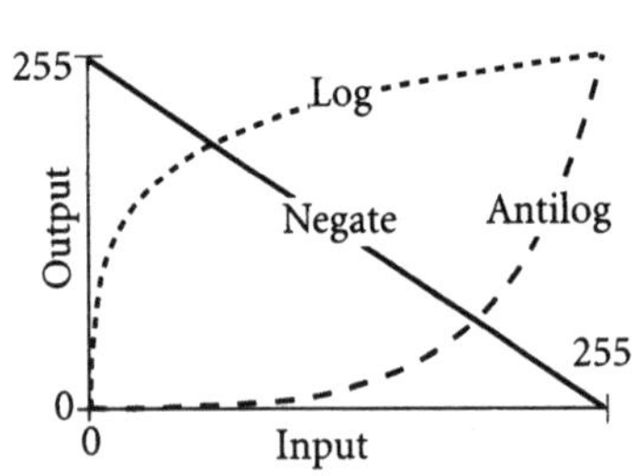

Fig. 5.1 Graphical LUT representation

For an 8-bit input grey scale, the required [8,8]-LUTs are tiny. For 10-bit and 12-bit inputs, the required [10,8] and [12,8]-LUTs are still of manageable size. Beyond this, one would likely use the LUT approach only when the computational formula is complex. LUTs are also useful for color recognition and the derivation of the HSI coordinates from the RGB values and vice versa. Three examples of LUTs are given in graphical form in Figure 5.1. The LUT implementation will almost certainly be much faster than direct computation for the nonlinear logarithm and antilogarithm functions. For a simple function, such as negation, the difference will not be as dramatic.

5.3 Implementation of dyadic pixel-by-pixel operations

By definition, a dyadic operation has two input images. Most of the common dyadic pixel-by-pixel operations involve simple arithmetic operations (add, subtract, multiply) or logical operations (AND, OR, Exclusive OR) which correspond to operations built into the CPU. For these, the speed is limited more by the time required to fetch pixel values from two images and write the result to a third image than by the actual computation time. This is especially true for systems in which the processor clock is much faster than the data bus clock. Therefore, there is not much that can be done algorithmically to speed up these operations.

At the other extreme are more complicated 2-operand functions, such as those needed to analyze the Fourier transform (Section 2.6.2). This generates images whose elements are complex numbers. These might conveniently be represented using two standard monochrome (i.e., grey-scale) images, holding the real and imaginary components, $a(i,j)$ and $b(i,j)$, respectively. Let us suppose that these images are to be transformed into another image pair, $c(i,j)$ and $d(i,j)$, respectively, which holds the magnitude and phase angle. Then, we would need to apply the following transformations:

Magnitude: $c(i,j) \Leftarrow \text{integer}(\sqrt{(a(i,j)^2 + b(i,j)^2)})$
Angle: $d(i,j) \Leftarrow \text{integer}(128 + (255/\pi)\times\arctan(b(i,j)/a(i,j))$

These functions require a significant amount of computation, which must be repeated for every pixel pair in the input images, and therefore is likely to be slow. Two [16,8]-LUTs, each with 65 536 8-bit cells, would allow these outputs to be computed very quickly. These large LUTs are still small compared with the vast amount of RAM provided in today's computers, so this approach is practical, even in this extreme case. For a further near doubling of the speed, the two LUTs could be merged into one [16,16]-LUT, with the magnitude $c(i,j)$ occupying the high byte and the angle $d(i,j)$ the low byte of the LUT output. In this case, only a single LUT access would be required.

As an intermediate example, consider the "divide" operation, defined as

$$c(i,j) \Leftarrow \text{integer}(0.5 + \text{sat}_{255}(16\times b(i,j)/a(i,j)))$$

where the function $\text{sat}_{255}(X)$ is equal to X if $X \leq 255$ and 255 otherwise. The computations here are fairly straightforward. Of course, one [16,8]-LUT could be used, as in the previous example. However, there is also a way to use two [8,8]-LUTs instead. First, reformulate the problem as

$$c(i,j) \Leftarrow \text{integer}(0.5 + \text{antilog}(\log(16) + \log(b(i,j)) - \log(a(i,j)))$$

One [8,8]-LUT is used twice to compute the two logarithms. These are subtracted and the result used as the address for a [9,8]-LUT, which implements the antilog and the other parts of the calculation. For most systems, this will be faster than a direct computation. Increased accuracy for small input values (where the logarithm function has a steep slope) could be obtained, at no sacrifice in speed, by using an [8,10]-LUT (log) and an [11,8]-LUT (antilog). Notice that one [16,8]-LUT is 128 times as large as two [8,8]-LUTs combined.

There is another approach that is applicable in some cases: partitioning the problem. For example, we may partition the data into two parts, containing most-significant bits ("high bits") and least-significant bits ("low bits"). This approach will be illustrated by applying it to an operation for which it is probably not needed in practice: ordinary pixel-by-pixel multiplication. Represent the two 8-bit inputs as

$a(i,j) = 16 \times a_{hi}(i,j) + a_{lo}(i,j)$ and
$b(i,j) = 16 \times b_{hi}(i,j) + b_{lo}(i,j)$

where $a_{hi}(i,j)$, $a_{lo}(i,j)$, $b_{hi}(i,j)$, and $b_{lo}(i,j)$ are all 4-bit numbers. Then, the product, $a(i,j) \times b(i,j)$, is exactly equal to

$$256 \times a_{hi}(i,j) \times b_{hi}(i,j) + 16 \times (a_{hi}(i,j) \times b_{lo}(i,j) + a_{lo}(i,j) \times b_{hi}(i,j)) + a_{lo}(i,j) \times b_{lo}(i,j).$$

Each of the 4-bit products can be obtained in turn from the same [8,8]-LUT, and the remaining steps can be achieved by shifting and adding. If only the high 8 bits of the product are wanted, then the last product term is omitted, the first term is not multiplied by 256, and the middle term is rounded to four bits. Notice the similarity of this decomposition to the "quarter-squares" method presented in Chapter 4.

A more meaningful example of bitwise partitioning will now be presented, showing that this approach can be extended to longer word lengths in a straightforward way. Consider some very complex function of two variables which is reasonably smooth. (There are no step or slope discontinuities, and slopes are limited to ±7.) Otherwise, the function is perfectly arbitrary. We wish to calculate the value of the function with 8-bit precision. As a rather simple example of a function meeting these conditions, consider the following transformation:

$$c(i,j) \Leftarrow \text{integer}(127.5 + 127.5 \times \text{cosine}(\pi \times (a(i,j) - b(i,j))))/255)$$

To implement this using LUTs, first divide input a(i,j) into two parts: $a_{hi}(i,j)$ (6 bits) and $a_{lo}(i,j)$ (2 bits). Similarly for input $b(i,j)$. Then, concatenate $a_{hi}(i,j)$, and $b_{hi}(i,j)$ to form a 12-bit address, which is applied to a [12,8]-LUT (called LUT_1). LUT_1 performs a transformation that can be expressed in the following way:

$$c_1(i,j) \Leftarrow LUT_1(a_{hi}(i,j) \oplus b_{hi}(i,j))$$

where $\oplus$ denotes concatenation of two bit strings. The 8-bit result, $c_1(i,j)$, is exact, but only at points on a grid spaced 4 units apart (i.e., $a(i,j) = 0, 4, 8, \ldots$ and $b(i,j) = 0, 4, 8, \ldots$). A second [12,8]-LUT (LUT_2) is used to estimate the *slope vector* $[s_i(i,j), s_j(i,j)]$ of the desired function at the point $[a_{hi}(i,j), b_{hi}(i,j)]$ with 4-bit precision for each vector component. Here is the transformation performed by LUT_2 expressed in symbolic form:

$$[s_i(i,j) \oplus s_j(i,j)] \Leftarrow LUT_2(a_{hi}(i,j) \oplus b_{hi}(i,j))$$

(Alternatively, LUT_1 and LUT_2 can be the high and low bytes, respectively, of a single [12,16]-LUT.) A third [12,8]-LUT (called LUT_3) calculates a correction value $c_2(i,j)$. Here is the transformation performed by LUT_3 expressed in symbolic form:

$$c_2(i,j) \Leftarrow \mathrm{LUT}_3(a_{lo}(i,j) \oplus b_{lo}(i,j) \oplus s_i(i,j) \oplus s_j(i,j))$$

LUT_3, in effect, multiplies the values $a_{lo}(i,j)$ and $b_{lo}(i,j)$ by the corresponding slopes $s_i(i,j)$ and $s_j(i,j)$ to provide the interpolating correction $c_2(i,j)$, which is added to the value $c_1(i,j)$ to give the final result. This will, in most cases, give an answer very close to or identical to the "exact" answer, rounded to 8-bits. Note that a [12,8]-LUT (4096 bytes) is one sixteenth the size of the [16,8]-LUT (65 536 bytes), which could be used instead to compute the overall result in one step.

These examples illustrate one other point: Speeding up a particular operation often depends on the use of "tricks" peculiar to that operation and are therefore not generalizable. However, the use of an approximate-value LUT and an interpolation LUT, as in the last example, is a good approach in most cases for which a good "trick" cannot be found. We emphasize, however, that the smoothness conditions, while not particularly restrictive, must be met.

5.4 Implementation of monadic neighborhood operations

A neighborhood operation is one in which the value of each output pixel is a function of the values of the pixels in the neighborhood of the corresponding pixel of the input image. The neighborhood may be of any size less than the overall image size, and may or may not be symmetrically placed with respect to the output pixel. Examples include binary erosion, Laplacian, and 3×3 average. The implementation of neighborhood operations present much greater difficulties than pixel-by-pixel operations. Consider a very simple neighborhood operation: the average grey level of the pixels in the 9×9 neighborhood centered on each input image pixel. A "brute-force" solution must generate 81 input pixel addresses, fetch 81 pixels from the input image, add them all together, divide by 81 (with rounding to the nearest integer), generate the output pixel address, and write the result to the output image. This must be repeated for each input image pixel (perhaps excluding a border 4 pixels wide around the edges of the image). Compare this to a simple monadic operation such as negation, which must generate one input pixel address, fetch one pixel from the input image, subtract the result from a constant, generate the output pixel address, and write the result to the output image. In this example, the neighborhood operation takes more than 81 times as long as a simple monadic operator.

Large-neighborhood operations are even slower. Since large-neighborhood operations are usually very slow on standard serial processors, small neighborhoods are used wherever possible. Most of the neighborhood operations in common use are defined on the smallest symmetrical neighborhood, 3×3 pixels. Examples include most edge detectors, gradient operators, and filters. But there are many operations, such as *Gaussian blur*[1] and binary or grey-scale morphology, for which large neighborhoods are absolutely essential, because the operating regions must be

1. This is the name given to a lowpass (blurring) filter in which the weights approximately follow values given by the expression $[k_0 \exp(-k_x(x-x_0)^2 - k_y(y-y_0)^2)]$.

comparable in size to features to be detected in the image. In these cases, an intermediate strategy is often used when the type of operation allows it: approximate the large-neighborhood operation with a sequence of small-neighborhood operations. These may be either identical or different. Large-neighborhood operations for which this can be done exactly will be said to be *sequentially separable.* (If all the small-neighborhood operations are identical, this is equivalent to the *repeated operations* procedure described in Chapter 3.) One operation which can be separated in this way is the neighborhood maximum (or minimum). A *3×3* maximum operator (PIP operators *lnb* and *snb*) requires 28 operations (10 addresses, 9 pixel "reads," 8 comparisons, and one pixel "write"). Applying this operation four times in sequence (112 operations) yields the maximum on a *9×9* neighborhood. Compare this with the "brute force" single-pass approach to *9×9*-pixel maximization, which needs 244 operations (82 addresses, 81 pixel "reads," 80 comparisons, and one pixel "write"). In this example, the sequentially-separated (repeated-operation) approach is more than twice as fast, but it is still very slow compared with typical monadic operations. Thus, the sequentially-separated approach helps only a little. Another very useful approach takes advantage of row–column separability. For those cases in which this kind of separation can be achieved, the overall neighborhood operation is obtained by first applying a row (or column) operation and then applying a column (or row) operation to the result. As we saw in Chapter 3, this kind of separation is possible for certain linear convolutions and some other operations. For example, any linear convolution function which uses a 2-axis symmetrical weight matrix can be decomposed into a column operation and a row operation (in either order). The separated form clearly has fewer operations, but two problems remain: even separated operations can be fairly slow; furthermore, many very useful operations cannot be separated in this simple way. Also note that, in some cases, both row–column separability and sequential separability can be applied to the same problem. One noted example is the Gaussian blur, in which a large Gaussian row operation (say *1×15*) can be implemented by repeated application of a small row operator (seven applications of a *1×3* operation). The Gaussian column operation can be decomposed similarly.

A so-called *serpentine memory* can be implemented in either software or special-purpose hardware and can be used in some cases to increase the speed of neighborhood operations. This applies only to situations in which the input image pixels are accessed in a raster-scan sequence, or some other regular row-by-row or column-by-column manner. Such a memory structure presents all of the pixels of the current neighborhood to the processor simultaneously, eliminating the need for pixel-address generation and CPU-mediated memory access. Hardware implementation is straightforward. For the *3×3* case shown in Figure 5.2, two delay lines (FIFOs) plus nine extra 1-pixel stores (simple SR registers) provide all nine neighborhood pixel values simultaneously. For software implementations, it would be very time-consuming to shift hundreds of pixel values for each pixel processed. Therefore, a *barrel buffer* is implemented, and only a set of pointers is incremented. Fixed offsets are programmed into the software, so that the address-generation steps are simplified. Furthermore, if the buffer is in fast memory and the image in slower memory, the extra time for slow repeated access to the same image pixels is eliminated. However, a serpentine memory does not eliminate any of the other computation steps, and it does require some overhead to maintain the serpentine buffer. Hence, the speed increase is generally moderate, leaving the basic problem of slow neighborhood operations unsolved. All of these difficulties with neighborhood operations help to explain why special-purpose image-processing hardware has been so important in the past. Indeed, it is still widely used in speed-critical applications. For the same reason, hardware accelerator boards are frequently added to desktop computers that are used for image processing.

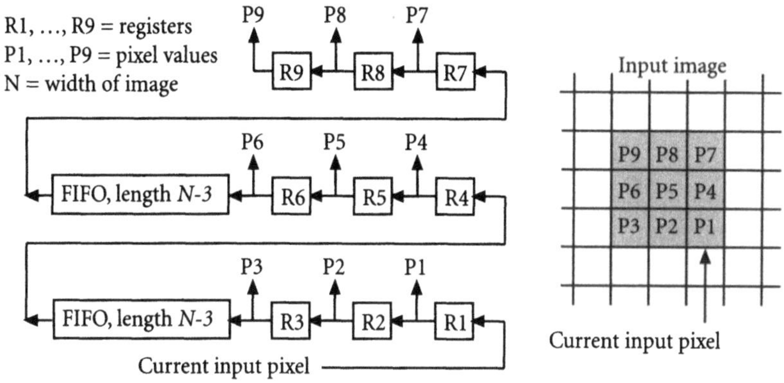

Fig. 5.2 Schematic diagram of a 3×3 serpentine memory

5.5 Implementation of monadic global operations

A "global" operation is one in which the value of each output pixel may depend on all of the pixels in the input image. The *discrete Fourier transform* (DFT) is one of the most widely used example of a true global image-processing operation. Not much can be done to speed up its execution, beyond that which has already been done in the so-called *fast Fourier transform* (FFT) implementation. The FFT is anything but fast when measured against the needs of industrial inspection. (Even this slow execution would be impossible if it were not for the fact that the DFT is row–column separable.) The DFT has a highly-developed theory and many attractive features. However, its slow execution is the main reason why spatial-domain filters (local operators, N-tuple filters, morphological operators) are used instead of frequency-domain filters in speed-critical Machine Vision applications.

There is another class of operations, which we will refer to as "pseudo-global" operations, in which the output pixel values depend on all of the input values in a complete row or column of the input image. Examples include run-length encoding on rows or columns [*rlc*], row or column sum [*rin, cin*], and row or column maximum and minimum [*rox, cox*]. Most of these can be done relatively quickly. The SKIPSM approach, described below and also in Chapter 3, is particularly good at executing these operators rapidly. The same operations on the diagonals of an image are also easily done with SKIPSM.

Certain global operations are input dependent and therefore cannot be performed exactly using only local neighborhood information. The "blob fill" operator [*blb*] is one example and performs the following operations on a binary image: Each pixel that is surrounded by a closed 8-connected curve of white pixels is set to white. Pixels that cannot be so enclosed become black. Pixels that were white in the input image remain white.

If we cannot see a contour in its entirety, we cannot always identify which pixels are inside and which are outside. Figure 5.3(a) shows a *9×9* pixel region from which we cannot resolve this ambiguity. The white contours here form part of a larger figure, as yet unspecified. Figures 5.3(b) and (c) show two possible examples. The crosshatching indicates "internal" pixels that are set to white by this procedure. Notice that "internal" pixels in one case are "external" in the other.

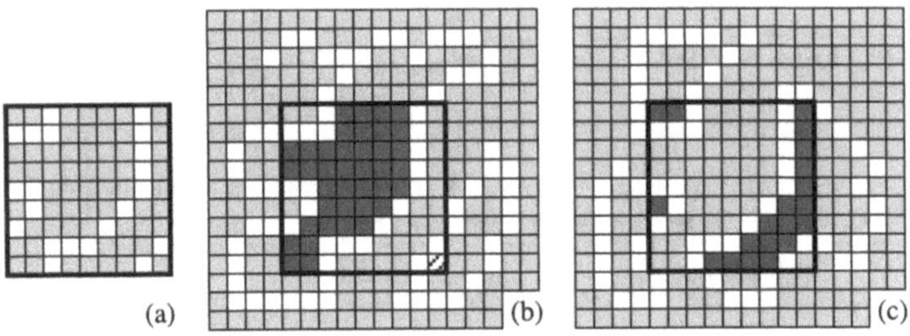

Fig. 5.3 (a) Pattern within *9×9* square. (b),(c) The same *9×9* pattern, with 2 possible surrounding patterns. Here, "inside" cannot be distinguished from "outside" using only the *9×9* region.

One approach to resolving this ambiguity is to carry out a CPU-directed traversal of the contour pixels, which requires random access to the image. Then, the blobs can be filled in various ways. One of the fastest is to use the CPU to assign special coded values to the contour pixels, and then use the SKIPSM approach, or some other raster-scan algorithm, to fill the areas based on the encoded values.

Another example of an "input-dependent" operation is the classical connected-component labelling problem [*ndo*]. As with "blob fill," no neighborhood less than the whole image suffices for all cases. Figure 5.4 shows a variation of Figure 5.3 (c) in which changing only two pixels (shown in black) changes it from a three-blob image into a one-blob image, assuming 8-connectivity. This inherent *structural* property makes this one of the more difficult binary image-analysis problems. Very little can be done to make it really fast, except to invest in expensive special-purpose hardware. However, a combined raster-scanned/random-access approach can provide some improvement: the (faster) raster-scanned part of the algorithm assigns temporary labels and identifies the relatively few pixels at which a merging of blobs could take place. The (slower) random-access part of the algorithm then processes a small list of possible blob connections, and reassigns labels as needed. The list of global or data-dependent operations could be extended, but these examples should suffice to indicate some of the inherent difficulties.

(a) 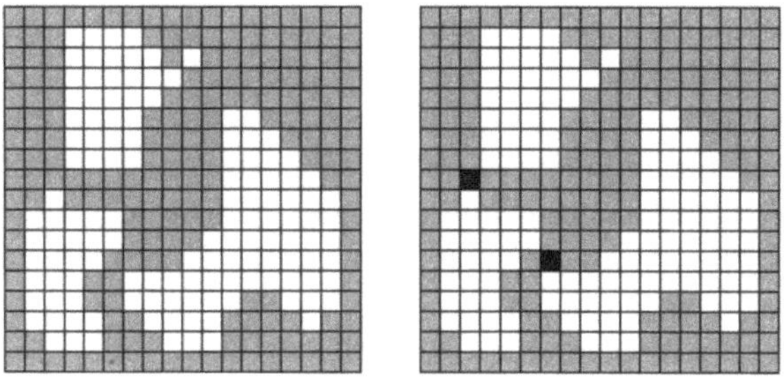 (b)

Fig. 5.4 Connectivity everywhere can be affected by a few pixels (e.g., the two black pixels here).

5.6 SKIPSM – a powerful implementation paradigm

In a series of papers starting in 1994, Waltz [WAL94a through WAL99b] described a powerful and radically new method of implementing many important neighborhood image-processing operations. These techniques, taken together, have been given the acronym SKIPSM (*Separated-Kernel Image Processing using finite-State Machines*). Although SKIPSM has not yet been used to carry out any operations that cannot also be done conventionally, the improvements in performance (either of speed or neighborhood size or both) are so great, in some cases, as to demand a radical rethinking of what is or is not practical. For example, binary erosion with large neighborhood (e.g., *51×51* pixels), can be carried out faster on ordinary desktop computers than these same computers can do a simple *3×*3 binary erosion (9 pixels) using conventional programming. Notice that the ratio of the areas of these neighborhoods, or *structuring elements* (SEs), as they are normally called, is 289:1. Speed improvements or neighborhood-size increases of over 100:1 have also been achieved for many other image-processing operations. The image-processing categories to which SKIPSM can be applied in software and/or special-purpose image-processing hardware implementations include but are not limited to the following:

- Binary erosion and dilation with large, arbitrary SEs. Structuring elements larger than *25×25*, some containing lakes ("holes"), and other nonconvex shapes can be applied in a single image pass, in software or hardware.
- Multiple SEs can be applied simultaneously in a single pass [WAL94a, 94b, 94c, 96a, 97b, HAC97a]. For example, six or more stages of the "grass-fire transform" [*gft*] have been carried out in one pass [WAL94c].
- Binary opening and closing in a single pass [WAL96b].
- Binary template matching with large, arbitrary templates [WAL94d].
- "Fuzzy" binary template matching [WAL94d].
- Binary correlation [WAL95a].
- Binary connected-component analysis, with assistance from a random-access processor [WAL99a].
- Grey-scale morphology [WAL94e, MIL97].
- Grey-scale template matching [WAL94e].

- Large-kernel Gaussian blurs, difference-of-Gaussian filters, etc. [WAL98e].
- Many standard linear neighborhood operations: edge enhancement, local averaging, horizontal, vertical, and diagonal gradients [WAL95b].
- Many standard nonlinear neighborhood operations: maximum, minimum, maximum of row minimums, Robert's edge detector, median and other ranked filters, largest gradient, direction of largest gradient, direction of brightest or darkest neighbor, pixels critical for connectivity, all possible functions defined on 3x3 binary neighborhoods, etc. [WAL95b, 98a, 98b].
- One-pass binary skeletonization [*ske*] [HUJ95a, 95b].
- Binary run-length encoding on rows [*rlc*], columns, or diagonals [WAL94f].
- Grey-scale run-length encoding on rows, columns, or diagonals [WAL94f].
- Generation of grey-level co-occurrence matrices for texture analysis [WAL99b].
- Certain operations previously thought to be impossible in pipeline systems, such as blob fill [*blb*] and "patterned blob fill" [WAL94f].
- Various "smearing" operations, including row, column, and diagonal summations and averages [*rin, cin*].
- Hough transforms [*huf*] [WAL94f].
- One-pass generation of certain standard images (e.g., grey-level wedges). Also, random and arbitrary image textures, either covering the whole image or masked arbitrarily. [WAL94f].
- Grey-scale morphology and other operations on color images. [WAL98d].
- Three- and higher-dimensional binary morphology. [WAL97a].

5.6.1 SKIPSM fundamentals

As mentioned in Chapter 3, SKIPSM is based on four key ideas:

1. The separation of two-dimensional (or *N*-dimensional) neighborhood operations into two (or *N*) one-dimensional operations, typically a *row operation* and a *column operation.*
2. The reformulation of these operations in a recursive manner.
3. The implementation of these recursive operations as finite-state machines, typically called the *row machine* and the *column machine.*
4. The automated generation of the finite-state machine configuration data.

Note that the separation of multi-dimensional neighborhood operations into multiple one-dimensional operations does not involve separability in the usual sense, as it is defined above for linear convolutions. All multi-dimensional operations, linear or non-linear, can be separated in a trivial way, roughly analogous to a serpentine memory. (This involves having the row machine simply encode the pattern of row pixel values in the current row into a single integer, and then passing this integer to the column machine. This requires the column machine to do the whole job of "remembering" the neighborhood pixel pattern and computing the overall result.) While always possible, this approach is in some cases not practical because rather large integers may be required to encode all the column-machine neighborhood information. The real power of SKIPSM becomes apparent when a certain condition, called the *compressibility condition* [WAL94a, WAL94b] is met, because this allows compact representations of the neighborhood pixel information. The compressibility condition is satisfied by most linear operations and some non-linear operations as well. In the latter category, we can include neighborhood minimum/maximum, but not the neighborhood median.

Finite-state machines (FSMs) are not machines in the usual sense; they are abstract autonomous input-driven sequential entities and are known to have some important analytical and computational properties. Their study is well-developed and forms a branch of mathematics, called *automata theory.* Hardware implementations of FSMs using flip-flops, multiplexing switches, or ASICs are widely used for such things as sequencers, timers, and computer disk I/O drivers. The FSM has an internal memory recording its state, which represents its "history." The input and state taken together are sufficient to calculate the output and next state. When it receives each new input, the FSM calculates its output and updates its state to assimilate the new information while retaining a sufficient record of the past. Thus, an FSM is a recursive device driven by a sequence of inputs and which embodies *rules* for determining its output and updating its state. It is usually represented visually as a directed graph and for computer or electronic hardware as a state-transition table, held in RAM or ROM. The most important step in defining an FSM is choosing the state in such a way that only the *necessary* information is retained. We have written C-language programs to create SKIPSM state transition tables automatically from a structuring element defined by the user. The details are covered in various papers [WAL94a, 94b, 96a, 97b] but are omitted here because they are a distraction from our main argument.

5.6.2 SKIPSM example 1

Two examples, one very simple and the other considerably more complex, will be used to illustrate the principles of SKIPSM design and use. The first example involves binary erosion using a diamond-shaped structuring element, contained within a 5×5 square (Figure 5.5). Much larger and more complicated SEs are handled just as easily by the SKIPSM approach, but these would be inappropriate as examples for the purposes of illustrating general principles. Furthermore, software programs [WAL94b, 96a, 97b] have been written to carry out the actual design steps in such a way that users do not need to understand all the complexities of the SKIPSM method.

Step 1 – Separation

In this case, there are three different kinds of row patterns to be detected by the row machine. (Figure 5.5) An "all-white" input pixel pattern is assigned a row machine output value of 3. An input pattern with three white pixels in the center and *either but not both* of the end pixels black is assigned a value of 2. An input pattern with one white pixel in the center and *either but not both* of the adjacent pixels black is assigned a value of 1. All other combinations of pixels are lumped together as the "null" pattern, and assigned a value of zero.

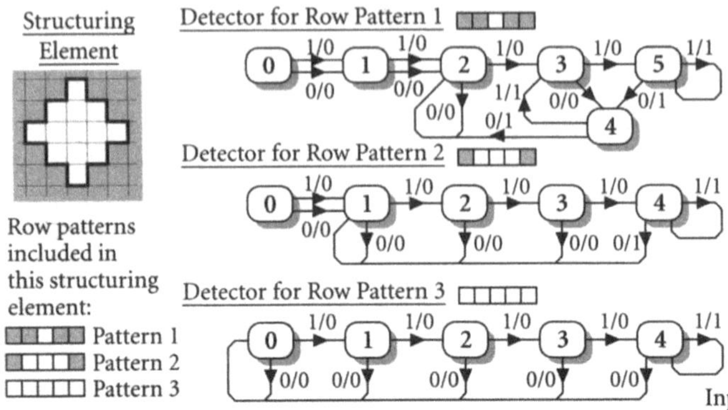

State Transition Table for Row Machine 2

In	CS	NS	Out
0	0	1	0
1	0	1	0
0	1	1	0
1	1	2	0
0	2	1	0
1	2	3	0
0	3	1	0
1	3	4	0
0	4	1	0
1	4	4	1

Fig. 5.5 A diamond-shaped structuring element, plus 3 row-machine state-transition diagrams and one of the three state-transition tables. Notation: CS = Current State, NS = Next State.

Step 2 – Defining state-transition tables for individual row machines

The overall row machine must be able to test for all three row patterns of interest simultaneously, as shown in Figure 5.5. As a preliminary to creating the single overall row machine, the FSM synthesizer program was used three times to create state transition tables for the three individual row patterns. One of the tables is given in Figure 5.5, which also shows the state transition diagrams for the three individual row patterns. These diagrams are simply conceptual aids for the reader, and are represented in the software by the corresponding tables.

Step 3 — Generation of the overall row machine configuration data

The row machine has only one output, which must indicate which, if any, of the three row patterns matches the five most-recent image pixels in the current row. Since we are concerned here with binary morphology, white pixels in the SE *must* be matched by white pixels in the image. However, black ("don't care") SE pixels match either white or black image pixels, by definition. Because of the "don't care" pixels, more than one of the SE row patterns can be matched simultaneously by the current set of 5 image pixels. The overall row machine provides an output of 0 (no match), 1 (Pattern 1 only), 2 (Patterns 1 and 2), or 3 (all three patterns).

A second program was used to combine these three row-pattern detectors into one FSM, whose state transition diagram and table are shown in Figure 5.6. A very important point about SKIPSM is illustrated here. From Figure 5.5, it is evident that the individual row-pattern detectors have 6, 5, and 5 states, while Figure 5.6 shows only 13. Thus, we have already achieved compression in two ways. Each machine could have as many as sixteen states (4 pixels to be remembered results in $2^4 = 16$ combinations). The combined row machine exhibits further compression. If the three row machines were independent, there would be 150 (= $6 \times 5 \times 5$) combinations of individual states, each one of which would be a state of the combined machine, thus requiring 8 bits. But they are not independent; there are only 13 states, and a 4-bit state representation suffices. This is the kind of data compression usually achieved in SKIPSM binary morphology applications. In fact, for larger neighborhoods, even greater compression is often achieved. This compression is the key to the power and wide applicability of SKIPSM.

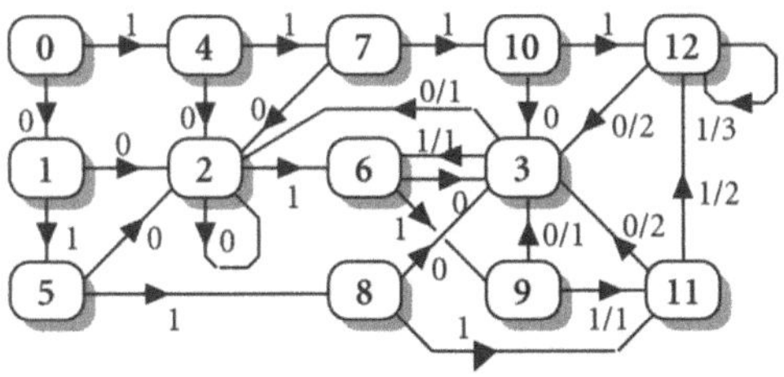

(a) Row machine state-transition diagram

In	CS	NS	Out	In	CS	NS	Out
0	0	1	0	0	6	3	0
1	0	4	0	1	6	9	0
0	1	2	0	0	7	2	0
1	1	5	0	1	7	10	0
0	2	2	0	0	8	3	0
1	2	6	0	1	8	11	0
0	3	2	1	0	9	3	1
1	3	6	1	1	9	11	1
0	4	2	0	0	10	3	0
1	4	7	0	1	10	12	0
0	5	2	0	0	11	3	2
1	5	8	0	1	11	12	2
				0	12	3	2
				1	12	12	3

(b) Transition table

Fig. 5.6 The combined row machine for the "diamond" example. State transitions not shown explicitly return to the zero state. Outputs not shown explicitly are zero.

Step 4 — Defining the state-transition table for the column machine

A third program was used to generate the column machine. Its state transition diagram and table are shown in Figure 5.7. It has only six states, which can be represented in 3 bits. Compare this with what would be required using a conventional serpentine memory: all 12 pixels (or 20 pixels, if "don't care" pixels are included) in the top four rows of the structuring element must be remembered. These are independent, so this corresponds to $2^{12} = 4096$ (or $2^{20} = 1\ 048\ 576$) states, and requires a 12 bit (or 20 bit) representation. Thus, the column machine by itself has achieved a data compression ratio of *4096/6 = 682:1* (or *1 048 575/6 = 174 762:1*).

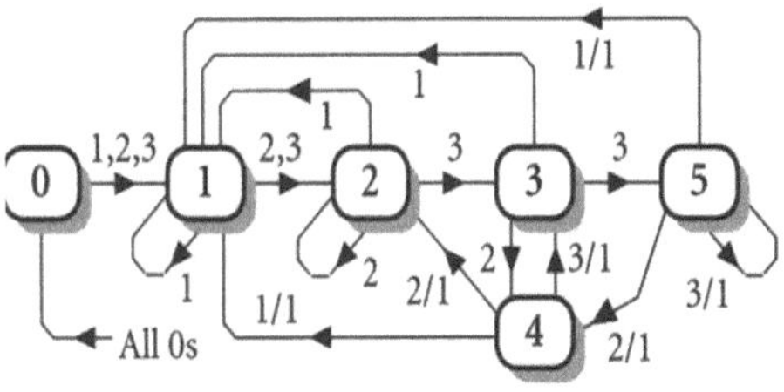

(a) Row machine state-transition diagram

In	CS	NS	Out	In	CS	NS	Out
0	0	0	0	0	3	0	0
1	0	1	0	1	3	1	0
2	0	1	0	2	3	4	0
3	0	1	0	3	3	5	0
0	1	0	0	0	4	0	0
1	1	1	0	1	4	1	1
2	1	2	0	2	4	2	1
3	1	2	0	3	4	3	1
0	2	0	0	0	5	0	0
1	2	1	0	1	5	1	1
2	2	2	0	2	5	4	1
3	2	3	0	3	5	5	1

(b) Transition table

Fig. 5.7 The column machine for the "diamond" example. All state transitions not shown explicitly correspond to zero inputs, and all of these return to the zero state.

There is also a program which merges many column machines (each performing a different operation) into one overall column machine, subject to the easily-satisfied condition that they share the same row machine. In this way, many different operations can be performed simultaneously, with *no increase at all* in execution time. A good example of this is the SKIPSM implementation of the grass-fire transform described below as Example 2 and in [WAL94c]. This applies six stages of erosion per pass, each pass requiring the same execution time as a single 3×3 SKIPSM erosion pass, which is itself very fast. The only significant limiting factor here is the combined size of the resulting column- and row-machine lookup tables. These can become rather large if incompatible operations are combined. The details of these limitations are given in the papers cited earlier.

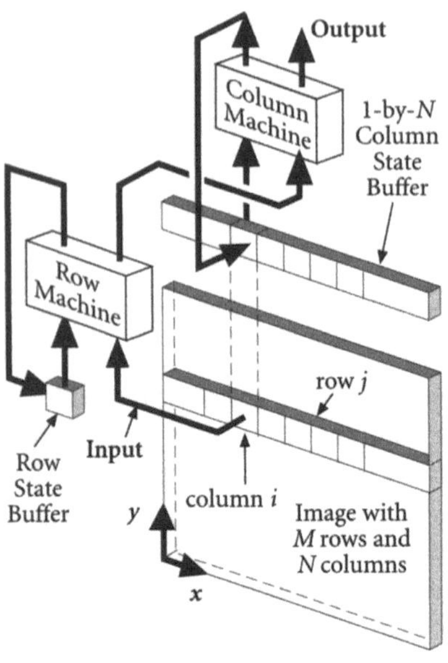

Fig. 5.8 SKIPSM flow chart.

One point remains to be explained: Where do the row machine and column machine retain their state information? Figure 5.8 shows a software implementation of the two machines operating on a raster-scanned image. One register (i.e., one memory location) is used to store the row state and one row containing *P* registers (*P* memory locations) is used to store the column state.

The pixel in row *j*, column *i* of the input image, labelled *Input* in Figure 5.8, is fed to the row machine. The row machine uses this pixel and the current contents of the *Row State Buffer* to calculate the next row machine state, which is loaded back into the *Row State Buffer*, overwriting its previous contents. The row machine also produces an output, which is fed to the column machine. The column machine uses this output and the current contents of column i of the *Column State Buffer* to calculate the next column machine state, which is loaded back into column *i* of the *Column State Buffer*. The column machine also produces the final output, labelled *Output*. This process is then repeated for the next pixel in row *j*, and so on. For pipeline hardware implementations, a delay line takes the place of the column state buffer.

5.6.3 SKIPSM example 2

Next, we will consider the grass-fire transform [*gft*], referred to in Sections 2.3 and 3.4.1. Diverse uses for this operation are described in this book. However, in many cases, and particularly in industrial applications, the slow operation of the algorithm is a problem: one pass through the image is required for each layer that is "burned away." For large blobs, this could require 50 or even more passes through the image. One of the earliest applications of SKIPSM [WAL94c] showed that six stages of "burning" could be done in one pass, thus producing a six-to-one speed-up. This is accomplished by applying six "circular" binary-erosion structuring elements simultaneously, as shown in Figure 5.9.

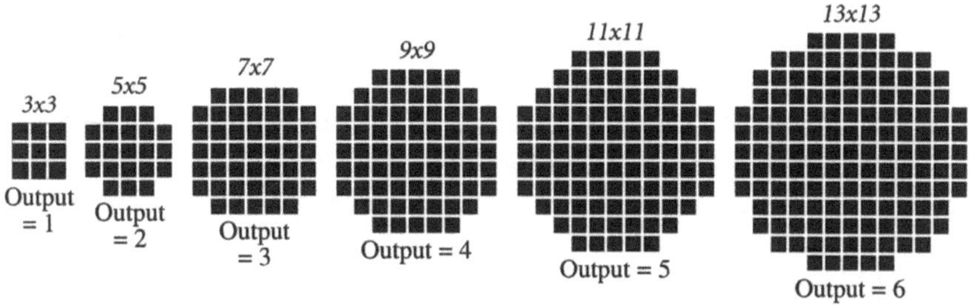

Fig. 5.9 The six "circular" SKIPSM SEs applied simultaneously in one image pass to perform six stages of grass-fire-transform erosion. When none of these "fit," the output is set to zero.

Because this operation generally requires more than one pass, even with the 6-to-1 speed-up provided by SKIPSM, the overall process must be iterative. Figure 5.10 shows the six-stage erosion operation embedded in a larger configuration which provides for progressive accumulation of the transform. On any given pass, the SKIPSM operator receives as its input a binary image containing only the binary

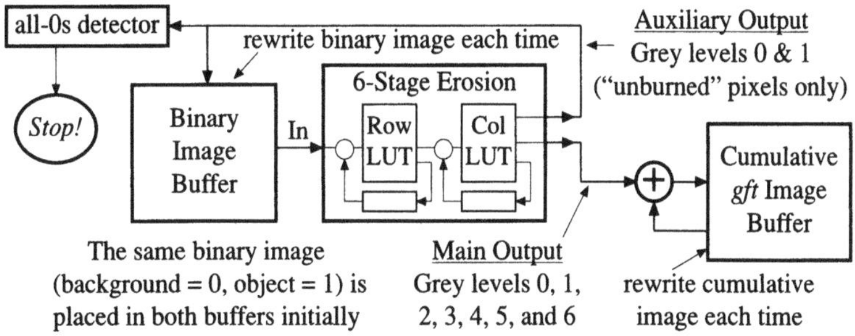

Fig. 5.10 Using a six-stage SKIPSM erosion operator recursively to create the grass-fire transform of a binary image. At the end of a frame in which the *ones detector* "sees" only black pixels, the process terminates and the cumulative buffer contains the *gft* of the original image.

pixels which are as yet "unburned." These are "burned" again, and the output is set to 0, 1, 2, ..., 6, according to these rules: If the largest SE fits, set the output to 6. If the largest SE does not fit, but the next largest SE does, set the output to 5. And so on. If not even the smallest SE fits, set the output to 0. (Providing multiple outputs in this fashion is standard practice for the SKIPSM LUT-creation software.) These values are added to the cumulative image, so that the "unburned plateaux" in the centers of the blobs can be processed in a subsequent pass. The cumulative image is combined with this new result. An auxiliary output with values 0 (already burned) and 1 (as yet unburned) is also applied as a mask to the binary image. This result then serves as the input for the next pass. The process ends when all the auxiliary output values are zero.

The choice of six stages of "burning" in the 1994 implementation [WAL94c] was determined by the limited amount of RAM memory available for lookup tables on machines of that era. Current machines have much more memory, so that eight or more *gft* stages can now be carried out in one pass.

5.6.4 Additional comments on SKIPSM implementations

The example above involved a morphological operation on binary images, for reasons of simplicity of explanation. It is important to notice that SKIPSM has been applied with excellent results to various grey-scale operations. For those grey-scale operations for which speed comparisons have been carried out, such as the *3×3* median filter, morphology, and Gaussian blurs of various sizes, the SKIPSM implementations have been shown to be significantly faster than the best conventional implementations. The details are given elsewhere [HAC97a, MIL97a, WAL98c].

SKIPSM implementations were first developed and implemented for binary image processing on special-purpose pipeline image-processing hardware. (See below.) The results can be summarized as follows: For those operations to which SKIPSM could be applied, execution times were 50 or more times faster than the identical hardware could do when programmed in the conventional way. For example, a SKIPSM implementation of a doughnut-shaped *27×27* binary morphology structuring element applied to a *512×485* image was performed one pass (*1/30* second). Using the identical set of hardware modules programmed conventionally required 81 passes (2.7 seconds). Recent developments in SKIPSM-based implementations have concentrated on developing software for standard desk-top computers, with particular emphasis on grey-scale operations. Although the speedups achieved here are not as spectacular as those demonstrated for binary processing, they are nonetheless significant and well worth pursuing.

5.7 Image-processing architectures

To be useful in other than well-chosen "niche" applications, a target vision system must be capable of performing a very diverse set of image-processing and feature-extraction operations. For real-time or near-real-time applications, it must also be fast. Most of the systems actually being used for such applications include either or both of the following:

- One or more fast microprocessors, with random access to all or a defined subset of the pixels in the image. In the discussion below, these are referred to as "software-based" systems. They use fast microprocessor chips: DSP chips, RISC chips, CISC chips with added signal-processing features, or Transputers®. In general, such systems can be programmed to perform a wide range of tasks in addition to image processing.
- Pipeline image-processing systems, which operate on a sequence of image pixels, usually in raster-scan order. In view of the highly parallel nature of their processing functions, these often perform these tasks much faster than software-based systems with comparable clock rates. However, such systems have limited capabilities for computations other than image processing. In the discussion below, these are referred to as "hardware-based" systems.

5.8 Systems with random access to image memory

The common factor in all of the systems described in this section is that images are mapped to the memory space of the CPU, so that it has programmable access to the individual image pixels. Pipeline systems do not provide this kind of access; one must wait until the appropriate moment in the raster scan to access a particular pixel.

5.8.1 Systems based on personal computers or workstations

The simplest Machine Vision systems consist of a computer, often an inexpensive personal computer, with a "frame grabber," either an external module or a card installed in one of the computer's slots, as in Figure 5.11. In the past, when computer memory size was limited and clock rates were slow compared with the

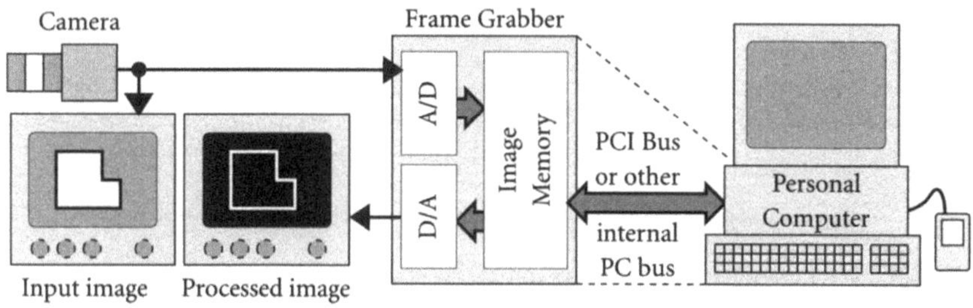

Fig. 5.11 Personal computer with frame-grabber card installed.

pixel rates of typical cameras, there was no choice but to build computer-based systems this way, because the computer could not keep up with the incoming pixel flow. Computer clock rates are now typically faster than the pixel stream from a video camera. Nevertheless, there are still very good reasons to use this approach: it frees the CPU to do other tasks, such as image processing, and eliminates the burdensome task of interfacing with the camera. If reasonable overall speed is to be achieved, more than just a fast CPU is required. The image should be memory-mapped, so that it effectively lies within the CPU's RAM space, enabling the latter to access it quickly and easily. However, even with fast microprocessors and memory mapping, pixel access can still be a bottleneck. This results from the fact that improvements in memory access times and bus speeds have not kept pace with those in CPU clock rates. At the time of writing (2000), it is common to find 60 nsec (16.7 MHz) RAM chips being used with 300 MHz microprocessors. This results in many CPU "wait states" for each "pixel fetch" operation. Some chip manufacturers offer some form of vector processing, a technique borrowed from parallel computing machines. It is normal to fetch a set of eight or sixteen pixels at one time, allowing the fast CPU to process these while the next set is being fetched. This is highly effective when the pixels needed are located sequentially in the image buffer, as they are in pipeline algorithms. However, if truly random access is needed, this merely adds to the computational overhead, without producing any speed increase at all. Another way to mitigate the problems caused by the image access bottleneck is to perform limited arithmetic capabilities (image addition, image averaging, grey-level mapping via lookup tables, or certain small-kernel convolutions) in the frame grabber. According to the classification scheme given above, this is a hardware solution. Even here, most of the image processing, feature extraction, decision making, external device control, and user interface processing must still be done by the computer's CPU. Therefore, this usually provides only moderate speed increases.

Another configuration, which addresses this data transfer bottleneck, is shown in Figure 5.12. So-called "multi-media" desktop machines have at least some moderately-fast memory, often called "video RAM". This allows the separate frame-

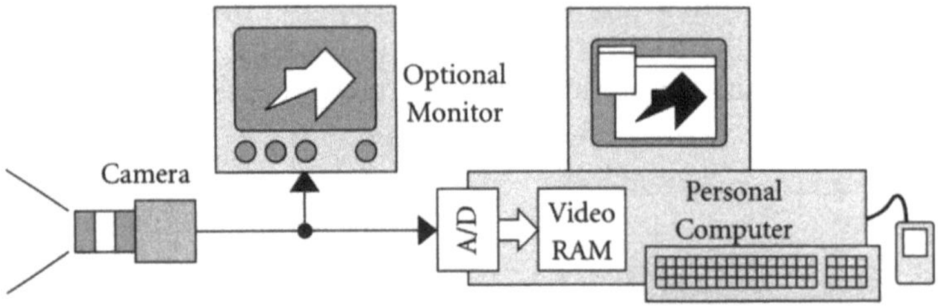

Fig. 5.12 A personal computer using built-in video RAM as a frame store

grabber image memory to be eliminated, by loading the input image directly into the fast RAM, accessible by the CPU. Since this avoids using the slower main RAM, the image access time is reduced. Furthermore, the main CPU does not have to bother about the image display, because the contents of the video RAM are automatically displayed on the computer screen, using a separate processor. Many of the features of multi-media computers are precisely those needed for Machine Vision. As the speed and power of desk-top computers continues to increase and multi-media capabilities become standard, Machine Vision systems based on such computers will, no doubt, be applied with increasing frequency to many significant inspection problems. However, these machines are still far short of what they could be. The ideal architecture for image processing on a conventional von Neumann computer would include a very large "back-side" cache, connected directly to the internal bus of the CPU and operating at CPU clock speed. Ideally, part of this cache would be free of the control exercised by the cache manager, so that it could be pre-loaded with user-specified data, such as lookup tables, reference images, etc. Such an architecture would allow even faster execution of LUT-based operations, including many of the SKIPSM operations. However, it appears unlikely that such capabilities will be incorporated into standard PCs, on the grounds that the demand is not large enough. An alternative would be to develop affordable memory chips capable of running at the CPU clock rate (i.e., trade storage capacity for operating speed), but this does not appear likely to happen either. Within the foreseeable future, it seems that speed increases will depend primarily on increased processor and bus speeds.

5.8.2 Systems based on bit-slice processors

By connecting together several chips, a bit-slice processor allows random-access processors to be assembled with virtually any word length. Each chip contains all the elements of a CPU: registers, arithmetic unit, etc., plus connections to higher and lower bits. These chips are concatenated to form processing elements with word lengths appropriate to the desired tasks, which are not necessarily the same throughout the machine. Bit-slice processors are usually programmed at assembly-language level, making it possible to create very fast software, but requiring high levels of programming skill. This is a parallel processor only in the sense that many chips are working on the data at the same time. The algorithms they implement are similar to those of conventional CPUs, although this kind of machine is intermediate between conventional random-access CPU implementations and true parallel machines. One successful product of this type is the ruggedized Intelligent Camera from Image Industries, Ltd. (now Cognex, Inc.). Figure 5.13 shows a cut-away view of an early version of this device. Inspection algorithms are created interactively on a desk-top computer (PC), with easy-to-use proprietary software. Many such algorithms can be downloaded to the system and saved in its internal non-volatile memory. During normal operation, the PC can be disconnected and the system run autonomously. The PC can be reconnected at any time to monitor the performance statistics, which the system collects automatically. When power is re-applied, the system starts automatically and reconfigures itself, as instructed by the user during programming. The programmer can control various external devices (solenoid drivers, lamp controllers, motor controllers, robots, etc.) via the I/O ports. (Both parallel and serial ports are provided.) These ports can be used to control the program, for example by adjusting parameters needed during its execution, or to choose one of a set of pre-defined inspection algorithms to be executed. Configurations with built-in camera, multiple remote cameras, grey-scale processing, and color recognition have been developed. Later models use multiple processors to improve program execution times.

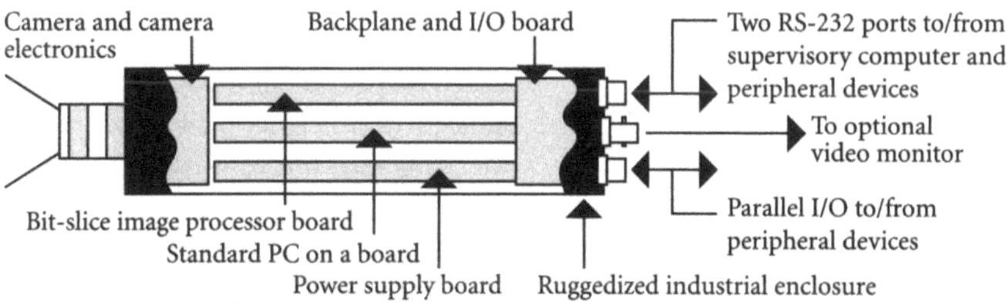

Fig. 5.13 A ruggedized self-contained vision system using a bit-slice processor.

5.8.3 Systems based on multiple random-access processors

When the available circuits are simply not fast enough for the task at hand, the only remaining solution is to use parallel processing, so that more than one chip operates on the image at the same time. Before this can happen, the problem must somehow be reformulated to separate the problem into parts ("parallelized"). Each part of the algorithm is then assigned to a separate processor. Parallel computers have been available from the very early days of machine-assisted computing: The wheels and gears of early mechanical calculators were, in effect, performing parallel computing. Analog computers are inherently parallel, which accounts, in part, for their speed advantage over digital computers. Some of the earliest large electronic computers were parallel machines. Numerous academic papers have been written about parallel computing, and numerous technical conferences are devoted to the subject every year. *The problem of programming them to operate efficiently remains!* Very little of this work on multi-processors has manifested itself in the industrial inspection context, except for array and vector processors and the widely-used pipeline systems, discussed below. The use of multiple CPUs in lower-priced (i.e., mass-market) desk-top computers is still in its infancy. The main reason for this is the difficulty in programming such devices using conventional step-by-step programming languages, even those having parallel-processing extensions. Except for the Transputer® and related devices, which are designed from the start to implement in hardware the software structures of the OCCAM parallel-processing language, the languages do not map well onto the chip architectures. The central issue, then, is parallelization: dividing up the overall task into manageable pieces.

One approach uses what could be called *sequential partitioning*: The processing elements are connected in sequence, called a "pipeline." All the pixels of the image "stream" through each element in turn. Each processing element does one sub-task, passes its result to the next processing element, and repeats its operation for all the pixels of the image. All of the processing elements are busy almost all of the time, resulting in very high performance. When measured in terms of operations per second, pipeline processors often outstrip even the fastest supercomputers. Pipeline systems are discussed in some detail in a later section.

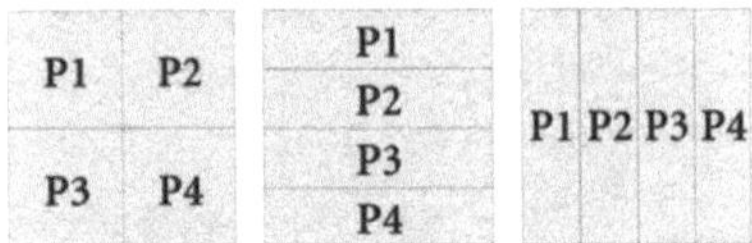

Fig. 5.14 Three partitioning schemes.

Another obvious parallelization technique, called *spatial partitioning*, divides the *image* in some way and assigns each part of the image to a different processor, which carries out all the steps of the overall algorithm. Figure 5.14 shows three of the many ways one could do this with four processors. Most of the difficulties associated with spatial partitioning arise when it is necessary to pass information between processors. Any processing, other than a monadic or dyadic pixel-by-pixel operations, requires that we pass some data from each processor to other processors concerned with adjoining areas of the image, and vice versa. The larger the processing neighborhood, the worse the problem becomes, because more information must be transferred between processors. This kind of overhead transaction can slow down the process significantly, and also greatly complicates the programming. Figure 5.14(a) has the additional disadvantage that data must flow in two orthogonal directions (up/down and left/right), to accommodate horizontal and vertical neighborhoods. This adds to both the overhead and programming difficulty. The partitioning shown in Figure 5.14(b) has been used to good effect in a parallelization of the connected-component labelling problem [*ndo*] (FLO86). This was chosen over the other possibilities because it minimizes the amount of information passed between adjacent areas. Many systems using image partitioning have been proposed and a few have even been sold as commercial products. An early system used DSP chips and partitioned the image as shown in Figure 5.14(c). Another system, using Transputers®, was designed for fingerprint analysis and identification. It has a distributed fingerprint database, local computing, and inter-site communications. One product from Image Inspection, Ltd. uses multiple Transputers to achieve very high processing rates.

There is yet another way to divide up the overall processing task, which could be called *temporal partitioning*, and is shown in Figure 5.15. Consider a production line in which objects moving on a conveyor belt must be inspected. Assume that the images arrive too fast for the preferred (and presumably inexpensive) standard Machine Vision system. Assume further that in all other respects this system is ideal for the task. Notice that the execution time of the inspection algorithm is longer than the elapsed time between the arrival of successive input images. To make the example concrete, suppose that the images arrive at a rate five times as fast as they can be processed by the preferred Machine Vision system. We could abandon this preferred system altogether, and instead use a fast hardware-based system, such as that described below. This would probably incur considerable cost, both for the purchase of the system and for retraining operating and maintenance personnel. Alternatively, we could instead use five identical systems of the preferred type. *System 1* receives an image (call it *Image 1*) and processes it until the analysis of that

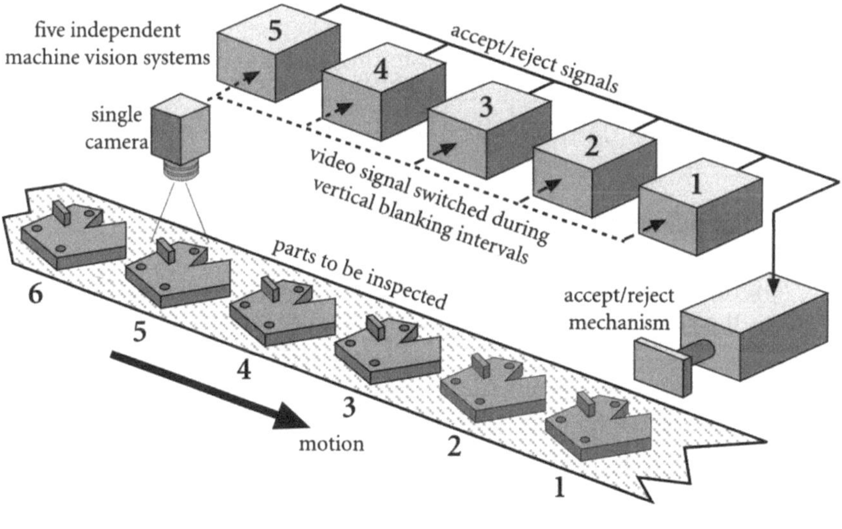

Fig. 5.15 Temporal partitioning of a task.

image is completed. During this time, four other images (*Image 2 – Image 5*) have arrived, each taken by one of the other systems. By the time *Image 6* arrives, *System 1* is available again. This process is repeated indefinitely. Since typical high-speed systems cost considerably more than standard PC-based systems, this multiple-system solution may be cheaper, just on hardware cost alone. In addition, there are savings in programming and training costs and in the reduced cost of stocking back-up systems. The redundancy provided by multiple identical systems also offers a reliability advantage: if one of the standard systems becomes inoperative, reduced-capacity inspection can be maintained, thereby avoiding a total shut-down. Clearly, when long latency cannot be tolerated, one must switch to a high-performance system. However, this intermediate approach should be considered before that step is taken.

5.9 Systems with sequential image memory access only

We now turn our attention to the pipeline systems mentioned in Section 5.2. Systems of this kind usually store intermediate results (i.e., images) in dedicated high-speed memory. It is not unusual to encounter systems that store ten or more images in dual-port memory. Furthermore, memory mapping is used so that these images lie within the extended address space of a conventional CPU, which acts as the system controller. Those operations which are within the functional capability of the dedicated hardware are then executed at very high speed, while other more complicated operations are executed in software by the CPU.

For certain operations, a pipeline system may be no faster than software-based systems. This can be a serious drawback in some applications, because typical hardware systems are rather limited in the range of operations they can perform. However, the reader will recall that in Chapter 3, we saw that some algorithms can be reformulated. Fast hardware can then be used to pre-process the data, allowing the tasks which require random access to the image (and hence which must be performed in software) to be far less demanding. The effective overall speed of a hardware–software system clearly depends on how well the algorithm can be mapped onto it.

5.9.1 Classification of operations needed for inspection

Early pipeline systems always operated on the full image, implying that each pass through the region to be processed required a full video frame-interval: 1/25 (Europe) or 1/30 second (USA and Canada). This happened even if only a small part of the image was of interest. Clearly, this is a serious drawback in some applications, since it wastes a lot of valuable time. Newer pipeline systems can be set up to operate on only a preselected rectangular region of interest (*ROI*), which may be much smaller than the complete image and hence much faster to process.

At the time of writing, pipeline processors are the principle form of high-speed hardware for the most demanding applications. In view of this, it is helpful to classify image-processing operations according to how easily they can be implemented in this way. Suppliers of pipeline systems provide different capabilities in their products. As a result, the classification scheme outlined in Table 5.1 gives an approximate but not an exact idea of what to expect in practice. Of course, *all* operations can be performed in a software-based system also, if lower speed can be accepted. Furthermore, those operations that are most easily implemented using a pipeline system usually lend themselves to efficient software implementation as well. Hence, the classification is also relevant to software implementation.

Table 5.1. Classification of image-processing operations according to the ease with which they can be performed in typical hardware-based pipeline systems.

Very easy: *single-pass monadic pixel-by-pixel operations*

Absolute value of intensity [abs]	Average intensity [avr]
Column or row integrate [cin], [rin]	Column or row maximum [cox], [rox]
Highlight a range of intensities [hil]	Negate [neg]
Obtain histogram [hpi]	Threshold [thr]

Easy: *single-pass monadic neighborhood operations*

Binary dilation (*3×3*) [dil]	Binary erosion (*3×3*) [ero]
Binary image edge [bed]	Convolutions (up to *8×8*) [con]
Isolated point removal [wrm]	Largest neighbor (*3×3*) [lnb], [snb]
Roberts edge detector [red]	Sobel edge detector [sed]

Very easy: *single-pass dyadic pixel-by-pixel operations*

Add/subtract two images [add], [sub]	Exclusive OR of two images [xor]
Logical AND of two images [and]	Logical OR of two images [ior]
Multiply/divide two images [mul], [div]	Maximum of two images [max]

Difficult: *operations requiring multiple passes due to large neighborhoods*

Binary closing and opening (> *3×3*)	Binary erosion and dilation (> *3×3*)
Convolution (> *8×8*) [*con*]	Largest neighbor (> *3×3*)

Difficult: *operations requiring multiple passes due to type of function performed*

Direction of brightest neighbor [*dbn*]	Direction of largest gradient [*dbn*]
Grass-fire transform [*gft*]	Grey-scale closing and opening
Grey-scale erosion [*snb*] and dilation	Largest gradient
Crack detector [*crk*]	Skeletonization [*ske*]

Very difficult: *operations which cannot be computed in a pipeline without CPU assistance*

Convex hull [*chu*]	Fetch value of specified pixel [*pgt*]
Freeman chain code [*fcc*]	Histogram equalization [*heq*]
Local-area contrast enhancement [*lce*]	Radius of curvature [*cnr*]

5.9.2 Pipeline Systems

Modular pipeline systems for image processing were introduced in the mid-1980s. A large family of compatible boards that could be plugged into the VME-bus were made available commercially and formed the basis for the standard approach to the

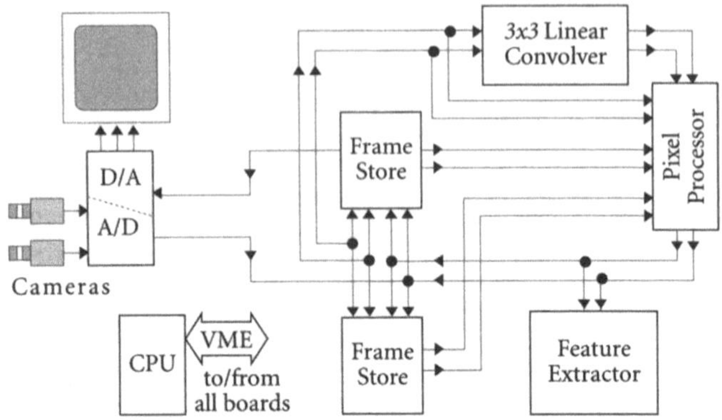

Fig. 5.16 A real-time system using seven VME boards

most demanding inspection applications. Figure 5.16 shows an example of such a system that can carry out many important monadic and dyadic pixel-by-pixel operations and 3×3 linear convolutions at high speed. However, it is relatively slow on morphological operations. Other systems could be configured to perform other types of operations at high speed. Theoretically, much larger system could be built to implement much more complicated algorithms. However there is a serious practical problem with this approach: in spite of great improvements in software support for these boards, optimized programming has always been extremely difficult and requires very great technical skill and detailed knowledge of each board. This continues to be the case. The difficulties are increased by the need to program a different architecture for each application.

In view of these programming difficulties, the industry goal then became to build a system with a fixed architecture, including all modules needed for general-purpose applications. Until about 1990, systems using this approach were very large, requiring 10–20 boards, and were very expensive. Since then considerable progress has been made to provide complete image-processing/feature-extraction systems on one board. At the time of writing, these boards provide better performance at a much lower cost than the boards they replace. The next section describes one commercial system of this type

5.9.3 Single-board system capable of many operations in one pass

The improved performance of the newer systems stems from the same factors that have improved the performance of desk-top computers: miniaturized chips, faster clocks, and larger/cheaper RAM. This has resulted in a continued redefining of the term "real-time." In the mid-1980s, a 7 MHz pixel clock, corresponding to an image with 485 rows and 384 columns in *1/30* second, was generally considered to be adequate for real-time processing. By the late 1980s, 10 MHz (485×512 pixels) was the accepted standard. The emerging standard for real-time systems in the late 1990s uses a 40 MHz pixel rate throughout, thereby processing 1024×1024-pixel images in *1/30* second.

Figure 5.17 shows a greatly simplified block diagram of a typical single-board vision system (Datacube MaxVideo 200) embodying many of these features. This has a 40 MHz pixel clock for image display and 20 MHz clock for most of the processing modules. It is a single-board system and occupies two VME-bus slots. It has literally hundreds of functional modules, many image buffers, and a very flexible system for interconnecting them. A complex series-parallel set of operations involving dozens of operations can be executed in one pass through the pipeline. Since the region(s) of interest (ROIs) being processed may be smaller than the full image, it is sometimes possible to pass data through the pipeline several times during a single frame period, thereby implementing a complex set of operations in a single video frame interval. To achieve even higher speeds, two or more of these systems can be connected together. Another version, the MaxVideo 250 board has somewhat reduced capabilities but occupies only one VME-bus slot.

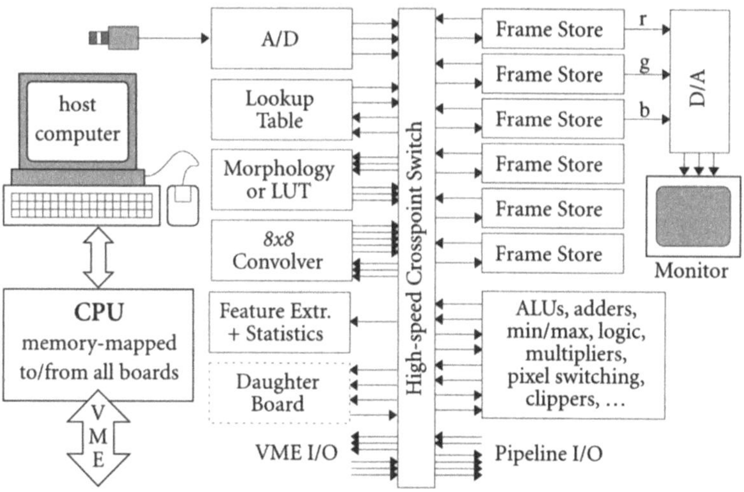

Fig. 5.17 Block diagram of a single-board real-time machine vision system.

However, the major difficulty with real-time pipeline systems remains. In spite of vastly-improved software, including programs to control system configuration, the programmer must know a very great deal about the details of the hardware before really efficient programs can be written. Graphical user interfaces have reduced the programming tedium somewhat, but have not eliminated the requirement that programmers be extremely knowledgeable about system details.

Modules written for other projects can occasionally be re-used to accomplish useful results. However, the cost of pipeline hardware cannot usually be justified, except for the most demanding applications, and it is precisely these applications for which existing modules are least likely to be useful. The shift away from specialized machines and expensive workstations, which was previously observed in low-performance software-based system (Figure 5.12), is now taking place for high-performance pipeline systems. Figure 5.18 shows a single-board image-processing system roughly comparable to that of Figure 5.17, installed in a personal computer. Such systems have been on the market for several years. As indicated above, the programming of these highly complicated systems remains a serious problem, which becomes progressively more difficult as system and problem complexity increases. The declarative programming style embodied in PIP offers one hope for lifting the heavy programming burden from the potential users of these systems, but little work has been done in this area to date.

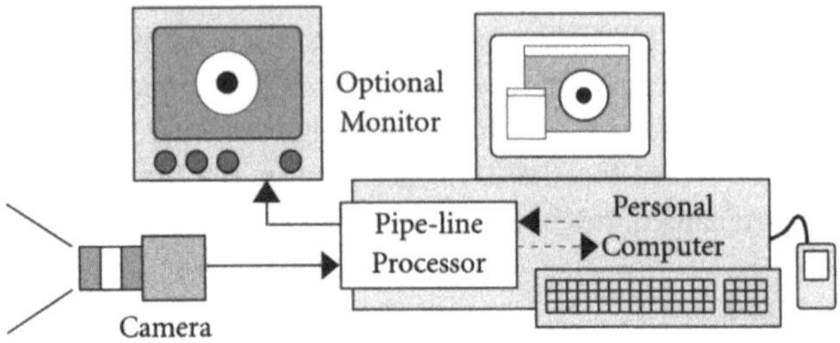

Fig. 5.18 A personal computer with a plug-in real-time pipelined processor.

5.9.4 Hardware implementations of SKIPSM

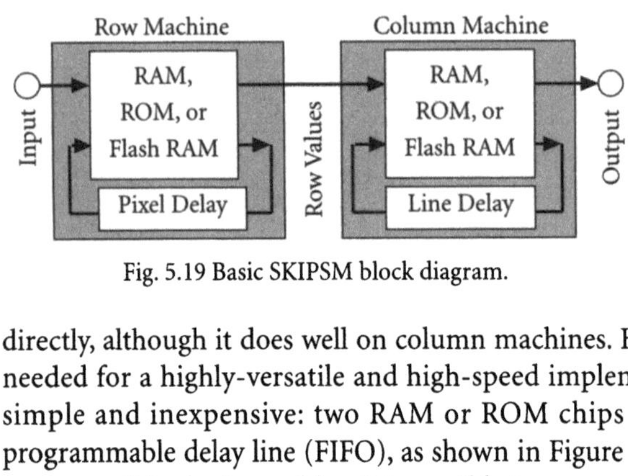

Fig. 5.19 Basic SKIPSM block diagram.

As noted above, the earliest SKIPSM implementations were in commercially-available pipeline hardware. Since this hardware introduces a processing delay (latency) of at least two pixel times, it cannot implement row machines directly, although it does well on column machines. However, the actual hardware needed for a highly-versatile and high-speed implementation of SKIPSM is very simple and inexpensive: two RAM or ROM chips (SIMMs or DIMMs) and a programmable delay line (FIFO), as shown in Figure 5.19. Merely by changing the contents of the lookup tables, it is possible to program this system to provide very fast execution of a wide range of large-neighborhood binary and grey-scale image-processing operations.

5.10 Systems for continuous web-based processing

Many mass-market products are made in the form of continuously-moving webs: paper products, lithographic plate, photographic film, sandpaper, magnetic disks and tapes, currency notes, adhesive-backed tape, etc. The production machines for these are very expensive, and profitability depends critically on running the machines as fast as possible, while maintaining the highest possible quality. Applications of this type confront us with very high data rates. In addition, because so much cost and potential profit are involved, rather expensive inspection systems can often be justified economically. The spatial partitioning approach of dividing the web into zones and assigning a separate system to each zone is sometimes effective. However, the most demanding applications require the fastest available systems, and sometimes more than one of these. It is here that high-performance pipeline systems, using high-resolution line-scan cameras, or ultra high-resolution laser scanners, currently have no competitors. The earliest pipeline hardware systems were designed around standard television timing patterns, and processed one video field or one frame at a time. This mode is inappropriate for continuously-moving webs, since about 40 lines of data are lost during each vertical blanking interval. Some newer systems dispense with video frames altogether, and use continuously-updated scrolling image buffers. Other systems achieve the same effect by using a region of interest of one scan line. The hardware manufacturers are devoting considerable effort to enhancing these products, both in the area of performance specifications and in software support. The price/performance ratio of these products will very likely continue to decrease, as it has in the past.

6 Adding intelligence

Even the vision of natural objects presents to us insurmountable difficulties.
Unknown
Oxford English Dictionary

6.1 Preliminary remarks

Writing code to implement just one of the simpler image-processing operators, such as negate (*neg/0*), threshold (*thr/2*), etc., presents no difficulties whatsoever, even to a novice programmer. However there is a large difference between writing code to implement a single isolated function and constructing a reliable software package that has a comprehensive command repertoire and a well-engineered user interface. In this chapter, we will explore some of the general issues that a programmer should bear in mind when implementing an interactive vision system for use in Machine Vision. We will not discuss the software implementation of specific image-processing functions at great length, since several excellent books containing C and Java programs for image processing, already exist [WHE00]. We will review the overall structure of several IVS packages, including:

a) VSP, a minimal but expandable system akin to PIP
b) PIP, hosted on a member of the Macintoch family of computers
c) WIP, running under Microsoft Windows
d) CIP, written in Java
e) JVision, a visual programming environment.[1]

6.1.1 Basic assumptions

There are several fundamental points which should always be borne in mind when designing or choosing an IVS for prototyping in Machine Vision.

Axiom: *In Machine Vision, there is no single image-processing operator that is so important that the implementation requirements of all others must be subjugated to its needs.*

This statement applies to any software (and/or hardware) development that purports to provide a multi-purpose system. It reflects the highly varied nature of

1. JVision was developed by Dr. Paul Whelan and his colleagues at Dublin City University, Dublin, Ireland.

image-processing procedures used in industrial Machine Vision systems. It is this variability that makes the writing of an efficient and comprehensive IVS package so challenging. Of course, if the algorithm is known beforehand, as it is when designing a target system (TVS), then the software (and/or hardware) can be optimized accordingly. Since our main concern here is the design of multi-purpose interactive systems, we must not allow ourselves to become preoccupied with optimizing the implementation of any one image-processing operator, to the possible detriment of all others. This is in marked contrast to the approach traditionally taken in both Image Processing and Computer Vision, where there is usually a concerted effort to optimize the performance and implementation of a single well-defined procedure.

We must bear this principle in mind, even when there is one algorithmic step (PIP command) that seems conceptually to be pivotal to the whole algorithm. An example of this is to be found in Section 9.7, where the operator *ctp/0* (Cartesian-to-polar coordinate transformation) is critical for finding the orientation of a metal grid. In the first program listed in Section 9.7, there are nine other PIP commands and we must not forget that these are of equal importance for the overall success of the project. Of course, most applications do not use *ctp/0* at all, so it would be folly to design an IVS in such a way that *ctp/0* alone is optimized. We must always be aware of both the temptation and danger of automatically translating conceptual importance into priorities for system design.

Assumption: *The basic operations (PIP commands) used within a set of programs for different applications must be assumed to occur in random order.*

While some combinations of commands are clearly useless (e.g. *[ctp, ptc]*, *[swi, swi]*, *[zer, raf]*, etc.), the number of potentially useful sequences is very large. (Since PIP currently has 274 different commands, the number of 10-operation sequences exceeds 2.3×10^{24}.) We cannot easily dismiss large numbers of these, since some seemingly strange combinations of commands can be surprisingly useful in practice.[2] We might, for example, apply operators that are normally used on grey-scale images to binary images instead. (The commands *lnb*, *snb*, *neg*, *rin*, *rox*, *raf*, *lpf*, *hil*, and *con* are all used in this way, occasionally.) Therefore, we must assume that, in practice, any combination of commands can occur. This has important repercussions for software implementation, since it means that the structure and range of the input and output of each command must be the same.

Axiom: *We must ensure that the image format, input range, and output range of all image-processing commands are the same.*

This principle has been adopted in each of the IVS packages from SUSIE onwards and provides the rationale for normalization, which was introduced in Chapter 2. It accounts in large part for the success that these packages have enjoyed. In an attempt to provide more facilities for the user, certain commercial image-processing packages allow different types of images to be stored and processed. This

2. Of course, no human being could possibly explore all of these possibilities. We are merely pointing out that no discernable pattern has been found that would help us to limit the input and output ranges of the image-processing operators in PIP.

can be counterproductive, because is exceedingly frustrating for a user to have a certain algorithm in mind, only to find that some component operations are prohibited. While it is often possible to perform an appropriate conversion of image type, this is distracting for the user and may introduce serious rounding errors or other information loss. This state of affairs has deliberately been avoided in PIP and its predecessors, by always using the same data structure to represent images.

Axiom: *The best data structure for representing (both grey-scale and binary) images is an M×N array of integers.*

Using the same format for all images can sometimes lead to inefficiencies, both in terms of reduced program execution speed and excess data storage. However, considered overall, a rectangular array of pixel intensity values is the most useful and effective format for representing images. While certain other formats may be more efficient as a basis for performing certain tasks (especially for processing binary images), a rectangular array is the only format for representing images that can reasonably claim to being "universal" in its range of application. This advantage greatly outweighs all negative considerations, which even when taken together, are of minor consequence by comparison. The rectangular-array representation is, of course, the most "natural" one to use for the first stages of any vision algorithm, since this is the format created by a video camera and initially contains all of the information that we can ever extract from that image.

In all of the software packages discussed below, the input and output images are always $M \times N$ arrays of 8-bit integers. (Typically, $M = N = 256$ to 1024. We do not assume that the images are square (M and N are not necessarily equal), even though some commands (e.g. *yxt* and *tur*) work properly only if they are. This permits 256 grey levels and it is customary to employ the convention that, in a grey-scale image, 0 represents black and 255 white. Some operators require a binary image and there are several conventions that we might adopt in this case (see Table 6.1). Whichever convention is chosen, it is imperative that it be applied consistently to all binary image-processing functions.

Table 6.1. Conventions for representing binary images in a 2-dimensional (grey-scale) array, whose elements can vary in the range [0,255].

Black	White	Comments
0	1	Other values have no defined effect.
0	1 – 255	
0	255	Values 1 – 254 have no defined effect.
0 – 254	255	
0 – 127	128 – 255	Most significant bit alone is used to determine the intensity.

Axiom: *The choice of algorithm or heuristic to implement each image-processing function (PIP command) should be made in the light of the overall architecture of the software, not vice versa.*

In Chapter 3, we explained that, in many cases, an image-processing function can be implemented using different algorithms, based on a variety of image formats. However, we have just suggested that a rectangular $M \times N$ array provides the best overall format for implementing a broad set of functions in a multi-purpose system. With this in mind, when considering the design of an IVS we can disregard some of the suggestions made in Chapter 3. (The points made there are nevertheless pertinent to target vision systems, where multi-function use is not contemplated but performance, speed and cost are of paramount importance.) The overall software structure of an IVS is therefore best chosen before the implementation of individual commands is studied in detail.

Axiom: *Many factors, most notably environmental protection and the machine–machine and machine–human interfaces, are just as relevant to the design of an IVS as the image-processing functions.*

These factors have a pivotal role in designing both an IVS and TVS and should therefore be given equal weight with the image-processing engine when designing an IVS. Under no circumstances should we ever allow interfacing to become a "detail" to be considered after the "real work" of designing the image-processing engine has been completed.

The best approach to designing an interactive Machine Vision system is far from obvious. Indeed, there can be no proven "optimal design," just good ones and poor ones, since there are so many intertwined factors to be considered. Let it suffice to say that 25 years ago a choice was made about the design of SUSIE, without much serious thought being given to alternative designs. SUSIE used what then seemed to be the obvious organization: the 2-image operating paradigm and 2-dimensional image arrays. Fortunately, that proved to be a good choice and, so far, no better organization has been found. PIP and its modern counterparts rely on this model also.

6.2 Implementing image-processing operators

In the limited space available here, it is not possible to provide more than a very brief introduction to the software implementation of image-processing operators, such as those embodied in PIP and its relatives. Our intention therefore is simply to demonstrate the basic ideas involved, referring the reader elsewhere for a more comprehensive review of the implementation of image-processing functions in C and Java [Chapter 8].

We present four Java routines, more properly called methods, illustrating the implementation of the commands *neg*, *cwp*, *cgr* and *con*. (The programs would be very similar in C.) At this stage, we do not explain how the images are displayed after processing. No user interface has yet been provided and there is no error checking. The image resolution is assumed to be *imageWidth* by *imageHeight* pixels. All intensity values lie in the range [0,255]. Since arrays in Java are 1-dimensional, it is necessary to calculate the array subscript (*subscript*) from the image-position indices, *i* and *j*, using the formula

$subscript \leftarrow i + j \times imageWidth$

The intensity in the current image at the point (i,j) is given by $currimage[i + j \times imageWidth]$. Similarly, the intensity in the alternate image at the same point is *altimage[i + j×imageWidth]*. (Note: In the computer code to follow, "*" indicates scalar multiplication.)

neg (negate)

The following method calculates the negative of the current image. Notice that there are two nested loops, scanning *i* (outer loop) and *j* (inner loop). There are no input or output parameters. The method *swi()* (switch images) has already been defined.

```
public void neg()
  {
  // Scan the image horizontally (i) and vertically (j).
  for (int i=0; i<imageWidth; ++i)
  for (int j=0; j<imageHeight; ++j)
  altimage[i+j*imageWidth] = 255 – currimage[i+j*imageWidth];
  // Interchange the current and alternate images
  swi();
  // Display the result. This method is not defined here.
  displayimages();
  }
```

cwp (count pixels with a given intensity value)

This method provides a slight enhancement on PIP's *cwp* operator, in that the number of pixels at any intensity level (*level*) is computed. This version of *cwp* takes either zero or one input parameters and returns the frequency (number of pixels at level level) as the value of *cwp(_)*.

```
// Case 1: no arguments. Count white points.
// Find number of pixels with intensity 255. (Equivalent to PIP's cwp/1 function.)
public int cwp()
  { return cwp(255); }

// Case 2: one argument. Find the number of pixels with intensity equal to "level"
public int cwp(int level)
  {
  int i,j;
  // Set counter "countpixels" to zero.
  int countpixels=0;
  // Scan image horizontally (i) and vertically (j), counting pixels at intensity "level."
  for (i=0;i<imageWidth;++i)
  for (j=0;j<imageHeight;++j) if (currimage[i+j*imageWidth]==level)
  countpixels++;
  // The result is returned
  return countpixels;
  }
```

cgr(centroid of all points at a given level in a grey-scale image)

This is a slightly generalized version of PIP's *cgr/2* operator, since it calculates the centroid of that region at which the intensity is at a specified level (*level*). (This region may be disjoint.) The operation is effectively equivalent to [*thr(level, level)*, *cgr(X,Y)*] in PIP. One argument is used by the following method to define the intensity-slicing level applied prior to calculating the centroid.

```
public int[ ] cgr(int level)
  {
  // cwp (already defined), calculates the number of pixels with intensity equal to "level"
  int area=cwp(level);
  int c[ ]= new int[2];
  // What to do if the area is non-zero
  if (area!=0)
    {
    // Set row and column counters to zero
    c[0]=0;
    c[1]=0;
    // Scan the image horizontally (i) and vertically (j)
    for (int i=0; i<imageWidth; i++) for (int j=0; j<imageHeight; j++)
      {
        // Check whether the pixel (i,j) has intensity equal to level
        if (currimage[i+j*imageWidth]==level)
        {
          // Increment row and column counters
          c[0]+=i;
          c[1]+=j;
        }
      }
    // Normalize counters by dividing by area
    c[0]=(int)(c[0]/area);
    c[1]=(int)(c[1]/area);
    }
  else
    {
    // What to do if there are no white pixels in the image
    cip.output.addtext("No pixels have the given intensity.\n");
    // Return "nonsense" values: (-1,-1)
    c[0]=-1;
    c[1]=-1;
    }
  // Return centroid coordinates, or (-1,-1)
  return(c);
  }
```

con (general purpose linear convolution 3×3 window)

This example implements a linear convolution operator that uses a 3×3 window. It requires nine weights: (*w1,w2,… , w9*). The normalization constants, *k1* and *k2*, are explained in Section 2.2.4 (see list item 7).

```
public void con(int w1,int w2,int w3,int w4,int w5,int w6,int w7, int w8,int w9)
  {
  float k1, k2;
  int i, j, res, res1, conResult, a, b, c, d, e, f, g, h, k;
```

```
// Sum of the absolute values of the weights (|w1| + |w 2|+... + |w9|).
res = (Math.abs(w1)+Math.abs(w2)+Math.abs(w3)+Math.abs(w4)+
Math.abs(w5)+Math.abs(w6)+Math.abs(w7)+Math.abs(w8)+Math.abs(w9));
// Sum of the weights (w1 + w 2 + ... + w9).
res1 = (w1+w2+w3+w4+w5+w6+w7+w8+w9);
// Test whether the sum of the weights is zero
if (res == 0) {k1 = 0; }
    else
    k1 = (float)(1 / (float)res);
if (res1 == 0) {k2 = (float)255/2; }
    else
    k2 = (float)(1 – ((float)k1 * (float)res1)) * (float)(255/2);
// k1 and k2 are normalization coefficients as discussed in Section 2.2.4.
// Simplify the notation for intensity values. See Figure 6.1
for (j = imageHeight – 2; j > 1; j —)
    {
    for (i = 1; i < (imageWidth – 1); i ++){
        a = currimage[(i–1)+(j–1)*imageWidth];
        b = currimage[(i–1)+j*imageWidth];
        c = currimage[(i–1)+(j+1)*imageWidth];
        d = currimage[i+(j–1)*imageWidth];
        e = currimage[i+j*imageWidth];
        f = currimage[i+(j+1)*imageWidth];
        g = currimage[(i+1)+(j–1)*imageWidth];
        h = currimage[(i+1)+j*imageWidth];
    // In Chapter 2, i was used instead of k, which is used here to avoid ambiguity
```

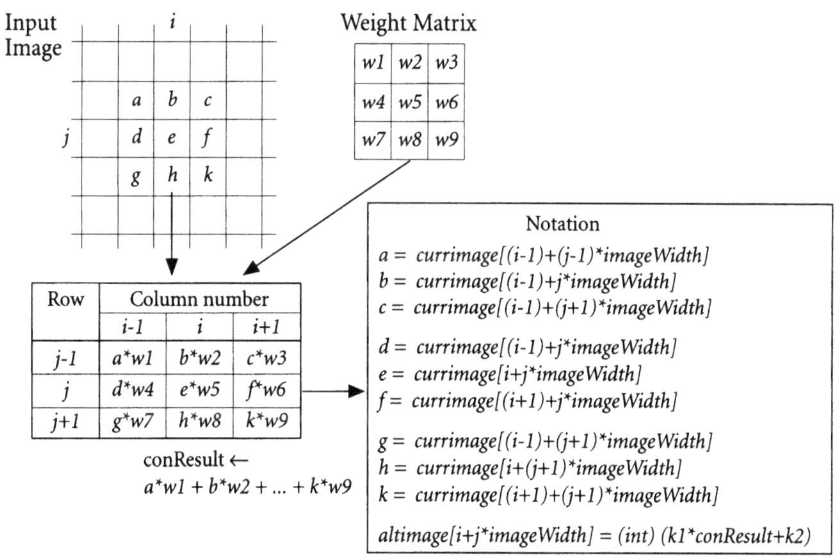

Fig. 6.1 Combining the intensity values in a 3×3 window *(a, b, ..., k)* with the weights *(w1, w2,... , w9)*, for the linear convolution operator. Notice that this is a slightly different notation from that defined in Section 2.2.

```
        k = currimage[(i+1)+(j+1)*imageWidth];
    // Now calculate the weighted sum of the pixels intensities in the 3*3 window
    conResult = ((a*w1)+(b*w2)+(c*w3)+(d*w4)+(e*w5)+(f*w6)+
            (g*w7)+(h*w8)+(k*w9));
    // Normalize the linear weighted sum so that the result is in the range [0,255]
    altimage[i+j*imageWidth] = (int) (k1 * conResult + k2);
    }
// Switch the current and alternate images
swi();
// Display the result. This method is not defined here.
displayimages();
}
```

6.3 Very Simple Prolog+ (VSP)

Prolog+ was the generic name given in the 1980s to predecessors of PIP that employ a Prolog program as the top-level controller for a separate image-processing engine operating as a slave [BAT86]. PIP is simply one of the latest implementations of Prolog+ and is the most advanced technically. VSP is a "minimal" Prolog+ system, although it is possible to extend it, using only Prolog-level programming, to provide a comparable set of facilities to those enjoyed by PIP. Figure 6.2 shows the architecture of a typical VSP system. Notice that all image-processing is performed within a hardware module that is quite distinct from the Prolog host. In the past, the VCS512 plug-in card [VCS] and Intelligent Camera [ICM] have both been used to perform the image processing in VSP systems designed at Cardiff. Much faster hardware, based on the Datacube MaxVideo [MaxV] series of image-processing boards was used in the same way, in a system developed for internal use within the 3M Company. Notice, however, that Prolog does not require high-bandwidth I/O facilities to control the image processor effectively, since speed is not a major requirement for an interactive prototyping system. It should also be noted that we never transfer whole images into or out of Prolog. Only very occasionally do we need to transfer a vector of numbers, such as the histogram or the intensity values in a whole row or column of the image. For the same reasons, Prolog's inability to operate in real time is not important. The data link between Prolog and the image-processing subsystem can be as simple as a standard RS-232 serial line, or a multi-bit parallel port if greater speed is needed. When Prolog and the image processor coexist on the same computer, data may be transferred between them via a standard inter-application communication mechanism built into the operating system. AppleEvents (Macintosh family of computers) have been used in this way in PIP. Since MacProlog (Macintosh operating system) and WinProlog [LPA] are all closely related, the programs presented in this section can be converted with minimal effort to run on machines which use the Windows operating system.

Our objective in this section is to explain how a simple but expandable "minimal" system may be implemented following the general approach outlined in Figure 6.2. It should be noted that VSP is not committed to using any one specific image processor. We must assume, however, that the image processor implements a significant proportion of PIP's basic commands.

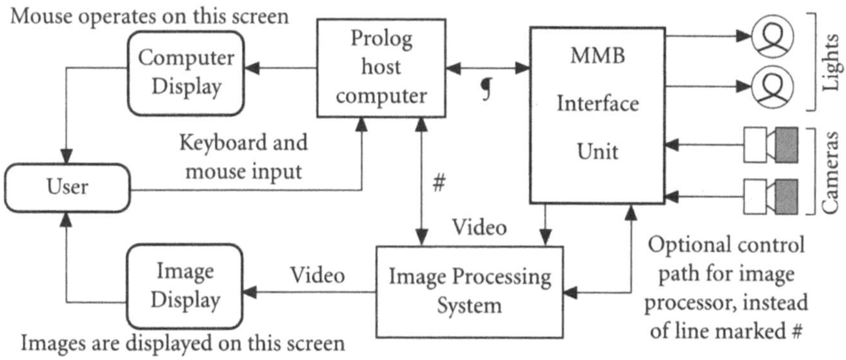

Fig. 6.2 Architecture of a VSP system.

VSP uses two operators:

a) # controls the image processor

b) ¶ controls any external hardware devices.

Figure 6.3 explains how the # operator works. For all commands except those that return parameters derived from the image, the effect of the # operator is simply

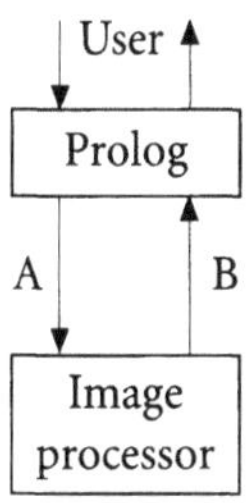

Command	Signal A	Signal B
# neg	neg	[0].
# thr(12,34)	thr(12,34) [0].	
# avg(X)	avg	[0,123].
# cgr(A,B)	cgr	[0,123,234].
X is 123, # thr(X)	thr(123) [0].	

Fig. 6.3 Flow of control and data signals for the # and ¶ operators.

to send the term following it to the image processor as an instruction to be executed. Hence, the subgoal

thr(12,34),

causes Prolog to despatch the string

"thr(12,34)"

to the image processor. The image processor then tries to execute that instruction. If it succeeds, a message is returned to Prolog, which causes the goal involving # to succeed. If the term following the # operator contains any variables, these are removed before the string is sent to the image processor. Then, when image measurement values are returned by the image processor, these variables are instantiated within the Prolog host program. If the instruction received by the image processor cannot be completed successfully, the goal beginning with # fails.

In order to understand the relationship between PIP (or Prolog+) and VSP, let us compare two programs, which check that the minimum and maximum intensity values and image contrast are all within prescribed limits.

PIP / Prolog+

```
intensity_limits:-
    repeat,          % Standard Prolog predicate. Succeeds at first, always on back-tracking
    grb,             % Digitize an image
    raf(11,11),      % Blur (lowpass filter)
    gli(X,Y),        % Calculate minimum & maximum intensities
    X > 0,           % There is no camera underload. i.e. Is Imin > 0?
    Y < 255,         % There is no camera overload. I.e. Is Imax < 255
    Y > 150+X,       % Is the contrast large enough? If not, backtrack to "repeat"
    avr(R),          % Calculate average intensity
    nl,              % Print a message for the user
    writenl('[Minimum, Average, Maximum] = ', [X,R,Y]).
```

VSP

```
intensity_limits:-
    repeat,              % Standard Prolog function
    # grb,               % Digitize an image
    # raf(11,11),        % Blur (lowpass filter)
    # gli(X,Y),          % Calculate minimum & maximum intensities
    X > 0,               % There is no camera underload
    Y < 255,             % There is no camera overload
    Y > 150+X,           % Is contrast OK?
    # avr(R).            % Calculate average intensity
    nl,                  % Print a message for the user
    writenl('[Minimum, Average, Maximum] = ', [X,R,Y]).
```

Comparing these two programs will reveal only one obvious difference: in VSP, all commands issued by Prolog for action by the image processor are prefixed by the # operator. This single feature greatly simplifies the implementation of VSP. However, it does not restrict us in any significant way, since we can use Prolog to extend VSP to implement a full version of PIP.

6.3.1 Defining the # Operator

Prefixing all image-processing commands with the # operator may at first seem a little clumsy, compared with PIP.[3] However, it greatly simplifies the software interface. Moreover, VSP is able to accommodate different types of image processor, with a minimal amount of reprogramming. The # operator may be defined using in the following way:

3. Some users have actually welcomed it, since it highlights the differences between VSP and standard Prolog. Obviously, when typing programs that incorporate the # operator, the user must press two extra keys for each command (including the space bar).

```
# A :-
    A =.. [B|C],          % Decompose the input term
    constants(C,D),       % Eliminate the variables from the input parameters
    variables(C,E),       % Eliminate the constants from the input parameters
    F =.. [B|D],          % Rebuild command containing only the constants from input A
    image_proc(F,[done|G]),   % Send reconstructed command F to image processor,
                          % which responds by returning "done" and parameter list G
    (append(E,_,G); append(G,_,E)),   % Instantiate as many input variables
                          % as we can use from list G
    !.                    % Succeeds once only; not resatisfied on backtracking
# A:- message(['Command ',A,' was not recognized by the image processor').
% Extracting a list of the constants included in the list A.
% Example: The goal "constants([a, b, 3, X, Y, 5, c,Z], Q)"
% instantiates Q to [a, b, 3, 5, c]
constants([ ],[ ]).
constants([X|Y],[X|Z]) :- not(var(X)), constants(Y,Z).
constants([X|Y],Z) :- constants(Y,Z).
% Extracting a list of the variables included in the list A. Example: The goal
% "variables([a, b, 3, X, Y, 5, c,Z], Q)" instantiates Q to share4 with [X,Y,Z]
variables([ ],[ ]).
variables([X|Y],[X|Z]) :- var(X), variables(Y,Z).
variables([X|Y],Z) :- variables(Y,Z).
```

The goal *image_proc(A,b)* sends command A to the image processor, which responds by returning a list consisting of

Head: the atom done and

Tail: a list of parameters calculated from the current image.

Notice that *image_proc/2* is directly responsible for all communication between the image processor and the Prolog-level controller. *image_proc/2* might, for example, direct its output to a serial port (e.g., Macintosh modem port) and collect whatever values are returned by the image processor through the same port. Notice that any number of parameters may be returned, irrespective of the number of variables included in the Prolog goal. (This will be discussed again later.) The following definition of *image_proc/2* has been modified slightly for ease of understanding and is valid for both MacProlog and WinProlog.

```
image_proc(A,b):-
    io_port(X),           % Consult the database to find which I/O port we are using
    write(a)~> X,         % Send command (a) to image processor via port X
    nl ~> modem),         % New line character sent to port X
    skip(10) <~ X,        % Look for RETURN character on port X
    read(Y) <~ X,         % Instantiate Y to incoming data on port X
    Y = [D|B],            % Get error no. (d) & list of returned values (B)
    skip(58) <~ X,        % Skip ahead looking for prompt character (':' ) on port X
    !.
```

4. In Prolog, two variables A and B are said to *share* when A is instantiated to B while B remains uninstantiated.

Table 6.2 Command formats permitted by the # operators. In examples labelled Mixture 2 – Mixture 4, the image processor returns three parameter values: res1, res2, res3.

Type of Arguments	Prolog+ Command	String sent to Image Processor	Instantiated variables
None	# neg	neg	None
Instantiated	# thr(12,34)	thr(12,34)	None
Uninstantiated	# avr(X)	avr	X
Mixture 1	# ndo(3,X)	ndo(3)	X
Mixture 2	# abc(1, 2, Q, 3, 4, W, 5, 6)	abc(1,2,3,4,5,6)	Q = res1, W = res2 (One parameter is not collected)
Mixture 3	# abc(1, 2, Q, 3, 4, W, 5, E, 6)	abc(1,2,3,4,5,6)	Q = res1, W = res2, E = res3
Mixture 4	# abc(1, 2, Q, 3, 4, W, 5, E, R, 6)	abc(1,2,3,4,5,6)	Q = res1, W = res2, E = res3 (R remains uninstantiated)
Error condition	No such command or parameter is out of range	Image processor returns error	Goal fails

The command formats allowed by this definition of the # operator are illustrated in Table 6.2. It is a trivial matter to define new predicates which do not require the use of the # operator, as we show in Table 6.3. In this way, it is possible to develop a system with PIP's full command set in terms of VSP, provided, of course, that an appropriate subset of commands is implemented directly by the image processor.

Table 6.3 Defining new mnemonics and expanding the VSP command set using Prolog-level programming.

Prolog programming	Function
thr(X,Y) :– # thr(X,Y).	Both arguments specified
thr(X) :– # thr(X,255).	Single argument option
thr :– # thr(128,255).	Default values for both arguments
threshold(X) :– # thr(X,255).	New name for an existing operator
thr(light_grey) :– # thr(192,255).	New argument type
thresh_neg(X) :– # thr(X), # neg.	New command

6.3.2 Device Control

Control of electro-mechanical devices, such as lamps, (X,Y,θ) table, pick-and-place arm, etc., may be achieved using another operator (¶), which is defined in a similar way to #. The ¶ operator is most effective if it operates through a multi-purpose interface device, similar to the MMB unit discussed in Chapter 4 (see Figure 4.24). In this case, all devices are controlled through a single serial port. A goal of the form

¶ device_command(a,b,c)

simply transmits the string "*device_command(a,b,c)*" to the device controller. As before, parameters obtained from the external hardware can be returned to the Prolog control program, simply by instantiating variables included in the command (Table 6.4).

Table 6.4 Device control using the ¶ operator.

VSP goal	Device controller receives	Function
¶ home	home	Move the (X,Y,θ) table to its home position
¶ move_to(45,67,8)	move_to(45,67,8)	Move the table to position/orientation (45,67,8)
¶ rotate(37)	rotate(37)	Rotate the table by 37 degrees
¶ pick	pick	Move the pick-and-place arm out, lower its gripper, grasp, and then raise it
¶ out	out	Move the pick-and-place arm to out position
¶ zoom(out)	zoom(out)	Operate the camera's zoom lens
¶ light(7,9)	light(7,0)	Set bulb number 7 to brightness level 9
¶ input(2,6,A)	input(2,6)	Instantiate variable A to the input signal on line 2 of port 6
¶ port(7,B)	port(12)	Instantiate variable B to the input signal on parallel port 7
¶ robot_at(X,Y,Z,Theta)	robot_at	Instantiate (*X,Y,Z,Theta*) to the position/orientation coordinates of the robot

6.3.3 VSP Control Software

While the VSP system described above is able to perform image processing and operate various electro-mechanical devices under the control of a Prolog program, it does not provide an adequate user interface. Throughout this book, we have emphasised that this is just as important as any other feature of an interactive vision system. We therefore complete this brief introduction to VSP by explaining how a range of other useful faciities can be added to the minimal system based on the # and ¶ operators. It is important to note that only Prolog-level programming is needed to build the user interface for VSP. The programs are straightforward but are too detailed to include here.

Interactive window

Prolog's standard dialog-box generators are used to program the interactive dialog box for VSP, which is similar to that provided in PIP. (Figure 4.3) The built-in predicates needed to build the dialog box are: *prompt_read/2* (MacProlog), or *call_modal/3* or *call_modeless/2* (WinProlog). The interactive dialog box should be designed so that the user does not have to type the # operator, although this should be tolerated, if he does. (The definition of the # operator given above allows for this.)

Journal

The Journal is simply a user-defined output window. Each time an image-processing operation is performed, the command string is added to that window. The easiest way to do this is to extend the # operator slightly:

```
# A :-
   A =.. [B|C],
   constants(C,D),
   variables(C,E),
   F =.. [B|D],
```

```
    image_proc(F,[done|G]),
    (append(E,_,G); append(G,_,E)),
    writeseqnl( [F,',']), % Command as sent to IP is printed in the Journal window
    (( E = [_|_], writeseqnl( ['% Command with variables instantiated:', A])); true),
                                  % User's original command with variables instantiated
    !.
# A:- message(['Command ',A,' was not recognized by the image processor').
```

(Since PIP was not designed using the # operator, the output it generates in the Journal window is slightly different from that produced by this program.)

Pull-down menus

We saw in Chapter 4 that pull-down menus form an important part of the user interface for an IVS. For this reason, they have also been integrated into VSP. However, they have been organized in such a way that the user can add extra menu items at will, rather than being limited to using a fixed set of preprogrammed procedures. This was done to avoid becoming dependent upon specialist help for reprogramming, while retaining the highest level of flexibility. The choice of actions that can be invoked by the menus is less important initially than the ability to extend the menus at will. Of course, this simply reflects the general spirit of VSP, which is "minimal" in its initial form but can be extended almost indefinitely.

It is easier to keep a separate record of the actions that can be invoked via the user-extended menus in a database, rather than trying to store and then recover this information from the menus themselves. The database might typically contain a series of facts of the general form *menu_item(Menu, Menu_item, Action)*.
Here are some typical examples:

```
menu_item('U', 'Digitize image from camera', grb).
                                  % 'U' menu. Digitize an image from the camera [grb]
menu_item('G', 'Threshold at mid-grey',thr).
                                  % 'G' (Grey-image)menu. Threshold. [thr(128)]
menu_item('B', 'Convex hull',neg).
                                  % 'B' (Binary image)menu. Convex hull [chu]
```

The dialog outlined in Figure 4.7 simply places an additional item in this database using Prolog's built-in predicate *assertz/1*. Once a set of facts like this has been established in the database, the menus can be built using *build_menus /0*, defined thus:

```
build_menus :-
    menu_item(A,B,C),           % Consult the database
    extend_menu(A,[B]),         % Add item B to menu A
    fail.                       % Force backtracking, to do same for all items in database
build_menus.
```

When the user operates selects an item X from a user-extended pull-down menu (say menu 'G'), the following program consults the database and performs the appropriate action.

```
'G'(X) :-                   % What to do when an item is selected from menu 'G'
    menu_item('G',X,Y),     % Find action Y associated with item X within menu 'G'
```

```
    writeseqnl('Journal', [Y, ', % was executed using the 'G' menu']),
                              % Save the command and a comment in the Journal window
    call(Y).                  % Try to satisfy goal X.
'G'(X) :- message(['Item', X, 'in the G menu failed']).
                              % So warn the user. (The goal succeeds though.)
```

In order to add a new menu, say 'M', defined by the list L = [*item1, item2, ..., itemN*], the user simply invokes the goal *install_menu('M',L)*. Short menu names (U,G,B,A,D,I, S) were chosen so that the maximum space was left on the computer screen for more menus, if needed.

While the "I" menu can be programmed in this way, there is a danger that the set of stored images might be changed without Prolog ever being aware of the fact. Hence, to be most effective, Prolog must be able to find the names of all of the stored images available to the image processor. This presupposes that the image processor has a command (call it *find_images*) which can identify the images stored in a given folder. Then , the 'I' menu is programmed as follows:

```
'I'(X) :-
    # find_images(L),         % Instantiate L to be the list of stored images
    member(X,L),              % Is image X still present?
    # loa(X).                 % Load image X
'I'(X) :- message(['Image',X, 'was not found']).    % Warn the user. (The goal succeeds.)
```

Of course, *build_menus/0* also has to be modified to build the 'I' menu. The changes are fairly obvious:

```
build_i_menu :-
    # find_images(L),         % Ask the image processor to find the list of images available
    member(A,L),              % Choose an image name (A)
    extend_menu('I',[A]),     % Add image name A to menu 'I'
    fail.                     % Force backtracking, to do the same for all images available
build_i_menu .                % Force the goal to succeed
```

System startup and error recovery

The fact that the VSP system consists of two distinct units can cause some difficulties, particularly if certain types of error occur. Hence, there is an occasional requirement to resynchronize the image processor and its Prolog host, in order to accommodate unusual and unexpected error conditions. This is essential at startup, when the state of the image processor is not necessarily known exactly. Since it is very difficult to anticipate all situations that can occur in practice, it is probably easier to build in a facility for re-initialization that can be triggered if the user suspects that synchronization has been lost. If the reader feels that this is admitting to poor engineering practice, he should bear in mind that VSP is a minimal Prolog program, which is expected to interface to an image processor of unknown type. Moreover, we may never know very much about the internal operation of the latter, since it is likely to be a proprietary system.

There are three problems that must be addressed:

a) Uncontrolled character sequences on the Prolog-to-image-processor communication path. (It should be understood that the user can place any sequence of characters on this line, since the # operator is completely indiscriminate.) There is a danger that the user might accidentally place the image processor in some strange state (e.g., test mode), so that it will never respond properly to further commands until that unit is reinitialized.

b) Unwanted characters on the communication path between the image processor and Prolog. Certain commands might unexpectedly produce a large amount of data. Again, this might be a consequence of the user having uncontrolled access to the image processor.

c) Image processor being in a strange and unknown state. This might occur for a variety of reasons, not necessarily connected with anything that the user has done.

Situations (a) and (b) can be helped considerably if the image processor has a software-interrupt mechanism. This might conveniently consist of some string of symbols, preferably involving at least one control character which would not normally be found in a command. Let us assume that this character is Ω and that the character string to start/restart the image processor is Σ*<IP restart string>*.

It is difficult to make any general statements about exactly what will happen next, except to say that we can be fairly certain that the image processor will respond by issuing a lot of textual information. (This is probably intended to be useful for a human being to read.) However, our non-intelligent VSP program has to reject all of these data and will probably wait for a certain character or combination of characters in order to be sure that the image processor is ready to receive instructions. Of course, it is easiest to cope with large volumes of "unknown" text if it is always terminated by a unique control character that never appears at any other time. Let us suppose that the terminating character is Ω. Then, the following error recovery programs are useful:

```
reinitialize :– writenl('Modem', 'Σ<IP restart string>').
flush_input :–
    % Removes all characters from the 'Modem' port, up to and including 'Ω'
    repeat,              % Repeat on backtracking
    get('Modem', X),     % Get one character from the I/O port
    X = 'Ω'.             % End of the character string? If not, backtrack
```

Given these facilities, the VSP system can be made to initialize itself at startup by using the *'<LOAD>'/1* predicate in MacProlog. This goal is invoked automatically every time the Prolog software is loaded. Here is the definition of the startup routine:

```
'<LOAD>'(_) :–
    delay(100),          % Allow image processor to settle down. Adjust time limit to taste
    reinitialize ,       % Initialize the image processor
    delay(100),          % Allow image processor to settle down. Adjust time limit to taste
    flush_input .        % Clear out the rubbish on the input channel
```

Although this is necessarily a very sketchy picture, it does provide a good basis for building a reliable VSP system. Let us end our discussion of this important, if slightly tedious, topic with one final remark: the system is made more reliable if we use the programmed versions of the commands (Table 6.3), rather than relying on the user to type each command correctly when using the # operator.

Image stack

The image stack can be programmed purely in Prolog, using named files to save images. Here are the relevant programs, which rely heavily on MacProlog's properties. This is faster and somewhat easier than using assert and retract.

```
% Defining the maximum size of the stack. Adjust to taste
image_stack_max_size(48).

% csk/0; Clear the image stack. This is one of the functions performed at startup
csk:– remember(image_stack,0).
% Place image on stack

pis :– psk.
psk :–
    recall(image_stack,X),        % Recall stack pointer
    image_stack_max_size(N),      % How big is the stack?
    Y is X + 1,                   % Increment the stack pointer
    Y ≤ N,                        % Is that value acceptable?
    concat(['image',Y],Z),        % Form image name (Z) from "image"
                                  % ... and the stack pointer (Y)
    wri(Z),                       % Save image in file Z
    remember(image_stack,Y),      % Remember the new value of the stack pointer
    !.                            % That is enough. Do not permit backtracking
psk :–                            % We had an error: stack is full
    message(['Image stack is full']),      % Tell the user
    abort.                        % Give up. Potentially dangerous: running out of memory

% Pop the image stack
pop :–
    recall(image_stack,X),        % Recall stack pointer
    X > 0,                        % Is X positive?
    concat(['image',X],Z),        % Form image name (Z) from "image" ...
                                  % ... and the stack pointer (X)
    rea(Z),                       % Read image file Z
    Y is X – 1,                   % Decrement the stack pointer
    remember(image_stack,Y),      % Remember the new value of the stack pointer
    !.                            % That is enough. Do not permit backtracking
pop :–                            % Stack is empty
    writenl('Image stack is empty'),       % Tell the user
    fail.                         % Force goal to fail

% View image stack
vsk :–vsk(top).                   % Default. Look at the top of the stack but do not alter it
vsk(top) :–                       % Look at the top of the stack but do not alter it
    recall(image_stack,A),        % Recall stack pointer
    concat(['image',A],B),        % Form image name (b) from "image" ...
                                  % ... and the stack pointer (A)
    rea(B),                       %Read image file B
    !.                            % Stop backtracking
```

```
vsk(bottom) :-                % Look at the bottom of the stack but do not alter it
    vsk(1),                   % "image1" the bottom of the stack
    !.                        % Stop backtracking
vsk(a):-
    recall(image_stack,B),    % Recall stack pointer
    A ≥ 1,                    % Is the input pointer positive
    A ≤ B,                    % Is the input pointer within range defined by stack pointer?
    concat(['image',A],C),
                              % Form image name (c) from "image" and the stack pointer (A)
    rea(C),                   % Read image file C
    !.                        % Stop backtracking
vsk(_) :-                     % Pointer value given by user was out of range
    message(['Argument for vsk out of range']),
    fail.                     % Force this goal to fail
    % View bottom image on stack. Defining a new synonym.
bsk :- vsk(bottom).
    % View top image on stack. Defining a new synonym.
tsk :-vsk(top).
```

Cursor

VSP's dual-processor structure system does not allow a cursor to be implemented easily. The reason is that the current image is displayed on the screen of a video monitor, whereas the pointing device (e.g., mouse) operates in association with the computer screen. However, an indirect approach has been devised, providing a rudimentary cursor function. Although this is not ideal, it is probably acceptable to many users, in view of the fact that the cursor is not used very frequently during interactive image processing. (The cursor is used principally to determine feature position and intensity values in an image and to control cropping and intensity plotting.) A blank "dummy" window, called '*My Graphic Window*' in the program below, is displayed on the screen of the Prolog host computer. Provided the user moves the cursor, controlled by the mouse, within the active central area of '*My Graphic Window*', the position coordinates are detected by the program and a white cross is displayed on the current image. (Even a system of modest speed enables the user to maintain good hand–eye coordination.) When the user clicks the mouse button once, the position and intensity are printed on the computer screen (in the '*Σ Output Window*'). When the user double-clicks the mouse button, the parameters A, B and C are instantiated to the cursor's current *(X,Y)* coordinates and the corresponding intensity value, respectively. Since he is watching the image display rather than the computer screen, the user may accidently move the cursor outside the active region within '*My Graphic Window*'. In this case, he temporarily loses control of the cursor In order to warn the user when this happens, the following program provides an audible signal (a simple "beep").

```
cur :- cur(A,B,C).                      %Default option
cur(A,B,c):-                            % A, B and C are uninstantiated initially
    wkill('My Graphic Window'),         % Kill "My Graphic Window"
    wgcreate('My Graphic Window', 40, 10, 350, 350,0,350,350,1,1),
```

```
                                      % Recreate "My Graphic Window"
   wfront('Σ Output Window'),         % Bring the default output window to the front
   wfront('My Graphic Window'),       % Now bring "My Graphic Window" to the front
   wsize('My Graphic Window',120,4,275,275),
                                      % Reposition and resize "My Graphic Window"
   add_pic('My Graphic Window',_,box(1,1,256,256)),
                                      % Draw a square of 256*256 pixels
   refresh_now('My Graphic Window'),  % Refresh the display
   single_click(A,B,C),               % Enter the interactive loop
   !.
single_click(A,B,c):-                 % Position = [A,B]. Intensity = C
   repeat,                            % Always succeeds on backtracking
   (get_mouse('My Graphic Window',V,U); fail),
                               % Find position of the cursor in "My Graphic Window"
   V1 is 257 – V,              % A bit of arithmetic to invert scale of the vertical axis
   range_check(U,V1,U2,V2),    % Check that [U,V1] is in range.
                               % If not use default values to define [U2,V2]
   # get_intensity(U2,V2,C),   % Find intensity (c) at point [U2,V2].
   # cross(U2,V2),             % Draw vertical cross on current image at (U2,V2)
   # cross(U2,V2),             % Remove the cross
   mouse_up('My Graphic Window', B1,A),
                               % (B1,a)are coordinates when mouse is released
   B is 257 – B1,              % A bit more arithmetic, again to reverse vertical axis
   writeseqnl(['Position = ', [U2,V2]),' Intensity = ', C]),   % Messages for the user
   mouse_up('My Graphic Window',_,_).
                               % User double-clicked to escape from cursor function
```

Help

The following program installs the H (Help) menu, using a file called called HELP which contains information on a range of topics. The file format is described later.

```
pip_help_file('My Computer:VSP:HELP for VSP').
                               % Identity and location of the Help file

build_help_menu :–
   pip_help_file(A),           % Find location of the Help file
   see(A),                     % Open the file
   build_help_list([ ],B),     % Build the list
   close(A),                   % Close the file
   install_menu('H',B).        % Install the 'H; menu

build_help_list(A,[D|B]) :–    % Builds a list of key words from the Help file
   skip(36),                   % Skip until we find the $ symbol in that file
   read_line(C),               % Read the remainder of the line beginning with $
   pname(C,D),                 % Convert between Prolog term (c) & atom ...
                               % ... representing its print name(D)
   not(C = `end_of_file`),     % End of file has been found
   build_help_list(A,B).       % Carry on until we reach end of the file

build_help_list(A,A).          % End of file reached. Keep list already calculated
read_line(c):–                 % Read the rest of the line in the Help file
   line_read([ ],B),           % This bit does most of the work ...
```

```
    string_chars(C,B),          % ... and this bit reformats the result
    !.
line_read(A,[C|B]) :-
    get0(C),                    % Get single character from the Help file
    not(((C = -1); (C = 13))),  % Are we at end of file or end of line?
    !,                          % We are not coming back, so freeze all instantiations
    line_read(A,B).             % Carry on until we get to the end of the line
line_read(A,A).                 $ End of line or end of file reacehed. Keep results
```

The following program activates the 'H' menu. When the user selects a topic, the appropriate portion of the *Help* file is displayed.

```
'H'(X) :-
    pip_help_file(A),           % Find location of the Help file
    close(A),                   % Close it if it isa already open
    see(A),                     % Open or re-open the Help file
    repeat,                     % Always repeat if we are backtracking
    skip(36),                   % Skip until we find a '$ ' character in the Help file
    read_line(B),               % Read the remainder of the line just found
    pname(B,C),                 % Convert between Prolog term (c) and atom ...
                                % ... representing its print name(D)
    C = X,                      % Do contents of the current line match the ...
                                % ... menu item selected by the user
    read_line(D),               % Read the next line: this is information we want
    writeseq(['PIP Command:', X, ' - performs the following operation:~M ']),
                                % Message for the user
    writenl(D),                 % Message for the user
    nl,                         % New line
    close(A).                   % Close the Help file
'H'(X) :- message(['Item', X, 'in the H menu was not found']).
                                % Catch all, safety clause
```

The structure of the HELP file is explained with reference to a simple example. Items in brackets {...} are not part of the file.

```
$'&'                        {Topic marker. Topic is the operator '&'}
Equivalent to Prolog's goal concatenation operator ','  {Information on topic '&'}
......
$aad – arity = [0,3]        {Topic marker. Topic is the command aad}
Aspect adjust               {Information about topic aad}
$abs – arity = 0            {Topic marker. Topic is the command abs}
Fold intensities about mid-grey ("absolute value or "modulus" function)
                            {Information about topic abs }
......
$yxt – arity = 0            {Topic marker. Topic is command yxt}
Interchange X and Y axes (transpose image).   {Information about topic yxt}
$zer – arity = 0            {Topic marker. Topic is command zer}
Black image                 {Information about topic yxt}
$end_of_file                {End of file}
```

6.4 PIP

PIP differs from VSP and other early implementations of Prolog+ in the fact that it runs on a single machine, namely an Apple Macintosh computer. Since PIP is based entirely on software, rather than a closed hardware-based system, a far more flexible approach to image display and processing can be taken. For example, PIP is able to function with a wide variety of operating paradigms, some involving the processing and display of a large number of images. This is useful, for example, when analyzing image sequences, color/multi-spectral images, or to keep the intermediate results in a long processing sequence.

The PIP software consists of two distinct parts:

a) Top-level control (LPA MacProlog) [LPA]

b) Low-level image processing (Symantec Think C).

In this section, we will describe PIP's basic design principles. In particular, the impact of the Macintosh operating system on the implementation of the image-processing functions, and the interface between these and the Prolog control software will be discussed. We also explain how individual image-processing commands and the system infrastructure have been implemented. Normally, the user works by interacting with the high-level (Prolog) controller, as we have explained in Chapter 4, and will not usually be aware of this intermediate level. However, it makes PIP potentially very flexible, by enabling a third level of programming to be used. Most PIP image-processing operators are programmed either directly in C, or by including sequences of commands within a Prolog program. However, the possibility exists for adding very significantly to PIP's functionality by programming at this intermediate level, as well.

6.4.1 System considerations

Using specialized equipment for image processing has the obvious advantage over a software implementation that the hardware is tailored to image processing and will often yield substantially higher processing speeds. If a software implementation is capable of providing adequate performance for a particular application, then such an implementation offers a number of benefits, including the following:

- Apart from initial image capture, little or no investment in specialized hardware is required. A complete interactive image-processing system may be assembled by merely purchasing a standard (CCIR/RS320, or RS170) video camera and installing the PIP software on a Macintosh computer that has been fitted with a video capture card. Alternatively, a low cost camera may be plugged into the serial (RS423) port, USB, or "firewire" (IEEE 1394) socket, without using any other hardware. Another option is to use scanned images or pictures downloaded from the World Wide Web (for demonstration and training purposes only), or transferred from the customer's experimental rig.

- As a user upgrades his computer, he will obtain a corresponding improvement in image-processing performance without incurring the additional cost of investing in new hardware.
- The software can be extended indefinitely, whereas image-processing hardware is typically packaged in a closed "black box," providing a predetermined range of functions and a fixed mode of operation.

It should be clearly understood that Prolog is not an appropriate language for implementing "low-level" image-processing operations, such as image addition, thresholding, filtering, skeletonization, convex hull, etc.(These are often described colloquially as "pixel pushing" operations.) A more conventional high-level language, such as C, or assembly language, is much better suited to rapid, iterative processing of large arrays of data.

The Apple Macintosh computer, LPA MacProlog32, and Symantec Think C were chosen as the platform for developing PIP. The software has been run successfully on the latest models of Apple computer and a stand-alone[5] version of PIP runs successfully under the Macintosh Application Environment 2.0 (Apple Computer, Inc.) on a Sun or Hewlett Packard workstation. Unfortunately, the PIP software will not run under the Windows 95 or MS-DOS operating systems. However, the promised enhancements to the Executor 2 software[6], which emulates a Macintosh computer, should make this possible soon.

We chose to use the Apple Macintosh for a number of reasons. Historically, our previous work on Prolog+ has been carried out on Apple Macintosh computers, due to the availability of a good implementation of Prolog. Macintosh computers are relatively cheap, and the authors find the operating system more aesthetically pleasing than the systems available on IBM PC-compatible machines (the main alternative). The choice of Prolog implementation for Macintosh computers is very limited, although LPA MacProlog is suitable for our needs. It provides a full implementation of Prolog, user-interface development facilities, the ability to call functions written in C or Pascal, and the ability to act upon low-level events, such as activation of a window, in a user-defined manner.

We are using THINK C because it was one of the first languages supported by LPA MacProlog. The former offers an integrated programming environment, which has proved useful in developing and testing image-processing functions, before attempting to integrate them into the PIP system.

6.4.2 Why not implement Prolog+ commands directly?

PIP usually displays the current and alternate images in dedicated windows on the Macintosh computer screen and a considerable amount of programming effort has

5. This does not have MacProlog's user interface, although Prolog programs can still be written, edited, compiled, and run.
6. The Executor 2 software is available from Ardi Software, Inc., Suite 4-101, 1650 University Boulevard, Albuquerque, NM 87102, USA. Also consult the following WWW site: WWW: http://www.ardi.com

to be applied to control these windows. We elected not to implement C routines that perform PIP functions directly. Instead, a new (*results*) image is created by each image-to-image mapping operator (e.g., *neg*, *add*, *lpf*, *chu*, etc.) and separate routines attend to general housekeeping tasks , such as

- Disposal of images which are no longer required. (Recall that the alternate image is discarded.)
- Creation of windows to display images.
- Association of a new image with a window. (A new window is associated with the newly-created results image.)

The principal reasons for this approach are as follows:

- It is a straightforward matter to implement the standard (i.e., 2-image) operating paradigm on top of this, by writing appropriate Prolog code.
- The idea of leaving the source images unchanged is more in keeping with the spirit of the Prolog language.
- It is not always desirable to have a continuous display of the images when the system is working. (The programmer may, for example, wish to hide intermediate results from a customer.)
- There is freedom to implement and explore other image-processing models, if desired. (For example, we may wish to define operations such as add, subtract, mutliply, edge detectors, etc. on 3-component color images.)

6.4.3 Infrastructure

LPA MacProlog allows the programmer to call "foreign code" functions written in C or Pascal. On the Apple Macintosh family of computers, files have two separate parts: the data fork and the resource fork. MacProlog requires that a new code resource be created containing the compiled foreign code. This can be likened to a library on other operating systems. Having opened the file containing the resource, the *call_c* (or *call_pascal*, as appropriate) predicate is used to invoke the required function. A collection of "glue" routines supplied with MacProlog must be linked into the foreign code resource, enabling the programmer to access arguments and manipulate the data structures supported by Prolog, such as lists.

In order to obtain a system which successfully coexists with the Macintosh Finder (the Graphical User Interface) and other applications, it was necessary to build an application within the framework provided by Apple Computer, Inc. There is a wide range of *Toolbox* routines for managing windows and menus. The *QuickDraw* toolbox makes it possible to create, manipulate, and display graphical data. A particularly useful feature is the ability to manipulate off-screen graphics worlds, which provide a means of storing bit maps and associated data. These features were used in a way that is entirely consistent with the Apple Computer company's recommendations, to ensure the future portability of the PIP software.

6.4.4 Storing, displaying, and manipulating images

It is generally best to avoid accessing the screen display directly on the Apple Macintosh computer. Instead, drawing is carried out using *QuickDraw* routines via a graphics port, which is normally a window. The operating system ensures that only visible parts of the window are drawn, and generates update events when part of a window needs to be redrawn, perhaps as a result of another overlapping window being moved. If a pixel map must be manipulated directly, then an off-screen graphics world (*GWorld*) may be used. *QuickDraw* may be used to draw into the *GWorld.* Alternatively, the *GWorld*'s pixels can be accessed directly. The result may be copied to the appropriate window, using the *QuickDraw copyBits* routine. In PIP, an off-screen *GWorld* is used to represent each image currently in use. Not all of these off-screen *GWorlds* necessarily have a corresponding window, but each image display window does have a corresponding off-screen *GWorld.*

Using off-screen *GWorlds* offers a number of benefits to the programmer:

- An off-screen *GWorld* can be associated with a window for future redrawing, as necessary.
- Whatever the pixel depth of the display, it is possible to make an off-screen *GWorld* of appropriate depth for the image. (So far, the images we have dealt with have had a depth of 8 bits per pixel, and the PIP program assumes this pixel depth when it creates a new off-screen *GWorld.*) If a display mode is selected in which not all the image colors are available, the copyBits routine will select the nearest possible color from the current palette.
- It is possible to associate a color lookup table (CLUT) with an off-screen *GWorld* which is different from the default. This is useful because the default Macintosh CLUTs assign white to a pixel value of 0, whereas PIP's grey-scale image-processing operations are based on the assumption that 0 represents black and 255 white. It is not necessary to change the screen CLUT, which would corrupt the colors of other items on the screen, since mapping between CLUTs is performed automatically by the *QuickDraw* software.
- Since the Macintosh operating system has its own memory-management routines, storage occupied by an off-screen *GWorld* may be released as soon as it is no longer needed.

Tables 6.5 and 6.6 illustrate how these functions are performed in PIP's C code. Table 6.5 contains annotated extracts from the *negate_image* routine, which indicates how a new off-screen *GWorld* is created and accessed, while Table 6.6 contains extracts from *update_window.* The latter indicates how an off-screen *GWorld* is associated with a window and how that window is updated. (Notice that error-handling code has not been included in these routines.)

6.4.5 Prolog–C interface

It is important to understand that information about the current and alternate images and windows is stored by the Prolog program, which provides overall mangement of their creation, use and eventual destruction. In this section, we explain how parameters are passed between the two parts of PIP: the Prolog top-level controller and the underlying C routines. We will also describe the Prolog predicates that call the C routines and provide the infrastructure mentioned above. Finally, we discuss how user events are handled.

Table 6.5 The *negate_image* routine. (C language)

Code	Comments
extern OSErr negate_image(GWorldPtr iml, GWorldPtr *im2)	iml:input image; im2: output image.
{...	
GetGWorld(&origPort, &origDev);	Store current GWorld for restoration later .
sourcePM=GetGWorldPixMap(iml);	Get input image pixel map (NB. This contains a reference to the memory where the pixels themselves are located, and other information.)
good=LockPixels(sourcePM);	Prevent it from being moved by the memory manager
boundRect=(*iml).portRect;	Get boundaries of pixel map.
cTable=GetCTable(129);	Obtain the greyscale CLUT.
errNo=NetGWorld(im2, 8, &boundRect, cTable, nil, 0);	Create new off-screen GWorld of depth 8, with our special CLUT.
DisposeCTab(cTable)	Free memory
SetGWorld(*im2, nil);	Drawing to occur in this new GWorld.
destPM=GetGWorldPixMap(*im2);	Obtain the new pixel map.
good=LockPixels(destPM);	Prevent the destination pixel map from being moved by the memory manager
srcAddr=(unsigned char*)GetPixBaseAddr(sourcePM);	Calculate where pixels are stored
srcRowBytes=(**sourcePM).rowBytes & 0x3fff;	Prepare to copy the pixels, negating.
destAddr=(unsigned char*)GetPixBaseAddr(destPM);	
destRowBytes=(**destPM).rowBytes & 0x3fff;	
width = boundRect.right – boundRect.left;	
height = boundRect.bottom – boundRect.top;	
for (row=0; row<height; row++)	Copy pixels negating. (We assume a greyscale image with pixel values between 0 through 255.)
{	
srcAddrl=srcAddr;	
destAddrl=destAddr;	
for (column=0; column<width; column++)	
*destAddrl++ = 255?(*srcAddrl++);	At last! Here is the negation operation.
srcAddr=srcAddr+srcRowBytes;	
destAddr=destAddr+destRowBytes;	
}	
UnlockPixels(destPM);	Allow pixel maps to be moved again.
UnlockPixels(sourcePM);	
SetGWorld(origPort, origDev);	Restore original GWorld (screen).
...}	

Table 6.6 The update_window routine.

Code	Comments
extern OSErr update_window(WindowPtr theWindow)	
{…	
GetGWorld(&origPort, &origDev);	Store current GWorld for later restoration.
theImage=(GWorldPtr) GetWRef-Con(theWindow);	Retrieve pointer to off-screen GWorld associ-ated with the window
SetPort(theWindow);	Drawing to occur in this window.
BeginUpdate(theWindow);	Indicate to OS that an update event is being processed; call the image drawing routine; indicate update is complete.
display_in_window(theImage, theWindow);	
EndUpdate(theWindow);	
SetGWorld(origPort, origDev);	Restore original GWorld (screen).
…}	

Passing parameters

Routines such as *negate_image*, which was described earlier, cannot be called directly from Prolog: Some pre- and post-processing is required, in order to retrieve and set the parameters passed between it and the C routine. This extra processing is not incorporated directly into each routine, since it is easy to write a stand-alone C program that allows individual image-processing functions to be tested and debugged separately.

A C routine is invoked from LPA MacProlog using a call of the form

```
call_c(<parameter list>, <resource type>, <resource id>)
```

In our case we have a single code resource of type '*MINE*' and resource *id 0*. We always use a parameter list of the form

```
[<input param. list>, <output param. list>, <function no.>, <err. code>]
```

As an example of the implementation of the image-processing predicates, consider the following extract from PIP's code:

```
neg_im(Im1, Im2):-                  % Implements the negation function ...
                                    % ... with help from a C routine
   call_c([[Im1],Var,1000,Err],'MINE',0),   % Invoke C routine (no. 1000) ...
                                    % ... which negates current image
   Err=0,                           % Check that there was no error
   Var=[Im2],                       % Pattern matching extracts identity of new image (Im2) ...
                                    % ... from the single-element list generated by C routine
   recd_new_im(Im2).                % Record the fact that we have a new image
```

Notice that *neg_im* succeeds only if the error number (*Err*) is zero. The output list should contain a single element (i.e., the value of a pointer to the new image, Im2). The Prolog program then records the fact that this new image has been created. An example of the program code, illustrating this, is given later in this section.

User events

We need to be able to handle two kinds of event:

- update events, generated when a window needs to be redrawn; and

- mouse_down events, when they occur in the close box of the window.

MacProlog provides a way of trapping these events and acting upon them, provided that the window's property *windowKind* is greater than 32. (All windows created by our C routines have windowKind equal to 33.) When an update event is received, MacProlog calls the *user-defined x_update* predicate. Similarly, a mouse-down event calls the *x_mousedown* predicate. *x_update* calls the C routine which we have written to process update events for the specified window and is defined thus:

```
x_update(Win) :- call_c([[Win], _, 2, _], 'MINE', 0).
```

When the Prolog system invokes this predicate, *Win* is instantiated to a value (i.e., a pointer) indicating which window is to be updated.

A cursor facility has been added to PIP recently. This provides fairly obvious functions, enabling the user to investigate the position, size, and intensity of a chosen feature in the current image, or select a rectangular region prior to cut, paste, or crop operations. Details of the cursor software will not be discussed here since they would be a distraction from our main topic.

Linking Prolog and C components together

The processing is completed by a library program, which is written in C and resides in a resource file that has been identified to Prolog. Suppose this file is named '*PIP lib*'. Then, the Prolog program should contain a headless clause of the form

```
:- res_open('PIP lib').
```

in order to establish the link between itself and the C library. (A headless clause specifies a goal that is satisfied whenever that part of the program code is compiled.) Loading the (Prolog) software has the same effect.

The main C routine within the Prolog-C interface (Table 6.7), receives the command from *call_c/3* and allocates the specified task to the appropriate module, for example, the routine *negate_image* (Table 6.5).

6.4.6 PIP infrastructure

Using the infrastructure facilities directly is rather laborious, since there is no automatic creation and disposal of images. Moreover, an image is only displayed in response to an explicit command to do so. The following example illustrates how the facilities provided by the infrastructure layer of PIP may be used.

```
strange_new_program(Im,Im4) :-        % Instantiate Im4 during satisfaction ...
                                      % ... of strange_new_program(Im,Im4)
    new_im(Im),                       % Read new image from file Im
    new_win_for_im_disp(Im,Win,"Original",1,50)   % Create new window, ...
                                      % ... titled "Original", top-left corner at (1,50)
    linop3_im([2,3,2,3,5,3,2,3,2],Im, Im1),  % Local operator ...
                                      % ... (blur, equivalent to [con(2,3,23,5,3,2,3,2)])
    new_win_for_im_disp(Iml,Win1,"New",320,50),
                                      % Create new window, titled "New"
    linop3_im([2,3,2,3,5,3,2,3,2],Im1,Im2),  % Local operator ...
                                      % ... (blur, equivalent to [con(2,3,23,5,3,2,3,2)])
```

```
new_win_im_disp(Im2,Winl),      % Display this new image
kill_im(Im1),                   % Dispose of previous image
sobel_im(Im2,Im3),              % Sobel edge detector [sed]
new_win_im_disp(Im3,Win1),      % New window (Win1) for image Im3
kill_im(Im2),                   % Kill window Im2
thresh_im(Im3,Im4,15,255),      % Threshold image Im3 between ...
                                % ... levels 15 and 255, to give image I4
new_win_im_disp(Im4,Win1),      % Create new window (Win1) for image Im4
kill_im(Im3).                   % Kill intermediate result image Im3
```

Table 6.7 The main C routine, interfacing with Prolog.

Code	Comments
bool main(long argc, void *link())	
{... ~ inListTag=get_arg(1);	Get 4 arguments passed by call_c
outListTag=get_arg(2);	
fNoTag=get_arg(3);	
errNoTag=get_arg(4);	
switch (get_int_val(fNoTag))	Recover the value of the function number; call routine which handles the specified function number.
{ ...	
case 1000;	
errNo=do_negate_image(inListTag, outListTag);	
break; ...	
}	
put_int_val(errNoTag, errNo);	Store error number.
return SUCCESS;	Return value indicating predicate succeeded
OSErr do_negate_image(cellpo inListTag, cellpo, outListTag)	
{ ...	
iml = (GWorldPtr)get_int_val(get_list_head(inList-Tag));	Get 1st (and only) input parameter (head of the input list).
errNo = negate_image(iml, &im2);	Call the negate routine.
outListTag = put_list(outListTag);	Create a list to hold output parame-ter(s).
put_int_val(get_list_head(outListTag), (long) im2);	
put_nil(get_list_tail(outListTag));	Tail of output list is empty, i.e. the list has only one element
return errNo;	Return error number to main routine.
}	

The Sobel edge detector function (*sobel_im*) is defined as follows:

```
sobel_im(Im1,Im2):-
   call_c([[Im1],Var,5001,Err],'MINE',0),
   Err=0,
   Var=[Im2],
   recd_new_im(Im2).
```

The parameter *Im1* identifies the source image and is passed to the C routine. This call binds *Var* to a single-element list which identifies the destination image. The Sobel edge detector routine is invoked on receipt of the control variable defined by the third element in the list (i.e., the number 5001). The C program binds *Err* to the error code, which is 0 if no error occurred. Table 6.7 contains extracts from the main C routine, which chooses the appropriate image-processing function based on the function number it receives, and passes parameters between the Prolog environment and that function.

It is normally undesirable for user programs to invoke the *call_c* predicate directly, for two reasons. Firstly, the semantics imposed upon the *call_c* predicate are opaque. Secondly, when an image-processing operation is called, additional processing is required at the Prolog level, involving tasks such as keeping a record of any new images and windows that have been created. The infrastructure layer provides an arcane facility but one which has considerable potential for building non-standard operating paradigms in PIP. We will encounter more examples in Section 6.4.8.

6.4.7 Defining infrastructure predicates

Unless the reader plans to implement an IVS from first principles, he can safely skip this section, in which we define the main predicates forming the infrastructure layer of PIP.

```
% "new_im/1" selects a new PICT-format file and draws that file's contents ...
% ... into a new image. Upon return, the variable "Im" points to this new image.
new_im(Im):–
    call_c([[ ],Var,0,Err],'MINE',0),    % The C program generates a file-selection ...
                                         % ... dialog box and then reads the file specified by the user.
    Err=0,                               % Check that there was no error.
    Var=[Im],                            % Simplify the result. This consists of a list (Var) ...
                                         % ... with a single element (Win)
recd_new_im(Im).                         % Add the new image to the list of open images
% "kill_im/1" disposes of the named image, freeing the memory it occupies.
kill_im(Im):–
    recd_dead_im(Im),                    % Delete the image Im from the list ...
                                         % ... of open images held by Prolog.
    call_c([[Im],[ ],1,Err],'MINE',0).        % The C program frees the memory ...
                                              % ... allocated to this image
% "new_win_for_im_disp/5" creates a new window for the image identified by ...
% ... by "Im". The new window title is "Name". Its top-left corner is at ...
% ... (OffH, OffV). "Win" points to the new window on return from this predicate.
new_win_for_im_disp(Im,Win,Name,OffH,OffV):–
    call_c([[Im,Name,OffH,OffV],Var,100,Err], 'MINE', 0),
                                         % Call C routine
    Err=0,                               % Check there was no error.
    Var=[Win],                           % Simplify the result: this consists of ...
                                         % ... a list (Var) with a single element (Win)
    call_c([[Im,Win],[ ],102,_],'MINE',0),  % Call C routine
    recd_new_win(Win,Im).                % Add a new window to the list of open windows.
% Associate the existing window "Win" with a new image "Im".
new_win_for_im(Im,Win,Name,OffH,OffV):–
```

```
    pip_max_window_size(H,W),          % Maximum size of window for displaying ...
    % ... images - one of several control parameters defined in program "header".
    call_c([[Im,Name,OffH,OffV,H,W],Var,100,Err], 'MINE', 0),
                                  % Call C routine
    Err=0,                        % Check that there was no error
    Var=[Win],                    % Reformat the result
    recd_new_win(Win,Im),         % Add a new window to the list of open windows
    recd_named_window(Name,Win),  % Add new window (Name) to the list of ...
                                  % ... open windows (Win). "recd_named_window/2" ...
                                  % ... is not defined, although its function is obvious.
    !.                            % Cut!
% Same as "new_win_im, except displays the window's contents immediately.
new_win_im_disp(Im,Win):-
    call_c([[Im,Win],[ ],104,Err],'MINE',0),    % Call C routine
    Err=0,                                      % Check that there was no error.
    call_c([[Im,Win],[ ],102,_],'MINE',0),      % Display; still waiting for update event
    recd_new_im_for_win(Win,Im).                % Associate a new image "Im" ...
                                                % ... with window "Win"
% Close the window, freeing the memory it occupied.
close_win(Win):-
    recall(win_open,Open),        % Get list of open windows and associated images
    member([Win,Im],Open),        % Find image ("Im") associated with window "Win"
    call_c([[Win],[ ],103,Err],'MINE',0),       % Call C routine
    ( Err=0 ; writenl('Warning, error in closing window') ),   % Do nothing if ...
                                     % ... there is no error. Otherwise warn user.
    recd_dead_win(Win),              % Remove window "Win" from list of open windows
    !,
    kill_im(Im),                     % Kill the image associated with that window
    unrecd_named_window(Win),        % Remove window from open window list.
                                     % Not defined here
!.
% Copy image "Iml" to a new image. "Im2" points to this new image on return.
copy_im(Im1, Im2):-
    call_c([[Im1],Var,19000,Err],'MINE',0),     % Call C routine
    Err=0,                           % Check new error
    Var=[Im2],                       % Reformat result
    recd_new_im(Im2).                % Add a new image, "Im2", to list of open images
% Dispose of all windows and images currently in use, freeing the memory they occupy.
kill_wins_and_ims:-                  % Dispose of all open windows
    recall(win_open,List),           % Get list ("List") of all open windows
    close_wins(List),                % Close all windows in "List". Based on "close_win/1"
    remember(win_open,[ ]),          % Remember that there are no open windows?
    fail.                            % Go on to clause 2
kill_wins_and_ims:-                  % Dispose of all open images
    recall(im_open,List),            % Get list ("List") of all open images
    kill_ims(List),                  % Close all images in "List".
    remember(im_open,[ ]),           % Remember that there are no open images
    fail.                            % Go on to clause 3
```

```
kill_wins_and_ims:- forget(named_im).        % Forget property "named_im"
kill_wins_and_ims.                      % Goal "kill_wins_and_ims" always succeeds
% Sample of the auxiliary predicates required for infrastructure layer just defined
% "recd_dead_im/1" de-installs an image from the list of open images.[7]
recd_dead_im(Im):-                      % What to do if a window is using the named image - ...
                                        % ... namely do not de-install it
   recall(im_open,List),                % Get list ("List") of open images
   recall(win_open,WList),              % Get list ("WList") of open windows
   member([Win,Im],WList),              % Find a member of Wlist
   writenl('Window using the image!'),       % Warn the user
   !,
   fail.                                % Force goal "recd_dead_im(_)" to fail
recd_dead_im(Im):-                      % Dispose of image not associated with open window
   recall(im_open,List),                % Get list ("List") of open images
   remove(Im,List,Remains),             % Eliminate "Im" from "List" to leave "Remains"
   remember(im_open,Remains).  % Retain record of what is left
% "recd_new_im/1"adds a new image ("Im") to the list of open images
recd_new_im(Im):-                       % What to do if some images are already open
   recall(im_open,List),                % Get list ("List") of open images
   remember(im_open,[Im|List]),  % Record updated list of open images
   !.                                   % Avoid continuing to next clause when backtracking
recd_new_im(Im):-                       % What to do if no images are already open
remember(im_open,[Im]).                 % Record that image "Im" is open
% recd_new_win/2 adds a new window–image pair to the stored list
recd_new_win(Win,Im):-                  % What to do if some windows are already open.
   recall(win_open,List),               % Get list ("List") of open windows
   remember(win_open,[[Win,Im]|List]),       % Record updated list of open windows
   !.                                   % Avoid continuing to next clause when backtracking
recd_new_win(Win,Im):-                  % What to do if no windows are already open
   remember(win_open,[[Win,Im]]).  % Record that window "Win" is open
% Remove a given window ("Win") from the list of open windows
recd_dead_win(Win):-
   recall(win_open,List),                  % Get list ("List") of open windows
   member([Win,Im],List,_),                % Window "Win" is associated with image "Im"
   remove([Win,Im],List,Remains),          % Remove [Win, Im] from "List"
   remember(win_open,Remains).             % Record what is left behind
```

---

7. *remember/2* and *recall/2* are built-in predicates that are peculiar to MacProlog. *remember(A,B)* records B as the currently stored value of property A. Any previously assigned value is lost. *recall(A,B)* retrieves the currently stored value (B) associated with property A. *recall(A,B)* fails if property A currently has no stored value. MacProlog properties are closely related to the properties employed in Lisp and provide a convenient way to handle parameters that are updated frequently. As far as the programmer is concerned, these predicates can be defined as follows: (In practice, they are implemented in quite a different way.)

```
remember(A,B):-
   retractall(property(A,_)),
   assert(property(A,B)).
recall(A,B):- property(A,B).
```

```
% "recd_new_im_for_win(Win,Im)" associates a new ...
% ... image ("Im") with window ("Win")
recd_new_im_for_win(Win,Im):-
   recall(win_open,List),              % Get list ("List") of open windows
   member([Win,OldIm],List,_),         % Image "OldIm" is associated with "Win"
remove([Win,OldIm],List,Remains),      % Eliminate that association
remember(win_open, [[Win,Im]|Remains]). % Append new window-image pair [Win,Im]
                                        % ... to the list "Remains" and store the result
```

6.4.8 Implementing PIP's basic commands

There are two distinct levels of basic operator in PIP:

a) "core" operators, defined directly in terms of the infrastructure just described.

b) higher level operators defined in Prolog, combining two or more of the "core" operators.

In this section, we will concentrate on the former category, by presenting several sample programs. These illustrate how the infrastructure is used to perform the necessary "housekeeping" operations, such as disposing of old images and maintaining information about the current and alternate images. In what follows, MacProlog properties record the identities of the current and alternate images and windows. Just before the execution of a command, these objects are identified as indicated in Table 6.8.

Table 6.8 Image and window identifiers for implementing PIP image-processing functions.

Property name	Identifies	Prolog variable
curr_im	Current image	CurrIm
alt_im	Alternate image	AltIm
curr_win	Current window	CurWin
alt_win	Alternate window	AltWin

During the execution of a command these property values are adjusted. The names may then seem anomalous. Simply remember that the four variables just mentioned are fixed at the instant just before the image manipulation begins.

swi/0 (switch images)

This operator does not perform any image processing at all. It simply interchanges the current and alternate images, which is achieved by interchanging the links between the images and windows. No images are moved or copied.

```
swi:-
   recall(curr_im,CurrIm),          % Find current image, stored in property "curr_im"
   recall(alt_im,AltIm),            % Find alternate image, stored in property "alt_im"
   recall(curr_win,CurrWin),        % Find current window, stored ...
                                    % ... in property "curr_win"
   recall(alt_win,AltWin),          % Find alternate window, stored in property "alt_win"
   new_win_im(AltIm, CurrWin), % Associate window "CurrIm" with image
   % "AltIm". That is, we associate the value held in the variable "AltIm"
   % with the window defined by the value held in the variable "CurrIm".
   new_win_im(CurrIm, AltWin), % Associate window "AltWin"
```

```
                                   % ... with image "CurrIm"
   remember(curr_im,AltIm),        % Record interchange. First part
   remember(alt_im,CurrIm).        % Record interchange. Second part
```

neg /0 (negate)

This predicate is one of the simplest image–to–image mapping operators. It uses no input parameters and generates no output results. It exemplifies the format for monadic operators.

```
neg:–
   recall(curr_im,CurrIm),         % Find current image
   recall(alt_im, AltIm),          % Find alternate image
   recall(curr_win,CurrWin),       % Find current window
   recall(alt_win,AltWin),         % Find alternate window
   neg_im(CurrIm,NewIm),           % Negate the image. Defined in Section 6.4.5
   new_win_im_disp(NewIm,CurrWin), % Display new image in current window
   new_win_im_disp(CurrIm,AltWin), % Display previous current image ...
                                   % ... in alternate window
   kill_im(AltIm),                 % Dispose of previous alternate image
   remember(curr_im,NewIm),        % Store references to new current ...
   remember(alt_im,CurrIm)         % ... and alternate images
```

thr/2 (threshold)

This is an example of a monadic operator which uses two input parameters and generates no output

```
thr :– thr(128,255).               % Default case. Arity = 0
thr(a):– thr(A,255).               % Default case. Arity = 1.
thr(Lo,Hi) :–                      % Arity = 2. Using the infrastructure
   recall(curr_im,CurrIm),         % Find current image
   recall(alt_im, AltIm),          % Find alternate image
   recall(curr_win,CurrWin),       % Find current window
   recall(alt_win,AltWin),         % Find alternate window
   thresh_im(CurrIm,NewIm,Lo,Hi),  % Pixels with intensities in range ...
                                   % ... [Lo, Hi] are set to white. All others are set to black.
   new_win_im_disp(NewIm,CurrWin), % Associate image "NewIm" ...
                                   % ... with window "CurrWin"
   new_win_im_disp(CurrIm,AltWin),% Associate image "CurrIm" ...
                                   % ... with window "AltWin"
   kill_im(AltIm),                 % Kill old alternate image
   remember(curr_im,NewIm),        % Remember where current image is
   remember(alt_im,CurrIm).        % Remember where alternate image is
thresh_im(Im1, Im2, ThrLo, ThrHi):– % Auxiliary predicate, useful for other PIP operators
   call_c([[Im1, ThrLo, ThrHi],Var,1003,Err],'MINE',0),
                     % Call C routine (no 1003) to perform thresholding operation
   Err=0,                          % Check that there was no error
   Var=[Im2],                      % Reformat the result. Var is a 1-element list. Im2 is s calar.
   recd_new_im(Im2).               % Record the new image
```

Notice the way that the control parameters (*ThrLo, ThrHi*) are passed to the C routine.

avg/1 (average intensity)

This is an example of an operator that generates an output parameter. Neither image is altered in any way.

```
avg:-                                  % Default case. Arity = 0
    avg(A),                            % Calculate average intensity and instantiate A
    write('Average intensity ='),      % A message for the user ...
    writenl(A).                        % ... in the default output window
avg(A):-
    recall(curr_im,CurrIm),            % Find current image
    MinMaxAvgArea(CurrIm,[_,_,A,_]),   % C routine calculates average intensity.
    !.                                 % Inhibit backtracking
% The following multi-function predicate forms the basis of "gli/2" and "cwp/1", ...
                                       % ... as well as "avg/1".
MinMaxAvgArea(Im1,[Min,Max,Avg,Area]):-
    call_c([[Im1],Var,9610,Err],'MINE',0), % C routine (no. 9610) calculates minimum,
        % maximum, and average intensities, plus area (no. of pixels with intensity ≥ 1)
    !,
    Err=0,                             % Check that no error occurred
    Var=[Min,Max,Avg,Area].            % Return all four values calculates
```

In this case, four parameter values calculated by the C program (*Min, Max, Avg, Area*) are passed to Prolog via the variable *Var*. However, *avg/0* needs only *Avg* and the remainder are discarded.

Other operators are implemented in a broadly similar manner. A complete list of Prolog+ operators currently available in PIP is given in Table 4.7.

add/0 (add)

This predicate exemplifies the implementation of the dyadic operators, which combine the current and alternate images to generate a third image. The (previous) current image is retained, becoming the new alternate image, while the (previous) alternate image is destroyed.

```
add:-
    recall(curr_im, CurrIm),               % Find current image
    recall(alt_im,AltIm),                  % Find alternate image
    add_im(CurrIm, AltIm, NewIm),          % Add the current and alternate images
    recall(curr_win,CurrWin),              % Find current window
    new_win_im_disp(NewIm, CurrWin),       % Display the new image ...
                                           % ... in the current window
    recall(alt_win,AltWin),                % Find alternate window
    new_win_im_disp(CurrIm, AltWin),       % Display the current image ...
                                           % ... in the alternate image display
    kill_im(AltIm),                        % Destroy the alternate image
    remember(curr_im, NewIm),              % Record image "NewIm" as being ...
                                           % ... the new current image
    remember(alt_im, CurrIm).              % Record image "NewIm" as being...
                                           % ... the new alternate image

add_im(Im1, Im2, Im3):-
```

```
call_c([[Im1,Im2],Var,1001,Err],'MINE',0),      % C routine (no. 1001) ...
                                                % ... adds the current and alternate images
Err=0,                                          % Check that no error occurred
Var=[Im3],                                      % Reformat result Im3 gives the identity of the new image
recd_new_im(Im3).                               % Record the new image identity
```

6.5 WIP

6.5.1 WIP design philosophy

We have already seen that the Macintosh family of computers provides a very satisfactory basis for building an interactive vision system. However, many people regard the Macintosh OS as an eccentric choice and would be happier to accept a system that uses the same hardware and operating system that they perceive "everybody else" as using. While we believe this to be a purely emotional and erroneous judgement[8], the authors are convinced that PIP would have received greater attention had it been based on the "standard PC" (i.e., an Intel (Pentium) processor and the Microsoft Windows operating system). With this in mind, WIP was developed, albeit with some reluctance, since PIP had already proved itself to be more than capable of fulfilling the needs of an IVS for Machine Vision prototyping. Thus far, WIP cannot match all of the facilities that PIP provides. In particular, PIP has a much larger command repertoire and WIP's user interface and machine interface are both weak in comparison. However, WIP can run with a range of host applications. To date, host applications have been developed for WIP in C, Visual C++, Visual J++, Java, Delphi, Visual Basic, and Prolog. Exactly the same mnemonic commands can be used in the top-level application, whatever host language is used. (For the sake of compatibility, WIP and PIP commands use the same mnemonics and parameters.)

WIP is based on a multi-layer software structure forming a Dynamic Link Library (DLL), written in standard C. The host application is interfaced to the top layer in this hierarchy and does not form part of the DLL. Each layer within the DLL communicates only with those layers immediately above and below it. For the sake of robustness, the protocol at each inter-layer interface is well defined. The host application can be changed at will, with no modifications being required within the DLL. The same DLL can run simultaneously under several host applications, written in different languages. In this respect, WIP is potentially more versatile than PIP, which provides an interface only to Prolog.

It should be noted that, at the time of writing, the authors have no plans to develop WIP further. The arguments which led us to begin work on WIP have more

8. Clear evidence for this emotional attachment to the "standard PC" can be found in the fact that numerous people queued (at midnight) to be among the first to buy the Windows95 operating system. This provides approximately the same facilities as those available in the Macintosh Plus computer available nine years earlier!

recently been seen as encouraging the development of CIP instead. Since the latter is based on Java, which offers multi-platform operation and easy access to the Internet, the authors believe that CIP has an even greater potential user base than either PIP or WIP.

6.5.2 Host application

The host application generates the user interface to WIP. This may provide an interactive dialog, or a programming environment that offers facilities for developing sophisticated programs, consisting of standard code plus embedded image-processing commands. Any host language that supports the use of DLLs and is Windows32-compatible may be used. As we have already indicated, simple host applications in C, Visual C++, Visual J++, Java, Delphi, Visual Basic, and Prolog have all been developed to prove the concept. However, none of these has yet been extended far enough to enable us to make a detailed comparison of the benefits they provide for an IVS. In each case, the interface is very straightforward, which we will illustrate by listing two Visual Basic programs. In the first example, we show how the WIP DDL may be interfaced to a standard program. In the second example, we provide the full code to generate an interactive dialog.

Example 1: Programmed Mode

The following defines a procedure called *Compoundfunction* which performs a sequence of 3 operations equivalent to the PIP program [*neg, thr(123,145), cwp*].

```
Private Sub Compoundfunction
   If (WIPModelCallBuffered("neg,thr(123,145), cwp", Result$) = 0)
                                  Then WIPReportError("Compoundfunction")
'____The function WIPModelCallBuffered is declared below.
   End If
End Sub
```

The procedure call *Compoundfunction* can then be inserted into any standard Visual Basic program. Clearly, a series of definitions of this type is needed to provide a comprehensive set of image-processing utilities. Although this is rather tedious to write, it is a straightforward exercise.

Example 2: Interactive mode

The following code generates the *Data Entry* window shown in Figure 6.4. The *Command Line* field has the internal name *txtCommand* and the *Result* field has the internal name *lblResult.*

```
'---- Register the application with WIP. Create a private memory space and image windows.
Private Sub Form_Load()
   WIPModelCreate
End Sub
'---- Create a results buffer ----------------------------------------------------------------
Private Sub cmdRun_Click(Index As Integer)
   Dim RString$
'---- Fill the buffer with spaces so that VB gives it enough space to store the results.
```

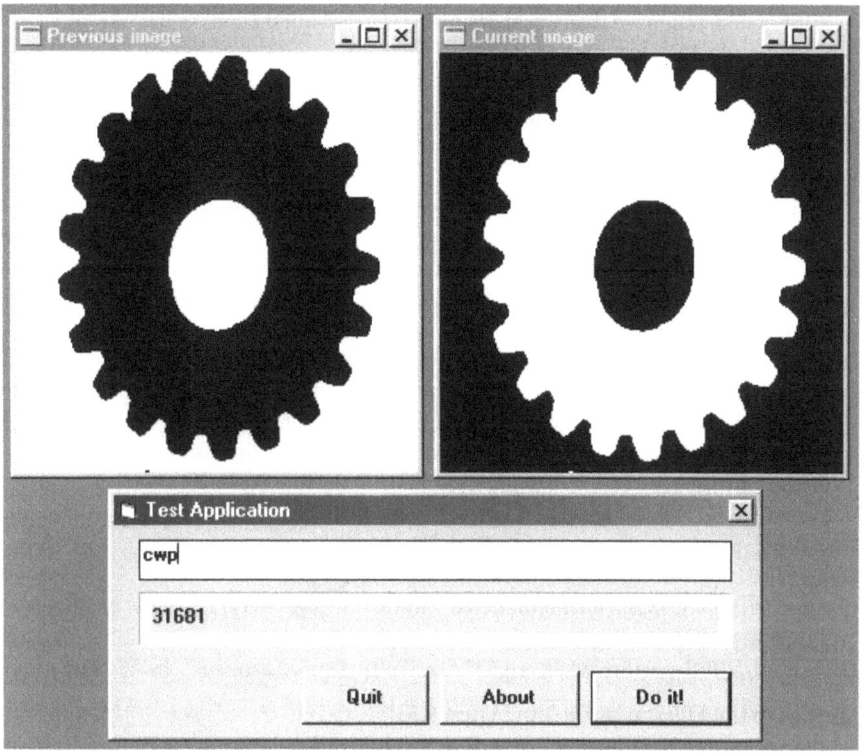

Fig. 6.4 Screen display for WIP, using the Visual Basic program discussed in the text.

```
   RString$ = String$(1024, 0)
' ---- Get command from input box. Ask WIP to execute it.
' ---- Then check returned value to see whether it worked.
   If (WIPModelCallFunctionBuffered(txtCommand.Text, RString$) = 0) Then
'----- An error! Display the error message.---------------------------------------------------
   WIPReportError ("VB Basic")
Else
' ---- It worked, so display any results returned in the results buffer. ----------------------
      lblResults.Caption = RString$
   End If
End Sub
' ---- Close any windows, release any private memory, unregister application with WIP.
Private Sub cmdClose_Click(Index As Integer)
   WIPModelDestroy
   End
End Sub
' ---- Unregister application with WIP, destroy private memory space & image windows.
Private Sub Form_Unload(Cancel As Integer)
   Cancel = 0
   WIPModelDestroy
End Sub
```

```
' ----------------------------------------------------------------------------------------------------
    A list of declarations so that Visual Basic can call relevant functions in the WIP DLL
' ----------------------------------------------------------------------------------------------------
Option Explicit
' ---- Declare the external functions in WIP. NOTE: The VB version of the function name
'      does not include the underscore character, because this seems to cause problems.
'
' ---- WIPModelCreate registers application with DLL, creates appropriate
'      windows and sets up private memory space.
Declare Sub WIPModelCreate Lib "wip32" Alias "WIP_ModelCreate" ()
' ---- WIPModelDestroy - Unregisters the application, deletes any windows,
'      and frees private memory.
Declare Sub WIPModelDestroy Lib "wip32" Alias "WIP_ModelDestroy" ()
' ---- WIPModelCallFunctionBuffered calls the WIP DLL with a given command line.
' ---- WIP carries out the command using the images in the image windows.
' ---- Any text results are returned in the buffer given in the function call.
' ---- The function returns TRUE (a non zero value) if the function is successful
'       and FALSE (zero) if an error occurred.
' ---- WIPReportError can be called to display the error.
Declare Function WIPModelCallFunctionBuffered Lib "wip32"
   Alias "WIP_ModelCallFunctionBuffered" (ByVal CString$, ByVal RString$) As Integer
' ---- WIPReportError - Display the last error which occurred.
Declare Sub WIPReportError Lib "wip32" Alias "WIP_ReportLastError" (ByVal Title$)
```

Output to the DLL (at point 1 in Figure 6.5)

We will trace the flow of control as WIP processes the following command sequence (as entered in the dialog box, Figure 6.4):

neg, thr(123,145), cwp

When the user clicks on the "*Do It*" button, the Visual Basic program listed above sends the following information to the Command Line Function

1) Address of the command string. This string consists of "*neg, thr(123,145), cwp*".
2) Address of the Result String. This consists of the empty string: "". (The results of any image-processing operation, such as *cwp*, *cgr*, *dim*, etc., which generate output values will be appended to this string.)

6.5.3 Command line function

The Command Line Function segments the incoming command string that it receives from the Host Application. Individual commands are then executed in turn within an iterative loop.

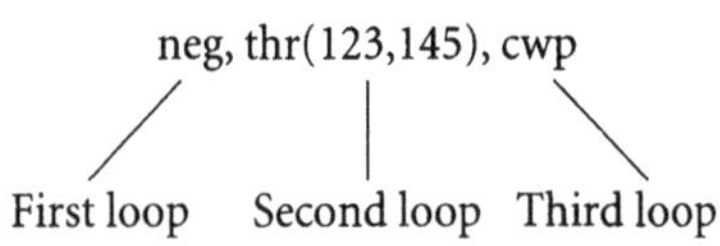

The Command Line Function is responsible for managing the creation, naming, and destruction of the image windows. This is a simple matter, if we adhere to the standard 2-image operating paradigm. In this case, the Command Function simply has to keep track of the identities of the Current and Alternate images. (As we will

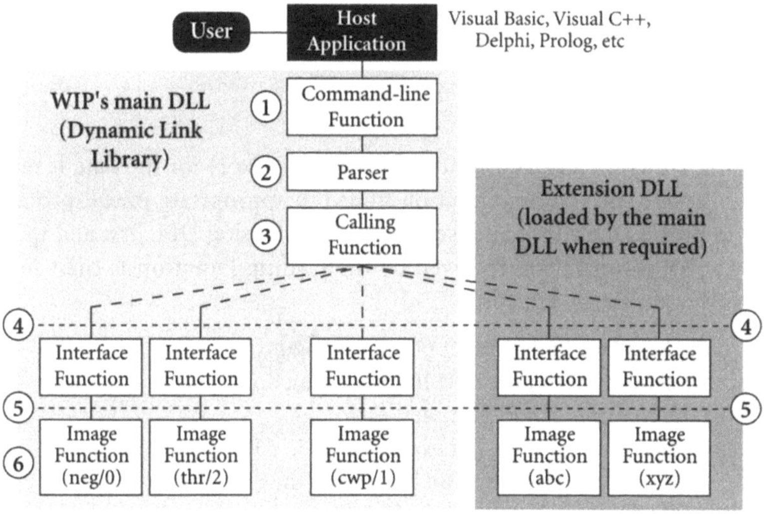

Fig. 6.5 Organization of WIP.

see later, modules operating at lower levels in the DLL hierarchy will generate other images which hold the results of the processing. The Command Line Function is aware of these and manages their creation, naming and destruction.) However, it is possible to program the Command Line Function to support other modes of operation, for example to maintain an archive of the most recent processing results. An image stack can also be supported at this level.

Output to the Parser (at point 2 in Figure 6.5)

This consists of the following four items:

a) Individual command string. This typically consists of items such as "*thr(123,145)*"
b) Address of the result string. This consists of the empty string ""
c) Address of the Current image
d) Address of the Alternate image

6.5.4 Parser

The Parser splits the command string received from the CommandLine Function (e.g., "*thr(123,145)*") into two parts:

a) Function name. e.g., "thr"
b) List of parameters. e.g., [123, 145]

Output to the Calling Function (at point 3 in Figure 6.5)

This consists of five items:

a) Name of the command. For example: "thr"
b) Pointer to the arguments list. For example, [123,145]
c) Pointer to the Results List. This still consists of the empty string ""

d) Address of the Current image
e) Address of the Alternate image

6.5.5 Calling function

The Calling Function acts as a router (or demultiplexor) for the data it receives from the Parser. The Calling Function finds the appropriate image-processing module to process the command, searching the extension DLL first and then the main DLL. All information received by the Calling Function is then sent on unmodified to the Image Function.

Output to the Interface Function (at point 4 in Figure 6.5)
Five items are sent to the appropriate Interface Function:
a) Name of the command. For example: "*thr*"
b) Pointer to the arguments list. For example, [*123,145*]
c) Pointer to the Results List. This still consists of the empty string ""
d) Address of the Current image
e) Address of the Alternate image

This is exactly the same data as that which the Calling Function receives from the Parser.

6.5.6 Interface function

The Image Functions in the bottom layer of the WIP hierarchy present a range of different interfaces standards to the higher levels. In order to standardize the format, each Image Function is provided with a tailored Interface Function, which decodes the parameter list received from the Calling Function. Each value in the parameter list is then associated with a variable used by the corresponding Image Function. In addition, the Interface Function verifies that the number of variables provided is correct and checks each parameter for type and range. It then calls its associated Image Function.

Output to the Image Function (at Point 5 in Figure 6.5)
There is no standard format, since each Image Function presents special requirements.

6.5.7 Image function

All processing, generation, analysis, loading, and saving of images is performed in the Image Functions. Each command mnemonic is associated with one Image Function module. If a given Image Function alters an image, it will create a new (third) image window, in addition to the Current and Alternate Images. This new image window will contain the results of the processing. Some Image Function modules, such as those responsible for image analysis, do not change either the current or alternate image. In this case, no "results" image is created and a special code is passed up the hierarchy to indicate this. Image Function modules which derive measurements append these to the Result List, which is also passed back up the DLL hierarchy.

6.5.8 Returning results and errors

Errors and results are passed up the hierarchy in a similar manner. A simple code indicates the outcome of obeying each command. See Table 6.9.

The integer variable representing this code can, of course, be modified by any function in the hierarchy that detects an error. For example, the Interface Function may detect that a parameter is out of range, or of the wrong type, or that too few parameters have been specified. The Calling Function must be able to signal an error if no appropriate Image Function exists. The Parser must be able to signal that it has detected a syntax error. While there is a single variable that indicates the presence/absence of an error, the type of error is indicated by the value stored in a global variable that any module can overwrite.

The histogram operators *hpi* and *hgc* return many values in the form of a list of lists of integer pairs. It is sometimes necessary to restucture complex strings such as these. This task is performed by the Parser, which, of course, receives the results generated by the appropriate Image Function via the Calling Function.

Table 6.9 Code indicating the outcome of WIP commands

Code	Meaning
0	An error has occurred somewhere.
1	The Image Function did not define a new image and no new images were created.
Integer > 65536	Address of the new image.

6.6 Concluding remarks

In this chapter, we have discussed in some detail the internal structure and operation of three systems: VSP, PIP and WIP. In addition, numerous "general purpose" image-processing packages, which could be used for problem analysis and algorithm prototyping in Machine Vision, are described elsewhere [BAT91a]. However, the advent of the Internet provides new opportunities that are beyond the capability of any stand-alone system. For this reason, we defer discussion of two further (Java-based)IVS packages, CIP/JIFIC and JVision, until the next chapter.

VSP provides a simple and rapid way to "add intelligence" and a wide range of other functions to an existing image processor. Systems of this type have been in use since the late-1980s and were among the first Prolog+ systems to be built. The same general approach could be adopted using different dialects of Prolog, possibly hosted on platforms other than a Macintosh computer. It would, for example, be a straightforward matter to build a VSP system using WinProlog and a stand-alone image processor. Indeed, a dual-processor system could easily be built combining virtually any "exotic" language, such as Lisp, Prolog, SmallTalk, etc., with image processing. The essential point to note is that, compared with the time needed to process every pixel in a large image, the delays introduced by the low-bandwidth link between the host program and the image-processing engine are tolerable. This is helped by the fact that interactive prototyping systems in Machine Vision do not normally demand high processing speed, provided user concentration can be maintained.

In terms of its functionality, PIP is by far the most advanced IVS that we or our immediate colleagues have developed to date. In addition to having an extensive image-processing repertoire, PIP is capable of performing a wide range of other functions, including the control of a variety of electromechanical, lighting, and other devices. PIP also provides a sophisticated user interface, even permitting limited processing of natural language. PIP was designed specifically to have a long working life. However, this goal was seriously jeopardized when technical support for MacProlog was withdrawn. Provided MacProlog continues to run satisfactorily under the ever-changing Macintosh operating system, PIP can be kept alive. Although the linkage between the host application, Prolog, and the image-processing subsystem is based upon standard Macintosh inter-application communication techniques, this is PIP's Achilles' heel and makes it impossible to transfer PIP easily to another platform.

WIP resembles VSP in the fact that the architecture allows a variety of systems to be built with relative ease. While VSP allows different image processors to be used in combination with a fixed top-level controller, in WIP the situation is reversed; the image processor is fixed and the language used within the top-level controller can be chosen at will. WIP is intimately tied to the Windows operating system and cannot therefore be transferred to other machines.

7 Vision systems on the Internet

Brain n. *An apparatus with which we think that we think.*
Ambrose Bierce

7.1 Stand-alone and networked systems

PIP and WIP are two of the latest developments in interactive vision systems that began in the mid-1970s, with SUSIE. All of the systems built by the authors since then operated in stand-alone mode. There have been brief and tentative incursions into networking, although these have never been fully developed, because at the time there was no universally accepted protocol for worldwide digital communication [BAT91a]. The advent of Internet technology[1] has made it possible to develop new avenues of approach that were not possible, nor even anticipated, in the mid-1990s. The Internet revolution is just beginning and already owes a great deal to the Java programming language, which was the chosen vehicle for developing CIP, the immediate successor to PIP. However, we will see in this chapter that our plans for a Web-based toolbox for vision system designers is much more ambitious than this.

When PIP was developed in the mid-1990s, it was envisaged that it would have a long working life, since it is based upon the standard Macintosh operating system and uses two well-established languages: C and Prolog. However, difficulties were encountered in updating PIP after the manufacturer's support for MacProlog faltered. (Commercial pressure on the company led it to pay greater attention to the market for Windows-based products, at the expense of Macintosh software.) In retrospect, it is clear that the dual-language approach used to implement PIP would eventually cause problems. It should be understood that the same difficulty is also inherent in WIP, since it also uses two languages. Hence, a completely new approach was sought that would avoid this problem in the future. It happened that our realization of this coincided with the introduction of Java. At first, we envisaged Java as simply another computing language for writing stand-alone applications. However, it soon became clear that its networking capabilities offered exciting possibilities that we had not previously even imagined.

The authors' latest Java-based language for interactive image processing is called CIP (*Cyber Image Processing*) and is written in Java. On its own, CIP is unremark-

1. When we use the term Internet, we imply both the public network and "closed" Intranet networks that are based on the same technology. The latter may interconnect different factories/offices within a large multisite company or other organization.

able, since it offers no more than either PIP or WIP. A package of device-control software is linked to CIP and is also programmed in Java. This is called JIFIC (*Java Interface to the Flexible Inspection Cell*). See Figure 4.26. The authors of CIP and JIFIC deliberately chose to implement all image processing, user interface, and device-control functions in Java. However, as we shall see later, this quest for homogeneity does not mean that we are discarding Prolog. In its rôle providing top-level control for an IVS, Prolog brings considerable benefits, which we are unwilling to forfeit. We will explain later how we envisage using a single language while retaining the benefits that Prolog brought to PIP and other Prolog+ systems. One of the main advantages that Java appears to offer is a promise of long-term stability. Although the language is very young and is still evolving, its very rapid rise in popularity is impressive, giving confidence that it will be available "by public demand" for many years to come. The development of Java is closely controlled by one of the world's major computer manufacturers, Sun Microsystems, Inc. However, numerous other companies are basing their products and services on Java. In addition, many university Computer Science departments have realized Java's potential and teach it, at both first-degree and postgraduate levels. The advent of Java is surely one of the most significant developments to take place in computing in recent years.

7.2 Java

In November 1995, Sun Microsystems, Inc. announced the first working ("beta") version of the Java programming language. Java's unprecedented rise in popularity since then is attributable to four main factors:

1. Java is the first really effective and widely accepted cross-platform computing language.
2. Java is able to support access to the Internet and World Wide Web.
3. The same language is able to operate at different levels of sophistication, ranging from free-standing (embedded) programs operating domestic appliances, to complex suites of software in which several users, possibly working many miles apart, cooperate interactively with one another.
4. There is a comprehensive and rapidly growing collection of facilities, called the *Class Libraries*, which are fully integrated with Java and provide a wide range of utilities.

It is interesting to note that Java was originally intended for applications such as industrial automation and controlling devices, including domestic equipment, automobiles, etc. using embedded processors. Its potential as a language for Internet programming was realized later.

7.2.1 Platform independence

One of Java's most important advantages over other computing languages is that it provides almost complete platform independence. This means, for example, that CIP is able to run on any combination of processor architecture and operating system, provided that it is equipped to provide a Java Virtual Machine. (JVM, Figure 7.1) Java guarantees that the results of executing a program are identical on all computer architectures and operating systems, although there may be insignificant cosmetic differences in the appearance of on-screen items such as windows, frames, and buttons.

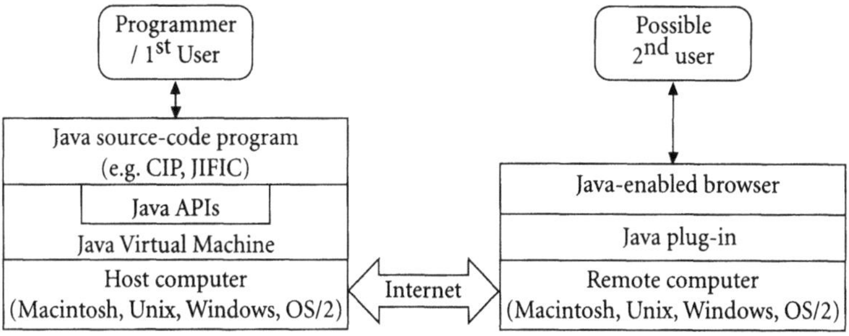

Fig. 7.1 Java Virtual Machine (JVM).

It is universally acknowledged that program development is both expensive and time consuming, especially in cases where the same program must be available to a variety of computing platforms. Even programs written in C or C++ are not truly transportable across different platforms. In many cases, transportability is not achieved even across different generations of the same family of computers. Java purports to overcome these limitations, once and for all! Once it has been compiled and debugged, any Java application can run on the JVM hosted on almost any make of computer, of any generation. This has the great benefit that programs will not need to be rewritten as often as they have in the past. Despite its youth, Java has already been proven to be stable, robust, and capable of building "real-world" applications. It is the Java Virtual Machine that has made this possible.

7.2.2 Applets, applications, and servlets

The majority of Java programs run via the Internet, through a standard Internet browser. Java programs that run in this way are called *applets*, and are commonly used to enhance the functionality of a Web page. On the other hand, Java programs which operate in stand-alone mode are called *applications* and require an interpreter for execution.

An applet may be linked to a web page, using a special HTML tag of the following form:

```
<html>
...
<applet code = "Myprogram.class" width = 250 height = 100>
</applet>
...
<html>
```

where *Myprogram.class* is the name of the compiled Java applet (bytecode). When this web page is loaded by a web browser, the latter also downloads the file *Myprogram.class* from the Web server and executes the program it contains on the client computer. Since a Java applet is associated with a frame managed on its behalf by the browser, the applet has access to all of the facilities that the browser enjoys, including graphics, drawing, text formatting, and event handling.

There is an important difference between applets and applications in their right to access resources. Local (i.e., client) resources are not normally available to an applet, whereas they are to an application. Hence, an applet cannot load or save a file on the client computer unless special privileges have been granted to it. Nor can an applet initiate the execution of another program on that machine. These restrictions are imposed for reasons of security. However, an applet can be granted the right to use local resources available on the client computer. To do this, the applet must be declared to be *trusted* by the client user. This declaration is certified as being so with a *digital signature.* Such an applet is said to be *signed* or *trusted.*

A Java application runs on a single stand-alone computer and cannot make use of the network as an applet does. An applet is stored on one computer (the server), is downloaded automatically to another (the client) when requested, via a WWW browser, and is then executed on the client machine.

A third category of Java program exists, called a *servlet.* This is a Java program that is stored on the server and is executed there, at the request of a browser running on the client computer. It has full right of access to all resources on the server computer.

Finally, it is interesting to note that a WWW browser or server can be programmed in a few lines of Java code.

7.2.3 Security

Although it is closely related to C and C++, Java is organized rather differently, with a number of C/C++ features omitted and other facilities borrowed from other languages. Java is intended to be a *production* language, not a *research* programming language. Hence, its designers avoided new and untested features. The nature of Java makes development of robust programs (applets or applications) relatively straightforward. Robustness is not normally determined primarily by logic errors or programming bugs; the main source of difficulties is usually faulty memory management. Java provides built-in mechanisms for memory management and does not allow the programmer to access memory resources directly. Pointers are the main source of run-time errors in a language such as C. In most cases, these cause either the application, or worse, the whole system, to crash. Elimination of pointer manipulation makes Java a more robust programming language than either C or C++. Furthermore, a Java applet will not stop running, even if a critical run-time error (called an exception) occurs. This is possible because Java runs on a virtual machine.

7.2.4 Speed

It is important to understand that Java is currently slower than most regular high-level languages, such as C or C++. The latter are compiled to produce target code in machine language. On the other hand, Java code is interpreted, although the term "compilation" is often used when speaking or writing about Java programming. There are many run-time checking operations to be performed and these limit the speed of processing of Java programs. However, *Just-In-Time* (JIT) compilation of an applet improves matters considerably. When a Java source code file is compiled, the compiler builds byte-code files that the JVM then interprets and executes. A JIT compiler is loaded automatically as soon as the byte-code files have been downloaded, and compiles them into the machine language of its host hardware. This means that the applet takes a longer time to start, after which it works considerably faster. However, even with this feature, Java is slow compared with a comparable well-designed C program.

Recent developments in Java technology offer the promise of further speed improvement. Furthermore, the advent of special-purpose integrated circuits ("Java chips"), designed to execute Java code directly, promise high-speed execution for all applets. Java is the lowest-level language that such a Java chip can execute. Since Java allows efficient networking to be achieved, it should be possible in the future to build high-speed parallel-processing arrays for such tasks as analyzing complicated visual scenes. This will, no doubt, be used in the future to achieve high processing speed in multiple Java-chip systems. Indeed, we should soon see a Java-based image-processing system being used in certain applications where real-time operation is required.[2]

7.2.5 Interactive vision systems in Java

Even with today's technology, the speed of processing for Java programs is acceptable for most tasks in prototyping vision systems. In other words, a Java application running on a standard desktop computer can successfully implement an IVS that has acceptable response times (Table 7.1).

Table 7.1 Execution times for CIP on various computing platforms. Image resolution: *300×300* pixels, 8 bits per pixel. Pentium Pro processor running at 180 MHz, with various operating systems and WWW browsers.

Operation	Time, ms	Operation	Time, ms
Average intensity	10	Lowpass filter	100–110
Binary edge detector	60–70	Negate	40–50
Centroid	10–20	Shift image	180–280
Compare images	200–240	Switch images	20
Enhance contrast	60–80	Threshold (fixed)	50–70
Euler number	30	Unsharp mask filter	510–550

2. Recall that an array of identical von Neumann processors can be used to produce a system that processes images at high throughput rates but with a high latency. (Figure 5.15)

There also arises the theoretical possibility of designing at least some shop-floor inspection/control systems based on Java that do not require the vision engineer to visit the factory where the target system will be used. At the time of writing, the relatively low execution speed of Java programs may suggest that this is merely a dream. However, the "Java chips" mentioned earlier may soon make this a reality.

CIP is available both as a signed applet and as an application. It is possible for either of these to use local resources. For example, CIP can read, store, and process image files on the client computer. It can also digitize an image from a camera without requiring other software.

7.3 Remotely-Operated Prototyping Environment (ROPE)

We turn away now from considerations of the technical issues of Java programming to discuss a new and fundamental question: What can we do for Machine Vision by using Internet technology?

ROPE (Figure 7.2) is the name given to a large suite of programs that is under development at Cardiff and which will eventually provide a comprehensive set of tools to assist a vision system designer. CIP is just one component within ROPE.

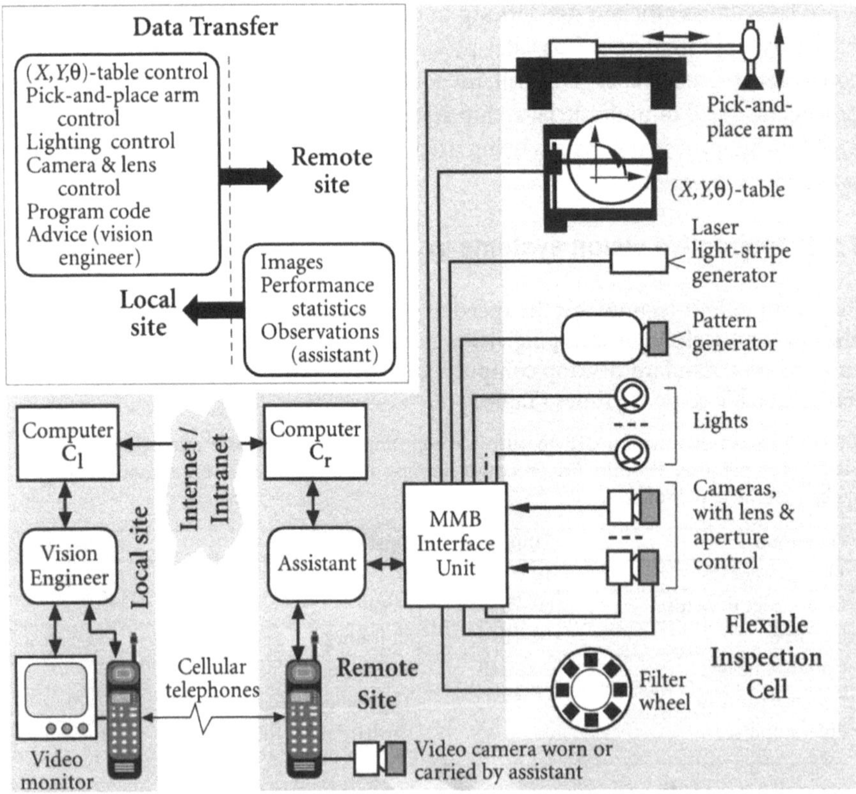

Fig. 7.2 Present organization of ROPE.

The design of ROPE is based on our belief that performing remote image acquisition (i.e., automatic configuration of the camera, optics, lighting, and handling mechanism) and image processing are likely to have a major impact on the productivity of vision systems engineers. One of the goals in our research is be to ensure that systems designers will be able to avoid traveling around the world unnecessarily. Site visits to factories are currently needed to examine the working environment for proposed vision system installations. They are both time-consuming and tiring. In many but clearly not all cases, we believe that the necessary background information could be acquired using Internet and/or mobile phone technology, to provide remote sensing and control.

7.3.1 What ROPE does

It is planned that ROPE will eventually perform the following tasks remotely (see Figure 7.3).

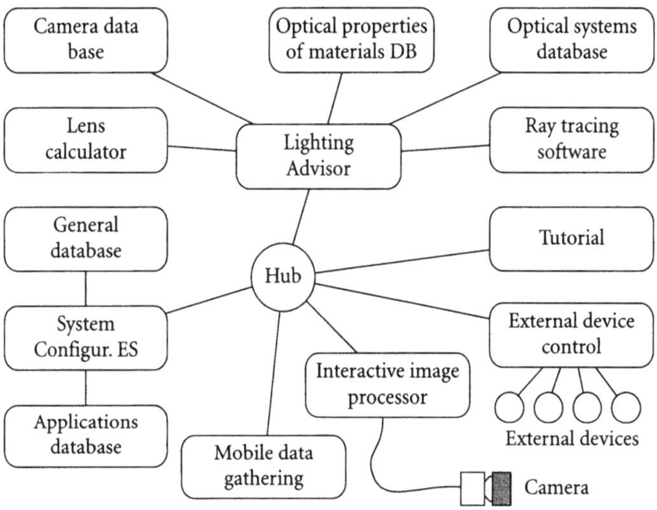

Fig. 7.3 a) Planned extensions and use of ROPE for designing an inspection system remotely: Overall view of the extended ROPE system.

1. Organize the collection of suitable product samples from a factory. This will be called the *remote site* and may be hundreds or even thousands of kilometers away from the vision system engineer's workplace. (The latter will be called the *local site.*)
2. Set up a suitable lighting-optics-camera system to view the objects being inspected, preferably *in situ*. That is, they should be examined during the feasibility study in the location and in exactly the same state as they would be inspected by the final on-line target system. It will be assumed that the prototyping image-acquisition subsystem will be located at the remote site. It will probably have the same general form as the Flexible Inspection Cell (FIC) described in Chapter 4, or the ALIS system described in Chapter 1.
3. Digitize a set of images and transfer them to the vision engineer's workstation.

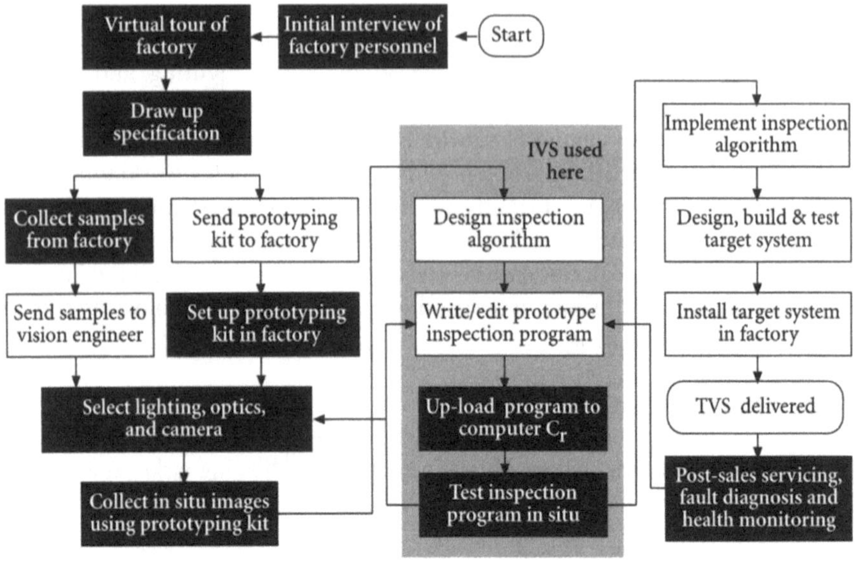

Fig. 7.3 (b) Planned extensions and use of ROPE for designing an inspection system remotely: Methodology. Black boxes indicate operations that are performed remotely. The grey rectangle indicates operations performed by the interactive image processor.

4. Analyze these digital images experimentally and, based on the lessons learned, write a program which performs the prescribed inspection task.
5. Download that program onto a computer attached to the image-acquisition subsystem at the remote site. In this way, the vision engineer sets up a prototype automated visual inspection system. (This will almost certainly not be as fast as the TVS whose design is the object of this exercise. However, the prototyping system will probably be much more versatile and be tailored for each new application very quickly and cheaply.)
6. Run that program, collect performance statistics as well as additional sample images and, if necessary, refine the program, repeating steps 4 and 5.

7.3.2 Assistant

To be realistic, bearing in mind our present state of knowledge and the communication difficulties caused by the Internet's limited bandwidth and transmission delays, the vision engineer will probably require a lot of help. The following points should be noted:

a. Help can usually be provided by an intelligent, but inexperienced, engineer/physicist. This person will be called the *assistant*, and might typically be a new graduate, with little or no specialized training in Machine Vision. However, he should be thoroughly familiar with the application requirements, as well as the working environment and practices of the factory where the TVS is to be installed.

b. The vision engineer will be helped considerably if he can receive a carefully selected and representative collection of products for inspection ("widgets") by post/courier. These should be collected by the assistant, working under the guidance of the vision engineer. It is important that these widgets be delivered to the vision engineer in a "fresh" condition; they must not corrode, oxidize, dry out, or otherwise change their appearance during transportation and storage.
c. The vision engineer must be able to talk freely with the assistant, to guide the latter as the latter walks around the factory, carrying a video camera mounted on a hard hat (Figure 7.4). The engineer should be able to view images of the factory environment and collect various other items of information about the application task and possible solutions. This involves the exchange of spoken and written text, sketches, scanned images, and video images. The authors are currently

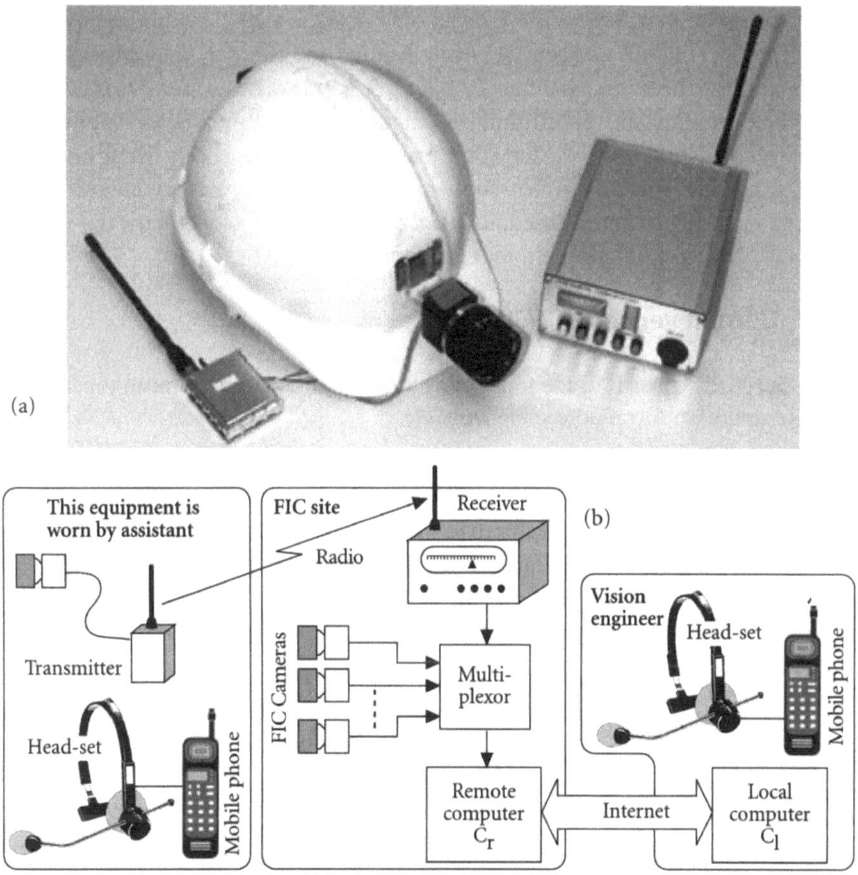

Fig. 7.4 Mobile data gathering in a factory. (a) Hard hat fitted with a small video camera and short-range radio transmitter (left). The latter is carried on a waist belt, together with a 2.5A-hr lead-acid accumulator (not shown). A 4-channel receiver is shown on the right. (b) Connection diagram. Speech is transferred by mobile phone, to facilitate real-time interaction between the vision engineer and his assistant. Video signals from the camera mounted on the hard hat are sent either via high-bandwidth mobile phone (not shown), or the Internet.

investigating the rôle and limitations of the Internet. The recent development of computers and video cameras that can be worn on the person, as well as radio/ IR/power-line connection to the Internet, are all relevant to this part of our research, which is still ongoing. We are also investigating the use of mobile phones as an alternative to, or to augment, the Internet (see Figure 7.4).

7.3.3 Vision engineers should avoid travel

For some applications at least, a vision engineer should eventually be able to avoid visiting the remote site during the design process. (It may, of course, be necessary for some representative of the vision equipment supply company to do so during the installation of the target inspection system but this is a separate issue and is outside the scope of our present discussion.) For remote information gathering and system design to be effective, there must, of course, be a good rapport between the vision engineer and his assistant. A clearly defined and properly understood system specification is therefore essential [BAT98a]. A good way to achieve this is via the use of a comprehensive questionnaire, such as the Machine Vision Applications Requirements Check List [MVAa]. This provides a comprehensive list of questions covering a wide range of topics, such as contract details, factory environment, working practices, materials used, range, size, and type of widget, how the widgets are manufactured and presented for examination, etc.

7.3.4 Current version of ROPE

ROPE currently consists of the following hardware and software, linked together via the Internet or an Intranet (see Figure 4.27).

Remote site

a. The Flexible Inspection Cell (FIC) mentioned in Chapter 4.
b. Java-enabled computer (C_r), fitted with a frame-grabber card.
c. Digital control unit providing the interface between the FIC and C_r.
d. Software to control the FIC.
e. Image digitization software running on C_r.
f. High-speed interface to the Internet.
g. Assistant.

Local site

h. Java-enabled computer (C_l).
i. Software for interactive image processing.
j. Program to control the FIC remotely. (This interfaces to the program mentioned under (d) via the Internet, and these together form the JIFIC software.)
k. Vision engineer.

The reader is asked to note the following comments relating to the present design of the Flexible Inspection Cell:

- The distribution of the lamps around the FIC is evident from Figs. 4.11 and 7.5.
- The laser light stripe generator (LS) is normally used in conjunction with the oblique camera (OC).
- The side camera (SC) is normally used in conjunction with the backlighting source (BI), to provide a bright background.

(a)

(b)

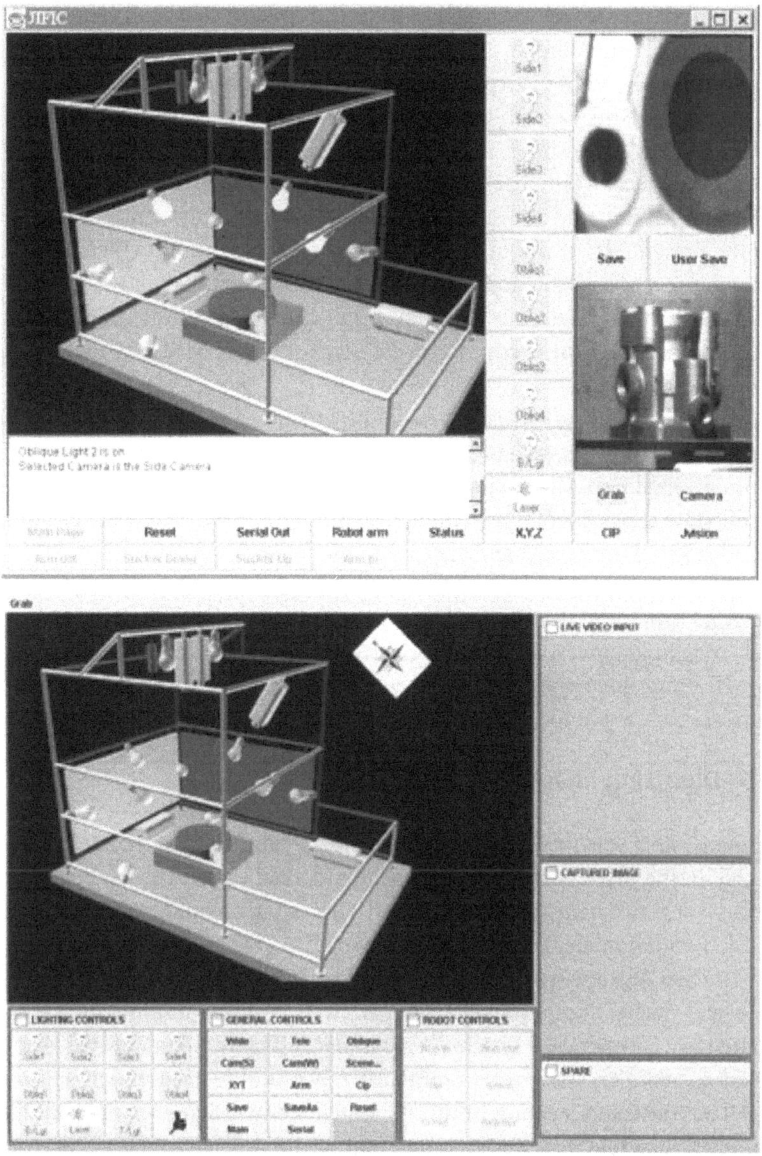

Fig. 7.5 Main control window for JIFIC. (a) Original design. Top left: Layout of the FIC. Bottom right: View from the currently selected camera. Top right: Image stored and thereby made ready for processing. (b) Revised design based on the Java Media Framework.

- The front camera (FC) faces a matte black screen, so that it views objects against a dark background.
- The wide-field (WC) and narrow-field (NC) overhead cameras. A single camera fitted with a motorized zoom lens would be more difficult to control precisely. However, the NC and WC optical axes are not aligned exactly. The offset must be taken into account when trying to register digital images derived from them.
- When the Flexible Inspection Cell was first conceived in the early 1980s, a motorized filter wheel was suggested for use with an overhead camera [BAT85]. Commercial units similar to our original design are now available. Although the present FIC does not incorporate such a device, the hardware and software interface mechanisms exist to control it. A filter wheel can, of course, hold color filters, polarizers, gratings, neutral-density filters, pinhole stops, etc.
- The MMB has already been used on a separate project to control a motorized pan-and-tilt unit and a zoom lens. Although neither of these mechanisms is used in the present FIC, it would be a simple matter to incorporate them as well.
- Many cameras available today have a serial input port that allows a computer to control such parameters as black level, gain, integration time, color balance, etc. Cameras such as these could also be controlled via MMB, which has several unused serial ports.
- The FIC lighting system could easily be enhanced by adding extra computer-controlled lamps, as well as UV and IR sources. There are numerous TTL signal lines available via MMB's parallel ports. These can be used to operate external solid-state relays, thereby switching mains power on/off. These are cheap devices (under $8/8 Euros each) and are easy to use.
- Numerous mains-power switches can be added very easily and might be used to operate low-speed devices, such as solenoids, relays, fans, air/water valves, heaters, indexing conveyors, motors, electromagnets, etc. There are also numerous MMB input lines available to sense external signals arising from binary transducers, such as photocells, thermostats, proximity sensors, relay contacts, etc.

7.3.5 Digitizing images

Originally, Java was unable to operate a video-capture card. For this reason, we decided, in the first instance, to use a standard "Webcam" software package, called *I-SPY*, which runs autonomously, digitizing a video image every few seconds [ISPY]. Two parameters needed by this program are set during system initialization. The first of these specifies the time interval between successive image-digitization cycles. Every time an image is digitized, it is stored in a file on a RAM-disc operated by C_r's operating system. (This is *file1* in Figure 7.6 (a).) RAM-disc should be used to avoid excessive wear on the main disc and to ensure high-speed operation.) The name and location of this file is defined by I-SPY's second configuration parameter. It should be noted that the input to the I-SPY program is derived from the output of the video multiplexor built into the MMB. This unit is operated by the JIFIC control software (about to be described), which periodically examines that file. JIFIC does this without interfering with the image-digitization process. Thus, I-SPY and JIFIC are asynchronous and C_r's operating system resolves conflicts of access to this file.

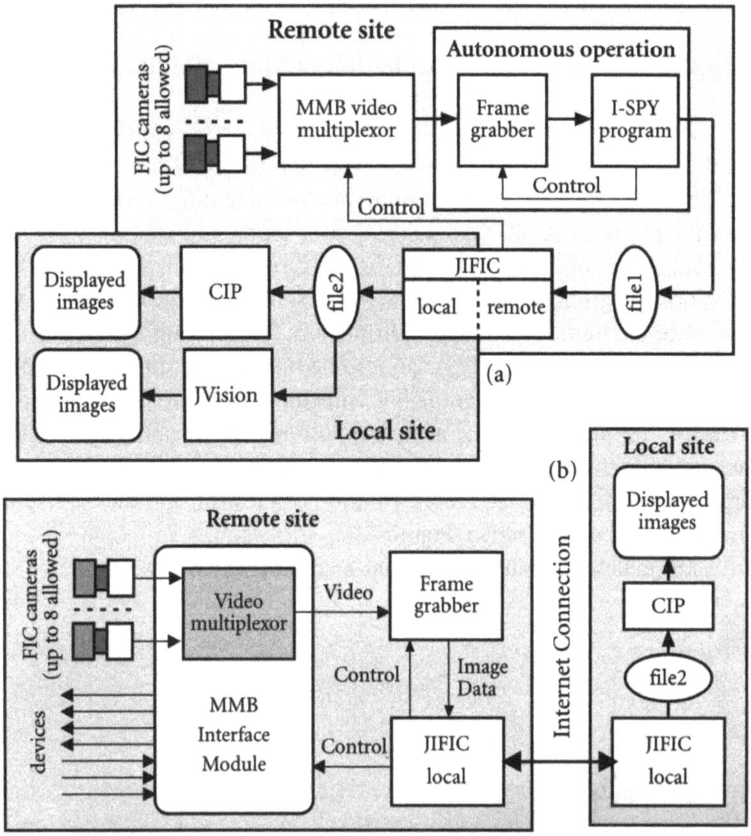

Fig. 7.6 Flow of image data in ROPE. (a) Original design using I-SPY for image capture. All cameras are connected to the MMB video multiplexor, whose output is taken to the frame-grabber card installed in C_r. The I-SPY software operates autonomously from JIFIC and digitizes a new image every few seconds. It saves the newly digitized image in a defined storage area (file1, on C_r). The FIC control program can read file1 and copy it to another defined storage area (file2), stored on C_l. This can, in turn, be read by CIP. (b) Revised design using Java Media framework for image capture.

Just as we were completing the manuscript for this book, Sun Microsystems, Inc. announced the introduction of Java Media Framework, which allows Java programs to control several standard video-capture cards [JMF]. This allows a Java program to digitize a video signal on demand, enabling this process to be synchronized with the rest of the ROPE software. Sample programs for image capture are provided at the WWW site just cited. The details are not given here because the Java Media framework is very new and therefore unlikely to remain static for long. The results that have been obtained so far are most promising; the modified version of ROPE is simpler. It runs smoothly and somewhat faster than the original version. The revised structure of ROPE, based on the Java Media Framework, is shown in Figure 7.6(b). It is worth noting that, like Java itself, the Java Media Framework is available free of charge. A commercial package is also available, providing an interface between a Java application and a TWAIN scanner or camera [TWA].

7.3.6 Java Interface to the Flexible Inspection Cell (JIFIC)

The FIC is controlled via a Java application running on C_l, which communicates with another application running on C_r. Together, these form the JIFIC software package. The main JIFIC control window shown in Figure 7.4 was designed with the following points in mind:

a. The user (either the vision engineer or his assistant) can control the FIC, by clicking on a diagram representing the physical layout of the cell. In this way, he can operate the lights and video multiplexor. The laser light-stripe generator (LS) and pattern generator (PG) can also be operated in this way. In addition, various buttons are provided in the FIC control window for operating the lights.
b. For clarity, the pick-and-place arm is not shown in the FIC layout diagram. However, when the user clicks on the appropriate button, the FIC layout picture is replaced by a diagrammatic representation of the arm, and the corresponding control buttons become active (Figure 7.7).
c. The (X, Y, θ) table is operated in a similar manner (Figure 7.8).

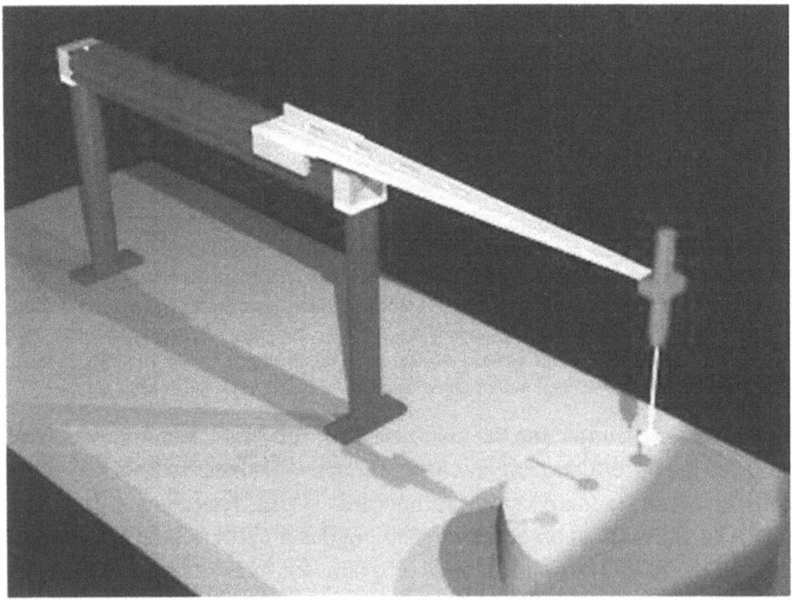

Fig. 7.7 Frame from the animation sequence showing the pick-and-place arm in operation. This is activated when the user clicks on one of the "Robot Controls" buttons (Figure 7.5(b)) and is displayed in lieu of the FIC diagram. (Reproduced by courtesy of R. J. Hitchell.)

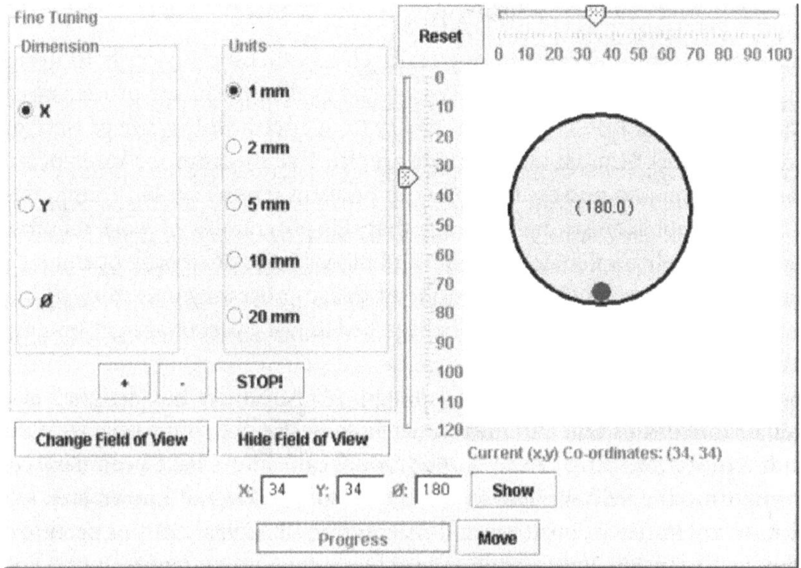

Fig. 7.8 A separate window used to control the (X,Y,θ) table and is displayed when the user clicks on the appropriate button. ("XYZ" button in Fig. 7.5(a); "XYT" in Fig. 7.5(b).)

d. Any of the cameras can be selected by the user, simply by clicking on the appropriate icon in the FIC layout diagram. The digitized image from the newly-selected camera is then displayed in the frame at the bottom-right corner of the JIFIC control window.
e. In order to provide an overall view of the FIC, a camera is mounted outside the frame. (This vantage point is equivalent to that used to draw the FIC layout diagram.) This is called the scene camera and is also connected to the MMB video multiplexor. The scene camera cannot always provide a clear image of the FIC, which contains naked light bulbs, polished metal, and matte black surfaces. Hence, the JIFIC control diagram is much easier to use in practice than the images derived from the scene camera. The latter merely reassures the user that the FIC controller is working properly.
f. Once the vision engineer obtains an image that he wishes to keep, he simply clicks on the Grab button in the main JIFIC control window. The image is then transferred to the frame at the top-right corner, where it is protected and can be saved for processing.
g. The JIFIC control window contains other buttons to transfer control to one of the image-processing packages.

7.3.7 Other features of ROPE

ROPE must also have the ability to process images. So far, we have given only scant attention to CIP, which will be discussed in Section 7.4. An alternative to CIP, called JVision, is also being considered. JVision is written in Java and possesses a similar command repertoire. While it has a command-line input mechanism similar to that used in PIP and its predecessors, it also provides a visual-programing interface.[3] JVision is described in detail in Chapter 8.

The long-term objective in our research in Machine Vision is to design a comprehensive and fully integrated IVS which has all of the image-processing capabilities available in PIP, as well as its ability to operate a wide range of electromechanical devices. It must also provide an effective medium for inter-personal communication and possess the ability to perform remote system configuration. We cannot emphasize strongly enough that this programme of work requires far more than simply implementing more and more image-processing operators. CIP already has a respectably large repertoire of image-processing operators, all implemented in Java. However, CIP must be regarded as just a part of a larger system that will eventually include:

a. Specialized calculators for optical design. M. A. Snyder has designed several calculators of this type but unfortunately these are not compatible with ROPE and require rewriting [SNY92]. Additional calculators have been devised for computing the field of view required to view an object of known size, with a known tolerance in position and orientation [BAT98a]. In this category we should also include more conventional ray-tracing programs for optical system design [KID].
b. Reference library, including such items as case study notes, glossary of terms, system design formulae, optical properties of materials, links to catalogues of light sources, optical components and cameras, etc.
c. Training package, particularly for the image processor.
d. Lighting Advisor. This an intelligent database of text, diagrams, and pictures, which is able to guide a vision engineer and/or assistant in the design of the image acquisition subsystem. Such a system already exists [BAT94a], albeit in a form that is incompatible with ROPE.
e. Expert system to help the assistant to perform as much of the design work as possible.

7.4 CIP

As we already indicated, CIP is the interactive image processor that has been devised specifically for use within ROPE. In broad terms, CIP is similar to PIP but differs in certain details, which reflect the nature of Java. The central design principle for ROPE has been maintained without compromise: all software must be written in Java.

Typically, CIP displays the windows, dialog boxes, and menus listed below. Also see Figure 7.9.

---

3. The leading author of JVision, Dr. Paul Whelan, has worked closely with BGB for several years. For this reason, the repertoire of JVision has been strongly influenced by that of PIP and its predecessors.

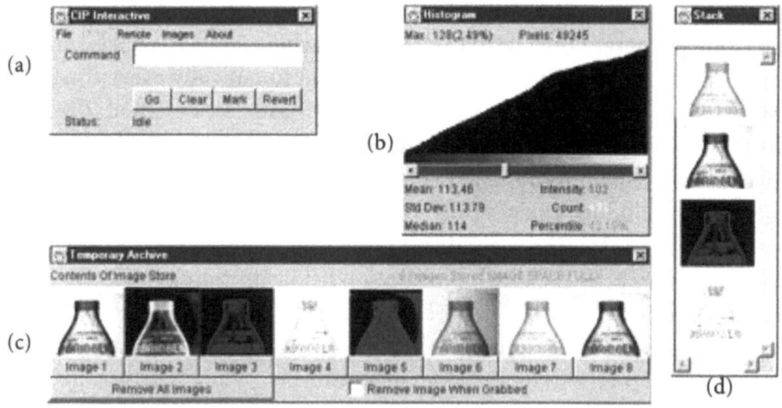

Fig. 7.9 Control windows for CIP. (a) Command-line entry window. Buttons labelled "Mark" and "Revert" perform push and pop operations on an image stack and allow the user to backtrack easily during interaction. Five pull-down menus provide a range of utilities, such as reading/storing images, accessing system documentation, etc. (b) Histogram window (in this case showing a cumulative histogram). (c) Image archive window. (d) Image stack window. This window is for display purposes only; it is not interactive.

a. **Interactive Dialog window** This is the central control tool that allows the user to issue commands to CIP. The user can enter a single command (i.e., a basic command, a script), or a short sequence of such commands. There is an array of buttons providing similar facilities to those available from PIP's command window (Figure 4.8). However, pulldown menus are attached to the Interactive Dialog window, rather than the main menu bar, as in they are in PIP. This window is shown in greater detail in Figure 7.10.

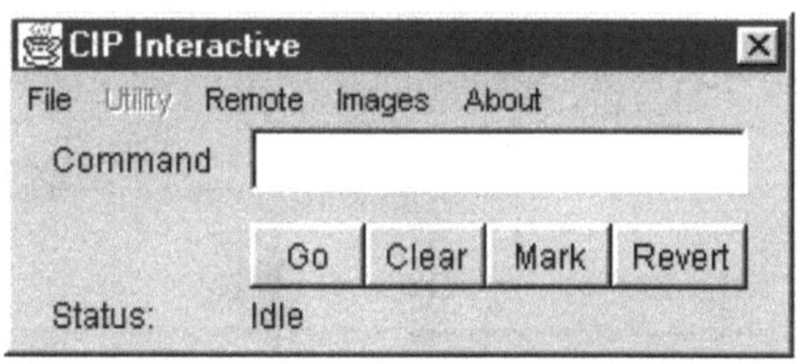

Fig. 7.10 A closer look at the Interactive window, shown at top left in Figure 7.9.

b. **Current Image window**
c. **Alternate Image window**
d. **Original Image window**
e. **Results window** This shows the results of any operations which compute numerical values. (e.g., average intensity, number of blobs, etc.) and optionally the execution time for each CIP command. (This has the same function as the Σ Output Window in PIP. Figure 4.3)
f. **Histogram window** Histograms are displayed in a dedicated window, as shown in Figure 7.9(b). (In PIP, they are drawn in the current image window, which

occasionally creates a poor display, due to scaling difficulties.) The user can examine the histogram interactively by moving a slider along the horizontal scale (intensity). Image statistics are (re-)calculated automatically every time a new current image is created; after each command has been executed, the mean, minimum, maximum, and median intensities, as well as the principal peak of the histogram, are all recomputed. (This feature is very popular with users.)

g. **Image Archive window** Images stored temporarily in RAM are displayed in the form of a strip containing several thumbnail images. The user can recover a stored image simply by clicking on the button below its thumbnail icon.
h. **Image Stack window** The top part of the image stack is displayed in the form of a strip of thumbnail images. Images near the bottom of the stack cannot always be seen, due to limitations of space on the screen.
i. **Pull-down menus** The user can load images, ready for processing, in two ways. By using the "Images" menu, he can select an image by name. Alternatively, under the "File" menu, he can choose the "Thumbnails" item, which creates an iconic display of the same files (Figure 7.11). The pulldown menus also enable windows (d)–(g) to be hidden/displayed and to control remote operation.

CIP does not yet have a top-level Prolog controller. Instead, a facility for writing and executing scripts has been provided. The user is able to write, modify, and save scripts (macros), using a standard text editor. When a command is entered in the Interactive Dialog window, the Command Processor (Figure 4.27) first consults the script file and tries to match the command name with that of a script. It does this before it attempts to interpret that command as a basic function. Hence, a script

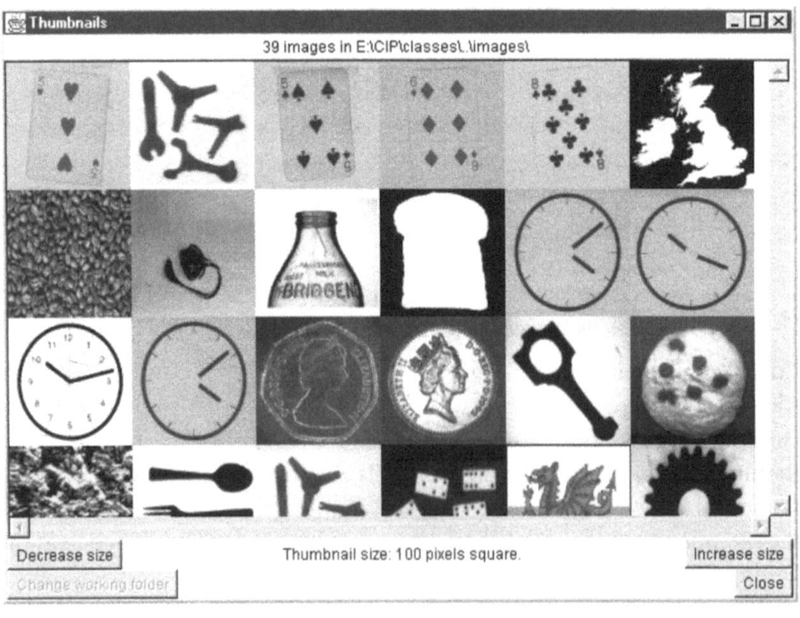

Fig. 7.11 Array of thumbnail images displayed when loading images.

may be used to redefine the operation of a basic function. Scripts may be nested and/or recursive. Moreover, they allow the usual arithmetic and conditional-command and program-control functions to be performed. (Table 7.2)

Table 7.2 Script commands for CIP. The characters a,b,c,d are identifiers representing integer variables,. Lower-case characters can be replaced with integer literals, as these are parameters passed to the function. Upper-case characters represent parameters returned by a procedure (variables). Variables do not have to be declared; they are assigned a value when parameters are returned. If a variable is passed to a function before it has been assigned a value, it is assigned the value 0 by default. Any legal string can be used to represent a variable. *string* represents any legal string. After G. E. Jones [Jon98].

Command	Description
+ (a,b,C)	$C = a + b$
– (a,b,C)	$C = a - b$
* (a,b,C)	$C = a * b$
/ (a,b,C,D)	$C = a/b$, $D = a \bmod b$
inc(A)	Increment *A* by 1
++(A)	Increment *A* by 1. Alternative format
dec(A)	Decrement *A* by 1
--(A)	Decrement *A* by 1. Alternative format
assign(A,a,B,b,C,c...........)	Assign *A* the value of *a*; *B*, the value of *b*, etc.
~~(A,a,B,b,C,c...........)	As above. Alternative format
owi(a)	Write the value of integer a to output window
ows("string")	Write string to output window
onl()	New line to output window
equ(a,b,C)	If $a == b$, C is assigned 1; else C is assigned 0
== (a,b,C)	As above. Alternative format
leq(a,b,C)	If $a <= b$, C is assigned 1; else C is assigned 0
<= (a,b,C)	As above. Alternative format
les(a,b,C)	If $a < b$, C is assigned 1; else C is assigned 0
< (a,b,C)	As above. Alternative format
label(string)	Labels position in code with identifier string
@(string)	As above. Alternative format
jump(string,a)	Jump to label "string" if $a >= 1$; else ignore
jump(string)	Unconditional jump to label string
^(string,a)/^(string)	Conditional/unconditional jump
j==(a,b,string)	Jump to label *string* if $a == b$; else ignore
j<=(a,b,string)	Jump to label *string* if $a <= b$; else ignore
j<(a,b,string)	Jump to label *string* if $a < b$; else ignore
j>(a,b,string)	Jump to label *string* if $a > b$; else ignore
j>=(a,b,string)	Jump to label *string* if $a >= b$; else ignore
j!=(a,b,string)	Jump to label *string* if $a \neq b$; else ignore
:string(any num of parameters)	Declaring the start of a procedure named string
return(return parameters)	Return parameters from procedure
#(return parameters)	As above. Alternative format
%% string	Comment, terminate by new line
%"string"%	Comment that can span multiple lines
run	Run the script displayed in the source code area

7.4.1 Structure and operation of CIP

Once again, the reader is referred to Figure 4.27, which shows how the major components of CIP are connected together.

CIP Launcher The CIP Launcher module loads the other CIP files and advises the CIP CORE module whether the source is trusted (i.e., *Is CIP implemented as an application or a signed applet?*)

CIP Core This is the heart of CIP and is responsible for system coordination and a wide variety of other actions, including loading/saving images and dispatching messages to other modules.

Command Processor The Command Processor builds an execution stack on behalf of CIP Core, and is responsible for storing variables used in CIP commands. Since certain commands are illegal in non-trusted applets (e.g., saving/storing images), the Command Processor must know whether or not CIP is trusted.

Client and Server Cores These provide facilities for remote operation, including the ability for one copy of CIP to utilize the resources available to another. Eventually, the user will be able to order the execution of operations in parallel on the local and remote machines, to speed up certain computational tasks.

Image-processing functions At the time of writing, CIP has a command repertoire of about 160 commands, which is somewhat smaller than that available in PIP. However, scripting facilities have only just become available in CIP and the pace at which new commands are added will accelerate from now on. This has happened in all of the previous IVS implementations, from SUSIE onwards. The ability to write high-level commands in terms of other "core" functions is the key element in this process. We might reasonably expect the CIP and PIP software packages to be broadly similar in the way they implement the image-processing primitives, since Java and C have many features in common. In fact, CIP and PIP are markedly different in the way that they generate and display images. PIP makes extensive use of the library of QuickDraw routines that help to define the Macintosh OS environment. Since CIP is platform-independent, it cannot make use of machine-specific toolboxes such as this. As a result, CIP image-processing functions are programmed in the obvious way, by simply manipulating an array of integers. Four samples of Java code implementing some of the simpler image-processing operators available in CIP were given at the beginning of Chapter 6.

Saving and loading images Originally, Java applets were unable to exploit local resources, such as the disc, on the client computer. This restriction was imposed in order to provide a high level of security. However, it imposed fundamental limits on what could be achieved in CIP and ROPE. Stand-alone Java applications have never been limited in this way and can, for example, read/write files to disc without hindrance. A trusted applet is needed if a Java program is to read and write files on the client's hard disc. The protocol for declaring Java applets to be trusted is well established and hence details will not be given here [JAV]. Let it suffice to say that, once an applet is trusted, it can read and write files in the same way that any normal program running on a stand-alone computer does.

Image acquisition and device control These functions are performed interactively in JIFIC. In order to control the camera in a CIP script, the commands *grb* and *ctm* can be used, as in PIP. Controlling the FIC's lights, pick-and-place arm, and (X,Y,θ) table requires the use of hardware that may not be available if CIP is used separately from ROPE. Programmed control of these electromechanical devices is being added at the time of writing.

7.5 Remarks

CIP and JIFIC are both still under development. New image-processing functions are being added to CIP just as fast as circumstances allow. At the time of writing, the script facility has just become available. This will make it possible to use "high-level" programming to add new functions rather more rapidly than hitherto. CIP's user interface and command repertoire are broadly similar to those provided by earlier systems, such as PIP. However, the potential to operate and configure an IVS remotely is important and is CIP's principal advantage over other systems. This feature is also under development.[4]

Although CIP does not yet possess a Prolog top-level control facility, it is possible to add this quite easily. In order to avoid violating the self-imposed design principle that only one implementation language is allowed, we are using a Prolog interpreter implemented in Java [CKI]. Of course, PIP is based on MacProlog, which has a rich repertoire of built-in predicates. These are, in turn, built on the (Macintosh) QuickDraw toolbox. The built-in predicates within MacProlog have been put to good use in extending PIP's facilities, most notably constructing its user interface. However, they are more than adequately matched by the large and sophisticated class libraries available in Java. Hence, by adding a "minimal" Prolog system to CIP, we believe that we will not be forgoing any important facilities. The general approach used in building PIP lends itself very well to implementation using a Prolog–Java system. Notice that the Prolog program, interpreted by a program written in Java, could easily be made to control another Java program, performing image processing, on a separate processor.

The present version of JIFIC is just the beginning of a much larger suite of software that will eventually form the basis of ROPE. JIFIC itself is currently being expanded to control the FIC's (X,Y,θ) table, servo motors, and other devices. An "uncommitted" version of JIFIC is currently being developed. This will be useful in those situations where the present graphical representation of a work cell is not appropriate. We have already indicated that various specialized calculators and advisory systems will be added to ROPE. The limitation on all of these developments is simply man–power.

Logically, Chapter 8 continues the theme of this chapter. The description of JVision is included in a separate chapter for the simple reason that material was

4. A major part of the programming effort for CIP and ROPE is provided by undergraduate and Master's degree students

contributed by a different author. JVision is an interactive image processor. It has a rich instruction repertoire that owes its origins to the same progenitors as PIP and CIP but uses visual programming. JVision is designed specifically for prototyping Machine Vision systems and is implemented using Java. We are currently evaluating JVision as a component for ROPE.

8 Visual programming for machine vision

by Paul F. Whelan[1]

This chapter will outline the development of a visual programming environment for machine vision applications, namely *JVision*[2] [WHE97a, WHE97b]. The purpose of JVision is to provide machine vision developers with access to a non-platform-specific software development environment. This requirement was realized through the use of Java, a platform-independent programming language. The software development environment provides an intuitive interface which is achieved using a drag-and-drop block-diagram approach, where each image-processing operation is represented by a graphical block with inputs and outputs which can be interconnected, edited, and deleted as required. Java provides accessibility, thereby reducing the workload and increasing the "deliverables" in terms of cross-platform compatibility and increased user base. JVision is just one example of such a visual programming development environment for machine vision. Other notable examples include Khoros [KRI99] and WiT [WIT99]. See [JAW96, GOS96] for details on the Java programming language.

8.1 Design outline

JVision is designed to work at two levels. The basic level allows users to design solutions within the visual programming environment using JVision's core set of imaging functions. (Currently JVision contains over 200 image processing and analysis functions, ranging from pixel manipulation to color image analysis. A full listing of all the functions available can be found at the JVision web site.) At the more advanced developers' level, JVision allows users to integrate their own functionality, i.e., to upgrade through the introduction of new image-processing modules.

The following sections outline the key issues involved in the development and use of a visual programming environment for machine vision. While many of the issues discussed are common to a wide range of visual programming environments, this chapter concentrates on the issues specific to a machine vision development system.

---

1. *http://www.eeng.dcu.ie/~whelanp/home.html*
2. JVision was designed and developed at the Vision Systems Laboratory, Dublin City University, Dublin 9, Ireland. See *http://www.eeng.dcu.ie/~whelanp/jvision/jvision_help.html* for more details, and for information on acquiring this package.

8.1.1 Graphical user interface (GUI)

The graphical user interface consists mainly of a "canvas" where the processing blocks reside. The processing blocks represent the functionality available to the user. Support is provided for handling positioning of blocks around the canvas or workspace, the creation and registration of interconnections between blocks, and support for double clicking. See Figure 8.1 for an example of the JVision GUI. The lines connecting each block represent the path of the image through the system, with information flowing from left to right. Some of the blocks can generate "child" windows, which can be used for viewing outputs, setting parameters, and selecting areas of interest from an image. If each block is thought of as a function, then the application can be thought of as a visual programming language; the inputs to each block are similar to the arguments of a function, and the outputs from a block are similar to the return values. The advantage in this case is that a block can return more than one value. The image-processing system can be compiled and executed as with a conventional programming language, with errors and warnings being generated depending on the situation. Warnings are generally associated with the failure to connect blocks correctly.

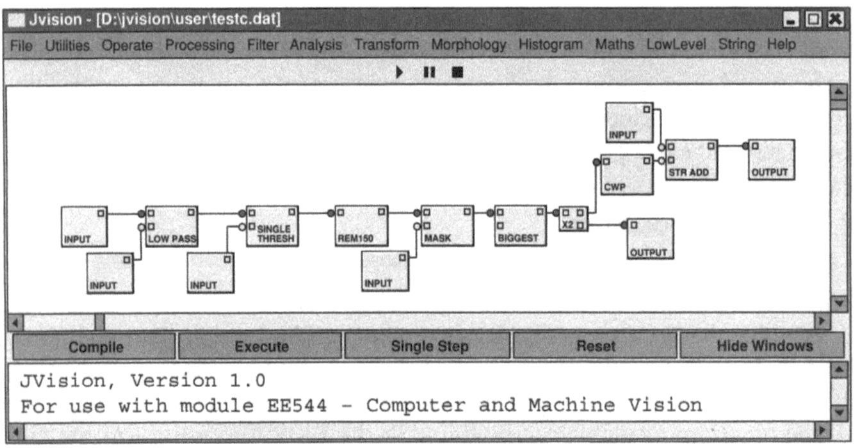

Fig. 8.1 A typical JVision canvas.

8.1.2 Object-oriented programming

The *Object-Oriented Programming* (OOP) paradigm allows us to organize software as a collection of *Objects* that consist of both data structure and behavior. This is in contrast to conventional programming practice that only loosely connects data and behavior. The object-oriented approach generally supports five main aspects: *Classes, Objects, Encapsulation, Polymorphism,* and *Inheritance.* Object orientation (OO) encourages modularization (decomposition of the application into modules), and the reuse of software in the creation of software from a combination of existing and new modules.

Classes allow us a way to represent complex structures within a programming language. Classes have two components: *States* (or data) are the values that the object has, and *methods* (or behavior) are the ways in which the object can interact with its data, i.e., the actions.

Objects are instantiated, in that they exist in the computer's memory, with memory space for the data. *Instantiation* is the creation of a particular object from the class description. An object is a discrete, distinguishable entity. For example, *my car* is an object, whereas the common properties that exist with *my car* and *every other car* may be grouped and represented as a class called *Car*. Objects can be concrete (a car, a file on a computer) or conceptual (a database structure), each with its own individual identity.

Encapsulation provides one of the major differences between conventional structured programming and object-oriented programming. It enables methods and data members to be concealed within an object, making them inaccessible from outside of the object. In effect, a module must hide as much of its "internals" as possible from other modules. Encapsulation can be used to control access to member data, forcing its retrieval or modification through the objects interface, which should incorporate some level of error checking. In strict object-oriented design, an object's data members should always be private to the object; other parts of the program should never have access to that data.

A *derived class* inherits its functions or methods from the base class, often including the associated code. It may be necessary to redefine an inherited method (i.e., alter the implementation) specifically for one of the derived classes. So, *polymorphism* is the term used to describe the situation where the same method name is sent to different objects and each object may respond differently. Polymorphism has advantages in that it simplifies the Application Programming Interface (API) [SUN99], and also in that it allows a better level of abstraction to be achieved.

Object-oriented languages allow a new class to be created by extending some other class, so *inheritance* enables the creation of a class which is similar to one previously defined. An inherited class can expand on or change the properties of the class from which it is inherited. In traditional programming, modification of existing code would be required to perform a function similar to inheritance, introducing bugs into the software as well as generating bulky repetitive source code. The level of inheritance can be controlled by the use of the key words *public, private,* and *protected.*

8.1.3 Java image handling

Java provides an *Image* class for the purpose of storing images. This class contains useful methods for obtaining information about the image. The most useful of these methods are described in Table 8.1. The image object can be displayed on the canvas using the *drawImage( )* method, which is a member of the *Graphics* class. This method allows the scaling of an image using a bounding box. The scaling function is particularly useful in performing basic magnification tasks.

Table 8.1 Methods of the image class.

Method	Description
getWidth(ImageObserver)	Returns the width of the image object. If the width is not yet known the return value is *–1*.
getHeight(ImageObserver)	Returns the height of the image object. If the height is not yet known the return value is *–1*.
getSource()	Returns the image producer which produces the pixels for the image. An *image producer* is sometimes used in the generation of a filtered image.
ImageObserver	An *image observer* is an object interested in receiving asynchronous notifications about the image as the image is being constructed.

An image consists of information in the form of pixel data. In the case of a Java image object, a pixel is represented by a 32-bit integer. This integer consists of four bytes containing an *alpha* or transparency value, a *red* intensity value, a *green* intensity value, and a *blue* intensity value. This representation of pixel data is referred to as the *default RGB color model.* The structure of a pixel using this color model is given in Figure 8.2. For the purposes of JVision, the *alpha* or *transparency* value is always set to its maximum value, because all images are required to be fully opaque for display purposes. Therefore, the alpha value can be ignored when processing an image. Also, for grey-scale images the values for each of the color planes will be the same; hence only the value for one of the color planes need be extracted in order to obtain the grey intensity value for that pixel. This approach increases the speed at which an image can be processed.

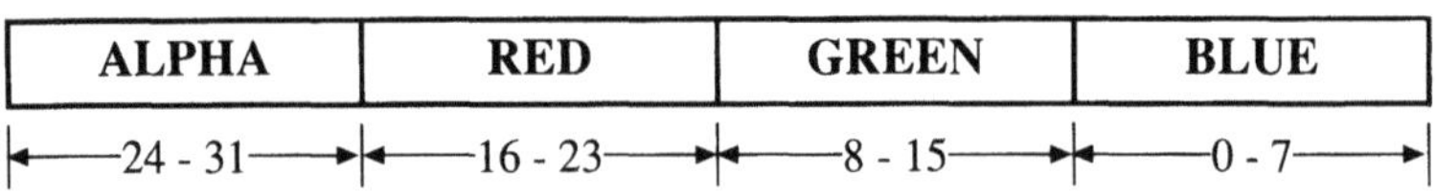

Standard Java 32-bit integer used to represent a default RGB pixel

Sample Java code for obtaining values for each of the constituent pixel planes

```
int rgb;

byte alpha = (byte)(rgb>>24);
byte red = (byte)((rgb>>16)&0xFF);
byte green = (byte)((rgb>>8)&0xFF);
byte blue = (byte)((rgb)&0xFF);
```

Fig. 8.2 The default RGB color model.

8.1.4 Image-processing strategy

The key requirement of the image processing and analysis strategy is generation of a robust method for image flow control. This is achieved through the introduction of the *sequential block-processing algorithm* or the *block manager.* Blocks appear as boxes, with a user-definable number of inputs and outputs depending on which image-processing function is being implemented. The only information the sequential block-processing algorithm requires is the actual state of each of the blocks in the image-processing system. Blocks request attention by setting an internal state variable, which is polled by the block manager, rather than generating an interrupt. A block can have one of several states, and the block manager will act according to the state of the block. The complete list of currently-available block states and their description are given in Table 8.2.

The purpose of the sequential block-processing algorithm is to bring all the processing blocks in a system to their steady state. This is achieved by monitoring for two main states. The operation of the algorithm can be summarized as follows: Any block which receives new information on any of its inputs eventually signals that it is WAITING_TO_PROCESS. When the algorithm sees a block signalling WAITING_TO_PROCESS it calls the *processImage( )* method of the relevant block, thus causing the block to signal WAITING_TO_SEND. When the algorithm sees a block signalling WAITING_TO_SEND it calls the *sendImage( )* method of the block in question, which results in the block returning to its STEADY_STATE. The algorithm loops through all the blocks in the linked list until they all signal STEADY_STATE. When this occurs the algorithm terminates.

Table 8.2 JVision block states.

Name of State	Description
STEADY_STATE	A block in STEADY_STATE is ignored by the block manger because it is assumed to have processed its image and requires no further attention.
WAITING_TO_PROCESS	A block which is WAITING_TO_PROCESS has received an image from one of its inputs and has verified that there are no images inbound on any other of its inputs. This checking back is required to avoid glitch-like operations evident in digital systems where asynchronous inputs to gates occur.
WAITING_TO_SEND	A block is WAITING_TO_SEND if it has processed an image and sent the image on to its output node.
WAITING_TO_RECEIVE	A block is WAITING_TO_RECEIVE only if it has more than one input. When an image is received on any input a check-back function is called to ascertain if there is any more inbound image data. If so, wait for that data before going in to the WAITING_TO_PROCESS state. Asynchronous inputs are caused by intensive processing taking place on one input to a block.
WAITING_TO_FEEDBACK	The only block which can implement the WAITING_TO_FEEDBACK state is the feedback block. Feedback can occur only when all the other blocks have settled into steady state.

In order to integrate the type of image processing required, certain changes to the GUI must be made. These changes are outlined below:

- The addition of image storage capabilities to the connectors, both male and female, is required because inter-block communications occur at the connector level. Thus the input and output images of a block must be stored at this level.
- Addition of *getImage( )* and *setImage( )* methods to each type of connector is required, so that communication of images between the male and female connectors is possible. If a block wishes to process an image, it reads that image

from its input connector and when the processing is complete it writes the new image to its output connector.

- The possible block states as outlined in Table 8.2 must be added to the block template so that the sequential block-processing algorithm may be incorporated into the GUI and thus control the flow of image information throughout the system.
- The definition of possible block types must also be added to the block template so it can be determined where an images originates and terminate.
- The addition of the *processImage( )* method to the block template is also required. This method is declared to be "abstract," and must be overwritten in any inherited versions of the block template class. The algorithm contained in the *processImage( )* method provides the only difference between each of the blocks available with JVision.

8.2 Data types

This section summarizes the various data types currently supported by JVision visual programming environment.

8.2.1 Image

Images are the best catered-for data types in the JVision library. Dark green nodes correspond to color images and red nodes correspond to grey-scale images. If a color image is specified for a grey-scale input then the grey-scale representation of the image is obtained and used in the consequent processing. Note that if the image viewer frame does not re-render properly after resizing, a refresh can be forced by clicking on the image. See Figure 8.3.

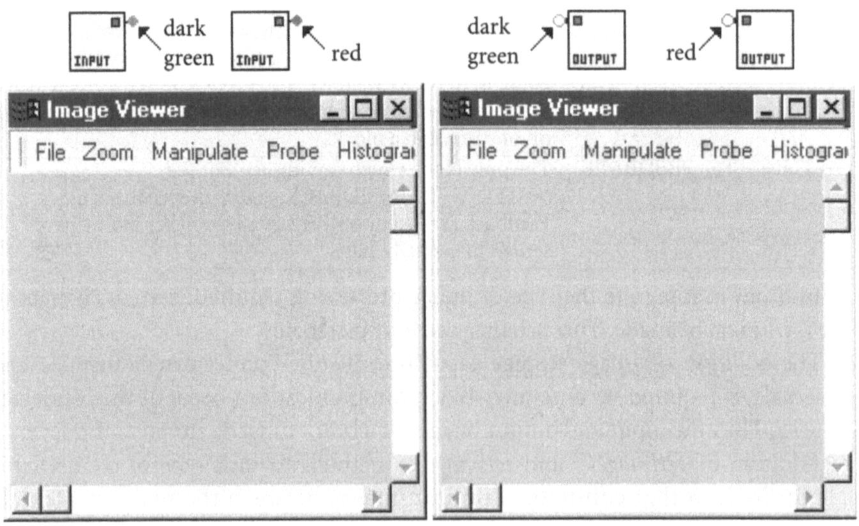

Fig. 8.3 Image data types.

8.2.2 Integer

Integer values may be used in conjunction with mathematics blocks or other image-processing blocks. The range of the Java integer is between *–2 147 483 648* and *2 147 483 647*. These are indicated by light green nodes. See Figure 8.4.

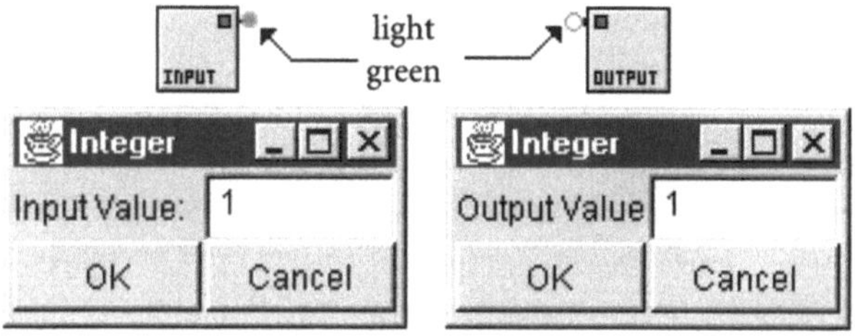

Fig. 8.4 Integer data types.

8.2.3 Double

Double values (double precision values) may also be used in conjunction with mathematics blocks or other image-processing blocks. The range of the Java double is between $-1.797\,693\,134\,862\,32 \times 10^{308}$ and $1.797\,693\,134\,862\,32 \times 10^{308}$. These are indicated by dark blue nodes. See Figure 8.5.

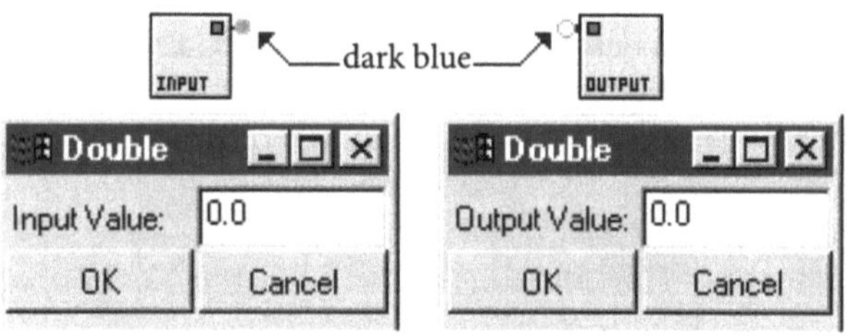

Fig. 8.5 Double data types.

8.2.4 Boolean

Boolean values are used in the interface between components of the mathematics library and the conditional processing blocks. The value of a Boolean variable may be either TRUE or FALSE. These are indicated by orange nodes. See Figure 8.6.

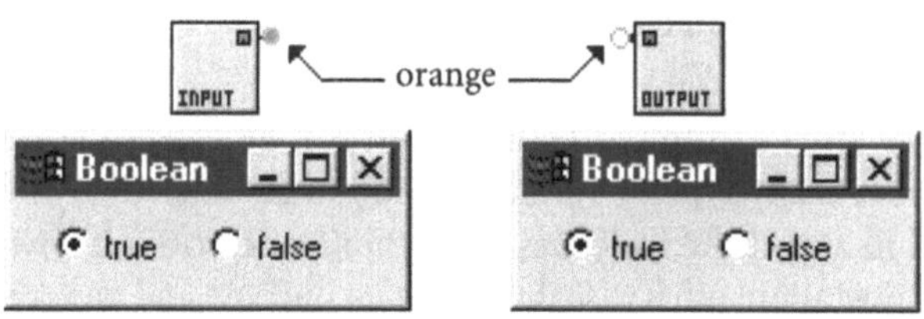

Fig. 8.6 Boolean data types.

8.2.5 String

String values may be used to alert the user to certain results. Any of the above variables integer, double, and Boolean, may be appended to a string using the *string add* block. Note that for file saving purposes, a string may not contain a new line. These are indicated by pink nodes. See Figure 8.7.

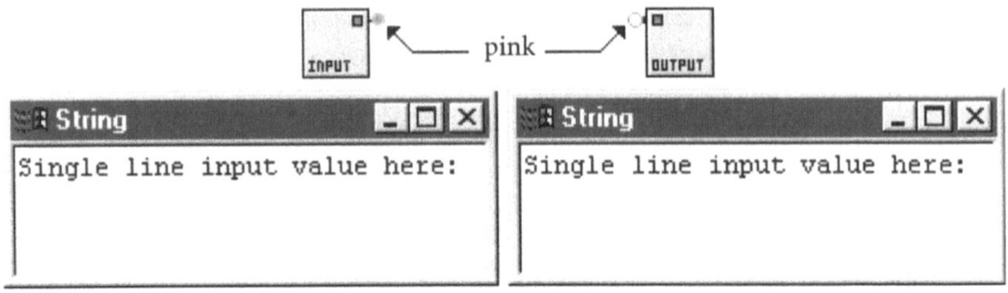

Fig. 8.7 String data types.

8.2.6 Integer array

Integer arrays are used to provide the mask input for the convolution filter or the structuring element for morphological operations. These are indicated by light green nodes. In the case of the morphological blocks, an 'x' in the structuring element corresponds to a *don't care* statement. (In fact, a non-valid integer has the same effect; this includes spaces.) See Figure 8.8.

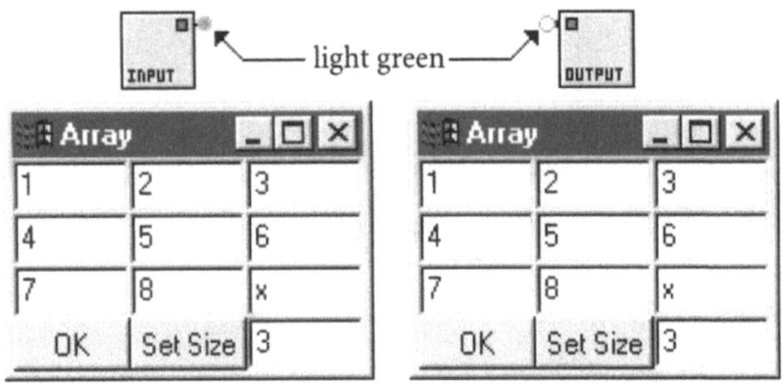

Fig. 8.8 Integer arrays.

8.3 Nonlinear feedback blocks

The nonlinear blocks supplied with JVision can be used to conditionally and repeatedly process any type of variable from images to strings. A feedback loop implemented in any of the instances of the blocks described below must contain some kind of operation; otherwise they serve no purpose and an error will occur.

8.3.1 Feedback

Feedback is an extremely useful mechanism for implementing a repetitive set of operations. In the example outlined in Figure 8.9, multiple dilations are applied to the input image (the top-most input on the left side of the feedback block). The number of times the dilation operation will be applied is specified by the integer input block. (This corresponds to the middle input in the feedback block). Once the image has been processed the specified number of times, it may be further processed. The result from the feedback loop is obtained from the top-most output on the right side of the feedback block (Figure 8.9).[3]

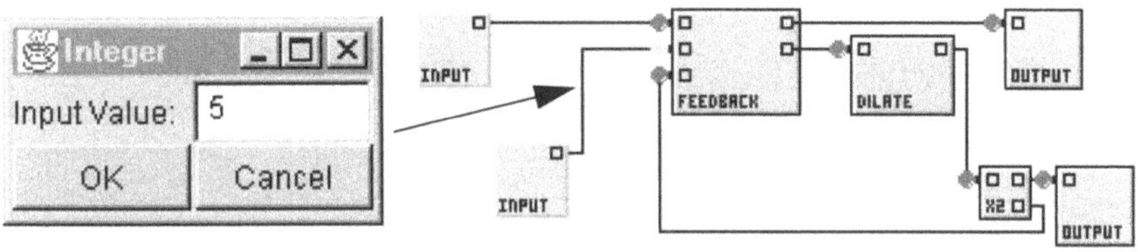

Fig. 8.9 (a) Feedback block diagram.

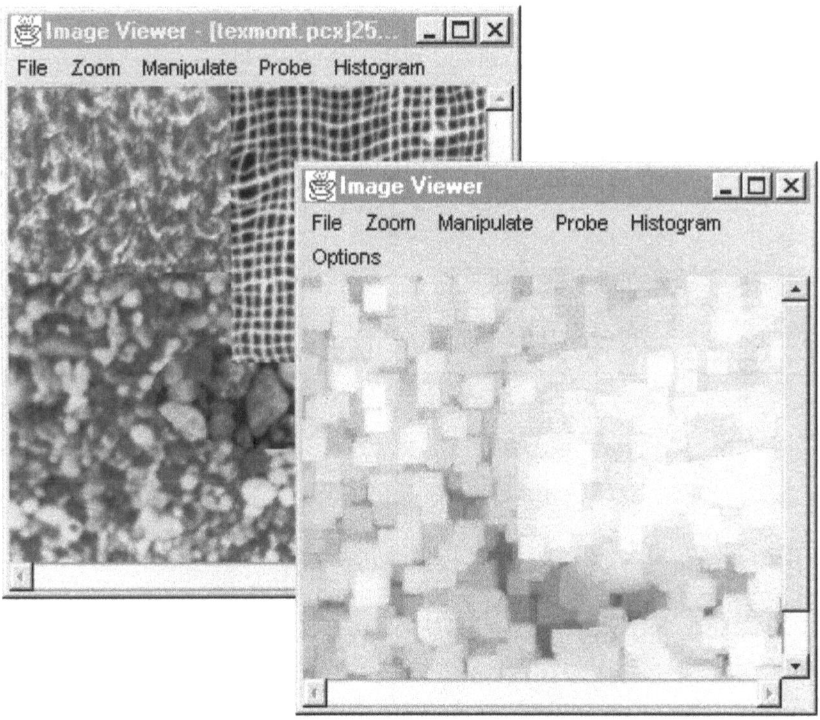

Fig. 8.9 (b) Feedback example. Multiple dilations of a grey scale image.

3. The image windows in this figure and in some of the later figures have been overlapped to allow them to be presented at a more readable size on the pages of this book.

8.3.2 FOR loop

Implementation of the FOR loop is similar to that described previously for the feedback structure, except in this case a loop variable is available to the feedback loop for further processing. Also, the single integer input used in Section 8.3.1 is replaced with three inputs, which represent the *start* value, *finish* value, and the loop *increment*. The implementation here is equivalent to the following C code:

```
For(x=start;x<finish;x+=increment)
```

where *start* is the top-most integer value, *finish* is the center integer value, and *increment* is the lower integer value in Figure 8.10.

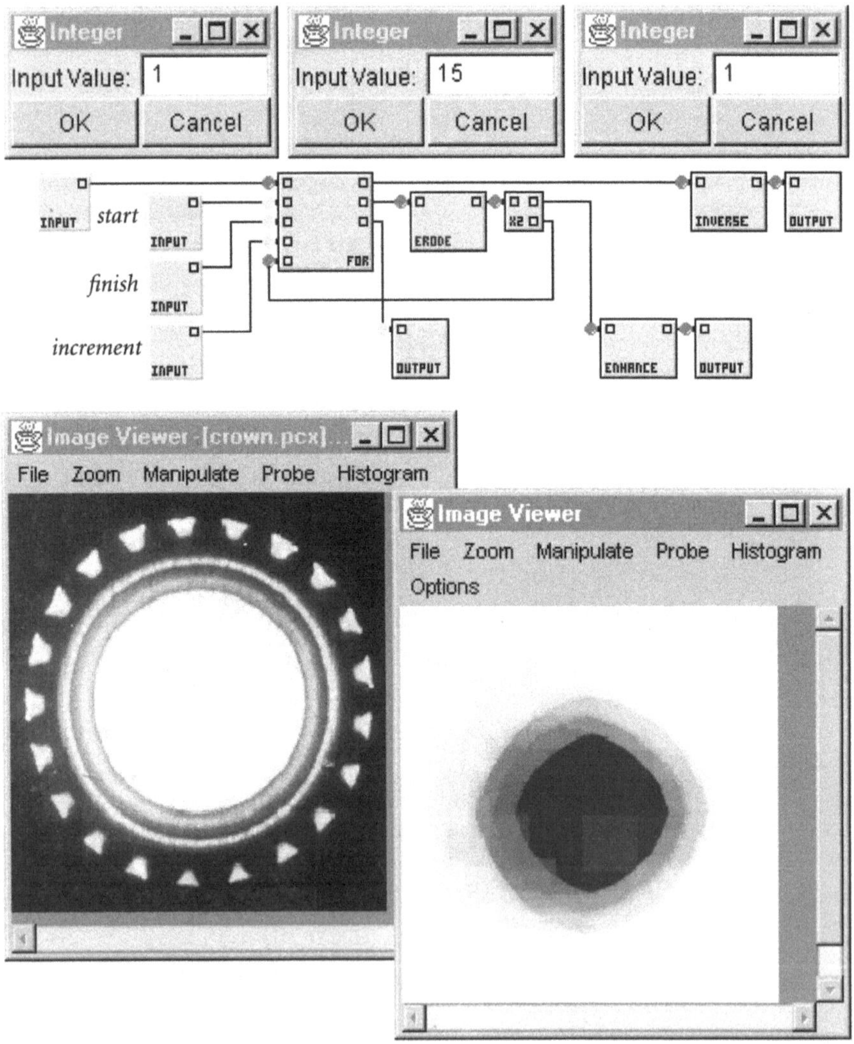

Fig. 8.10 FOR loop example: Setup and results for multiple erosion of a grey-scale image.

8.3.3 IF ELSE

The *IF ELSE* structure is yet another implementation of the feedback mechanism. In this case the feedback operation is applied only once, if at all. The decision whether processing of the data should be performed using the IF or the ELSE path is based on the status Boolean input variable. In the case of the JVision canvas illustrated in Figure 8.11, if the Boolean input is TRUE then the *Smallest-Intensity Neighbor (SIN)* filter is applied to the image. If the Boolean input is FALSE then the *Largest Intensity Neighbor (LIN)* is applied.

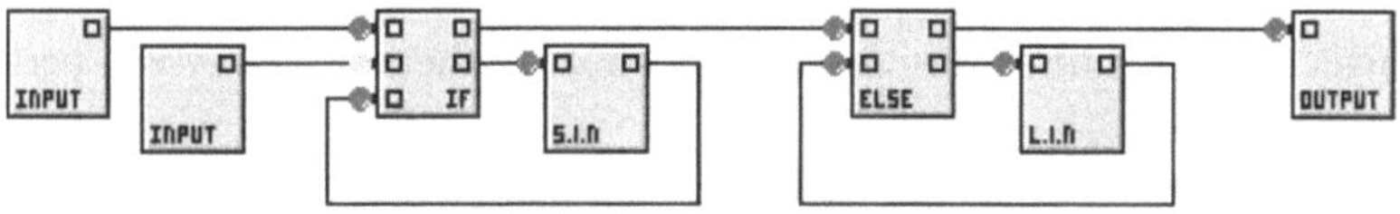

Fig. 8.11 IF ELSE example

8.3.4 Nesting

Another key feature of JVision is that it allows multiple nesting of feedback structures. For example the operation implemented in Figure 8.12 is low-level *image negation*, which performs a pixel-by-pixel raster scan of the image. The two FOR structures represent the vertical index and the horizontal index respectively. The vertical index block ranges from 0 to the image height in steps of 1, i.e., one pixel at a time.

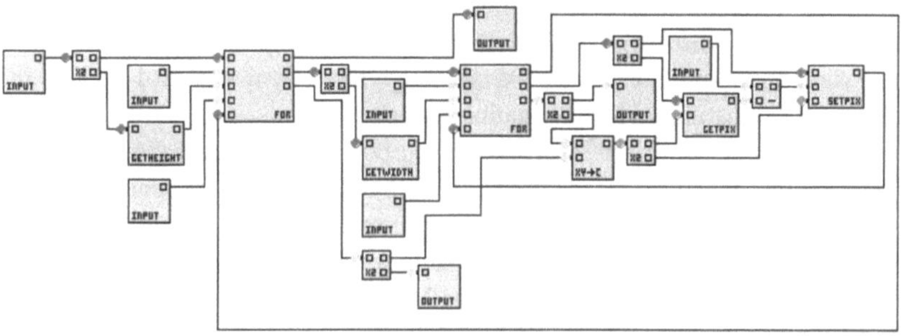

Fig. 8.12 Nesting. Pixel-by-pixel negation.

8.4 Visual programming environment

Blocks are selected from the menu system as required, then placed in the workspace. Interconnections are made between blocks following a strict 45 degree snap-to-grid rule system. When the processing system is set up, input images must be specified. This is done by double-clicking on the relevant input block; this causes an image viewer frame to appear. The *open* option is then selected from the file menu and the appropriate file name is chosen by navigating the file dialog box directory structure. If the selected file contains raw image data, then the dimensions of the image must be specified before the image can be loaded. There is currently support for nine input graphics file formats. See Section 8.4.1. Images can be saved in either Microsoft Windows Bitmap (BMP) or PC Paintbrush (PCX) formats.

Initialization of certain block parameter may be required throughout the program. If they are not set, the parameters will assume their default values. A program may be compiled using the *compile* button. If successful, then the program can be executed. Program parameters can be adjusted and the program may be reset and executed again until the desired response is obtained. At any stage, blocks may be added or removed from the program. A block can be deleted by clicking on it in order to highlight it, then pressing the delete key. If the delete key is not available, which may be the case with certain keyboards, then the 'd' key will perform this same operation.

Once a program is executed, each block is highlighted as it is processed. This enables novice users to get a sense of the relative speeds of the various processing options available within the JVision environment.

8.4.1 Interpretation of graphics files

The first problem which needs to be addressed in the development of any image-processing software is interpretation of image resource files containing bitmap or raster information. Java provides support for the two most common file formats found on the Internet, Graphics Interchange Format (GIF) and Joint Photographic Experts Group File Interchange Format (JPEG). Unfortunately, these file formats are not used for computer vision applications, because the compression techniques which they employ distort image information. The pixel error introduced by either compression technique is typically 1%. Although this may not be visually apparent, it does result in unacceptable information loss.

In order to preserve image information, support for several non-corrupting file formats has to be implemented in Java; the three main formats required are raw image data (BYT), PC Paintbrush (PCX) and Tagged Image File Format (TIFF). See below for the definitive list of file formats supported by the JVision [MUR96].

- *BMP:* Microsoft Windows Bitmap (BMP) is a bitmap image file format using *Run Length Encoding* (RLE), and supports a maximum pixel depth of 32 bits. The maximum image size is 32K×32K pixels.
- *BYT:* Raw image data, grey-scale only, with a maximum pixel depth of 8 bits. The data is stored with the first byte of the file corresponding to the top left-hand corner of the image. No header is contained within the image and details about the dimensions must be supplied by the user. This file format preserves image data; no information is lost because no compression is used.
- *GIF:* Graphics Interchange Format (GIF) is a bitmap file format which utilizes Lemple–Zev–Welch (LZW) compression and is limited to a maximum color palette of 256 entries and a maximum image size of 64K×64K pixels.
- *JPEG:* Joint Photographic Experts Group (JPEG) File interchange format is a bitmap file format utilizing JPEG compression and an encoding scheme based

on the discrete cosine transform. It has a maximum color depth of 16.7 million colors and a maximum image size of 64K×64K pixels.

- *PCX:* PC Paintbrush (PCX) is a bitmap file format using either no compression or RLE, and allows image data to be kept intact. The maximum color depth of 24 bits is available and a maximum image size of 64K×64K pixels.
- *PGM:* Portable Greymap Utilities (PGM) is a bitmap file format uses no compression. Hence, it allows image data to be left intact. It has a maximum grey-scale depth of 256.
- *RAS:* Sun Raster Image (RAS) is a bitmap file format using either no compression or Run Length Encoding (RLE). It supports a maximum of 16.7 million colors and has an undefined maximum image size.
- *RAW:* Raw image data. This has a similar specification to the BYT format described except that color image data is also supported. The image data is stored in RGB triplets, with the first triplet representing the upper left hand corner of the image.
- *TIFF:* Tagged Image File Format (TIF) is a bitmap file format using a wide range of compression techniques, uncompressed, RLE, LZW, International Telegraph and Telephone Consultative Committee (CCITT) Group3 & Group4, and JPEG. It supports a maximum pixel depth of 24 bits (16.7 million colors) and a maximum image size 4 G pixels.

8.4.2 Plug-in-and-play architecture

The main objective here is to allow the user to write code for a new block and integrate it into JVision without the need to recompile the source code. To accomplish this, a dynamic class loader must be written to allow a class (in this case an inherited version of the main *Block* class) to be imported into the application after it has been compiled. The name of the class referred to above (or, more precisely, the list of names of block classes which together create JVision) must be made known to the application, so that it may import them as required. This is implemented in a similar manner to the method in which the resource file information is made available to the file dialog box; i.e., through the use of a scripting language. In this case the file called *menubar.rc* contains the list of processing blocks available to the user. The block names are categorized into functional groups, e.g., filters, transforms, etc. A limitation associated with this approach is that the name of the block must be an exact match for the name of the class which implements the functionality of that block. This is easily overcome because the name of the class usually gives a good description of the function which it performs. The scripting language uses the four tags outlined in Table 8.3 in the generation of the main JVision menu bar.

When the JVision application is initialized, it reads the data from the *menubar.rc* file. With this data it constructs the menu bar for the main window automatically. Each menu in the menu bar corresponds to a menu block in the *menubar.rc* file, the name of the menu being specified by the first string in the menu block and the menu contents being specified by the remaining strings in that menu blocks. This method of setting up the menu bar performs two tasks: It means the user does not need to recompile the code to update the menu bar, and hence the inventory of processing blocks. It also means that when a menu event is handled, the argument specifying the event is the name of the new block to be added to the workspace. This block name is used in conjunction with the class loader to load the selected block, dynamically import it, and then add it to the linked list and eventually to the workspace. The method by which the menu bar is automatically updated is outlined in Figure 8.13. Note that the dashed lines in the script file correspond to menu separators in the actual menu.

Table 8.3 Tags for the automatic generation of the menu bar.

Tag	Meaning
<number of menus>	The value directly after this tag specifies the number of menus described by the file. This number is used to initialize an array of Java menu objects and must be an integer.
<menu>	The new line terminated string following this tag is used as the menu name. The strings following the first line specify the menu items i.e., the individual blocks.
</menu>	This tag marks the end of a menu, the file interpreter now waits for another menu or the end of file tag.
<end>	This tag informs the file interpreter that all the menus and menu items have been updated and to close the input stream.

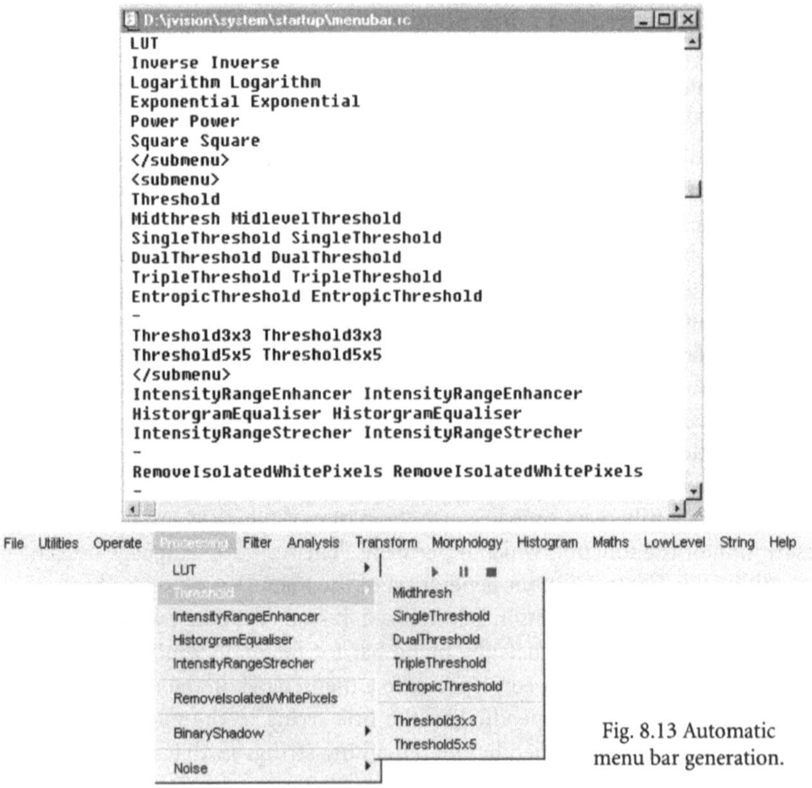

Fig. 8.13 Automatic menu bar generation.

8.5 Image viewer and tools

JVision provides several image investigation tools, horizontal and vertical intensity scans, normal and cumulative histograms, pseudocolor tables, and a 3-D profile viewer. These functions are common to both the color and grey-scale image viewer frames.

8.5.1 Horizontal and vertical scans

The horizontal and vertical scans tools provide cross-sectional intensity maps of the average intensity of the displayed image, this means that it works with both grey-scale and color images. The profile lines may be removed by closing the Horizontal and Vertical Scan windows. Both scans can be applied simultaneously, giving the crosshair display illustrated in Figure 8.14.

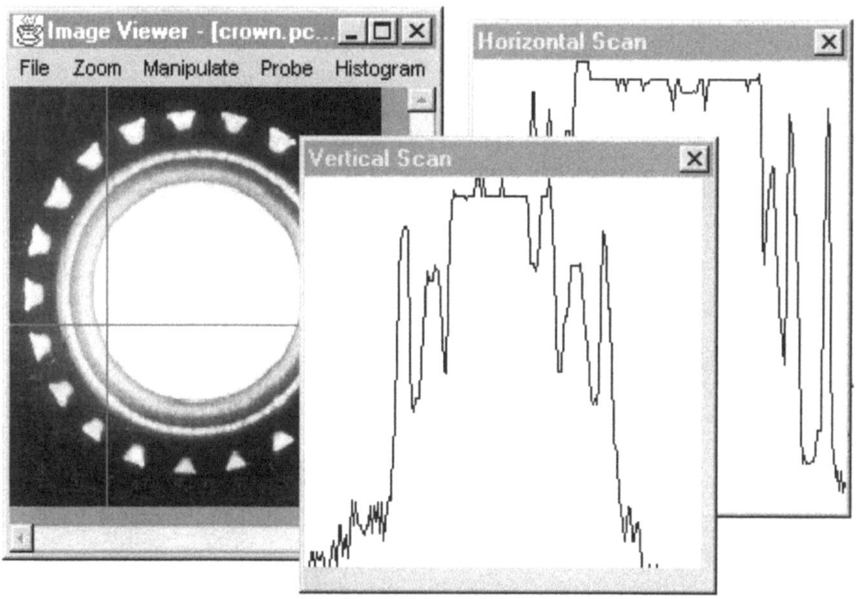

Fig. 8.14 Horizontal and vertical scans.

8.5.2 Histograms

The displayed image may be represented using the histogram function in either normal or cumulative mode.

8.5.3 Pseudocolor tables

JVision provides a selection of pseudocolor tables which may be used to re-render an image. This is especially useful when it is required to distinguish between several grey levels of similar intensity. A random pseudocolor table is also provided. This has been found to be useful when dealing with images which have been processed using the *label* operation. See Plate 19(TR) for color renditions of these tables.

8.5.4 3-D profile viewer

The 3-D profile viewer can be launched from the image viewer frame only when a valid *Region of Interest (ROI)* is selected. A ROI can be selected only if neither of the other two probing utilities (horizontal and vertical scan) are in operation, because they have precedence over ROI selection. Once the 3-D viewer is launched, then the profile may be manipulated using the menu system or the 3-D navigator, which can be launched using the *navigate* menu. The 3-D profile viewer may be applied to a color image, in which case only the intensity of the image is represented.

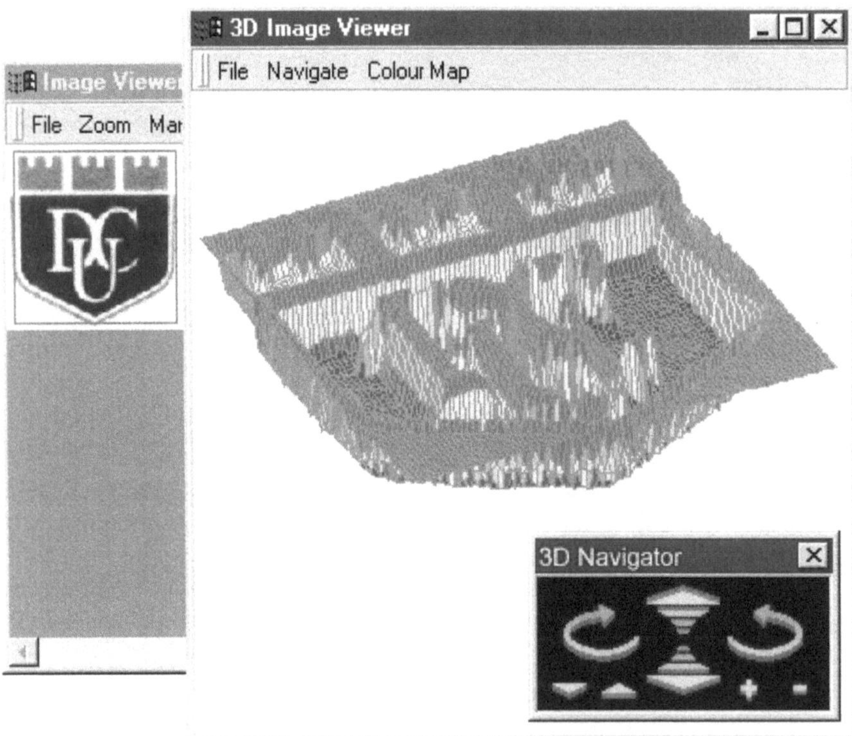

Fig. 8.15 Grey-scale representation of 3-D image viewer. The 3-D navigator is shown on the bottom right of this diagram.

In addition to the default grey-scale color table used with the profile viewer, we may also use any of the pseudocolor tables described previously in order to distinguish between differing grey levels, as well as adding extra depth information to the image.

In addition to using the 3-D navigator or the menu bar, we may manipulate the profile using the keyboard, as described below. Note that the profile must be brought to the foreground by clicking on it before any keystrokes are valid.

- UP ARROW – tilt backwards
- DOWN ARROW – tilt forwards
- RIGHT ARROW – spin anticlockwise (counterclockwise)
- LEFT ARROW – spin clockwise
- MULTIPLY – intensity multiply
- DIVIDE – intensity divide
- PLUS – zoom in
- MINUS – zoom out

8.6 Sample problems

JVision provides an image analysis software development environment that can work at several levels. For example, at a relatively low level we can manipulate individual pixels (Section 8.6.1). Alternatively, we can use JVision's built in functionality to generated solutions to complex machine vision tasks (Section 8.6.5).

8.6.1 Low-level programming

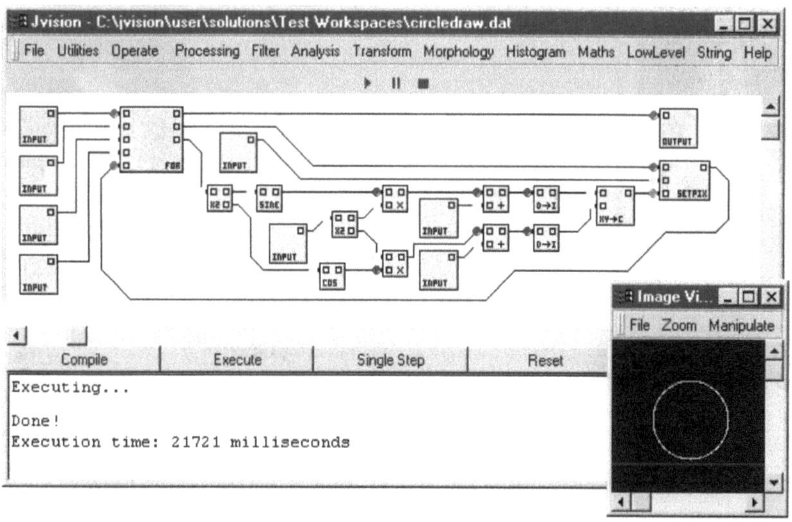

Fig. 8.16 Low-level functionality of JVision.

The solution outlined in Figure 8.16 illustrates this for the case where pixels in a circular arrangement are superimposed on a black image. This solution implements the following equivalent code:

```
int xcent = 128;
int ycent = 128;
int radius = 20;
int colour = 255;
for (int i=0;i<360;i++)
```

```
{
int x = radius*sin(i);
int y = radius*cos(i);
input.setxy(xcent+x,ycent+y,colour);
}
```

8.6.2 High-level programming

Working at a higher level maximizes the use of the predefined image-processing algorithms. For example, the *label by location* operator is used to implement a naive blob-fill algorithm.

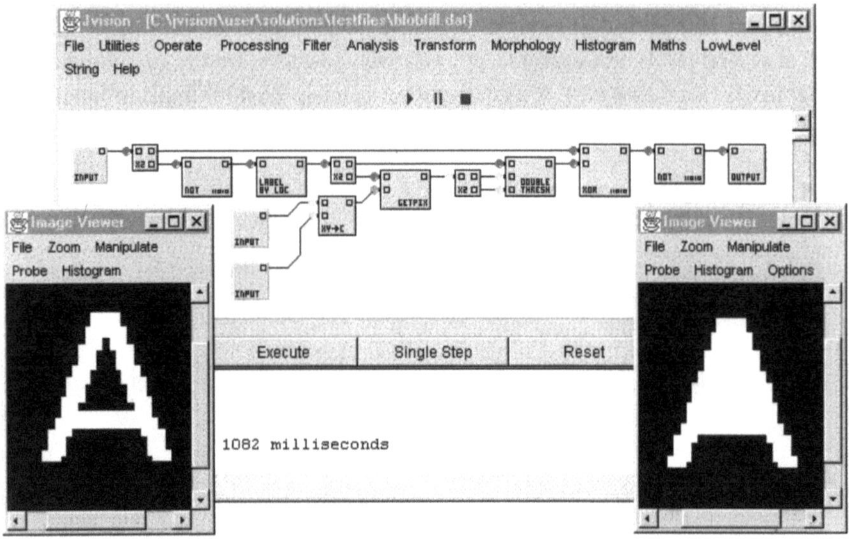

Fig. 8.17 Blob-fill algorithm implementation.

8.6.3 Convolution

Figure 8.18 illustrates the operation of user-defined convolution masks of sizes 3×3 and 5×5 in extracting horizontal and vertical information from an image.

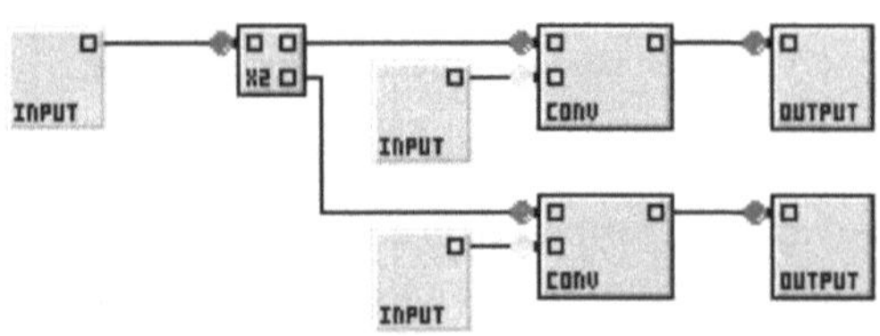

Fig. 8.18 (a) Convolution: JVision canvas.

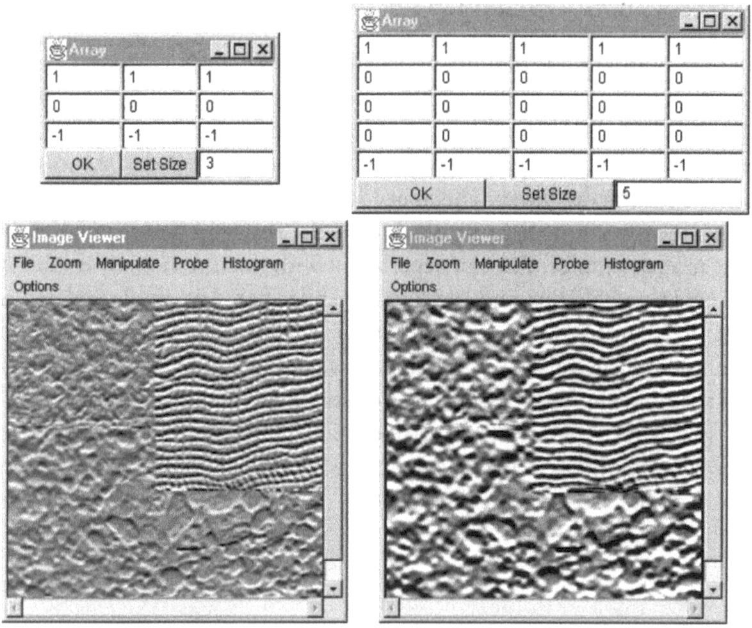

Fig. 8.18 (b) Convolution: Results from the 3×3 and 5×5 convolution operators. The input image is the same one used as the input in Figs. 8.9 and 8.19.

8.6.4 Fourier transform

At the highest level, complex abstract concepts such as Fourier analysis can be implemented; for example, the bandpass filter illustrated in Figure 8.19.

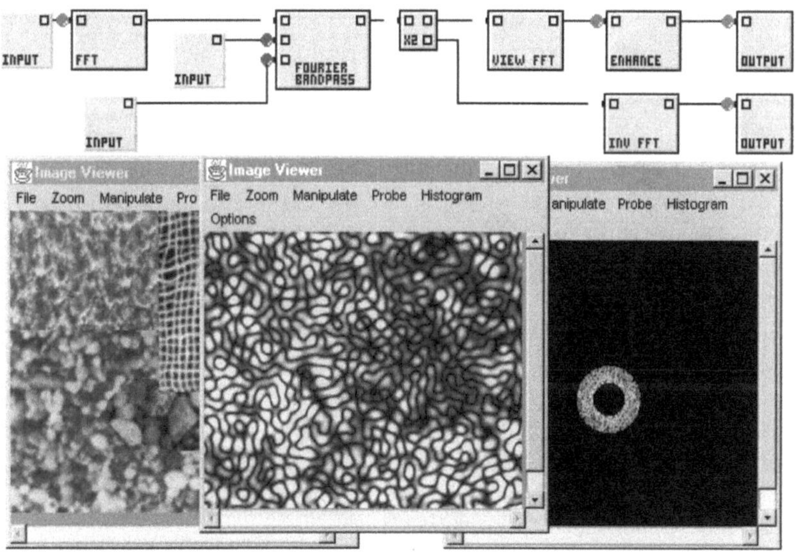

Fig. 8.19 Fourier analysis: Bandpass filter. The original input image is indicted on the lower left. The resultant filtered image is illustrated on the lower right.

8.6.5 Isolate the largest item in the field of view

The aim of this program is to find and isolate the largest white region in the scene and indicate the area (in pixels) of this region in an embedded text message. The input image is in PCX format. This is loaded by double-clicking the input image box and selection the *crown.pcx* image. A low pass filter is applied to the image, with the effect of blurring the image. The image is then thresholded at grey scale 200. All pixel values below 200 will go to black, and the rest will be assigned to white, producing a binary image. You can experiment with the threshold by double clicking the *single threshold* box and moving the slider bar. Any single isolated white pixels are then removed and a 3-pixel-wide black border is drawn around the image. This is implemented to eliminate any white pixels touching the boundary, a necessary requirement for the *biggest blob* operator. See Figure 8.20.

The largest white region is then isolated, and its area calculated by counting the number of white pixels. This is assigned to an integer variable which is then combined with the predetermined text string "The area in pixels is: " and displayed to the user.

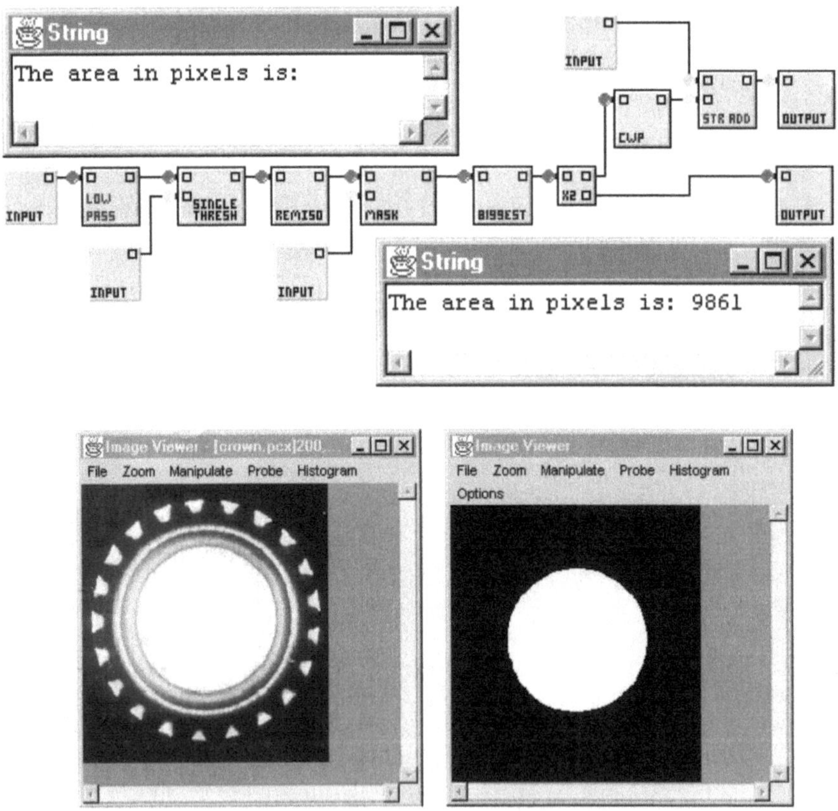

Fig. 8.20 Find and isolate the largest white region in the scene.

8.6.6 Character detection using the N-tuple operator

Figure 8.21(a) and (b) illustrate how the N-tuple operator can be used to highlight the Arial font character "t" in a sentence for a given orientation. The original image is threshold at the mid grey level and inverted. This image is then convolved with an N-tuple representing the Arial font character "t", resulting in a output image in which the location of the desired character is highlighted by a white point. The threshold operator is then applied to isolate these points and the detected characters are emphasized by overlaying them with a small square box. The JVision canvas for this is illustrated in Figure 8.21(c).

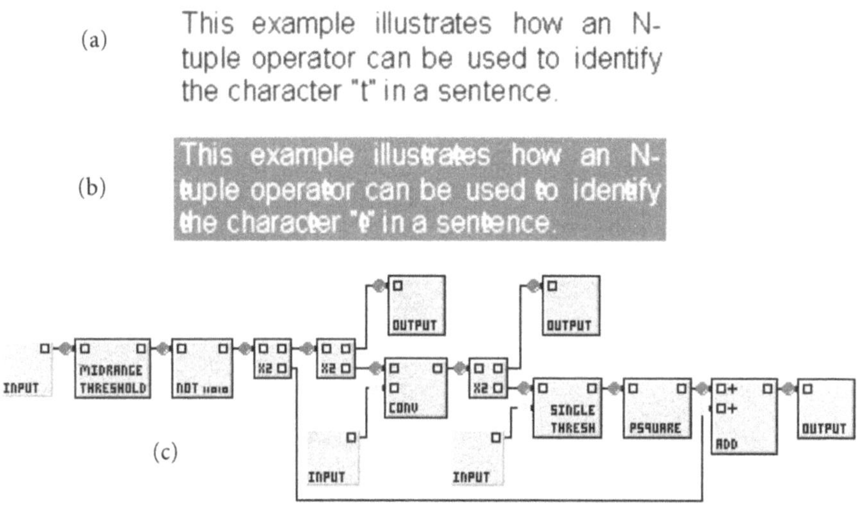

Fig. 8.21 (a) Original image. (b) Detection of the Arial font "t" character using an N-tuple operator. (c) The associated JVision canvas.

8.7 Summary

Visual programming environments such as JVision provide a versatile platform which facilitates high-level access to a wide range of machine vision and general data processing algorithms. These can be combined to create a visual program by specifying data interdependencies between the individual algorithms. In addition, the path of data through a visual program can be dictated using special flow-control components. These techniques provide a fast and simple alternative to conventional text-based programming, while still providing much of their power and flexibility.

JVision provides a fully integrated software development solution, with several levels of user interaction. It is extendable, portable, and most importantly, easy to use. The choice of Java as the development language for the application not only improves portability but also introduces the developer to a very powerful and straightforward approach to object-oriented programming for image processing and analysis, which Sun Microsystems is committed to supporting and improving for the foreseeable future.

9 Application case studies

If it's green, we reject it.
If it's too ripe, we reject it.
If it's bruised, we reject it.
If it's diseased, we reject it.
If it's dirty, we reject it.
*If it's just right, we squash it.**
Advertisement: McDonald's Restaurants, December 1992
* Tomato used to make ketchup

9.1 Preliminary remarks

Machine Vision has been studied in universities, commercial companies, and various other research organizations for over 20 years. The world-wide market for vision-based products and services for industrial, medical, security, and other applications is already about US$4000 million *per annum* and is growing rapidly. However, it must be admitted that industrial Machine Vision is still in its infancy, in the sense that the technology is not widely understood by the general engineering and business fraternity. As a result, it is often mistrusted, misused, and underused. Recall that in Chapter 1 we pointed out that most people, including technologists, erroneously believe that they can explain how human beings see. It has been found by experience that "obvious" solutions to Machine Vision applications almost invariably do not work. Indeed, even an experienced vision engineer can only rarely predict correctly what algorithm will be most appropriate, if he is limited to viewing an object or scene by eye. The applications studies described in this chapter are intended to demonstrate this point, plus two others:

a) The algorithmic/heuristic techniques described at length earlier in this book are capable of achieving technically-feasible solutions to a very wide range of applications. (Demonstrating *commercial* feasibility is a completely different question, and is beyond the scope of our present discussion.)
b) Interactive image processing, as exemplified by PIP, is able to analyze a large proportion of interesting and commercially-important applications in a surprisingly short period of time.

Our objective in this chapter is simply to give credence to the ideas discussed earlier. This book is not intended to provide a comprehensive review of important applications of Machine Vision. It is not even possible to describe the details of each application mentioned below or illustrated in the halftone plates, since to do so would greatly increase the size of this book and would, in many cases, violate commercial confidence. The approach that we shall take is therefore to use specific applications to illustrate a number of general points and, in a few cases, explain in some detail how PIP or one of its predecessors has been used to discover the broad nature of a possible solution.

9.2 Taking a broad view

The reader is referred to Plates 4 through 18, which illustrate a number of general lessons for would-be designers/installers of Machine Vision systems. The most important point to note is that the image-processing subsystem can be designed only after a good image-acquisition subsystem has been chosen. However, we cannot rely on having good-quality images to process. For the sake of robustness and system reliability, we must design our computational procedures on the assumption that the images will be of indifferent quality.

9.2.1 Automobile connecting rod (conrod)

Plate 6 shows the results obtained by applying various methods of analyzing a single isolated object in a binary image: the silhouette of an unfinished conrod. (Also see Plate 7.) On its own, Plate 6 does not provide a convincing demonstration that Machine Vision is fundamentally different from Digital Image Processing as it is described in the many excellent textbooks on that subject. However, let us go back to consider other aspects of this application, which is concerned with manipulating a dusty dull-grey ferrous component or detecting faults in it. This may have a few rust spots and will almost certainly be viewed against a dirty background. Obviously, matters are not quite so simple as Plate 6 suggests. We can see that this is so if we examine Plate 1(TR). This shows the original grey-scale image from which Plate 6(TL) was derived by processing with PIP. In many applications, the image processing begins with a picture of low-to-moderate quality, not with a clean binary image. We must not assume that a high contrast necessarily exists between an object and its background. The nonuniform background illumination and the dirt on the right of Plate 1(TR) both render the "obvious" image-processing operations (e.g., thresholding, *thr/2*), unreliable. In practice, we must use a more complex procedure, such as *[enc, thr, neg, big]*, to obtain a reliable segmentation of this image. Experimentation with PIP and related systems over a period of 25 years has shown clearly that even backlit subjects are not as easy to analyze as we might imagine.

In many instances, there are other objects within the camera's field of view (Plate 7). Somehow, the most important ones have to be isolated before they can be recognized/analyzed. This might be done on the basis of physical size (use *big/1*), position (use *fil/5* to mask off unwanted regions of the image), or some higher-level attribute, such as shape *(shf/1)*. The author's experience has demonstrated quite clearly that PIP has more than sufficient power of expression for this type of application [BAT97]. Obviously, the TVS needed to implement a complicated algorithm will be slower/more expensive than one based on a simple procedure. Anything that we can do to separate the components as they move in front of the camera can have very beneficial effects in reducing TVS cost and/or increasing its speed. We might, for example, use guide rails over a conveyor belt to do this. Even simple mechanical devices such as this can have quite dramatic effects on the complexity of the image-processing algorithm/heuristic and its implementation.

9.2.2 Coffee beans

Plate 1(CL) shows a collection of coffee beans. From this picture, people can identify individual beans, but it does not provide a reliable basis for inspecting individual beans. It is not particularly useful either for counting beans or for identifying types of beans. (Identifying types of beans/seeds is important commercially, as a check on the honesty of suppliers of bulk products.) This image, although seemingly "obvious" as a means of examining large numbers of beans simultaneously, is quite unsuitable for Machine Vision (or human vision). A better approach would be to separate the beans and use backlighting, so that their individual images can be isolated easily. Although it is easier if the beans do not touch, image processing can readily separate those that just "kiss." Another approach for counting beans uses front lighting, using a point source placed very close to the camera. This causes one point on the top of each bean to glint, making it very easy to separate and then count the beans. Using such a lighting arrangement permits dense packing like that shown in Plate 1(CL) to be used. Although it would be very convenient to base our inspection process on the image shown in Plate 1(CL), it is unsuitable for the purposes of Machine Vision. Despite long periods of study by an experienced vision engineer using PIP, it has not been possible to extract any really useful information reliably from this image. In general, failure to find a solution with a powerful tool such as PIP should be taken as a strong indication that the given image is inappropriate for the given application.

9.2.3 Table place-setting

See Plate 1(CR). This provides a convenient conceptual analog for a range of industrially relevant applications. Here, we will ignore the problem of processing the initial grey-scale image, discussed in Section 9.2.1. At first sight, the image in Plate 1(CR) does not appear to be very different from that shown in Plate 7(TL). In both cases, there are several separate objects, viewed in silhouette. However, these two applications have quite different requirements and, as a result, require completely different solutions. This is an illustration of the general maxim that two applications which appear to be very similar to an uninitiated person may, in fact, be quite different in practice.

The question to be answered in relation to the table place-setting is quite simply expressed: *Is this a well-laid place-setting?* What constitutes an acceptable place setting can be expressed by a person in natural language (e.g., English), using a collection of statements such as

"The soup spoon is to the right of the butter knife."

Although seemingly very straightforward, this application unexpectedly leads us into the realm of Artificial Intelligence, since it involves our manipulating abstract symbolic relationships relating objects in physical space: *left*, *right*, *above*, *beside*, etc. Although these concepts seem harmless enough to the uninitiated, they can be difficult to express in conventional computer languages. However, Prolog provides a very much more natural way to explain what a relationship such as *"left of"* means [BAT91a]. This illustrates the need to be able to integrate vision with AI techniques. PIP is designed to do just that.

Despite Prolog being a natural way to represent spatial and other abstract relationships, users of a TVS would probably not accept a system that appears to rely on programming in this language. There are therefore two courses of action:

a) Reprogram the algorithm using a different language. For example use C to recode Prolog programs.
b) Use Prolog in the TVS but do not tell the user, and hide the fact by employing a user-friendly dialog. In Chapter 4, we emphasized the use of Natural Language for this very reason.

9.2.4 Hydraulics manifold

See Plate 9(TL), (TR), and (CL). Employing the correct view is clearly essential for a component with a complicated geometry, since some features are visible only from certain vantage points. As we saw in Chapter 1 when considering another complicated object (Plate 9(CR)), several cameras might be needed, together with sophisticated parts-handling and computer-controlled lighting. For this type of application, the Flexible Inspection Cell, described in Chapters 4 and 7, is ideal, since it allows a variety of lighting and viewing conditions to be set up quickly and easily under software control. A few years ago, the author encountered a somewhat similar application that required the inspection of a zinc die-casting of comparable complexity that was then being inspected manually. This involved a total of 257 measurements, and the inspection procedure took *four days*! Automating this task is far from trivial. However, there are few potential applications of Machine Vision that better demonstrate the need to automate the inspection process. The FIC is exactly the type of facility that is needed for this task. Here, as in similar applications, versatility rather than speed is the prime requirement. While neither PIP nor the FIC was designed for use on the factory floor, such a combination would be appropriate within a TVS for this application.

While PIP has the potential to examine objects of this level of complexity, the programs can be rather long and tedious to explain. For this reason, they are not listed here. The art of writing programs for applications like this is to break the task into a number of smaller subproblems, each of which can be solved by a single (Prolog) clause. This "divide and conquer" approach is normal practice in Prolog programming.

9.2.5 Hydraulics cylinder for automobile brake system

See Plate 9(BL). The inspection task required in this instance is to examine the smooth polished internal surface of the bore. Prior to our involvement with this application in the late 1970s, a special-purpose laser scanner had been designed to obtain the image data [BAT79a].

Images obtained using this scanner show the scratches as dark streaks against a light background, which has a faint criss-cross pattern on it. Isolating the scratches was found to be straightforward, using the crack detector operator, *crk/1*. In fact, this operator was discovered by the author (BGB) experimenting with SUSIE in the 1970s. *crk/1* precedes the more general grey-scale morphology techniques by several years. Apart from contributing to the success of the project, this operator was subsequently incorporated into SUSIE and has since become one of the author's favorite filters for detecting scratches and cracks. A highpass filter such as *[pis, N•lpf, pop, sub]* also performs quite well in this application, demonstrating the point that there are usually several algorithms that are potentially able to perform a given inspection function. Helping the user to choose the best one is a task that is particularly well suited to PIP.

The reader is reminded of the possibility of designing specialized image-acquisition hardware, such as a laser scanner, for certain demanding tasks. Although the cost of designing this equipment may seem daunting, it may well save a great deal of effort and expense in the end. There is much that an IVS such as PIP can do to predict and quantify the benefits of such a venture. (Recall the possibility of using PIP to simulate a line-scan camera. See Section 4.4.7 and Plate 17(TR).)

9.2.6 Electrical connection block

See Plate 9(BR). Four faults can be seen. If the blocks are unconstrained in space. inspecting them is relatively difficult. On the other hand, it should be borne in mind that these connectors are manufactured by first molding a long plastic strip. The brass collars and screws are then inserted automatically into a length of this strip, which is held in a jig during assembly. Finally, the strips are cut up into smaller sections before packaging and shipment. It therefore seems sensible to examine the long strips while they are still being held in a known position. The inspection algorithm is then quite straightforward, although we will not describe the minutiae here. It should be noted that PIP finds no difficulty at all in expressing even the more complicated algorithm required to solve the general problem in which there are no constraints imposed on the position of the blocks. However, the program is much larger, slower in operation, and is inherently less reliable than one designed to examine strips that are held in fixed position and orientation. In general, the ability to constrain the problem is an important part of reducing the complexity and cost of the image-processing procedure. Its implementation, either in hardware or software, also benefits in terms of reduced cost and increased speed/ throughput rate.

9.2.7 Electric light bulb

See Plate 10(TL). There are very real dangers caused by wires protruding from malformed light bulbs. This problem has been known for many years; the author is aware of three companies in the United Kingdom that have been obliged to publish warning notices about their products. (The reader should bear in mind that the high power-supply voltages used in Europe make the danger more acute than, for example, in the USA.) The speed of production is high enough to make human inspection unreliable. As a result, this is an application in which there is no serious competitor to Machine Vision.

The presence of both bright shiny brass and glass on the light bulbs makes the design of the lighting/viewing system critical. This is a demanding application which provides a chance for an experienced illumination engineer to demonstrate his skill. Expressed another way, this is an application where a sloppy unprofessional approach to lighting will result in disaster!

It should be understood that wires can protrude either from the connecting pads or squeezed between the brass cap and the glass envelope. A camera is needed to inspect the base of the bulb. It is also necessary to examine all around the glass–metal joint, since the exact position of the wires is unpredictable. To obtain an all-around view, the light bulbs can be rotated through 360° and viewed by a line-scan camera (see Section 4.4.7). Alternatively, three area-scan cameras can be placed at 120° intervals around the glass–metal joint. If we are not very careful, the lighting for one camera may interfere with the view of another, so considerable care has to be taken to block all unwanted light. One possible approach is to modify the lighting pattern by switching the lamps on/off as the video output from each camera is sampled in turn. Another approach uses narrow-band filters in front of the lights and cameras. However, there are many other optical tricks that a good vision engineer can employ to obtain good images. Let it suffice to say that it is possible to obtain good images from all around the light bulb. Once the mechanical/lighting problems have been overcome, choosing the inspection algorithm is straightforward. Once again, the crack detector *crk/1* is likely to form the basis of the inspection algorithm.

9.2.8 Analysis of industrial X-rays

See Plates 11 and 18. As well as being concerned with applications in which sensing of the physical world is based on the visible part of the electromagnetic spectrum, Machine Vision also encompasses Automated X-ray Inspection (AXI). This offers considerable untapped potential for examining safety-critical parts, the internal structure of assemblies of components, and detecting foreign bodies in food and pharmaceutical products. While user safety is of paramount importance when building industrial x-ray systems, many of the other design factors are common to Automated Visual Inspection systems. In AXI, the image structure is often quite complex, because the sensing process integrates the x-ray absorption through the whole object being examined (Plate 11(BL) and (BR)). One particular feature of AXI is that the image contrast may be quite low, as is evident in Plates 11(TL) and (CR). This places an even greater emphasis on image-segmentation algorithms and on the need for holding the object to be examined in a known position. The latter can greatly simplify the analysis procedure. Later in this chapter, we discuss the use of PIP to examine the mains power plug.

9.2.9 Highly variable objects (food and natural products)

See Plate 12. Loaf shape analysis is important for the baking industry, since unattractive misshapen loaves are unpopular with customers. What constitutes an acceptable loaf of a given type? Such a question is far from easy to answer, since personal preference cannot be defined in any objective way. The difficulty is compounded by the fact that a consensus opinion, derived from many individual customers, must somehow be determined. Bakers may claim to know what a "good" loaf is but, on close examination, their opinions are only a rough guide to what the "average customer" really thinks. The conventional approach to this kind of situation is to employ machine learning and try to emulate the opinions of a panel of people. The operational difficulties here may well dwarf the technical ones. However, we will assume that such problems have been overcome and that there is a definitive way to classify loaves as *"acceptable"* or *"non-acceptable."* This forms the *teacher* for a learning system, which must also be given a set of numeric and/or symbolic values describing each loaf. These measurements/observations can be derived using a PIP program. First, however, we must obtain one or more images from which to derive these values.

It is easy to obtain a description of the shape of the top and side surfaces of a loaf using a laser light-stripe system, similar to that explained in Plate 17 [BAT93a]. An image in which the "intensity" indicates the height of an object surface is called a *range (or depth) map* (Section 4.4.8). Such an image can be constructed in a variety of other ways, although details need not concern us here. Alternatively, an off-line inspection system can be built which derives and combines shape descriptors for several (vertical) slices, taken along the length of the loaf.

Deciding whether a range map represents an "acceptable" loaf can be determined by first deriving a set of numeric shape descriptors, then applying these to a Pattern Recognition system. The latter uses self-adaptive learning to distinguish between *"acceptable"* or *"non-acceptable"* loaves. However, there is a rather severe snag: almost all conventional learning techniques are designed to learn to distinguish between representative samples of two or more classes. Bakeries do not make "bad" loaves! Most bakers will admit to making only a small number (if any) of "unacceptable" loaves. (Notice the subtle choice of words used here.) However, just as soon as they discover that "unacceptable" loaves are being made, they stop doing so by either altering the mixing/baking machine settings or by simply stopping the production line altogether. The result is that very few "unacceptable" loaves are made on which a system can learn. It must therefore learn on samples of just the "acceptable" class. Certain specialized types of learning algorithm must therefore be used. Some of these have been coded in Prolog and hence can be used in PIP [BAT91a].

Another application that requires the use of learning is bread-texture analysis [BAT93a]. Texture is a very complex issue, although PIP has all of the essential features needed to study it in detail: the ability to derive a wide range of descriptive measurements and to represent a range of complex analysis procedures. One severe problem is always encountered when we try to analyze texture: deciding what we are trying to do! How do we recognize the texture of an "acceptable" loaf. The problem is compounded by the fact that the texture is not constant over the whole area of a typical slice of bread. Plate 12(CL) shows another example of a texture surface: a cork floor tile. The problems of analyzing its texture are also difficult. However, an interactive system such as PIP probably provides a suitable way to study this type of application too.

Numerous other food products (e.g., pizza, quiche, Plate 12(CR)) present similar problems, as do many natural products such as nuts, fruit, vegetables, etc. These applications are typified by being associated with a high level of object variability. This is in stark contrast to molded metal and plastic components, which are usually very similar to one another. As the level of variability increases, ever more intelligent analysis techniques are needed. PIP's reliance on Prolog therefore becomes even more important in the study of natural and food products.

One recent study which used Prolog+ (PIP's immediate predecessor) concentrated on inspecting cake decoration patterns (Plate 12(B), [BAT91b]). Another project, which again used Prolog+, was concerned with directing a robot to dissect very small plants [BAT89]. Both rely on AI techniques for their solution.

9.3 Cracks in ferrous components

Objective

Detect cracks in a nonferrous metal workpiece.

Illustrations

Plate 13(TL), (TR), and (CL)

Sample preparation

Clean the surface of the object first. Then, immerse it in a low-viscosity, high surface-tension, fluorescent or vividly-colored dye. Rinse and dry the workpiece. Finally, coat it with fine, white, water-absorbent powder. This causes any dye that has penetrated a crack, by capillary action, to spread out, forming a visible colored line around it.

Lighting and optics

Fluorescent dye method: Illuminate using ultraviolet light, with a UV-blocking filter in front of the camera lens.

Colored dye method: Illuminate using colored or white light, with a narrow-band filter in front of the camera lens.

Camera

Area-scan camera. This must be sensitive enough to cope with the low light levels produced by fluorescence.

Image processing

Two viable solutions have been found using PIP:

Highpass filter: [*raf,sub,thr(A)*], control parameter *A*
Crack detectors: [*crk(B), thr(C)*], control parameters *B* and *C*

To eliminate small spots due to camera noise, minor surface blemishes, and dirt, use [*big(1)*] or [*kgr(N)*], where *N* is some predefined integer control value. It also may be necessary to use additional processing to join segments of "broken" cracks. This may consist of simple morphological closing [*M•exw, M•skw*] (*M* is some predefined, integer, control parameter.) Alternatively, the skeletonization operator [*ske*] may be applied prior to extending the ends of the arcs so formed, using *egr/[0,4]*.

Remarks

Exposure to ultraviolet light can cause eye damage and skin cancer. The dye may be toxic or carcinogenic. Airborne dust may contaminate exposed optical surfaces and possibly affect breathing. There are various alternative sample-preparation techniques for ferrous components.

Commercially, this is a very important application, with profound implications for safety in automobiles, aircraft, bridges, and similar structures.

9.4 Aerosol spray cone

Objectives

a) Measure the spray-cone angle and symmetry.
b) Detect fine jets within the spray (manifest as bright streaks).

Illustrations

Plate 13(CR), (BL), and (BR)

Lighting and optics

Place the light source above the spray cone. Light must be cold (i.e., contain little or no infrared), to avoid heating the spray droplets and causing them to evaporate prematurely. The camera should face a black nonilluminated background.

Camera

Area-scan camera. Must be sensitive enough to cope with low light levels.

Image processing

Intensity contours (isophotes) are generated directly by [*ict*].

The following program sequence plots the cone width:

```
loa('Spray'),     % Load image
yxt,              % Rotate image (Spray now points downwards)
wim(a),           % Save image in file a
rox,              % Row maximum
csh,              % Make all columns the same as RHS
hin,              % Halve intensities
rim(a),           % Recall original (i.e., rotated) image
sub,              % Subtract images
```

```
thr,                    % Threshold (Edges shown in curve B)
rin,                    % Row integration
csh,                    % Make all columns the same as RHS
plt,                    % Plot width of spray (image is rotated by 90°
yxt.                    % Final result (see curve A in Plate 13(BR))
```

We can use [*neg,crk*] to detect any fine jets. (Not illustrated.)

Remarks

The spray may contaminate exposed optical surfaces. An air curtain and enclosure may be required to keep them clean and avoid any possible health hazard.

9.5 Glass vial

Objectives

a) Measure the radius of curvature of the shoulder. (This is a critical factor determining the vial's mechanical strength. The shoulder must be properly formed to avoid shattering the vial when the cap is fitted.)

b) Measure the neck diameter.

c) Identify points of interest; for example, where a robot can place its fingers to lift the vial.

Illustrations

Plate 14

Lighting and optics

Dark-field illumination. That is, the camera faces a large bright screen that has a small opaque black patch on it. This patch fills the camera's field of view. The vial is placed between this patch and the camera.

Camera

Area-scan camera.

Image processing

The following program calculates the radius of curvature (*Rad*) and draws the circle which fits three points on the shoulders of the vial.

```
loa('Vial'),            % Load image of the vial
yxt,                    % Interchange X and Y axes
enc,                    % Enhance contrast
lnb,                    % Largest neighbor
sub,                    % Subtract – this & previous line form edge detector
thr(135),               % Threshold – adjust to taste
dim(A,_,_,_),           % Position of vial top
X2 is A + 100,          % First scan line defined in relation to vial top
X1 is X2 – 20,          % Second scan line is a little way to left of first one
X3 is X2 + 20,          % Third scan line is a little way to right of first one
scan3(X1,X2,X3,Y1,Y2,Y3), % See below
fcd(X1,Y1,X2,Y2,X3,Y3, X0,Y0,Rad).% Circle parameters
```

The auxiliary predicate *scan3/6* finds the coordinates of the top-most white points along three vertical scan lines and is defined thus:

```
scan3(X1,X2,X3,Y1,Y2,Y3):-
pis,                      % Push image onto stack
dgw(L,T,R,B),             % Image size
bve(X1,T,X1,B,_,Y1,_,_),  % Extreme point along vertical scan line
tsk,                      % See top of stack
bve(X2,T,X2,B,_,Y2,_,_),  % Another extreme point
tsk,                      % Image at top of stack
bve(X3,T,X3,B,_,Y3,_,_),  % Another extreme point
pop.                      % Restore original image
```

Once the vial image has been converted into a binary format, the neck can be found easily by using [*rin*] (Plate 14(CL)). By applying [*bed, cox*] again to the binary image, the sides of the main body, the neck, and the mouth can all be identified (Plate 14(CR)) and (BL)). The concavities formed by the neck can be identified using the convex deficiency operator, [*cvd*]. The centroids of these concavities can provide a useful cue to indicate where a robot should place its gripper to lift the vial.

Remarks

A variation of dark-field illumination can be used in conjunction with a line-scan camera [BAT94a].

9.6 Coin

Objective

Determine the orientation of a bright "silver" coin as a prelude to detecting faults.

Illustrations

Plate 15

Lighting and optics

Coaxial illumination and viewing from a diffuse source [BAT94a]

Camera

Area-scan camera.

Image processing

This program finds the orientation of the coin by locating the tip of the crown:

```
loa('Coin'),              % Load image
enc,                      % Enhance contrast
thr,                      % Threshold at mid-grey
big,                      % Select biggest blob
chu,                      % Convex hull
blb,                      % Fill holes – Coin outline
cgr(X,Y),                 % Center of coin
```

```
xor,                    % Exclusive OR. Monarch's head isolated
big,                    % Select biggest concavity
blb,                    % Fill holes. Monarch's head is "solid"
ctp(X,Y),               % Cartesian-to-Polar coordinate transform
rin,                    % Row integration
csh,                    % Make all columns same.
gli(_,B),               % Find maximum intensity
thr(B,B),               % Threshold at maximum intensity
big,                    % Select biggest blob (i.e., widest stripe)
cgr(_,C),               % Y-coordinate of center of widest stripe
Z is -360*C/256,        % Rescale to degrees
dsl(X,Y,Z),             % Draw line through (X,Y) at angle Z
loa('Coin'),            % Reload image
max.                    % Superimpose onto original
```

Remarks

Finding the orientation is a useful prelude to recognizing the pose (i.e., deciding whether the coin lies "heads" or "tails" up). It is also helpful if we are to use model fitting to inspect the coin for defects. (That is, we compare the image of the coin with a stored reference image and thereby compensate for the wide intensity variations that exist even in a "good" coin.)

The lighting is of crucial importance if we are to obtain good, consistent results from a simple algorithm.

9.7 Metal grid

Objective

Determine the orientation of a metal grid. (The grid illustrated in Plate 16(TR) consists of a flat disc, similar to a fine "woven" kitchen sieve.)

Illustrations

Plate 16

Lighting and optics

Backlighting.

Camera

Area-scan camera.

Image processing

The program has two parts. First, a line indicating the orientation of the wire grid is calculated quickly. However, it may not be very accurate and is therefore refined later.

```
loa('Grid'),            % Load image
ect,                    % Threshold at middle intensity
wim(binary),            % Saved for second half of the program
ctp,                    % Cartesian-to-polar coord. Plate 16(TR)
rox,                    % Row maximum
```

```
csh,                       % Make all columns the same as RHS
neg,                       % Negate
dim(_,A,_,_),              % Find top-most white line
Z is 90-360*A/256,         % Convert its Y-coordinate to angle
mim(X,Y),                  % Middle of the image
dsl(X,Y,Z)                 % Draw line through (X,Y) at angle Z
```

The line produced in this way is shown in Plate 16(CR). Once a row of spots has been established tentatively in this way, the estimate of the orientation can be refined, using the following program.

```
rim(binary),               % Binary version, created earlier in program
mim(X,Y),                  % Middle of the image
balloon(X,Y,1,X2,Y2)       % [X2,Y2] is the white point closest to the image center1
dsl(X2,Y2,Z,Z),            % Draw line through [X2,Y2] at angle Z,Z
rim(binary),               % Reload binary image
max,                       % Superimpose line onto it
kgr(50),                   % Keep line and all points lying along it2
rim(binary),               % Reload binary image
min,                       % Mask to keep only points along that line
dpa.                       /* Draw principal axis – refined line
                           Use [lmi] to obtain numeric values. */
```

Remarks

This method, based on the *ctp/0* operator, is akin to the Radon transform [BAT91a]. Algorithms based on the Hough transform *(huf/0)* and *fan /7* predicate (Figure 4.18 and Table 4.12) can be used instead.

9.8 Toroidal metal component

Objectives

a) Identify the component type.
b) Determine its orientation
c) Measure its 3-dimensional profile.

Illustrations

Plate 17

Mechanical handling

Slow continuous or indexed rotation in the horizontal plane.

---

1. The *balloon/5* predicate is one of a class of so-called *Gauge Predicates* that have been incorporated into PIP, See Figure 4.18, Table 4.12, and [BAT91a]. *balloon(X,Y,R,X1,Y1)* searches for the white point that is closest to [X,Y]. The newly found point is at [X1,Y1] and must be not closer than a distance R from the point [X,Y], otherwise the goal will fail. The calculation effectively simulates inflating a circle of initial radius R and centered at [X,Y] until it encounters a white point. The circle is scanned in an anticlockwise direction.
2. The obvious way to achieve the same result would be to use [*big*]. However, this operator has difficulty with images containing more than 255 spots.

Lighting and optics

Two different image-acquisition schemes can be employed, depending on the function to be performed:

1) Uniform illumination along a single radial line. The light may be projected vertically downwards, or at an oblique angle.

2) A fan-shaped beam of light generated by a laser which is fitted with a cylindrical lens. The laser beam lies in a vertical plane. This method is used to measure the surface profile (see Section 4.4.8).

Camera

Two different image-acquisition schemes are employed, depending on which illumination system is used:

1) Line-scan camera, placed vertically above the toroidal object. This type of camera is used in conjunction with lighting method 1 (see Section 4.4.7).

2) Area-scan camera, used in conjunction with the fan-shaped beam. The camera looks down obliquely (at about 45°) onto the toroidal object and sees only the bright line of light. See Plate 17(TL).

Image processing

1) The simulated line-scan image, Plate 17(TR), was produced by repeatedly sampling the central row in the digitized signal from an area-scan camera. Meanwhile, the metal component was rotated beneath it. This process is discussed in Chapter 4. The pronounced sinusoidal "wiggle" can be removed by the following program.

```
ptc,              % Polar-to-Cartesian mapping, Plate 17(CL)
wim(a),           % Save image for a little while
ect,              % Threshold image to isolate object
chu,              % Draw its convex hull
blb,              % Fill any holes
cgr(X,Y),         % Find centroid of resulting disc
mim(X1,Y1),       % Center of the image
X2 is -X + X1,    % Calculate offset of object centroid along X axis
Y2 is -Y + Y1,    % Calculate offset of object centroid along Y axis
rim(a),           % Recover the transformed image saved earlier
psh(X2,Y2),       % Shift image
ctp               % Cartesian-to-polar coordinate mapping
```

Following this, processing the resulting image to find the object orientation is easy: we simply locate the notches on the right-hand side (corresponding to the outer edge of the component). The vertical position coordinate is proportional to the orientation of the component.

2) Generating the range map is discussed in Section 4.4.8. The "wiggle" can be removed in the same way as before. The object's top surface consists of three annular rings. The average height of each one can be estimated using the following program:

```
yxt,              % Interchange X and Y axes
rin,              % Integrate intensity
```

```
csh,                    % Make all columns the same as RHS
plt                     % Plot intensity
```

9.9 Mains power plug (X-ray)

Objectives

a) Check that there are no loose strands of copper wire.
b) Check that the internal cable insulation has not been stripped back too far.
c) Check that the fuse has been fitted correctly.

This is a contrived application, intended to demonstrate the potential for applying Automated X-ray Inspection to safety-critical assemblies consisting of materials with very different levels of x-ray absorption.

Illustrations

Plates 11(BL) and 18

Mechanical handling

For safety's sake, any repositioning of the component while the x-ray source is switched on must be accomplished automatically. (It may be appropriate to use MV techniques, based on visible light, to assist in this process.)
Accurate alignment of the object by mechanical means is important if model-based vision is to be used.
Care should be taken to align the brass connecting pins with the axis of the x-ray beam, to avoid obscuring the internal structure of the plug.

X-ray

A fine-focus low-energy x-ray system with the ability to penetrate the plastic casing is used. The beam does not need to penetrate the metal components.
The beam occupies a cone, expanding from a point source. This causes a variety of parallax problems. For example, off-axis component edges lying parallel to the central axis of the x-ray beam will appear to be out of focus.

Image processing

Simple thresholding can separate the plastic and metal parts of the assembly with reasonable accuracy. It may be necessary to adjust the threshold parameter in different parts of the image. Detecting the metal components is straightforward, since they are effectively opaque to x-rays (Plate 18(BL)). Features on the plastic molding can also be seen, though slightly less reliably, since the intensity contrast is lower (Plate 18(CR)).
In this particular application, sharp intensity gradients exist at the edges of both plastic and metal (Plate 18(TR)). Hence, an edge detector can locate a range of features easily and reliably. Once two or more "anchor points" have been found, a model-fitting program (written in PIP) can be used to locate specific parts of the plug assembly. Other features can then be examined in detail. It is important to obtain accurate positioning, since it makes model fitting more straightfor-

ward and reliable. Thin, bright streaks, caused by non-insulated copper wire can be detected most easily using a model-based inspection technique.

The very thin bright x-ray shadows created by loose strands of copper (Plate 18(CL)) and the insulated multi-strand wires (Plate 18(BR)) can be detected, independently of the alignment of the plus, using morphological operators, such as the crack detector, (*crk/1*).

9.10 Conclusions

In a short chapter such as this, it is impossible to describe more than a very small fraction of all the applications of Machine Vision that have been studied to date. While some applications have critical implications for product and/or operator safety, other can only be described as bizarre. However, they are all important for one or more good reasons: to increase the efficiency of a production process, reduce scrap, make better use of natural resources, or improve product appearance or process/product safety. Some of the applications that have been studied by the authors in the past are impossible to solve in the light of our present knowledge, while some others are trivial. Sometimes, Machine Vision may offer a solution but it must always compete with other technologies. When we are given a few sample "widgets" and are asked to assess what the potential for Machine Vision really is, we must be able to do so without a great deal of effort or expenditure. The applications illustrated above clearly show that *outline solutions* can often be expressed in terms of the PIP command language. What they do not show is how easy it is to derive those first tentative solutions. The author's experience with final-year undergraduate and Master's degree students, learning to use PIP for the first time, shows that they can begin to derive solutions for a range of varied applications within a few hours of study. On the other hand, an experienced user can often derive and demonstrate outline solutions within a few *minutes*. The point is that solutions can often be discovered quickly. If they cannot, then we must question whether the problem is solvable at all. An IVS such as PIP enables this question to be answered quickly, easily, and cheaply.

A very large group of applications is composed of problems that are tractable, provided that the image-acquisition subsystem has been carefully designed beforehand. In general terms, the tighter the control that can be achieved on object form, color, position, and orientation, the easier the task of designing an image-processing algorithm will be. In this book we have discussed only image-processing algorithms and their implementation in detail. However, a vision system must also incorporate devices for mechanical handling, lighting, optics, and an image sensor, all properly integrated, with a great deal of sound engineering applied to the design process. The only way that a proper sense of proportion can be maintained during the design of a system is to understand the image-processing component and be able to balance its needs and abilities with those of the other parts of the system. The real value of an IVS lies in its ability to provide the engineer with a tool to remove the mystery from image processing and provide firm objective evidence for difficult design decisions. Every design engineer knows the importance of being able to answer "*What if ....?*" questions quickly and easily. This is just as important for image processing as it is for electrical or mechanical design. In this short chapter, we have attempted to demonstrate that IVS can very often provide the kind of information that is crucial for achieving a balanced design.

While it is clear from the foregoing pages that PIP and its predecessors have been able to demonstrate outline solutions to a wide variety of applications, a large number of other examples have been given in earlier publications [BAT91a, BAT94b, BAT97]. However, it is impossible to convey a true impression of the real power of an interactive vision system relying only the printed word; it really has to be seen in operation, or better still used in earnest, to appreciate it fully.

10 Final remarks

"Everybody is an expert on vision"

"The following inequality is always true:
Vision_system ≠ PC + Framegrabber + Software"

When the authors began working (separately) in Machine Vision research in the mid-1970s, the subject was still very much in its infancy. The limited computing power then available made it very difficult to apply image processing to any but the simplest and least demanding inspection tasks. Since then, the subject of Machine Vision has been transformed by several key innovations, which have greatly improved:

(a) The range of off-the-shelf lighting units specifically designed for Machine Vision
(b) General and specialized optical devices
(c) Image sensors
(d) Algorithmic/heuristic techniques
(e) Design tools, including but not restricted to interactive image processing
(f) Standard computer hardware and software
(g) Turnkey software packages for image processing
(h) Dedicated hardware for high-speed image-processing.

Within the limited scope of this book, we have concentrated on only part of the story, by discussing topics (d), part of (e), (g) and (h); we must emphasize that only part of the overall picture relating to Machine Vision as a whole is described in these pages.

As a result of these technical developments, both engineers and customers have gained far greater confidence in Machine Vision, so that it is no longer regarded as an "exotic" technology. This increase in confidence is also due in large part to our improved understanding of the vital rôle that Systems Engineering plays in the design of vision systems. In particular, our awareness of the importance of the image-acquisition subsystem led to more reliable systems being built. This, of course, improved confidence in later systems.

At the moment, Machine Vision technology is developing rapidly. Numerous companies are selling a wide range of equipment, which is being installed in a very wide range of manufacturing industries. However, the overall coverage of potential and feasible applications is depressingly small and is likely to remain so, until great improvements can be achieved in the design process.

Predicting the future is always difficult in rapidly moving areas of technology, such as electronics and computer technology. The reason is, of course, that history does not provide any effective clues about what we can expect to happen, even in the near future. Emergent technologies, as yet invisible except to the *cognoscenti* in a particular subject, are liable to thwart even the most learned scholar in another area. Technology is rather like an iceberg; at any given time, we see only part of the situation, since our limited minds cannot properly assimilate more than a very narrow section of all available knowledge. Prophesy is therefore a dangerous pastime and liable to produce acute embarrassment for those bold enough to try anticipating the future. Despite this, there are some clear signs that allow us to make some fairly firm predictions for the future development of Machine Vision, at least covering the next 5-year period (i.e., from the time of writing up to 2006).

10.1 Interactive prototyping systems

It is clear to anyone who has followed this subject for a number of years that interactive prototyping systems have proved their worth many times over. This does not mean that their development is complete. For example, there are many features that we would like to see included and/or improved in PIP:

(a) SKIPSM interface
(b) Color image processing
(c) Analysis of image sequences
(d) Image transforms, most notably those based on
    Fourier transform
    Walsh/Haar transforms
    Wavelets
    Fractals
(e) Fuzzy reasoning
(f) Neural networks
(g) Ability to operate on different types of image representation, such as run- and chain-code.

In some cases, of course, facilities already exist for some of these functions but are not yet fully integrated into a system that operates harmoniously.

The command repertoire for a system such as PIP is never complete. A large body of ideas and procedures have been developed by researchers in Computer Vision. Many of these techniques will eventually find an appropriate rôle in Machine Vision, particularly in those situations where there is a high level of variability in the objects to be inspected. Improved computing power makes some of these ideas more attractive. Despite the arguments put forward in earlier chapters, the boundary between Machine Vision and Computer Vision is both fuzzy and dynamic.

The user interface will, no doubt, continue to develop over the next few years. Visual programming represents a (relatively) new approach and clearly has a rôle that is quite distinct from that represented by PIP's Prolog top-level programming. WIP has shown that it is possible to operate the same image-processing engine beneath any one of a set of different top-level controllers (Figure 6.5). Perhaps this will be the model for future generations of interactive vision systems. Providing a choice of "plug in" top-level controllers has many advantages, because each user can then choose the language/programming mode that best suits his particular skills and activity. This also provides the advantage that a "slow" top-level controller, such as Prolog, can be replaced by code written in a faster language, once the appropriate algorithm has been developed. This approach will eventually lead to a blurring of the distinction between interactive prototyping and target systems. If an IVS is fast enough, there is really no objection to it being used in the factory.

Every step we can make to enhance the speed of an IVS makes it less likely that a separate TVs is needed. In fact, the distinction between them is an admission of our failure to make an IVS that is fast enough for use in the factory. SKIPSM is an important development, because it provides a substantial improvement in operating speed over a wide range of functions.

As long as IVS and TVS exist separately, there is a need to generate code for a fast factory-floor system from an algorithm specified as a sequence of (PIP) commands. The IVS should be able to generate object code for a software-based system, or instructions defining the inter-module connections in a hardware system. At the moment, this requires long periods of work involving a highly skilled programmer/ hardware designer.

10.2 Target vision systems

Target vision systems may be based solely on standard computer hardware and software, or incorporate dedicated electronics to facilitate certain computationally intensive image-processing operations. For obvious commercial reasons, many manufacturers of vision systems prefer to concentrate on the former type. These are, of course, suitable for only those tasks where very high processing speed is not required. Through the use of good lighting/viewing techniques, clever algorithm design and good programming practice, it has often been possible to make software-based systems that are fast enough to cope with modern speed of production. However, for such applications as those involving real-time control and web inspection, much higher processing speeds are needed, in which case it is essential to use dedicated electronic hardware. It is a universal and undeniable truth that no system is ever really fast enough; improvements in processing speed are always welcome! Simply because there is such a demand, we can confidently predict that we will see a continuing improvement in the speed of target vision systems.

The continuing rise in the speed of standard computers will, as in the past, have a beneficial effect on Machine Vision. Within the short-to-medium term, improved systems software will have a welcome but probably smaller effect than that due to the introduction of new hardware. The recent rise in the power of desktop computers, up to the point where they can provide a cheap and effective facility for editing and processing home videos, will probably have a significant impact on industrial Machine Vision. Systems that use USB or IEEE 1394 ("firewire") interfaces are already being used to a small extent in industrial applications [SMI99]. The cost of such a system is very attractive indeed. However, it is the fact that any reasonably well-paid person can afford such a system for use at home that will surely have the biggest psychological impact on would-be customers? A customer will inevitably question why he should pay 10 – 20 times more for an industrial system than what he considers to be a comparable home video system. We can look forward to the day when many industrial systems will use exactly the same camera, electronics, and computer (protected in a rugged environmental housing), as the customer uses to record his wedding, or his child's first steps. This will have a mixed effect:

(1) Rising popular demand for "home" video facilities will effectively subsidize the development of future industrial systems.
(2) Processing speed will rise as costs fall. The rate of change will probably exceed anything experienced previously, because customers will not tolerate anything less than they perceive to be provided by their computers at home.

It is clearly imperative that, in the future, vision equipment suppliers should explain with greater force than hitherto that "systems issues," relating to the design of the image acquisition subsystem, user interface, and protective housing for the optics and electronics, do not represent mere luxuries to be reduced to an absolute minimum by haggling over price. We must argue with utter conviction that quality of design costs money and that quality produces reliability of operation. We must also remind customers that improving product and process quality is what motivates our subject. So, while great benefits will undoubtedly be obtained by adapting "home" video products for industrial use, there will be forceful arguments about cost along the way. We must be prepared to defend our corner; remembering the two "proverbs" given at the top of this chapter will provide us with the ammunition to do so effectively.

Greater processing speed in dedicated systems will come about through greater use of integrated circuits that are specifically designed for image processing and analysis. "Chips sets" were introduced recently for certain types of calculation, such as image filtering, histogram generation, and other forms of feature extraction. So far, relatively few systems have used them, although we anticipate that this situation is likely to change quite quickly. Dedicated image-processing systems can be further subdivided into those that are free-standing, where a computer is used for control and final decision making, and those in which the additional electronics takes the form of a plug-in card. In the latter case, the host computer is probably more deeply involved in certain parts of the calculation. Notice how close this architecture is to that needed to implement a VSP system (Section 6.2). A high-speed version of VSP is technically feasible if a suitable plug-in card were developed. This would possess the great advantage that the effort of rewriting a PIP command sequence would be eliminated, thereby reducing the overall system development cost by a substantial amount.

Two embryonic technologies are also worth mentioning here. Optical processing, which has for so long been seen as the "Holy Grail" for machine vision and image processing, still seems to be a remote and wonderful dream. If this is proved to be an incorrect prediction, the benefits will be enormous and Machine Vision will receive a great boost in its capabilities. However, the commonly held view is that this goal still seems to be a long way off and it would be unwise to rely on this happening before 2006. A much more promising development is, we believe, the possibility of building really large arrays of von Neumann processors, all programmed in a suitable high-level language. A few years ago, the Transputer was seen as the technology that would transform image processing. Multi-Transputer networks were shown to be very powerful but the company that made these devices failed to make a commercial success out of the product. Multi-chip arrays, programmed in Java, seem to offer another chance to achieve broadly similar objectives. (The reason why we look forward to using chips programmed in Java is explained later.) However, as far as the authors are aware, there are no vision systems yet available that use this technology.

10.3 Design tools

To date, a variety of tools have been devised for Machine Vision system design [BAT94b], including software for
(a) Designing the optical subsystem (ray tracing)
(b) Planning the lighting/viewing subsystem
(c) Selecting the appropriate lens
(d) Searching a database of equipment and service suppliers
(e) Giving advice about the overall system design (development is incomplete).

In addition, we should also mention the ALIS 600 and Micro-Alis lighting system development kits and the Machine Vision Application Check List. However, these various design tools are not yet integrated so that they work together as a unified "seamless" entity. Harmonizing and expanding the existing design tools is essential, if we are to exploit Machine Vision technology to its full potential. The present piecemeal development of systems by competing companies that are afraid of each other is not helping the intellectual growth of the subject and is having a distinctly detrimental effect on the market. In view of the scale of the development program needed, inter-organization rivalries must be put aside temporarily, in order to achieve the greatest overall benefit for Machine Vision, both as an academic and commercial activity. Whether there exists the collective will to do this is questionable. However, there is one approach that offers hope and relies on the Internet providing a unifying standard.

The Internet allows the same software and database services to be made available anywhere in the world. A company that wishes to sell its products or services globally cannot ignore its potential to reach customers literally anywhere on Earth. Companies may feel that it is in their interest to provide tools to encourage would-be customers to understand and apply their technology better. For example, a company selling illumination equipment would probably find a system such as the Lighting Advisor helpful in promoting its products. Similarly, a company selling high-end image-processing hardware may provide complimentary interactive software to enable future customers to learn about the technology and even use it to obtain an initial design for an algorithm. The reader can, no doubt, envisage several other ways in which the Internet might be used to encourage potential customers to learn about and gain greater confidence in Machine Vision. Of particular note is the possibility that small groups of companies providing complimentary services can form (and reform) alliances to their mutual advantage. Thus, a lighting company might form a working partnership with another organization that designs mechanical-handling equipment and a supplier of cameras, in order to provide a complete service for designing and building image-acquisition subsystems. Since this type of liaison is easily set up and advertised via the Internet, there are grounds for holding the optimistic view that refined design tools will emerge naturally and that they will be harmonized through the medium of the Internet.

10.4 Networked systems

It is in the use of networked vision systems that we envisage perhaps the greatest changes occurring within the next 5 years. Hitherto, vision systems have been regarded as necessarily being complicated, expensive devices that fulfill just one task, albeit one that has great technical or commercial value. The falling cost of image processing and the ability to communicate effectively via the Internet open up exciting new possibilities for building multi-camera systems that consist of a number of interconnected semi-autonomous vision modules. Many of these devices might be used for undemanding tasks, such as making sure that

(a) the doors of an oven are closed,
(b) the factory floor walkways are free of debris,
(c) a material-feed hopper does not have an unhygienic build-up of perishable material,
(d) a flame is lit,
(e) there is an adequate supply of raw feedstock and materials for the manufacturing systems,
(f) there is no "log jam" on a production-line conveyor, and
(g) an extrusion nozzle is not blocked.

The essence of such a system is that it consists of several or many low-cost modules scattered around the factory and that some of them may be quite slow. There will probably be other faster units, also connected to the network and checking product size, shape, integrity, etc., as we have described elsewhere in this book. Each module connected to this network would normally report to a central computer on an occasional basis, so that performance statistics may be collected. Sometimes, two or more of these vision modules might cooperate together more closely. For example, one module placed "up stream" on a factory production line might report what it sees to another sensor placed further "down stream", so that the latter knows what to expect later.

There are several important points to note about such a networked multi-camera system:

(1) The incremental cost of adding a new camera should be small.
(2) While the speed of some modules might not be very high, others might need to operate much faster. All of these units should employ the same network protocol.
(3) Individual modules should be programmable via the network.
(4) Some modules may be connected to the network via a wireless link, which should be transparent to the user. The use of wireless connections is important, so that new ideas can be tried easily and cheaply, without the expense of installing cabling.
(5) Each module should be capable of performing simple control of external devices such as a camera, lens, lights, accept/reject mechanism, etc.
(6) A manager, perhaps working many hundreds of kilometers from the production plant, may need to monitor what is happening there. This does not mean, of course, that he should be able to interfere with the minutiae of the manufacturing process, but instead be able to assure himself that production is running smoothly and see where bottlenecks occur.
(7) The network must be reliable and secure. It is important that the information flowing around the network should not reach anybody who is unauthorized to see it. Nobody, either inside or outside the company, should be able to disrupt the information flow by malicious action. The system should also degrade gracefully if individual vision modules fail.

At the moment, cooperating multi-camera vision systems are not used to any large extent, although there are many potential benefits to be obtained. By incorporating the appropriate cabling and other infrastructure elements into a new factory, the full benefits of a multi-camera vision network would be obtained later.

10.5 Systems integration

The adoption of widely accepted standards for hardware and software has already had a major impact on the cost effectiveness of vision systems. The IEEE 1394 ("firewire") standard allows real-time data capture of medium-resolution color video signals and is therefore likely to provide the mainstay for vision systems in the immediate future.

There is less agreement about the operating system. There are many objections from professional software engineers to using a "closed" software system that cannot be changed to accommodate special requirements. This attitude favors Linux and Unix over the Windows and Macintosh operating systems but does move away from standardization. An important issue that must be faced squarely in the future is that of long-term stability of software. At the moment, if a computer breaks down and has to be replaced, we might well find that the software will not run properly on the new machine, since the operating systems are sometimes not backwardly compatible. While new operating systems are introduced for commercial reasons, the changes, however subtle, can have disastrous consequences for a software package that was designed to run on an earlier version of the OS. Engineering systems are susceptible to being made obsolete overnight by the dangerous policy of constantly updating the operating system. Paradoxically, an "open" operating system, such as Linux, might overcome this problem, since all aspects of the overall system are then under the software engineer's control. However, this does impose an even greater intellectual burden on the design team. During its short but impressive period of development, Java has not been stable to the point where programs written in an early version will run successfully later. The difficulties in ensuring the long-term stability of computer operating systems is, of course, not restricted to Machine Vision but the problem is just as troublesome here as elsewhere.

10.6 Algorithms and heuristics

It is particularly difficult to anticipate developments in computational procedures with any great accuracy, since algorithms and heuristics often develop in a revolutionary rather than an evolutionary manner. On the other hand, hardware and software normally evolve "smoothly" and are unlikely to change beyond recognition overnight, as a result of discovering just one seminal idea. One moment of inspiration is all that it took to develop the germ of the idea embodied in SKIPSM, although it took several years to understand its scope and develop it as a useful tool for a wide range of applications. Despite the comment made in the first sentence in this section, the authors will venture to make some remarks about SKIPSM, about which they feel they can write with particular authority.

As is made clear in earlier chapters, SKIPSM lends itself to high-speed implementation of a broad range of image-processing operators. It can provide a very significant speed increase in both software and hardware systems. SKIPSM achieves this by using RAM/ROM to implement large lookup tables. Since many such tables can normally be held on a modern RAM/ROM chip, there is the potential to build a fast multi-function image-processing engine which relies heavily on algorithms that can be implemented in LUTs. All monadic operators, dyadic operators, PCF color recognition procedures, and many local and morphology operators can be implemented in this way.

Speed increases in excess of 100:1 have already been achieved for individual algorithms, compared with conventional implementation methods. Despite this, SKIPSM is not used in present-day industrial vision systems. How can such a large speed-improvement factor be ignored? Surely, once one company decides to take the initial step to convert SKIPSM from an academic research topic into a factory inspection system, it will quickly gain much wider acceptance. We envisage that the first practical uses of SKIPSM will be seen in the next five years and that thereafter it will rapidly rise in popularity. SKIPSM also lacks the wide interest among academics that is needed to broaden its scope and appeal. Once the proven merits of this approach are understood by the "right people" in academic circles, it seems inevitable that greater attention will be paid to this idea by theoreticians.

10.7 Concluding comments

Machine Vision is currently being applied to a huge variety of industrial inspection and control tasks. Inspection systems have been built for such diverse applications areas as clothing, timber, agricultural produce, minerals, food, pharmaceuticals, toiletries, paper, packaging materials, electronics, automobile and aircraft components, assemblies and even complete engines, munitions, steel, sheet materials, etc. Machine Vision has an important rôle too in the monitoring and control of manufacturing processes. In each of these areas, it is clear that far more unsolved applications still exist than have been studied to date. The authors would suggest that the following will be among the most important ones:

- Design of an inspection system for the æsthetic appearance of highly variable items, such as food products (e.g., cakes), clothing, plants, etc.
- Learning by showing from a "Golden Sample". This will, of necessity, require the use of meta-knowledge. To suggest that a person or machine accepts that a sample should "look like this example that I am showing you," presupposes that there is higher-level knowledge about what constitutes acceptable similarity.
- Declarative programming involving natural language will be refined to a much higher level.
- The user interface will continue to develop and improve, through the use of multimedia techniques. Prolog+, for example, should be provided with at least rudimentary graphics facilities, to aid feedback to the human operator.
- Multi-camera/multi-processor systems will become more commonplace and networking will develop to allow really effective cooperative action between vision systems.
- Closed-loop process control, through the use of Expert Systems with visual inputs, will become more common in manufacturing industry.

One of the main factors limiting the adoption of Machine Vision technology remains the high cost of the skilled labor needed to design and build a system for each new application. The ideas discussed in earlier chapters were developed specifically with this point in mind. Experience has shown that when a system like PIP is used, it is often possible to discover/invent a suitable image-processing procedure for a new application within an hour or so. (Where no effective solution is possible, an IVS can often demonstrate the inherent difficulty of an application in a short time, thereby saving a large amount of wasted engineering effort later.) The major bottleneck currently limiting the application of Machine Vision lies in the large amount of engineering effort needed to design the system. Improved design tools are essential if we are ever to exploit the vast potential market for Machine Vision. Interactive vision systems form one of the main tools used to aid the design process. While IVSs are highly developed, there is still room for considerable improvement. Over the last 25 years, interactive vision systems have demonstrated time and again the technical feasibility of applying this technology, only to find that the commercial will does not exist to build a complete system. This unfortunate state of affairs is changing, as more people accept that Machine Vision has an important rôle to play in improving the efficiency and safety of manufacturing. The future for our subject seems to be bright, provided we continue to maintain the distinct identity and ethos for Machine Vision; this is a practical engineering subject where many factors must be combined to bring success.

Appendix A Programmable color filter

Representation of color

It has long been known that by mixing together red, yellow, and blue paint in different proportions many other colors can be created. Later, it was discovered that by adding red, green, and blue light, a similar effect can be obtained. This phenomenon is known as *trichromaticity.* It is also known that the mammalian eye contains three differently-colored pigments. These observations and various physiological and psychological experiments provide strong evidence to support the view that the human eye has three different color sensors. These are commonly attributed to the detection of red, green and blue light. However, they each have a broad spectral response and these names can be misleading. In an attempt to emulate human vision and provide acceptable color reproduction, most color video cameras are provided with three outputs, known as the *RGB* signals. Each of these may be regarded as representing a monochrome image (Plate 19).

High-quality color printing commonly relies on four imprints using inks based on different pigments: cyan, magenta, yellow, and black (Plate 20). This has been found to give superior quality of reproduction when colored inks are superimposed onto white paper. This uses what is known as the *CMYK* representation. However, it is not normally used when working with video, since the latter relies on a light-additive process, whereas printing is a light-subtractive process. For this reason, we need not concern ourselves further with the CMYK representation.

Given the practical limitation that video cameras almost invariably use three colored optical filters to separate the RGB signals, this representation provides the starting point for our work on color recognition. It must be understood that trichromaticity places bounds on what an organism or a vision system can see. Within these limitations, the RGB representation provides a good starting point for work on color recognition. It should be understood that most other representations of color that are described in the literature on color theory can be derived by transforming the RGB signals. Hence, any recognition rule that uses one of these representations can be mapped into RGB color space. No representation that can be obtained by transforming RGB signals can possibly provide any *additional* information not already contained in the RGB signals. Thus, it is pointless to argue that any of these alternatives is better than RGB for color recognition. However, one particular representation is of special interest to us, since it is closely associated with the programmable color filter (PCF) that is useful in Machine Vision systems. This is the *HSI (Hue, Saturation, Intensity)* representation. However, before we discuss that, let us briefly introduce another idea, the *color triangle*, which provides the conceptual link between the HSI and RGB representations.

Color triangle

James Clerk Maxwell adopted the idea of the *color triangle* (also known as the *Maxwell triangle*). It is based upon the idea that three highly saturated colored lights (red, green, and blue) are located at the vertices of an equilateral triangle. These are known as *primary colors*. Other points in this triangle correspond to mixtures of these three primary colors. Now consider Figure A.1. The cube in this diagram is called the *color cube* and represents the limits of the camera outputs along each of the RGB channels. Each signal is assumed to lie within the same finite range $0 \leq R, G, B \leq W$, where W is a positive constant. The color triangle is defined by the intersection of the plane defined by the equation

$$R + G + B = W$$

with the cube. This plane intersects the points (W,0,0), (0,W,0) and (0,0,W).

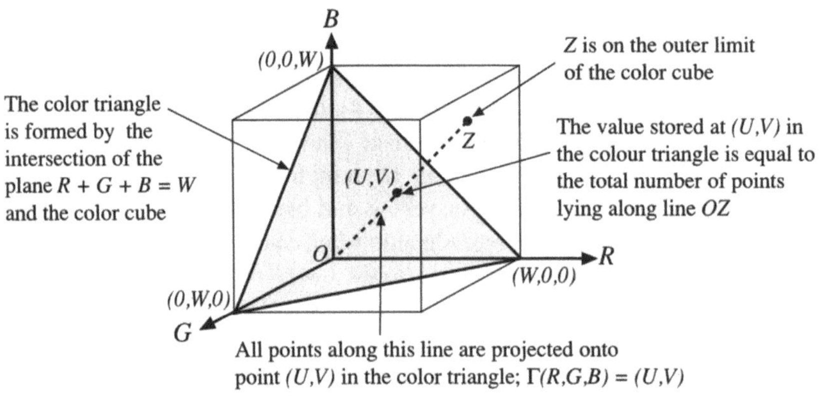

Fig. A.1 RGB space, showing the color cube and color triangle.

Both hue and saturation can be identified by the position of a point within the color triangle. (However, intensity information cannot be so identified, because it requires movement in an orthogonal direction.) Hue defines the intrinsic nature of the color (Figure A.2). In other words, hue is related to the *name* of a color that a human being might assign to a surface. When we name the colors of the rainbow/spectrum, we are simply responding to the changes in hue that we observe for different wavelengths of light. (It should be noted, however, that some hues cannot be represented on the spectral scale. Purple and "gold" are two examples.) There is no universal classification (i.e., naming) of colors. For example, the boundary distinguishing *green* and *blue* (English) is different from that between *gwrdd* and *glas*, their nearest equivalents in Welsh. Language, culture, illness, and drugs can all influence a person's naming of colors. There is no agreement about the names of colors, even among healthy drug-free individuals of the same sex, ethnic, and social group who all speak the same first language. Given this fact, we have to design a color-recognition device or program that can *learn* to copy a given person's color-recognition capability. The PCF is able to learn from a human being how to classify colors.

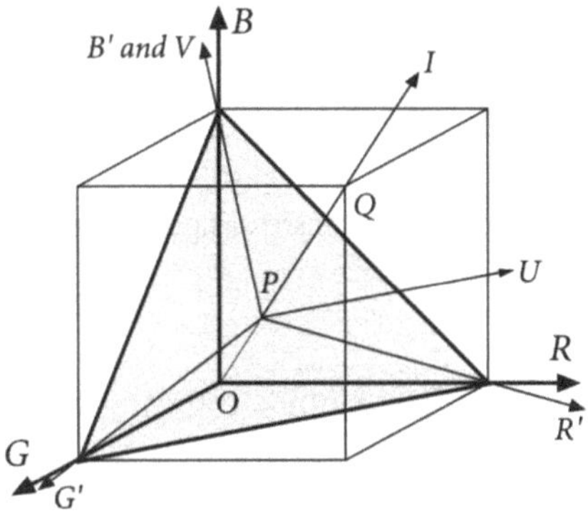

Fig. A.2 Coordinate axis transformation. (The axes *R'*, *G'* and *B'* are useful for deriving the formulae for *U* and *V*.)

Another important feature about a color is its *saturation* (Figure A.3). A saturated color is deep, vivid, and intense, due to the fact that it does not contain colors from other parts of the spectrum. Weak or pastel colors have little saturation. For example, pink is non-saturated red and can be produced by mixing approximately equal low levels of blue and green light with a much higher level of red. Fully saturated mixtures of two primary colors are found on the outer edges of the triangle. On the other hand, the center of the triangle, where all three primary components are balanced, represents *neutral* tones (i.e., white and greys). Other unsaturated colors are represented as points elsewhere within the triangle.

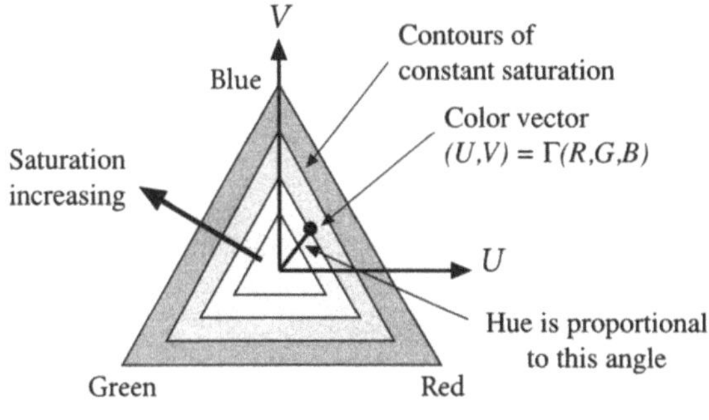

Fig. A.3 Hue and saturation.

Mapping RGB to HSI

Hue is represented as the angular position of a point relative to the center of the color triangle, while the degree of saturation is measured by its distance from the center (Figures A.2 and A.3). The following equations relate the hue (H) and saturation (S) parameters to the RGB representation and are derived in [GON92].

$$H = \cos^{-1}((2R - G - B)/(2\sqrt{((R - G)^2 + (R-B)(G-B))})) \qquad (A1)$$

$$S = 1 - 3\,MIN(R,G,B)/(R+G+B) \qquad (A2)$$

The observed brightness of a point in a scene is related to the quantity

$$I = (R+G+B)/3 \qquad (A3)$$

This a more useful estimate of the intensity, as perceived by a human being, than the length of the RGB vector, given by

$$\sqrt{(R^2 + G^2 + B^2)} \qquad (A4)$$

This is a reflection of a principle known to psychologists as *Grassman's Law* [HUT71].

It should be noted that it is possible to (re-)calculate the RGB values from HSI. Thus, equations (A1) – (A3) are non-destructive mappings.

Programmable color filter (PCF)

The Programmable Color Filter (PCF, Figure 4.13) uses a standard RGB video input from a color camera and digitizes each channel with a resolution of n bits. Typically, $n = 6$. Thus, a total of $3n$ (18) bits of data are available about each pixel and together they form the address for a random access memory (RAM). This RAM is assumed to have 8 parallel output lines and to have been loaded with suitable values, thereby forming a lookup table (LUT). By means that we will discuss shortly, the contents of this LUT can be modified, enabling it to recognize any desired combinations of the incoming RGB signals. Thus, the filter can be programmed to recognize one or more colors. The LUT has a total of capacity of 2^{3n} bytes (256 Kbytes) of data. Since the output of the LUT consists of 8 parallel lines, its output can be taken as defining the intensities in a monochrome video image. Notice that there is no attempt to store a color image. The digitized RGB video signal is processed *in real time* by the LUT, the output of which can be

a) redisplayed as a monochrome image, or
b) passed through a set of three further Lookup Tables, providing a pseudocolor display (Plate 19), or
c) digitized, stored, and then processed by PIP.

The starting point for training the PCF is the generation of the color scattergram. This is then processed to remove outliers (Figure 4.14(a)). The resulting image is then thresholded, probably producing small compact blobs, which are then labeled (i.e., shaded using *[ndo]*) and expanded to achieve some degree of color generalization (Plates 22 and 24). The resulting image resembles one of those shown in Plate 24. The blob size is chosen by the user to achieve the appropriate degree of generalization. Finally, the so-called *back-projection* method is applied to map the contents of the color triangle after processing, in order to calculate the LUT values. Here is a detailed description of each of these processes.

Creating the color scattergram

This process is explained in Figure A.1. Let us consider a point (i,j) in the input image. Furthermore, let us suppose that this generates a point at $(R_{i,j}, G_{i,j}, B_{i,j})$ in the color cube. We will use $\Gamma(R,G,B)$ to denote the mapping function for projecting a point (R,G,B) from the color cube onto the color triangle. That is, $\Gamma(R_{i,j}, G_{i,j}, B_{i,j})$ generates the coordinate pair $(U_{i,j}, V_{i,j})$[1], where

$$U_{i,j} = (R_{i,j} - G_{i,j})/[\sqrt{2}\,(R+G+B)] \quad \text{(A5)}$$

$$V_{i,j} = (2\,B_{i,j} - R_{i,j} - G_{i,j})/[\sqrt{6}\,(R+G+B)] \quad \text{(A6)}$$

Consider the point (i,j) in the input image. Its color vector, $(R_{i,j}, G_{i,j}, B_{i,j})$, maps onto the point $(U_{i,j}, V_{i,j})$ in the color triangle. The value stored at $(U_{i,j}, V_{i,j})$ is increased by 1. To generate the complete color scattergram, this operation is performed for all points *(i,j)* in the input image. (We begin by storing zeroes throughout the color triangle.) Notice that, as we scan through the input image, multiple "hits" may occur at certain values of *(U,V)*, causing a progressive increase in the value stored at those points. The result of applying this mapping process to a picture containing only well-defined and well-separated colors is typically a pattern in the color triangle consisting of several bright diffuse spots. In addition, there will inevitably be a number of outliers forming "clouds" around these spots. Outliers must be removed (step 2 below), since they represent noise effects and will lead to erratic behavior of the PCF.

Processing the color scattergram

The color scattergram is then filtered to remove outliers (Figure 4.14(a)). Three possible way to do this are

1) Apply a low-pass (blurring) filter, such as [*N•lpf*], where *N* is a (small) positive integer.
2) Apply a grey-scale morphology filter of the form *[N•lnb, 2•(N•snb), N•lnb]*. The resulting image is then thresholded, probably producing a number of small compact blobs. Sometimes it is necessary to merge clusters of small blobs. Again, a grey-scale morphology filter (dilation/closing) might be appropriate. It should be noted, however, that there is no prescription for this process. By experimenting with PIP, an experienced vision engineer can usually find a reasonable way to

1. To see how these equations can be derived, view the color triangle normally (i.e., along the line *OQ*, the diagonal of the color cube, Figure A.2). When the vector *(R,0,0)* is projected onto the color triangle, the resultant is a vector V_r of length $R/\sqrt{2}$ parallel with the *R'* axis. In a similar way, when the vector *(0,G,0)* is projected onto the color triangle, the result is a vector V_g of length $G/\sqrt{2}$ parallel to the *G'* axis. Finally, the vector *(0,0,B)* projected into the color triangle forms a vector V_b of length $B/\sqrt{2}$ parallel to the *B'* axis. *U* and *V* are then found simply by resolving V_r, V_g, and V_b along these axes.

process this binary version of the color scattergram. (This remains one of those areas in Machine Vision system design where human intuition cannot yet be fully automated or replaced.)

In this exercise, the goal is to obtain a number of blobs, one for each large area of nearly constant color in the input image. The blobs created thus are then labeled (e.g., shaded using *[ndo]*) and possibly expanded to achieve some degree of color generalization (Plates 22 and 24). That is, the blobs in the processed color triangle are somehow enlarged. The resulting image resembles that shown in Plate 24(TL), (CL), or (BL). Again, the vision engineer must be the final arbiter about the level of generalization that is most appropriate for a given application.

3) *Back-projecting each point from the color triangle into the color cube:* This is an automatic process. Every point (R,G,B) within the color cube that maps onto a given point (U,V) is assigned a label (i.e., an integer value) equal to the intensity stored at (U,V) (see Figure A.4). In order to represent this process in more formal terms, let us use $\Phi(U,V)$ to define that set of points such that $\Gamma(R,G,B)$ is equal to (U,V). That is:

$$\Phi(U,V) = \{(R,G,B) \mid \Gamma(R,G,B) = (U,V)\}$$

If the intensity of the point (U,V) in the color triangle is $T(U,V)$, then all (R,G,B) in $\Phi(U,V)$ are given the value same value, $T(U,V)$. This operation is repeated for all (U,V) in the color triangle. Since $\Gamma(R,G,B)$ is responsible for projecting points throughout the color cube onto the color triangle, the reverse process is called *back-projection.* The result of back-projection is a defined value for each point in the color cube. The lookup table in the PCF stores these values. Thereafter, the color recognition process can be performed by a single reference to the LUT for each pixel. Given the speed of modern RAM/ROM devices, it is possible to perform this operation in real time on a digitized video signal. (No further processing is needed.)

The result of back-projecting a solid blob-like region of the color triangle whose pixels all have the same value (Q) is to create a cone in RGB space in which all points also have value Q. Thereafter, any pixel whose RGB coordinates define a point within that cone will be associated with the color label Q. While the values

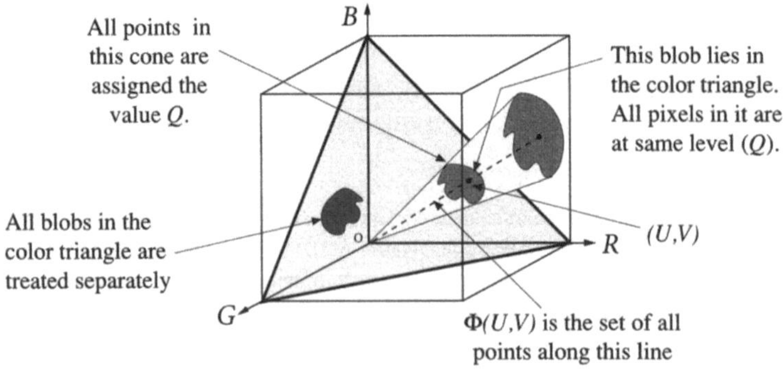

Fig. A.4 Back projection

stored in the PCF LUT use an 8-bit code, these should properly be regarded as abstract symbols, not necessarily as numbers. However, numeric operations (i.e., PIP image-processing operators) can often be applied to the PCF output to good effect. This is reasonable if colors that are classified by a human being as similar are mapped to approximately equal color codes stored in the LUT.

Software

In order to calculate the color scattergram, a single PIP predicate called *color_scattergram/1* has been written. This possesses an important facility not yet described: one extra step is included to reduce quantization effects. The color scattergram is computed for only those pixels in the input image that are brighter than a threshold value defined by the user. Since a human being finds it difficult to attribute color names to very dark areas of a color image, this prevents pixels of very low intensity from influencing the contents of the LUT. Indirectly, this also provides a facility for applying a mask to the input image, so that only a limited area influences the color scattergram and hence the contents of the LUT. The user can therefore select the interesting regions of a complex input image in order to "teach" the PCF, while ignoring others. A mask image might be drawn/selected by the user or generated automatically.

The color scattergram can be processed using any suitable operator sequence, since the color triangle is drawn within a standard PIP image. The user can also employ any of PIP's image-generation operators (e.g., *wgx/1 hic/2, cir/3,* etc.) to generate artificial patterns in the color triangle. This facility is useful for a variety of tasks, such as creating filters which measure hue, saturation, yellowness, redness, etc. (Plates 21 and 23). Operators such as *fil/5* and *cir/3* can be used to mask off certain clusters in the color scattergram. This is useful when some of the colors in an image are known to be more important than others.

The back-projection procedure is embodied within the predicate *create_filter/0.* Understanding the details of how *create_filter/0* works is not essential for designing or using the PCF, since the procedure is fully automatic.

Appendix B A brief introduction to Prolog

Prolog is different

The first thing to learn about Prolog is that it is completely different from almost all other computer languages. Indeed, it violates most of the established principles of programming. Experienced programmers often find this strange, even bewildering. Despite its unconventional approach, Prolog is a powerful tool for certain types of application, most notably those requiring the use of symbolic learning and other Artificial Intelligence techniques. It provides a very natural way to express complex abstract relationships, that is not available through the use of languages such as C, Basic, Java, etc. However, Prolog does not perform well on certain types of computational task, for example those requiring a large number of numeric calculations. The lesson is simple: do not expect Prolog to conform to previous experience.

It is easy to learn how to program in Prolog, provided that all previous experience with other computer languages is set aside. Indeed, complete novices often find it easier to write Prolog code than to begin programming in other languages. While it is possible to learn *how* to program in Prolog after just a few hours of study, it is much more difficult to understand *why* the language is significant.

Prolog cannot sensibly be classified as either a high-level or low-level computing language, in the way that assembly code, Basic, C, and Java can. Prolog is so different from these other languages that it cannot be related to the conventional pyramid of complexity.[1] [2] [3]

Declarative programming

A Prolog *application* consists of *facts* and *rules*; it does not contain any instructions at all. A Prolog program may be regarded as a description of (part of) the world.

---

1. Prolog system software may be written in assembly code, C, Java, Lisp, Perl, etc. However, this does not mean that is is a higher level language than these, since it ignores most of their most basic facilities. In addition, it possesses many novel features that conventional languages do not have. While Prolog's core facilities may be provided via a language such as C, much of its system software is usually written in Prolog.
2. The correct term is *application*, since a program is, strictly speaking, a sequence of instructions. However, we shall often use the term "program," since this is more familiar to most readers. For the same reason, we also use the term *commands*, particularly when refering to sequences of image-processing operators. To forestall criticism from Prolog purists, we point out that most PIP applications are procedural, since image-processing utilities are always satisfied on the first encounter but fail on backtracking.
3. One of the earliest programs used by a commercial dating agency was written in Prolog.

For this reason, Prolog is referred to as a *declarative* language, whereas most other computer programs, military orders, knitting patterns, automobile repair manuals, musical scores, cookery books, etc. are expressed in terms of *imperative* languages. (That is, they consist of sequences of instructions or commands.) This feature distinguishes Prolog from almost all other computer languages. To use Prolog, a person poses a *query. (Answering a query* using Prolog is often expressed as *satisfying a goal.)*

An excellent application which demonstrates the nature and power of declarative programming is that of finding a marriage partner. A person wishing to find a spouse might, for example, list the physical and personal characteristics, interests, etc. that he/she hopes to find in his/her future wife/husband. Such information might include details about such factors as age, height, weight, hair color, racial group, religious belief, personality, occupation, etc. Given such a set of "requirements," Prolog then sets about trying to match the individuals listed in its database with the given specification, so that all conditions are met. Notice that the programmer does not have to tell Prolog how to do the search. A program implementing this idea can be fully listed on two A4-sized pages.

Facts

Simple relationships can be stated in Prolog, as the following examples show

```
parent(susan,alison).        % susan is a parent of alison
parent(peter,graham).        % peter is a parent of graham
parent(lucy,peter).          % lucy is a parent of peter
parent(angus,peter).         % angus is a parent of peter
```

The numbers of arguments is called the *arity.* We indicate the arity of a relationship such as this in the following manner: *parent/2.* There is no reason why definitions for the same relationship cannot simultaneously exist with different arities. For example, *parent* might sensibly be defined with 1, 2 or 3 arguments:

```
parent(elaine).              % Indicates that elaine is a parent (of unknown children)
parent(david, andrew).       % ... that david is a parent of andrew
parent(helen, stephen, susanna).% ...that helen and stephen are both parents of susanna.
```

The syntax requires a short explanation. The names of all items (*parent, peter, alison,* etc.) begin with lower-case initial letters because they are Prolog *atoms.* Atoms and numbers are types of *constants.* Our first Prolog program consists of four *clauses,* all of which are *facts* (i.e., none of them contains the ':–' operator). Each clause is terminated by a period ('.'). Single line *comments* are preceded by '%', while *multi-line comments* are enclosed between the symbols '/*' and '*/'.

Simple queries

The results of four typical queries are given below. (The database consists of the collection of four clauses for *parent/2* listed above.) The search for a solution always begins at the top of the database.

Query no. 1:	parent(lucy, peter).
Remarks	The goal is included explicitly in the database
Prolog's response:	Yes % The goal matches clause 3.
Query no. 2:	parent(peter,X).
Remarks	*X* is a variable, since it begins with an upper-case letter. The user indicates that he is willing to accept a single solution.
Prolog's response:	No. 1 X = graham % Clause 2 Yes
Query no. 3:	parent(X, peter).
Remarks	X is a variable. Multiple solutions are requested by the user.
Prolog's response	No. 1 X = lucy % First solution. Clause 3 No 2 X = angus % Another solution. Clause 4 No more solutions
Query no. 4:	parent(X,Y).
Remarks	X and Y are both variables. Multiple solutions are requested by the user.
Prolog's response	No. 1 X = susan Y = alison % First solution. Clause 1 No. 2 X = peter Y = graham % Second solution. Clause 2 No. 3 X = lucy Y = peter % Third solution. Clause 3 No. 4 X = angus Y = peter % Fourth solution. Clause 4 No more solutions

Rules

In Prolog, it is a straightforward matter to make general statements about abstract concepts. For example, it is possible to define a relationship, *child/2* in terms of the existing relationship, *parent/2:*

In English:	A is a child of B if B is a parent of A.
In Prolog	child (A,B) :– parent(B,A).

Read ":–" as "*is true if*", or more simple as "*if*".

Clearly, we can define the relationship *older(A,B)* using three clauses each of which accommodates a different situation. Any one of the following classes can be used to prove that *A* is *older* than *B*:

```
older(A,B) :– parent(A,B).        % Parent (A) is older than his child (B)
older(A,B) :– child(B,A).         % Child (A) is younger than his parent (A)
older(A,B)) :–
    born(A,X),                    % Object A is/was born in year X
    age(B,Y),                     % Object B is was born in year Y
    Y > X.                        % Test whether Y > X
```

A collection of facts and rules such as this is said a to define a *predicate*. This is simply an entity with a binary truth value (i.e., either *true* or *false*) that has yet to be determined. Sometimes a predicate is defined using both facts and rules. Once again, the order in which the clauses are listed is important, since Prolog always tries to match the goal to items at the top of its database first.

Queries involving rules

Let us now consider further queries, this time involving rules. Consider the goal

older(lucy, peter).

As we have already explained, Prolog tries to satisfy this goal by scanning through its database, beginning at the top. In this case, a tentative and as yet unproven match is obtained with the first clause of *older/2*. Prolog temporarily binds the variables *A* to the value *lucy* and *B* to the value *peter*. This process is called *instantiation*.[4] Prolog then tries to prove the body of this clause (i.e., its right-hand side) by showing that *parent(lucy, peter)* is true. It is able to do this directly, by matching with the third clause for *parent/2*. Since *parent(lucy, peter)* has now been proved to be true, Prolog deduces that *older(lucy, peter)* is also true, thus answering the query.

Backtracking and instantiation

Now, consider, the relationship *grandparent/2* defined thus:

```
grandparent(A,B) :-
    parent(A,Z),          % Sub-goal 1; Z is instantiated here
    parent(Z,B).          % Sub-goal 2; Value of Z is used here to find a value for B
```

and the query

grandparent(A,B).

To satisfy this goal, we must satisfy two sub-goals. The proof then proceeds as follows:

a) In the first instance, the sub-goal *parent(A,Z)* is taken to match the first clause for *parent/2*, thereby instantiating *A* to *susan* and *Z* to *alison*.
b) These values are then carried forward to try to satisfy the second sub-goal *parent(alison,B)*, which *fails*.
c) Prolog then backtracks to the first sub-goal and uninstantiates both *A* and *Z*.
d) The first sub-goal is then resatisfied, this time by matching it to the second clause for *parent/2*, and thereby instantiating *A* to *peter* and *Z* to *graham*.
e) The second sub-goal, (*parent(graham,B)*), is then investigated and fails.
f) Prolog then backtracks to the first sub-goal and uninstantiates *A* and *Z* again.
g) The first sub-goal is then matched to the third clause of *parent/2*, thereby instantiating *A* to *lucy* and Z to *peter*.
h) This time, the second sub-goal is *parent(peter, B)* is investigated and matches the second clause of *parent/2*, thereby instantiating *B* to *graham*.
i) Since both sub-goals have been satisfied, the original goal (*grandparent(A,B)*) is proved true, by assuming that *A* = *lucy* and *B* = *graham*. On the way, we found a suitable value for Z (= *peter*) but this is discarded.

---

4. *Instantiation* is derived from the same linguistic root as *instance*. By making certain assumptions about the values of the variables in a Prolog application, we may be able to prove some postulate (i.e., satisfy a query). That is, we find an instance which ensures that the given query is true.

Notice how the program goes both forward and backwards (*backtracking*), as it tries to prove the original goal. Needless to say, the "flow of control" can be very complicated in a large practical application. Fortunately, the programmer does not normally have to think about this; he simply defines a specification of what constitutes an acceptable solution and does not need to give much thought to backtracking. (The programmer needs to consider backtracking only to avoid long searches generating lots of unwanted solutions.)

A special mechanism exists to control backtracking. This called the *cut* and is denoted by the exclamation mark (!). The effect of the cut is to prevent backtracking but it has no effect whatsoever when the flow is proceeding in the forward direction. The reader is urged to read about the cut in the specialized textbooks on Prolog, because its effect is profound and its behaviors can be puzzling to the uninitiated. No program that works with a cut will fail to do so if it is removed; more solutions will be produced.

Recursion

Prolog makes very extensive use of recursion. While many other languages also *allow* recursion, it is an *essential part* of Prolog. The relationship *ancestor* can be defined recursively in the following way:

```
% Simple case: the ancestor is a parent.
% This clause is included to terminate the recursion.
ancestor(A,B) :-          % A is an ancestor of B if ...
    parent(A,B).          % A is a parent of B
                          % Defining ancestor when the two individuals involved are...
                          % ... at least two generations apart
ancestor(A,B) :-          % A is an ancestor of B if ...
    parent(A,Z),          % A is a parent of Z ... AND ...
    ancestor(Z,B).        % Z is an ancestor of B.
```

Recursion, backtracking, cut (!), and the conditional evaluation operator (–>, described below are the only flow-control mechanisms that exist in Prolog. This is probably the most perplexing aspect of Prolog for experienced programmers, who yearn to use the familiar tools for controlling flow.

Lists

Prolog is able to handle *lists* easily. Prolog can handle lists just as easily as LISP, which is one of its main competitors. The two languages use quite different mechanisms. Prolog's ability to manipulate lists can be explained with reference to a simple example. Consider the list:

[a,b,[c,d,[e,f]],g]

This has the following characteristics:

No. of elements: 4
Head (1st element): *a*

Second element: *b*
Third element: Sub-list *[c,d,[e,f]]* whose 3rd element is a sub-sub-list: *[e,f]*.
Fourth element: *g*
Tail: *[b,[c,d,[e,f]],g]*

The head and tail of a list can be extracted from a given list by matching it to *[Head|Tail]*, as in the following example:

[a,b,[c,d,[e,f]],g] = [Head|Tail]

In this case, *Head* is instantiated to *a* , while *Tail* is instantiated to *[b,[c,d,[e,f]],g]*.

Other features

The connective operator ',' (comma) implies that two sub-goals are ANDed together. Another operator, ';' (semicolon), exists which ORs two sub-goals together. Hence, the following are exactly equivalent:

```
parent(A,B) :– mother(A,B) ; father(A,B).
```

and

```
parent(A,B) :– mother(A,B).
parent(A,B) :– father(A,B).
```

The ';' (semicolon) operator is not essential and it is regarded as good style to avoid using it wherever possible, since it is usually easier to read and understand several short clauses than one long one.

Prolog is extensible virtually without limit; new syntax rules can be defined easily, as can a large repertoire of utilities for a given area of application. One of the prime ways to do this is through the use of *operators*. For example, we can define operators *if* and *&* (or *AND*), so that we may write Prolog programs in the following way:

```
grandparent(A,B) if
parent(A,B) AND               % This sub-goal is ANDed with ...
parent(B,A).                  % ... this one
```

The operator '§' is used in PIP to perform the same image-processing operation on all blobs in an image. The operator 'Δ' is used to mark PIP commands that are to be reversed on backtracking. (Actually, the operator 'Δ' stores images, so that they can be read on backtracking.) Another example is provided by *N•G*, which performs operation *G* a total of *N* times. (In more formal terms, *N•G* tries to satisfy goal *G* a total of *N* times.) Here is the definition of this operator:

```
0•G :– !.                     % Finished, so do nothing and satisfy the goal
N•G :–
   G,                         % Satisfy G once
   M is N – 1,                % Decrement the counter
   !,                         % Do not permit backtracking
   M•G.                       % Satisfy G a total of (N – 1) more times
```

An entirely new language can be defined in such a way that it retains all of the facilities inherent in Prolog but does not restrict the user to employing standard Prolog syntax. (He may not even be aware that Prolog even exists!)

Prolog permits *meta-level programming*, as we can see in the definition of the Prolog version of the familiar IF...THEN...ELSE construction:

```
if_then_else(A,B,C) :– A, !, B.    % If A is satisfied then evaluate B as well
if_then_else(A,B,C) :– C.          % Otherwise evaluate C
```

Notice that variables (*A*, *B* and *C*) are included as undefined items yet to be evaluated. This and the ability that Prolog possesses to add and delete clauses from its own database (*assert/1* and *retract/1*) make it an extremely powerful language. The subtlety and use of these facilities are beyond the scope of this brief review of the language.

The conditional evaluation operator '–>' can be defined, as the following cryptic bit of magic demonstrates.

```
A –> B :– A, !, B.
```

This states that *(A –> B)* is true if *A* is true and *B* is true. However, if *B* fails, we do not permit another solution to be found for *A*. (Notice that *A –> B* implements the familiar IF...THEN construction.)

While Prolog is a small language with few syntax rules, most practical implementations also provide a wide range of facilities in the form of *built-in predicates.* Printing and reading / writing files is handled in this way, as is access to the system clock and control of the I/O ports. List membership can be tested using *member/2*. While this may be defined by the user as follows, it is usually made available as a built-in predicate.

```
member(A, [A|_]).                % A is a member of any list whose head is A.
% Underscore, '_', represents a variable whose value is unimportant and can be discarded.
member(A,[_,B]) :– member(A,B).
% A is a member of any list whose tail is B if A is a member of B
```

The author of this Appendix (BGB) advises his students that they cannot program in Prolog properly until they can write *member/2* from first principles.

Prolog contains a facility for defining grammars, parsing, and extracting meaning from simple Natural Language statements. Section 4.3.7 explains the rudiments of definite-clause grammars in Prolog.

Further reading

- A. Gal, G. Lapalme, and H. Somers, "Prolog for Natural Language Processing", Wiley, Chichester, 1991, ISBN 0 471 93012 1.
- C. McDonald and M. Yazdani, "Prolog Programming: A Tutorial Introduction", Blackwell, Oxford, 1990, ISBN 0 632 01246 3.

- G. Gazdar and C. Mellish, "Natural Language Programming in Prolog," Addison-Wesley, Wokingham, 1989, ISBN 0 201 18053 7.
- G. L. Lazaraev, "Why Prolog? Justifying Logic Programming for Practical Applications," Prentice-Hall, London, 1989, ISBN 0 13 959040 4.
- H. Coelho and J. C. Cotta, "Prolog by Example," Springer, Berlin, 1988, ISBN 3 540 18313 2.
- I. Bratko, "Prolog Programming for Artificial Intelligence," Addison-Wesley, Wokingham, 1986, ISBN 0 201 14224 4.
- L. Sterling and E. Shapiro, "The Art of Prolog: Advanced Programming Techniques," MIT Press, Cambridge, MA, 1987, ISBN 0 262 19250 0.
- P. Deransart, A. Ed-Dbali, and L. Cervoni, "Prolog: The Standard Reference Manual," Springer, Berlin, 1996, ISBN 3 540 59304 7.
- W. F. Clocksin and C. S. Mellish, "Programming in Prolog," Springer, Berlin, 1981, ISBN 3 540 11046 1.

Appendix C
PIP commands and their implementation

Classification of commands by input type and output result

The letters A–T in the following table are the codes used in the "Type" column (column 3) of the main table. It is not possible to include all commands in this classification scheme, and some operators can claim to span two or more categories.

<table>
<tr><th rowspan="2">Output / result</th><th colspan="6">Input</th></tr>
<tr><th>Grey</th><th>Grey x 2</th><th>Binary</th><th>Binary x 2</th><th>Data</th><th>Disk</th></tr>
<tr><td>Grey image</td><td>A</td><td>B</td><td>C</td><td></td><td rowspan="2">D</td><td rowspan="4">E</td></tr>
<tr><td>Grey image, altered geometry and/or features</td><td>F</td><td></td><td></td><td></td></tr>
<tr><td>Binary image</td><td>G</td><td></td><td>H</td><td>I</td><td rowspan="2">L</td></tr>
<tr><td>Binary image, altered geometry and/or features</td><td>J</td><td></td><td>K</td><td></td></tr>
<tr><td>Data</td><td>M</td><td></td><td>N</td><td></td><td>O</td><td>P</td></tr>
<tr><td>Program control</td><td></td><td></td><td></td><td></td><td>Q</td><td></td></tr>
<tr><td>Disk file, RAM image store</td><td>R</td><td></td><td>R</td><td></td><td>S</td><td>T</td></tr>
</table>

Classification of commands for implementation using a LUT/SKIPSM approach

The following codes are used in the "SKIPSM" column of the main table.

Code	Operator type
X	Function implemented by a simple lookup table
Y	Function implemented by a single pass through the image using a SKIPSM machine
Z	Function implemented by multiple passes through the image using a SKIPSM machine

PIP command mnemonics, arities, and functions

Mnemonic/Arity	Function	Type	SKIPSM
#	Backtracking operator for image-processing commands	Q	
&	Equivalent to Prolog's goal concatenation operator ","	Q	
if	Equivalent to Prolog operator ":–"	Q	
\	\ *X* performs operation *X* and then saves result in a file named *X*	Q	
§	*A* is equivalent to [*dab(A)*]; operation *A* is performed on all blobs	Q	
•	*N*•*G* repeats goal *G* a total of *N* times	Q	
æ	Issue AppleEvent. Interfacing MacProlog to other program modules	Q	

Mnemonic/Arity	Function	Type	SKIPSM
aad/[0,3]	Aspect adjust	F	
abs/0	Fold intensities about mid-grey ("absolute value" function)	A	X
acn/1	Add a given constant to all intensities	A	X
add/0	Add intensities at corresponding points in two pictures	B	
and/0	Logical AND of two binary images	I	
ang/6	Angle formed by line joining 2 given points and horizontal axis	O	
avr/[0,1]	Average intensity value (synonymous with *avg[0,1]*)	A	
bay/0	Isolate bays (edge concavity)	P	
bbb/[0,3]	Coordinates of centroid of biggest blob and its intensity.	P	
bbt/1	Is biggest concavity (bay or hole) above second largest?	P	
bcl/[0,1]	Binary closing (mathematical morphology)	H	Y
bed/[0,1]	Edge detector (binary image)	H	Y
bic/1	Set given bit of each intensity value to 0	A	X
bif/1	Flip given bit of each intensity value	A	X
big/[0,1]	Select the ith biggest blob	H	
bis/1	Set given bit of each intensity value to 1	A	X
blb/0	Fill all holes ("blob fill")	H	Y
blo/1	Increase contrast in center of intensity range	A	X
blp/6	Calculate six parameters for all blobs in image	N	
bop/[0,1]	Binary opening (binary morphology)	H	Y
box/5	Draw a rectangle with given corner points and intensity	L	
bpt/[0,2]	Bottom-most white point	N	
bsk/0	View bottom image on stack	C	
bug/0	Corner detection function and edge smoothing (bug) function	C	Z
bve/8	Find coordinates of extrema along defined vector	N	
cal/1	Copy all pixels above given grey level into another image	A	
cbl/[0,1]	Count blobs	N	
ccmb/[0,1]	Combine red, green, blue images to form a color image	–	
ccp/0	Cursor-controlled crop	–	
cct/[0,1]	Concavity tree of object in a binary image	N	
cgr/0[0,2]	Centroid coordinates	N	
cgrb/[0,2]	Grab color image and separate into [*R, G, B*] component images	–	
cgsg/3	Set up a LUT (used with *clve, cpsu*)	–	
chf/[0,1]	Flip horizontal axis if longest vertical chord is left of image center	K	
chu/0	Convex hull	K	Z
cin/0	Column integration, top to bottom	F	Y
cir/5	Draw circle inside given bounding box	L	
circ/9	Parameters of circle intersecting 3 given points	O	
clc/0	Column run-length coding (travelling from top to bottom)	F	Y
cloa/[0,1,2]	Load color image into 3 separate pixel planes	Color	
clve/[0,1,2,3]	Show live image in pseudocolors	–	
cnw/0	Count neighbors	C	Y
cob/0	Corners of blobs	C	Y
com/[0,1]	Compare two images	B	
con/[0,9]	General linear convolution operator, 3×3 pixels	A	Y
cox/0	Scan columns top-to-bottom to find column maximum	F	Y
cpsu/[0,2]	Set pseudocolor	Color	
cpt/2	Set pixels along vector to values defined by a given list	D	

Mnemonic/Arity	Function	Type	SKIPSM
cpy/0	Copy the current image into alternate image	A	
crk/[0,1]	Crack detector (morphological grey-scale filter)	A	Y
crp/4	Crop current image to rectangle specified by user	F	
csca/[0,1]	Color scattergram	Color	
cscm/[0,1]	Mask color scattergram	Color	
csh/0	Copy right-most column of current image to all columns	F	Z
csk/0	Clear the image stack	S	
ctm/0	Camera to monitor (set up camera)	U	
ctp/2	Cartesian-to-polar transformation	F	
cua/4	Use cursor to find the position of a rectangular area defined by user	Cursor	
cur/[0.3]	Cursor (point defined by user with mouse, synonymous with *cup/0,3*]	Cursor	
cvd/0	Convex deficiency	K	
cvr/[0,1]	Find set of non-black points in rectangular area defined using cursor	Cursor	
cwd/[1,2]	Set the current image to white	D	X
cwp/[0,1]	Count white points	N	Y
dab/1	Do given task for all blobs in image (synonymous with §)	Q	
dbn/0	Direction of brightest neighbor	F	Y
dcg/[0,1]	Draw centroid and print coordinates	K	
dci/0	Draw center of the image	L	
dcl/[0,2,3]	Draw cross lines	L	
dcn/1	Divide current image intensities by constant	A	
dgw/[0,4]	Get image size	M	
dif/0	Absolute value of difference between current and alternate images	A	
dil/1	Dilate white areas of binary image in direction specified by user	H	Y
dil4/0	4-neighbor dilation (takes place in 4 directions simultaneously)	H	Y
dil8/0	8-neighbor dilation (takes place in 8 directions simultaneously)	H	Y
dim/[0,4]	Bounding box of the white pixels in binary image	N	
din/0	Double all intensity values	A	X
disc/4	Draw a white disc given its center and radius	L	
div/0	Divide intensity in current image by intensity in alternate image	A	
dlp/2	Difference of lowpass filters	F	Y
dpa/[0,2]	Draw principal axis	K	
dpi/[0,1]	Copy current image into passive image display	–	
dsl/3	Draw straight line given one point on it and its slope	L	
eab/1	Evaluate given goal for each blob	Q	
ect/0	Threshold midway between min. and max. intensity	G	
edd/0	Edge detector (grey-scale morphology)	A	Y
edg/[0 – 2]	Set the picture boundary of given width to defined grey-level	A	Y
egr/[0,4]	Grow ends of limbs of skeleton-like objects in binary image	H	Y
enc/0	Enhance contrast; set darkest pixel to black, brightest to white	A	X
ero/1	Erode white regions in binary image in given direction	H	Y
ero4/0	4-neighbor erosion	H	Y
ero8/0	8-neighbor erosion	H	Y
eul/[0 – 3]	Euler number, area and perimeter	N	
exp/0	Exponential of all intensities in current image	A	X
exw/0	Expand white regions	H	Y

Mnemonic/Arity	Function	Type	SKIPSM
fac/0	Flip about vertical line through the centroid	K	
fbr/0	Find blobs touching border and remove	K	
fcb/4	Fit circle to blob	K	
fcd/9	Fit circle to data: coordinates of 3 points	L	
fil/5	Fill given rectangle shape with intensity value defined by user	D	
fld/4	Fit line to data: coordinates of 2 points	L	
frz/0	Digitize an image	F	
fsr/1	Remove blobs below given size (synonymous with *kgr*)	H	
gcl/[0,1]	Grey-scale closing (mathematical morphology)	A	Z
gdh/0	Extract measurements from intensity histogram	M	
gft/0	Grass-fire transform (also called "prairie-fire" transform)	C	Z
gis/[0,1]	Grab (digitize) an image sequence	–	
gli/[0,2]	Minimum AND maximum intensities	M	Y
gob/0	Get one blob	K	
gop/[0,1]	Grey-scale opening (mathematical morphology)	H	Z
gpx/1	List of all non-black pixels and their intensities	M	
gra/0	Intensity gradient (edge detector)	A	Y
grb/[0 – 3]	Grab (digitize) an image from the camera	–	
gri/2	Extract regions of grey level above given value	F	
gry/[0,1]	Set all intensities in current image to given grey level	D	X
hfl/0	Fill holes (synonymous with blb)	H	Y
hgc/	Cumulative histogram, output as a list	M	
hgi/[0,1]	Intensity histogram, output as a list	M	
hgr/0	Horizontal gradient function	A	Y
hid/0	Horizontal intensity difference	A	Y
hil(/[0,3]	Highlight given range of intensity values	A	X
hin/0	Halve all intensity values	A	X
hlp/0	Help – PIP system documentation	P	
hmx/[0,2]	Find peak in the intensity histogram	M	
hpf/0	Highpass filter	A	Y
hpi/0	Plot histogram	J	
hsm/0	Horizontal smoothing	A	Y
huf/0	Hough transform	C	Z
hyp/0	Hyperbolic intensity mapping	A	X
ict/[0,1]	Intensity contours, smoothed or unsmoothed. Also called isophotes	J	
iht/0	Inverse Hough transform (principal peak only)	J	
imx/[0 – 3]	Center point and intensity of largest region with max intensity	M	
inv/0	Invert each pixel value in a binary image	C	X
ior/0	Logical OR of two binary images	I	
isd/1	Display sequence of images in Source or Destination folder	E	
isg/1	Digitize image sequence from camera; store on disc	S	
isi/0	Display interactive control window for image sequence processing.	Q	
isp/1	Apply the given operator on the image sequence in the Source folder.	E	
isr/1	Read image from Source Folder, given its relative address	E	
isu/2	Read image, given its absolute address (numeric index)	E	
isv/2	Write image, given its absolute address (numeric index)	R	
isw	Write image into Destination Folder (relative address mode)	R	

Mnemonic/Arity	Function	Type	SKIPSM
isx(A)	Delete all files from named folder	S	
itv/0	Interactive mode	Q	
jnt/[0,1]	Isolate and count joints (of skeleton-like figure)	K	Y
kgr/1	Keep blobs with area greater than defined limit	H	
ksm/1	Keep blobs with area smaller than defined limit	H	
kwi/1	Kill all windows and images	–	
lak/0	Isolate lakes	K	
lat/1	Local averaging with thresholding	G	
lav/[0,1]	Local average (blurring filter)	A	Y
lgr/0	Largest gradient	F	Y
lgt(/[0,2]	Get pixel intensities lying along given vector	M	
lhq/1	Local-area histogram equalization	A	
lin/0	User defines point which is then used to invert Hough transform	J	
ljt/1	List of all skeleton joints	N	
lle/1	List of all skeleton limb ends	N	
lme/[0,1]	Isolate and count limb ends of skeleton-like figure	N	
lmi/[0,3,4]	Principal-axis parameters	N	
lnb/[0,1]	Largest neighbor in 3×3 neighborhood (local operator)	A	Y
lni/[0,1]	Load image and associate with a new name	E	
loa/[0,1]	Load image from disc (path is defined by user)	E	
log/0	Logarithm of all intensities	A	X
lpc/[0,1]	Laplacian filter (spot detector)	A	Y
lpf/0	Lowpass filter	A	Y
lpt[0,2]	Left-most white point	N	
lrt/0	Left-to-right (flip horizontal axis)	J	
mar/[0,4]	Minimum area rectangle enclosing white points	K	
max/0	Maximum intensity in two images (dyadic pixel-by-pixel operator)	B	
mbc/[0,3]	Minimum bounding circle	K	
mcn/1	Multiply all intensities by given constant	A	X
mdf/1	Median filter (also implements rank filters)	A	Z
mim/[0,1]	Coordinates of the center of the image	O	
min/0	Minimum intensity in two images (dyadic pixel-by-pixel operator)	B	
mul/0	Multiply current and alternate images	B	
ndo/0	Numerate distinct objects (blob labelling)	C	Z
neg/0	Negate image	A	X
neg_log/0	Negative logarithm intensity mapping	A	X
nmr/[0,2]	Normalize [*X,Y*] position of minimum–area rectangle	K	
not/0	Logical inversion (negation) of binary image	C	X
npo/[0,3]	Normalize position and orientation	H	
nxy/[0,2]	Normalize [*X,Y*] position of centroid	H	
pcc/[0,2,3]	Find white point closest to the center of image and its distance	N	
per/[0,1]	Perimeter	N	
pex/[0,2]	Picture expand	A	
pfx/3	Draw one given pixel at defined grey-level	A	
pgn/[0,1]	Draw polygon or arc using cursor, construct list of its nodes	Cursor	
pii/3	Test whether given point is inside the image	O	
pis/0	Push image onto the stack (synonymous with *psk/0*)	R	

Mnemonic/Arity	Function	Type	SKIPSM
plt/1	Plot intensity	J	
pop/0	Pop the image stack	E	
ppi/2	Print the intensity of a given point	M	
psh/2	Shift image, no wrap-around	A	
psi/1	Process a sequence of images	Q	
psk/0	Push image onto the stack (synonymous with *pis/0*)	R	
psq/[0,2]	Picture squeeze	A	
psw/[0,2]	Shift image, with wrap-around	A	
ptc/0	Polar-to-Cartesian transformation	F	
pth/2	Percentage threshold	G	
put/3	Draw one given pixel at defined grey-level (synonymous with *pfx*)	A	
raf/[1]	Blurring filter (repeat *lpf* many times)	A	Z
rbi/0	Recover current and alternate images from the stack	E	
rea/[0–2]	Read named image from RAM	E	
red/0	Roberts edge detector	A	Y
ria/1	Read image from Archive Images folder	E	
rim/[0,1]	Read image from the Temporary Images folder	E	
rin/0	Row integration, left to right	F	Y
rip/[0,1]	Remove white isolated points	H	Y
ris/0	Read frame of an image sequence from disc store	E	
rlc/0	Row run-length coding (scanning from left to right)	C	Y
rnd/0	Pseudorandom number generator (uniform distribution)	D	
rni/1	Read image, name assigned by lni	E	
roa/0	Rotate anticlockwise by 90 degrees	A	
roc/0	Rotate clockwise by 90 degrees	A	
rox	Scan rows left to right to find row maximum	F	Y
rpi/[0,1]	Read image from passive image display	E	
rpt[0,2]	Right-most white point	N	
rsh/0	Copy bottom row to all rows	F	Z
sbi/0	Save both current and alternate images on stack	E	
sca/1	Remove least significant bits of each intensity value	A	X
sco/0	Circular wedge (Intensity increases with angle relative to vertical axis)	D	
set/0	White image	D	
sgb/0	Initialize image grabber	–	
shf[0,1]	Shape factor (=area/(perimeter×perimeter))	N	
shp/[0,1]	Sharpen image	A	Y
sim/[1,2]	Open window with a copy of current image	A	
sio/0	Show image sequence (original)	E	
sip/0	Show image sequence after processing	E	
ske/0	Skeleton	K	Z
skw/0	Shrink white regions	H	Y
slt/1	Select LUT for image mapping	Q	
snb/[0,1]	Smallest neighbor in 3×3 neighborhood (local operator)	A	Y
sqr/0	Square all intensity values	A	X
sqt/0	Square root of all intensity values	A	X
ssi/0	View temporary stored images	E	
sto/[0 – 2]	Write current image picture to disc file	R	
sub/0	Subtract images	B	

Mnemonic/Arity	Function	Type	SKIPSM
swi[0,2]	Switch images (Current and alternate images are switched by default.)	B	
tbt/0	Top-to-bottom (flip vertical axis)	A	
thp/[1,2]	Threshold at given centile values	G	X
thr[0,1,2]	Threshold between two given intensity values	A	X
thx([0,1]	Threshold at maximum intensity (Isolate points with maximum intensity)	A	X
tia/0	List files in Temporary Images folder	P	
tia/0	List of temporary images available	P	
tpt/[0,2]	Top-most white point	N	
tsk/0	View top image on stack	E	
tur/1	Turn (rotate) image about its center	A	
usm/[0,1]	Unsharp masking	A	
vgr/0	Vertical gradient function	A	
vgt/2	Instantiate list to values in right-hand column of image	M	
vid/0	Vertical intensity difference	A	
vpl/5	Draw a line with between two given points	L	
vpt/2	Set right-hand column of image to values defined in given list	D	
vsk/0	View image stack	E	
vsm/0	Vertical smoothing	A	Y
vsw/0	Variable frequency sine-wave generator	D	
wdg/[0,1]	Intensity wedge	D	Y
wgx/[0,1]	Intensity wedge (synonymous with *wdg*)	D	Y
wia/1	Writing to the Archive Images folder	R	
wim/[0,1]	Write image to the Temporary Images folder	R	
wis/0	Write frame of an image sequence to disc store	R	
wri/[0,1]	Write image into named RAM file	R	
wrm/[0,1]	Remove isolated white or black pixels in binary image	H	Y
xor/0	Logical exclusive OR of two binary images	I	
yxt/0	Interchange *X* and *Y* axes (transpose image)	A	
zer/0	Black image	D	X

References

ALIS ALIS 600 and Micro ALIS, Application Lighting Investigation Systems, Dolan Jenner, Woburn, MA, USA.

BAT74 B. G. Batchelor, "Practical Approach to Pattern Classification," Plenum, London and New York, 1974, ISBN 0-306-30796-0.

BAT79a B. G. Batchelor and G. A. Williams "Defect detection on the internal surfaces of hydraulics cylinders for motor vehicles," Proc. Conf. on Imaging Applications for Automated Industrial Inspection and Assembly, Washington DC, April 1979, SPIE, Bellingham, WA.

BAT79b B. G. Batchelor, "Interactive image analysis as a prototyping tool for industrial inspection," Computers and Digital Techniques, Proc. IEE, part E, vol. 2, no. 2, April 1979, pp. 61–69.

BAT80 B. G. Batchelor, "Two methods for finding convex hulls of planar figures," Journal of Cybernetics and Systems, vol. 11, 1980, pp. 105–113.

BAT81 B. G. Batchelor and B. K. Marlow, "Converting run code to chain code," Cybernetics and Systems, vol. 12, 1981, pp. 237–246.

BAT82a B. G. Batchelor, D. H. Mott, G. J. Page and D. N. Upcott, "The Autoview interactive image processing facility," in N B Jones, Ed, "Digital signal processing," Peter Peregrinus, London 1982, ISBN 0-906-04891-5, pp. 319–351.

BAT82b B. G. Batchelor, "A laboratory-based approach for designing automated inspection systems," Proc. Int. Workshop on Industrial Applications of Machine Vision, Research Triangle, NC, May 1982, IEEE Computer Society, 1982, pp. 80–86.

BAT83a B. G. Batchelor and S. M. Cotter, "The automatic inspection of aerosol sprays using visual sensing," Sensor Review, vol3, no. 1, Jan.1983, pp12–16.

BAT83b B. G. Batchelor and S. M. Cotter, "Detection of cracks using image processing algorithms implemented in hardware," Image and Vision Computing, vol. l, no. l, Feb. 1983, pp. 21–29.

BAT85 B. G. Batchelor and A. K. Steel, "A flexible inspection cell," Proc. Int. Conf. on Automation in Manufacturing, Part 4: Automated Vision Systems, Singapore, September 1985, Singapore Exhibition Services Pte, pp108–134.

BAT86 B. G. Batchelor, "Merging the Autoview Image Processing Language with Prolog," Image and Vision Computing, vol. 4, no. 4, November 1986, pp.189 – 196.

BAT89 B. G. Batchelor, I. P. Harris, J. R. Marchant, and R. D. Tillett, "Automatic dissection of plantlets," Proc. Conf. on Automated Inspection and High Speed Vision Architectures II, Cambridge, MA, USA, Nov. 1988, SPIE, Bellingham, WA, USA, vol. 1004.

BAT90 B. G. Batchelor, "Tools for Designing Industrial Vision Systems," SPIE Conf. on Machine Vision Systems Integration, 6–7 Nov. 1990, Boston, MA, SPIE vol. CR36, 1991, ISBN 0-8194-0471-3, pp. 138 – 175.

BAT91a B. G. Batchelor, "Intelligent Image Processing in Prolog," Springer–Verlag, Berlin, 1991, ISBN 0-540-19647-1.

BAT91b J. P. Chan, B. G. Batchelor, S. R. Broderick, R. A. Lambert, and A.W.E. Weeks, "Expert system to aid learning of new products for inspection," Proc. Conf. on Machine Vision Systems: Architectures, Integration and Applications, Boston, MA, Nov, 1991.

BAT93a B. G. Batchelor, "Automated inspection of bread and loaves," Proc. Conf. on Machine Vision Applications, Architectures and Systems Integration II, Boston, MA, Sept. 1993, SPIE, Bellingham, WA, USA., vol. 2064, ISBN 0-8194-1329-1, pp. 124 – 134.

BAT93b B. G. Batchelor and F. M. Waltz, "Interactive Image Processing," Springer Verlag, New York, NY 1993, ISBN 3-540-19814-8.

BAT94a B. G. Batchelor, "HyperCard lighting advisor," Proc. Conf. on Machine Vision Applications, Architectures and Systems III, Boston, MA, Nov. 1994, SPIE, Bellingham, WA,

USA., vol. 2347, ISBN 0-8194-1682-7, pp. 180 – 188.
Also URL: http://bruce.cs.cf.ac.uk/bruce/index.html

BAT94b B. G. Batchelor and P. F. Whelan, Eds., "Industrial Vision Systems," SPIE Milestone Series, vol. MS 97, SPIE, Bellingham WA, ISBN 0-8194-1580-4.

BAT94c B. G. Batchelor and P. F. Whelan, "Machine vision: proverbs, principles, prejudices and priorities," Proc. Conf. Machine Vision Applications, Architectures and Systems III, Boston, MA, USA., Nov. 1994, SPIE, Bellingham, WA, vol. 2347, ISBN 0-8194-1682-7. pp 374 - 385.

BAT97 B. G. Batchelor and P. F. Whelan, "Intelligent Vision Systems for Industry," Springer Verlag, London and Berlin, 1997, ISBN 3-540-19969-1.

BAT98a B. G. Batchelor and J.-R. Charlier, "Machine Vision is Not Computer Vision," Keynote paper, Proc. SPIE Conf., Machine Vision Systems for Inspection and Metrology VII, Boston, MA, November 1998, vol. 3521, pp. 2–13, ISBN 0-8194-2982-1.

BAT98b B. G. Batchelor, "Knowledge-based Program to Assist in the Design of Machine Vision Systems," Proc. SPIE Conf., Machine Vision Systems for Inspection and Metrology VII, Boston, MA, November 1998, vol. 3521, pp. 312–324, ISBN 0-8194-2982-1

BAT99 B. G. Batchelor, M. W. Daley, R. J. Hitchell, G. J. Hunter, G. E. Jones, and G. Karantalis, "Remotely Operated Prototyping Environment for Automated Visual Inspection," Proc. IMVIP99 – Irish Machine Vision and Image Processing Conference 1999, Dublin City University, Dublin.

BOR86 G. Borgefors, "Distance transformations in digital images," Computer Vision, Graphics and Image Processing, vol. 34, 1986, pp. 344–371.

BRA86 I. Bratko, "Prolog Programming for Artificial Intelligence," Addison-Wesley, Wokingham, England, 1986.

CIP Cyber image processing software, Cardiff University, Cardiff, Wales.
URL: http://bruce.cs.cf.ac.uk/bruce/index.html

CKI CKI Prolog, S. van Otterloo,
URL: http://www.students.cs.uu.nl/people/smotterl/prolog/

CLO81 W. F. Clocksin and C. S. Mellish, "Programing in Prolog," Springer Verlag, Berlin, 1981.

DAV00 E. R. Davies, "Image Processing for the Food Industry," World Scientific, Singapore, 2000, ISBN 981-02-4022-8.

DOU92 E.R. Dougherty, "An Introduction to Morphological Image Processing," Tutorial Text Vol. TT9, SPIE Press, 1992.

DUF73 M. J. B. Duff, D. M. Watson, T. M. Fountain, and G. K. Shaw, "A cellular logic array for image processing," Pattern Recognition, 1973.

FLO86 A Floeder, "A connectivity algorithm for parallel processors," Master's dissertation, Electrical Engineering Department, University of Minnesota, Minneapolis, MN 1987

GAZ89 G. Gazdar and C. Mellish, "Natural Language Processing in Prolog," Addison–Wesley, Wokingham, England, 1989.

GON87 R.C. Gonzalez and P. Wintz, "Digital Image Processing," Addison-Wesley, Reading MA, 1987.

GON92 R. C. Gonzalez and R. E. Woods, "Digital Image Processing", Addison-Wesley, Reading, MA,1992, ISBN 0-201-50803-6.

GOS96 J Gosling, B Joy, and G Steele, "The Java Language Specification (Java Series)," Addison–Wesley Pub Co. 1996, ISBN: 0201634511.

HAC97a R. Hack, F. M. Waltz, and B. G. Batchelor, "Software implementation of the SKIPSM paradigm under PIP," Proc. SPIE Conf. on Machine Vision Applications, Architectures, and Systems Integration VI, vol. 3205, no. 19, Pittsburgh, PA, Oct. 1997

HAR79 R. M. Haralick, "Statistical and structural approaches to texture," Proc. IEEE, vol. 67, no. 5, 1979, pp. 768 – 804.

HAR87b R. M. Haralick, "Image analysis using mathematical morphology," IEEE Trans. Pattern Analysis and Machine Intelligence, vol. 9, no. 4, 1987, pp. 532 – 550.

HAR92 R.M. Haralick and L.G. Shapiro, "Computer and Robot Vision: Volumes I and II," Addison-Wesley, Reading , MA, 1992.

HEI91 H. J. A. M. Heijmans, "Theoretical Aspects of Grey-scale Morphology, IEEE Trans. Pattern Analysis and Machine Intelligence, vol. 13, no. 6, 1991, pp. 568 – 582.

HUJ95a A. A. Hujanen and F. M. Waltz, "Pipelined implementation of binary skeletonization using finite-state machines," Proc. SPIE Conf. on Machine Vision Applications in Industrial Inspection, vol. 2423, no. 2, San Jose, CA, Feb. 1995.

HUJ95b A. A. Hujanen and F. M. Waltz, "Extending the SKIPSM binary skeletonization implementation," Proc. SPIE Conf. on Machine Vision Applications, Architectures, and Systems Integration IV, vol. 2597, no. 12, Philadelphia, PA, Oct. 1995.

HUT71 T. C. Hutson, "Colour Television Theory," McGraw-Hill, London, 1971.

ICM Intelligent Camera, past product of Image Industries Ltd, now part of Cognex Ltd.,Contact: Mr. Peter Neve, Image Industries Ltd., Unit 7, First Quarter, Blenheim Road, Epsom, Surrey, England.

IMA "Image" image processing software, National Instiutes of Health, USA. Anonymous FTP: zippy.nimh.nih.gov

ISPY I–SPY "Webcam" software, Surveyor Corporation, Inc.
URL: http://www.ispy.nl/ or URL: http://www.surveyor.com/index.html

ITI Lighting Science Database, Sensor Center for Improved Quality, Industrial Technology Institute, P.O. Box 1485, Ann Arbor, MI 48106.

JAV Java, The Source for Java Technology, Sun Microsystems Inc.
URL http://www.javasoft.com

JAW96 J. Jaworski, "Java Developers Guide," Sams Net, 1996. ISBN: 1-57521-069-X.

JMF Java Media Framework, API, The Source for Java Technology, Sun Microsystems Inc., URL http://java.sun.com/products/java-media/jmf/

JON98 G. E. Jones, "A Scripting Language, Interpreter and User Interface to Extend CIP, an Image Processing Tool," MSc dissertation, Department of Computer Science, University of Wales Cardiff, Cardiff, Wales, 1998.

KID Sigma 2100, Optical Design Software, Kidger Optics Ltd/, 9a High Street, Crowborough, East Sussex, TN6 2QA, England.

KRI99 Khoral Research, Inc. – http://www.khoral.com/core.html

LPA MacProlog and WinProlog, Logic Programming Associates Ltd., Studio 4, Royal Vistoria Patriotic Building, Trinity Road, London, SW18 3SX, U.K.

MAR80 B. K. Marlow and B. G. Batchelor, "Improving the speed of convex hull calculations," Electronics Letters, vol. 16, no. 9, 24 April 1980, pp319–321.

MCC80 A. J. McCollum, C. C. Bowman, P. A. Daniels, and B. G. Batchelor, "A histogram modification unit for real-time image enhancement," Computer Graphics and Image Processing, vol. 42, 1988, pp. 387 – 398.

MIL97 J. W. V. Miller and F. M. Waltz, "Software implementation of 2-D grey-level dilation using SKIPSM," Proc. SPIE Conf. on Machine Vision Systems for Inspection and Metrology VII, vol. 3205, no. 18, Pittsburgh, Oct. 1997

MUR96 J. D. Murray & W. Van Ryper, "Encyclopaedia of Graphics File Formats," O'Reilly, 1996.

MVAa Machine Vision Application Requirements Check List, Machine Vision Association of the Society of Manufacturing Engineers, Dearborn, MI.
Available to SME members at URL:http://www.sme.org/cgi-bin/new-gethtml.pl?/mva/mvanarc.htm&GROUP&MEMBNUM&MVA

MVAb Machine Vision Lens Selector, Machine Vision Association of the Society of Manufacturing Engineers, Dearborn, MI, 1992.

OPL OptiLab and Concept Vi, Graftek France, Le Moulin del'Image, 26270 Mirmande, France.

OPT OPTO*SENSE, Machine Vision Database, Visual*Sense*Systems, 314 Meadow Wood Terrace, Ithaca, NY 14859.

PEN Lighting Advisor Expert System, Penn Video, Inc., a subsidiary of Ball Corporation, Inc., Industrial Systems Division, 929 Sweitzer Avenue, Akron, OH 44311. (No longer available commercially.)

PHO Photoshop, Adobe Systems, Inc., 1585 Charleston Road, PO Box 7900, Mountain View, CA 94039-7900.

PIT93 I. Pitas, "Digital Image Processing Algorithms," Prentice–Hall, Englewood Cliffs NJ, 1993.

PVB "Machine Vision: Proverbs, Opinion and Folklore." URL: http://www.eeng.dcu.ie/~whelanp/proverbs/proverbs.html

RUS95 J. C. Russ, "The Image Processing Handbook," CRC Press, Boca Raton, FL, 1995, ISBN 0-8493-2516-1.

RUT78 D. Rutovitz et al., "Pattern Recognition Procedures in the Cytogenetic Laboratory," in "Pattern Recognition: Ideas in Practice," B. G. Batchelor, ed, Plenum, London and New York, 1978, ISBN 0-306-31020-1, pp. 303 - 329]

SER82 J. Serra, "Image Analysis and Mathematical Morphology Vol. 1," Academic Press, New York, NY, 1982.

SER86 J. Serra, "Image Analysis and Mathematical Morphology Vol. 2," Theoretical Advances," Academic Press, New York, NY, 1988.

SIM00 S. Simpson, "For the bees: glowing paint may highlight the forces that make insects fly," Scientific American, May 2000, p 15. Also see URL http://www.sciam.com/2000/0500issue/0500scicit3.html

SMI99 T. E. Smith, D. F. Britton, W. Daley, and R. Carey, "Integration of USB and Firewire cameras, in machine vision applications," Conf. Machine Vision Systems, for Insoection and Metrology VIII, Boston, MA, Sept 1999, Proc. SPIE, vol. 3836, 1999, pp. 216 – 225.

SNY92 M. Snyder, "Tools for Designing Camera Configurations," Proc. Conf "Machine Vision Architectures, Integration and Applications," Boston, MA, Nov., 1992, Proc. SPIE, vol. 1615, 1992, pp. 18- 28.

SON93 M. Sonka, V. Hlavac, and R. Boyle, "Image Processing, Analysis and Machine Vision," Chapman and Hall," 1993.

STE78 S. R. Sternberg, "Parallel architectures for image processing," Proc. IEEE Conf. Int. Computer Software and Applications, Chicago, IL, 1978, pp. 712 – 717.

STE86a L. Sterling and E. Shapiro, "The Art of Prolog," MIT Press, Cambridge, MA, 1986.

STE86b S. R. Sternberg, "Grey-scale morphology," Computer Vision, Graphics and Image Processing," vol. 35, 1986, pp. 333 – 355.

SUN99 Java™ Technologies from Sun Microsystems: Products and APIs. http://java.sun.com/products/index.html

SUP "SuperVision" interactive image processing software, fomer product of Image Industries Ltd, now part of Cognex Ltd., Unit 7, First Quarter, Blenheim Road, Epsom, Surrey, England.

TWA TWAIN camera interface. URL: internet.sk/gnome/index.htm

VCS "VCS," Vision Control System, interactive image processing software, Vision Dynamics Ltd., Suite 7a, 1, St. Albans Road, Hemel Hempstead, Herts, HP2 4XX, England.

VIN91 L. Vincent, "Morphological transformations of binary imnages with arbitrary structuring elements," Signal Processing, vol. 22, 1991, pp. 3 – 23.

VOG89 R. C. Vogt, "Automatic generation of morphological set recognition algorithms," Springer-Verlag, 1989.

WAL88 F. M. Waltz, "Fast Implementation of standard and 'fuzzy' binary morphological operations with large structuring elements," Proc. SPIE Conf. Intelligent Robots and Computer Vision VII, vol. 1002, 1998, pp. 434- 441.

WAL94a F. M. Waltz, "SKIPSM: Separated-Kernel Image Processing using finite-State Machines," Proc. SPIE Conf. on Machine Vision Applications, Architectures, and Systems Integration III, vol. 2347, no. 36, Boston, MA, Nov. 1994.

WAL94b F. M. Waltz and H. H. Garnaoui, "Application of SKIPSM to binary morphology," Proc. SPIE Conf. on Machine Vision Applications, Architectures, and Systems Integration III, vol. 2347, no. 37, Boston, MA, Nov. 1994.

WAL94c F. M. Waltz and H. H. Garnaoui, "Fast computation of the Grassfire Transform using SKIPSM," Proc. SPIE Conf. on Machine Vision Applications, Architectures, and Systems Integration III, vol. 2347, no. 38, Boston, MA, Nov. 1994.

WAL94d F. M. Waltz, "Application of SKIPSM to binary template matching," Proc. SPIE Conf. on Machine Vision Applications, Architectures, and Systems Integration III, vol. 2347, no. 39, Boston, Nov. 1994.

WAL94e F. M. Waltz, "Application of SKIPSM to grey-level morphology," Proc. SPIE Conf. on Machine Vision Applications, Architectures, and Systems Integration III, vol. 2347, no. 40, Boston, MA, Nov. 1994

WAL94f F. M. Waltz, "Application of SKIPSM to the pipelining of certain global image processing operations, " Proc. SPIE Conf. on Machine Vision Applications, Architectures, and Systems Integration III, vol. 2347, no. 41, Boston, MA, Nov. 1994

WAL95a F. M. Waltz, Application of SKIPSM to binary correlation, Proc. SPIE Conf. on Machine Vision Applications, Architectures, and Systems Integration IV, vol. 2597, no. 11, Philadelphia, PA, Oct. 1995

WAL95b F. M. Waltz, SKIPSM implementations: morphology and much, much more, Proc. SPIE Conf. on Machine Vision Applications, Architectures, and Systems Integration IV, vol. 2597, no. 14, Philadelphia, PA, Oct. 1995

WAL96a F. M. Waltz, Binary openings and closings in one pass using finite-state machines, Proc. SPIE Conf. on Adv. Sig. Proc. Algorithms, Architectures, and Implementations VI, Denver, CO, Aug. 1996

WAL96b F. M. Waltz, Automated generation of finite-state machine lookup tables for binary morphology, Proc. SPIE Conf. on Machine Vision Applications, Architectures, and Systems Integration V, Boston, MA, Nov. 1996

WAL97a F. M. Waltz, Implementation of SKIPSM for 3-D binary morphology, Proc. SPIE Conf. on Machine Vision Applications, Architectures, and Systems Integration VI, vol. 3205, no. 13, Pittsburgh, PA, Oct. 1997

WAL97b F. M. Waltz, Binary dilation using SKIPSM: Some interesting variations, Proc. SPIE Conf. on Machine Vision Applications, Architectures, and Systems Integration VI, vol. 3205, no. 15, Pittsburgh, PA, Oct. 1997

WAL98a F. M. Waltz, "The application of SKIPSM to various 3x3 image processing operations, " Proc. SPIE Conf. on Machine Vision Systems for Inspection and Metrology VII, vol. 3521, no. 30, Boston, MA, Nov. 1998

WAL98b F. M. Waltz, R. Hack, and B. G. Batchelor, "Fast, efficient algorithms for 3x3 ranked filters using finite-state machines, " Proc. SPIE Conf. on Machine Vision Systems for Inspection and Metrology VII, vol. 3521, no. 31, Boston, MA, Nov. 1998

WAL98c F. M. Waltz, "Automated generation of efficient code for grey-scale image processing," Proc. SPIE Conf. on Machine Vision Systems for Inspection and Metrology VII, vol. 3521, no. 32, Boston, MA, Nov. 1998

WAL98d F. M. Waltz, "Image processing operations in color space using finite-state machines," Proc. SPIE Conf. on Machine Vision Systems for Inspection and Metrology VII, vol. 3521, no. 33, Boston, MA, Nov. 1998

WAL98e F. M. Waltz and J. W. V. Miller, "An efficient algorithm for Gaussian blur using finite-state machines," Proc. SPIE Conf. on Machine Vision Systems for Inspection and Metrology VII, vol. 3521, no. 37, Boston, MA, Nov. 1998

WAL99a F. M. Waltz and J. W. V. Miller, Connected-component analysis using finite-state machines,Proc. SPIE Conf. on Machine Vision Systems for Inspection and Metrology VIII, vol. 3836, Boston, MA, Nov. 1999

WAL99b F. M. Waltz, Grey-scale co-occurrence matrix generation using finite-state machines, Proc. SPIE Conf. on Machine Vision Systems for Inspection and Metrology VIII, vol. 3836, Boston, MA, Nov. 1999

WHE97a P. F. Whelan, Remote Access to Continuing Engineering Education RACeE, IEE Engineering Science and Education Journal, Oct 1997, pp205-211, 1997.

WHE97b P. F. Whelan, EE544 - Computer and Machine Vision – Online Course, http://www.eeng.dcu.ie/~whelanp/vsg/outline.html, 1997.

WHE00 P. F. Whelan and D. Molloy, "Machine Vision Algorithms in Java: Techniques and Implementation," Springer, London, 2000, ISBN 1-85233-218-2.

WIL89 G. R. Wilson and B. G. Batchelor, "Convex hull of chain-coded blob," Computers and Digital Techniques, Proc. IEE, pt. E, vol. 136, no. 6, Nov. 1989, pp. 530 - 534.

WIT99 Logical Vision - http://www.logicalvision.com/default.htm, 1999

ZAY93 I. Y. Zayas, J. L. Steele, G. Weaver, and D.E. Walker, "Breadmaking Factors Assessed by Digital Imaging," SPIE Conf. on Machine Vision Applications, Architectures and Systems Integration II, vol. 20, Boston, MA, Sept. 1993.

ZHU86 X Zhuang and R. M. Haralick, "Morphological structuring element decomposition," Computer Vision, Graphics and Image Processing, vol. 35, 1986, pp. 370 – 382.

Further reading

- A. D. Marshall and R. R. Martin, "Computer Vision, Models and Inspection," World Scientific, 1992.
- A. K. Jain, "Fundamentals of Digital Image Processing," Prentice-Hall, 1989.
- A. P. Pentland, "From Pixels to Predicates: Recent Advances in Computational and Robotic Vision," Ablex, 1986.
- I. Alexander, "Artificial Vision for Robots," Kogan Page Ltd., 1983.
- "AVA Machine Vision Glossary," Automated Vision Association, 1985.
- B. G. Batchelor, D. A. Hill and D. C. Hodgson, "Automated Visual Inspection," IFS Ltd/North Holland, 1984.
- B. G. Batchelor, "Pattern Recognition, Ideas in Practice," Plenum Press,London, 1978.
- B. K. P. Horn, "Robot Vision," MIT press, 1986.
- A. Browne and L. Norton-Wayne, "Vision and Information Processing for Automation," Plenum Press, New York, NY, 1986.
- C. A. Lindley, "Practical Image Processing in C," Wiley, 1991.
- C. H. Chen, L. F. Pau, and P. S. P Wang, "Handbook of Pattern Recognition and Computer Vision," World Scientific, 1993.
- C. R. Giardina and E. R. Dougherty, "Morphological Methods in Image and Signal Processing," Prentice Hall, 1988.
- D. Marr, "Vision," W. H Freeman and Company, 1982.
- D. H. Ballard and C.M. Brown, "Computer Vision," Prentice-Hall, 1982.
- E. L. Hall, "Computer Image Processing and Recognition," Academic Press, 1979.
- E. R. Davies, "Machine Vision: Theory, Algorithms, Practicalities," Academic Press, 1996.
- E. R. Dougherty and P. A. Laplante, "Introduction to Real-time Imaging," SPIE/IEEE Press, SPIE vol. TT19, 1995.

- E. R. Dougherty, "An Introduction to Morphological Image Processing," Tutorial Text vol. TT9, SPIE Press, 1992.
- G. Dodd and L. Rossol, "Computer Vision and Sensor-Based Robots," Plenum Press,1979.
- G. A. Baxes, "Digital Image Processing: A Practical Primer," Prentice-Hall/Cascade Press, 1984.
- G. A. Baxes, "Digital Image Processing: Principles and Applications," John Wiley, 1994.
- H. Barlow, C. Blakemore and M. Weston-Smith, Eds, "Images and Understanding," Cambridge University Press, 1990.
- H. Bassman and P. W. Besslich, "AdOculos – Digital Image Processing," International Thomson Publishing, 1995.
- H. Freeman, Ed., "Machine Vision for Inspection and Measurement," Academic Press, 1989.
- H. Freeman, "Machine Vision for Three Dimensional Scenes," Academic Press, 1990.
- H. Freeman, "Machine Vision: Algorithms, Architectures, and Systems," Academic Press, 1987.
- H. R. Myler and A. R. Weeks, "Computer Imaging Recipes in C," Prentice Hall, 1993.
- H. R. Myler and A. R. Weeks, "The Pocket Handbook of Image Processing Algorithms in C," Prentice Hall, 1993.
- J. Hollingum, "Machine Vision: The Eyes of Automation," IFS, 1984.
- K. R. Castleman, "Digital Image Processing," Prentice-Hall, 1996.
- K. S. Fu, Ed., "Syntactic Pattern Recognition and Applications," Springer-Verlag, 1981.
- K. S. Fu, R. C. Gonzalez, and C. S. Lee, "Robotics, Control, Sensing, Vision and Intelligence," McGraw-Hill, 1987.
- L. Uhr, "Parallel Computer Vision," Academic Press, 1987.
- L. F. Pau, "Computer Vision for Electronic Manufacturing," Plenum Press, 1990.
- L. J. Galbiati, "Machine Vision and Digital Image Processing Fundamentals," Prentice-Hall, 1990.
- M. Brady and H. G. Barrow, "Computer Vision," North-Holland, 1981.
- M. Eijiri, "Machine Vision: A Practical Technology for Advanced Image Processing," Gordon and Breach, 1989.
- M. Sonka, V. Hlavac, and R. Boyle, "Image Processing, Analysis and Machine Vision," Chapman and Hall, 1993.
- M. A. Fischler, "Readings in Computer Vision: Issues, Problems, Principles and Paradigms," M. Kaufmann, 1987.
- M. D. Levine, "Vision in Man and Machine," McGraw-Hill, 1985.
- N. J. Zimmerman and A. Oosterlinck, Eds., "Industrial Applications of Image Analysis," DEB, 1983.

- A. Pugh, "Robot Vision," IFS/Springer-Verlag, 1983.
- R. Chellappa and A. A. Sawchuk, "Digital Image Processing and Analysis: Volume 1: Digital Image Processing," IEEE Computer Society, 1985.
- R. Chellappa and A. A. Sawchuk, "Digital Image Processing and Analysis: Volume 2: Digital Image Analysis," IEEE Computer Society, 1985.
- R. Jain, R. Kasturi, and B. G. Schunck, "Machine Vision," McGraw-Hill, 1995.
- R. Nevatia, "Machine Perception," Prentice-Hall, 1982.
- R. C. Gonzalez and P. Wintz, "Digital Image Processing," Addison-Wesley, 1987.
- R. D. Boyle and R. C. Thomas, "Computer Vision: A First Course," Blackwell, 1988.
- R. J. Schalkoff, "Digital Image Processing and Computer Vision," Wiley, 1989.
- R. M. Haralick and L. G. Shapiro, "Computer and Robot Vision: Volumes I and II," Addison-Wesley, 1992.
- R. O. Duda and P. E Hart, "Pattern Classification and Scene Analysis," John Wiley, 1973.
- A. Rosenfeld and A. C. Kak, "Digital Picture Processing," Academic Press, New York, NY, 1982.
- S. L. Robinson and R. K. Miller, "Automated Inspection and Quality Insurance," Marcel Dekker, 1989.
- T. S. Huang, "Image Sequence Analysis," Springer-Verlag, 1981.
- C. Torras, Ed., "Computer Vision, Theory and Industrial Applications," Springer-Verlag, Berlin, 1992.
- D. Vernon, "Machine Vision," Prentice-Hall, 1991.
- W. B. Green, "Digital Image Processing - A Systems Approach," Van Nostrand Reinhold, 1983.
- W. K. Pratt, "Digital Image Processing," Wiley, 1978.

Index

N

O

P

X

Y

Z

Plates

The twenty-four plates on the pages following these captions are used to illustrate many of the concepts of this book. In the captions below, (TL) refers to the image at "top left," (CR) refers to the image at "center right," etc. With respect to Plates 19 through 24, the reader should realize that it is impossible, using available printing inks and papers, to reproduce the full range of color saturation achievable on an RGB color monitor. Therefore, when RGB colors are being described in the text or in the captions, the colors shown in the figures are merely approximations to the colors one would see on a monitor.

Plate captions

Plate 1: Typical test images for machine vision. Other examples are shown in the following plates.

(TL) A human face is useful for demonstrating a wide range of image processing operators. A person can very often detect subtle changes in a picture of a face that would be difficult to see if another subject were used.

(TR) Silhouette of an automobile connecting rod (conrod). Many industrial inspection applications employ back lighting to create a high-contrast image. Notice the nonuniform intensity of the background.

(CL) Coffee beans. This image is included as an example of texture, rather a view of a set of discrete objects.

(CR) Table place setting. Some applications generate images that consist of several objects that have to analyzed both separately and in relation to one another.

(BL) Test card, consisting of an intensity wedge [wdg(0)], a "staircase" providing equal-contrast steps [wdg(0), sca(3)] and a grid [wdg(0), sca(3), sed, thr(1), yxt, max].

(BR) Uniformly distributed random noise. [rnd]

Plate 2: Basic image processing operations demonstrated on the face image in Plate 1(TL). In some cases, the contrast has been enhanced by an unspecified operator *enhance* and the image negated to emphasize the effect.

(TL) High-pass filter [raf, sub, enhance , neg]

(TR) Intensity contours (isophotes) [ict] or [raf, heq, sca(2), sed, thr(1), neg]

(CL) Square all intensities [sqr]

(CR) Sobel edge detector [sed, enhance, neg]

(BL) Largest neighbor [lnb]

(BR) Edge detector, obtained by subtracting (BL) from the original [lnb, sub, enhance , neg]

Plate 3: Basic image processing operations demonstrated on the face image in Plate 1(TL).

(TL) Intensity histogram. (Abscissa: intensity, black on LHS. Ordinate: frequency.)

(TR) Crack detector [crk(5), enhance]

(CL) Horizontal gradient [hgr, enhance]

(CR) Vertical gradient [vgr, enhance]

(BL) Histogram equalization [heq]

(BR) Sharpening [con(-1, -1, -1, -1, 9, -1, -1, -1, -1), enhance]

Plate 4: Basic image processing operations demonstrated on the face image in Plate 1(TL).

(TL) Aspect adjust. [aad(0,3,5)] The vertical axis can be modified by using [yxt, aad(0,3,5), yxt].

(TR) Errors introduced by image rotation [wri, original, tur(37), tur(-37),rea original, sub, blo]

(CL) Picture shift with no wrap-around [psh(40,50)]

(CR) Picture shift with wrap-around [psw(40,50)]

(BL) Intensity profile [plt(98)] superimposed onto the original image

(BR) Making all columns the same as a given column [psh(98), csh]

Plate 5: Varying the resolution of the face image in Plate 1(TL)
(TL) 16 grey levels [sca(4)]
(TR) 8 grey levels [sca(3)]
(CL) 64*64 pixels [2●psq,2●pex]
(CR) 32*32 pixels [3●psq,3●pex]
(BL) Effects of image compression using the JPEG algorithm. The image restored after compression seems to the human eye to be almost exactly the same as the original. (Data storage: PICT format (lossless), 72788 bytes; JPEG (lossy), 19901 bytes. Image size: 256*256 pixels.)
(BR) Comparing the JPEG image and the original. Difference image, enhanced to show detail. (The standard deviation of the intensity difference is about 5.)

Plate 6: Binary image processing operations, performed on a binary version of Plate 1(TR). In each case, the object/outline has been superimposed to improve understanding.
(TL) Grass-fire transform [gft]
(TR) Convex deficiency (black). The convex hull consists of the union of the black and white areas. Its edge is the shape taken up by an elastic string stretched around the given figure.
(CL) Skeleton [ske], its joints [jnt], and limb ends [lme]. Notice that the skeleton forms a loop immediately surrounding the lake (hole). A better skeletonization algorithm would place it midway between the edge of the lake and the outer edge of the conrod.
(CR) Centroid [cgr(X,Y)] and principal axis [dpa]. Parameter values are found using [lmi(X,Y,Theta)]
(BL) Shading indicates how far it is from the most-recently-encountered edge point, travelling from left to right [rlc, neg] This operator produces an image from which it is possible to derive the run-length code using [rlc, wri temporary, swi, bed, rea temporary, min]
(BR) Row integration [rin, neg]

Plate 7: More binary image processing operations.
(TL) A set of four objects
(TR) Blob shading [ndo, neg]
(CL) Third largest blob [big(3)]
(CR) Minimum bounding circle, applied to (CL) [mbc]
(BL) Minimum area rectangle, applied to (CL) [mar]
(BR) The second largest blob, after it has been normalized for both position and orientation [big(2), npo] (The cross-lines [dcl] indicate the center of the circle.)

Plate 8: Image transformations
(TL) Test image. Also see Plate 1(BL).
(TR) Polar-to-Cartesian coordinate transformation [ptc]
(CL) Test image (generated using PIP)
(CR) Cartesian-to-polar coordinate transformation [ctp]
(BL) Silhouette of a slice of bread. The diagonal line indicates the direction of the principal linear feature found by applying [iht] to (BR).
(BR) Hough transform [huf], applied to (BL). (Horizontal axis: orientation. Vertical axis: distance from the center of the image. The bright spots A, B, and C correspond to the nearly-straight sides in (BL). Notice that features B and C have approximately the same orientation.)

Plate 9: Application of machine vision, emphasizing the need for accurate positioning of the object during inspection.
(TL) Hydraulics manifold. A view from an arbitrary angle can often give a person a better overall impression of what an object "looks like" than a pair of orthogonal views can. However, a general view, such as this, is usually unsuitable for machine vision, because it requires a very much more complicated image-processing algorithm.
(TR) Plan view of the hydraulics manifold. Lighting: omni-directional light source.
(CL) Side view of the hydraulics manifold. Lighting: omni-directional light source.

(CR) The automotive component discussed in Section 1.4.4 (supplied by J.-R. Charlier)
(BL) Automobile brake cylinder. The inspection function to be performed requires carefully inserting an optical probe inside the bore (bottom right).
(BR) Electrical connector. The inspection task is moderately difficult if the component is not restrained during inspection. However, if it is the problem becomes almost trivial.

Plate 10: Application of machine vision, emphasizing the need for well-controlled lighting.
(TL) Light bulb. Notice the thin vertical dark line (indicated with an arrow), which shows a piece of wire protruding from the cap.
(TR) Hallmark, stamped letter on English silverware. (The letter *M* is about 2 mm tall and indicates the year of manufacture.) Inspection task: locate and read the date letter. The quality of the hallmark is highly variable in practice, and locating it on antique silverware can be very difficult.)
(CL) Coin. Lighting: multi-directional grazing illumination, provided by a ring illuminator projecting light inwards.
(CR) Coin. Lighting: coaxial viewing and illumination from a diffuse light source. A processed version of this image is shown in Plate 15.
(BL) Glass bottle. Lighting: back illumination. Notice that this highlights both the crack and the embossing.
(BR) Stress patterns in a glass bottle. Lighting-viewing: two linear polarizers, one on each side of the bottle. Their easy axes are orthogonal. (Another image obtained from a glass vial using dark-field illumination is shown in Plate 14.)

Plate 11: Machine vision applications based on x-ray sensing
(TL) Plan (top) and side view (bottom) of six glass jars mounted on a cardboard tray (not visible). A twin-beam x-ray system was used. Notice that the two views are registered exactly along the horizontal axis. There are three foreign bodies present here.
(TR) Glass jar containing tomato sauce, a thin piece of a screw and, just above it, a short, thin piece of wire. Conventional x-ray system
(CR) Integrated circuit. Fine-focus x-ray system. Notice that the wires connecting the chip to the terminals are just visible to the eye but are difficult to detect automatically.
(BL) Power plug (UK and Irish electrical system). Conventional x-ray system. (Processing of this image is illustrated in Plate 18.)
(BR) Aerosol-spray nozzle assembly. Conventional x-ray system.

Plate 12 Machine vision applications which produce considerable image variability and hence require a high level of intelligence in the software.
(TL) Baked goods. Shape and texture are both important.
(TR) Plant. Vision task: guide a robot to dissect the plant.
(CL) Cork floor tile. Lighting: grazing illumination. Vision task: Check the aesthetic appearance of the tile by analyzing its texture.
(CR) Quiche. Inspection task: determine whether it is aesthetically pleasing by checking the distribution of vegetables on its top surface. Also see Plates 19 and 20.
(B) Four examples of cake decoration patterns. (These images have already been processed to produce high-contrast patterns, which would be difficult to display otherwise.) Inspection task: determine whether these patterns are aesthetically pleasing. (Do they conform to what we expect?)

Plate 13: Cracks and aerosol sprays
(TL) Image obtained from an automobile component. Lighting: ultra-violet, no color filter used.
(TR) Same scene as in (TL) but with a narrow-band color filter placed between the work-piece and the camera. (Pass band: 475 - 525 nm)
(CL) Crack detected using [crk(3)] followed by removal of binary noise.
(CR) Unprocessed image of an aerosol spray. Illumination: lamp located directly above the spray. Vertical axis expanded by a factor of two, using [yxt, aad(0, 1, 2), yxt].

(BL) Intensity contours (isophotes) [ict]
(BR) A - Graph of the peak intensity in each column of the image. B - Two contours showing where the intensity in each column falls to 75% of the peak intensity in that column. C - Position of the peak intensity in each column.

Plate 14: Glass vial
(TL) Original image. Compare this with the images shown in Plate 10. Lighting: dark-field illumination.
(TR) Radius of curvature of the shoulder is found by fitting a circle to three edge points.
(CL) Neck. Found by measuring the horizontal chord length for each row, applying [rin] to the grey region in (TR), and then thresholding.
(CR) Mouth, neck, and parallel sides of the vial, found by integrating the white curve in (CL) vertically. [yxt, rin, csh, thr(15), dab(ske), yxt]
(BL) Segmenting the edge contour by using the 6 white lines in (CR) to mask the edge contour. [rim(stripe_image, 2•exw, rim(outline_image), min, kgr(25)]
(BR) The vial could be picked up by a robot gripper placing two fingers at the positions marked by the crosses. These points are the centroids of the bays of the vial silhouette.

Plate 15: Coin. Original image is shown in Plate 10
(TL) The four white lines show the angular limits of the arcs of embossed lettering.
(TR) Cartesian-to-polar coordinate transformation. [ctp(X,Y)] ((X,Y) is the centroid of the coin silhouette.) A vertical slice through the center of this image can be used to find the limits of the lettering and hence the angular positions indicated in (TL).
(CL) The monarch's head was isolated, then ctp/2 applied. [loa('Coin'), enc, thr, big, chu, blb, cgr(X,Y), xor, big, blb, ctp(X,Y)] The arrows indicate features that might be employed to find the orientation of the coin. A - Tip of the crown. B - Bottom of the neck (front)
(CR) The slopes of these lines are determined by the vertical positions of features A and B in (CL).
(BL) Highpass filtering highlights the surface texture. The region to the left of the monarch's head has no visible texture because the original is saturated (i.e., all pixels are white.)
(BR) Monarch's head and lettering isolated. [loa('Coin'), enc, thr, big, chu, blb, cgr(X,Y), xor, big, blb] isolates the head.

Plate 16: Metal grid.
(TL) Original image. Lighting: with back-illumination.
(TR) After thresholding, negation, and applying the Cartesian-to-polar coordinate transformation. [ect,neg,ctp] Vertical axis: orientation. Horizontal axis: radius from the center of the image.
(CL) After integrating the intensity along each row [rin]. Arrows point to four bright streaks, which indicate the orientation of the mesh.
(CR) Line through the center of the image whose orientation is indicated by the vertical position of the top-most bright streak in (CL).
(BL) Spots lying along the line in (CR). Although this is quite a good fit, there is a slight error in its slope.
(BR) Drawing the principal axis through this set of spots produces a better estimate of the orientation of the grid [dpa].

Plate 17: Toroidal metal component; constructing images from multiple scans.
(TL) Machine part. Lighting: Natural, with simulated laser light stripe. (This was drawn manually to enhance clarity.) Notice that there is a raised ring 8 mm high with steep cliff-like sides in the center of the top surface. This component is light grey and has a diameter of 210 mm.
(TR) Simulated line-scan image. Lighting: Normal front illumination. This image was obtained by rotating the component while repeatedly scanning it with an area-scan camera. However, only the central row of each image was kept. Horizontal axis: radius. Vertical axis:

angle. The edges would be straight and vertical if the object were mounted concentrically on the turntable. The human eye can never see this component in the way shown here.
(CL) Image (TR) after polar-to-Cartesian coordinate transformation [ptc]. The cross-lines indicate the center of rotation.
(CR) Depth map.
(BL) Image (CR) after polar-to-Cartesian coordinate transformation [ptc]. Bright parts of the image represent surfaces that are raised. Notice the notch on the top surface and that the cliff-like sides of the raised ring are partially obscured (black regions).
(BR) Image (BL) after applying a noise removal filter [op(2)]. The cross-lines indicate the center of the component and the center of rotation.

Plate 18: X-ray image of domestic electrical power plug. Original image is shown in Plate 11(BL).
(TL) Intensity histogram. The centers of the two rather poorly-defined valleys correspond to the threshold values used for (CR) and (BL).
(TR) Intensity profile along the section shown by the vertical line. [plt(64)]
(CL) Loose strands. [neg, crk(2), thr(100), swi, thr(64), exw, kgr(50)]
(CR) Simple thresholding highlights the plastic molding and insulated sleeving quite well [thr(74,141].
(BL) Thresholding also isolates the metal quite well [thr(142)].
(BR) Grey-scale morphology (crk/1 and gop/1) is able to isolate the multi-strand copper wire and its immediate plastic insulation, except where the wires cross [neg, crk(5), thr(64), kgr(200)].

Plate 19: Pseudocolor and RGB color separation
(TL) Pseudocolor image. Original grey-scale image is shown in Plate 1(TL)
(TR) Various pseudocolor mappings applied to an intensity wedge (top).
(CL) A different pseudocolor mapping, applied to the image shown in Plate 1(CL).
(CR) Quiche: red component. The original image is shown in Plate 20(TL).
(BL) Quiche: green component
(BR) Quiche: blue component

Plate 20: CMYK color separation and pseudocolor
(TL) Original color image of a quiche. Also see Plate 12(CR).
(TR) Quiche: cyan component
(CL) Quiche: magenta component
(CR) Quiche: yellow component
(BL) Quiche: black component
(BR) Pseudocolor image, derived by performing [heq] on each of the CMYK color channels

Plate 21: Graphics overlay and color recognition
(TL) Colored graphics overlaid on a monochrome image is probably acceptable, if the graphics can be switched on/off at will.
(TR) A PIP command sequence superimposed on the image. Neither can be seen clearly, despite the use of color.
(CL) Original image for color recognition (a young child's jig-saw puzzle).
(CR) Test pattern, created to train the PCF. (Compare to Plate 23(CR).)
(BL) The PCF was trained on the image in (TL). The histogram was then computed.
(BR) Output of the PCF trained on the pattern in (CR). Pseudocolor switched on.

Plate 22: Color recognition, continued
(TL) Color scattergram derived from Plate 21(CL). Pseudocolor switched on.
(TR) Thresholding applied to (TL).
(CL) Blob shading. [exw, skw, ndo, enc, acn(-1)] applied to (TR). Pseudocolor switched on.
(CR) Output of the PCF trained on the pattern in (CL). Pseudocolor switched on.

(BL) Watershed operator applied to the image in (CL). (This divides the space using a nearest neighbor decision rule.)
(BR) PCF trained on the image in (BL). Pseudocolor switched on.

Plate 23: Color recognition, second example.
(TL) Original image. (Flag of the Principality of Wales, against a yellow background.)
(TR) The PCF was trained on the image in Plate 21(CR). The color scattergram was then computed for the flag image.
(CL) Two-dimensional scattergam derived from the flag image. Abscissa, green. Ordinate, Blue. Pseudocolor switched on.
(CR) Test pattern, created to train the PCF. Pseudocolor switched on. (This differs from Plate 21 (CR) in creating only 7 discrete color classes. The pseudocolors were chosen to approximate the "average" real color within each class.)
(BL) Output of the PCF trained on the pattern in (CR). Pseudocolor switched off.
(BR) As (BL), with pseudocolor switched on.

Plate 24: Color recognition, second example, continued. In each case, pseudocolor is switched on. Notice how the color recognition becomes more accurate as the blobs in the color triangle are expanded.
(TL) Color scattergram followed by thesholding, noise removal, and blob shading.
(TR) Result of applying the PCF trained on (TL) to the flag image in Plate 23(TL).
(CL) Enlarged blobs in the color triangle. [10•lnb] applied to (TL).
(CR) Result of applying the PCF trained on (CL) to the flag image in Plate 23(TL).
(BL) Watershed operator applied to the image in (TL).
(BR) Result of applying the PCF trained on (CL) to the flag image in Plate 23(TL). Pseudocolor switched on. (Ignore the change of pseudocolors, because this is an artifact of the watershed software and can be corrected easily.)

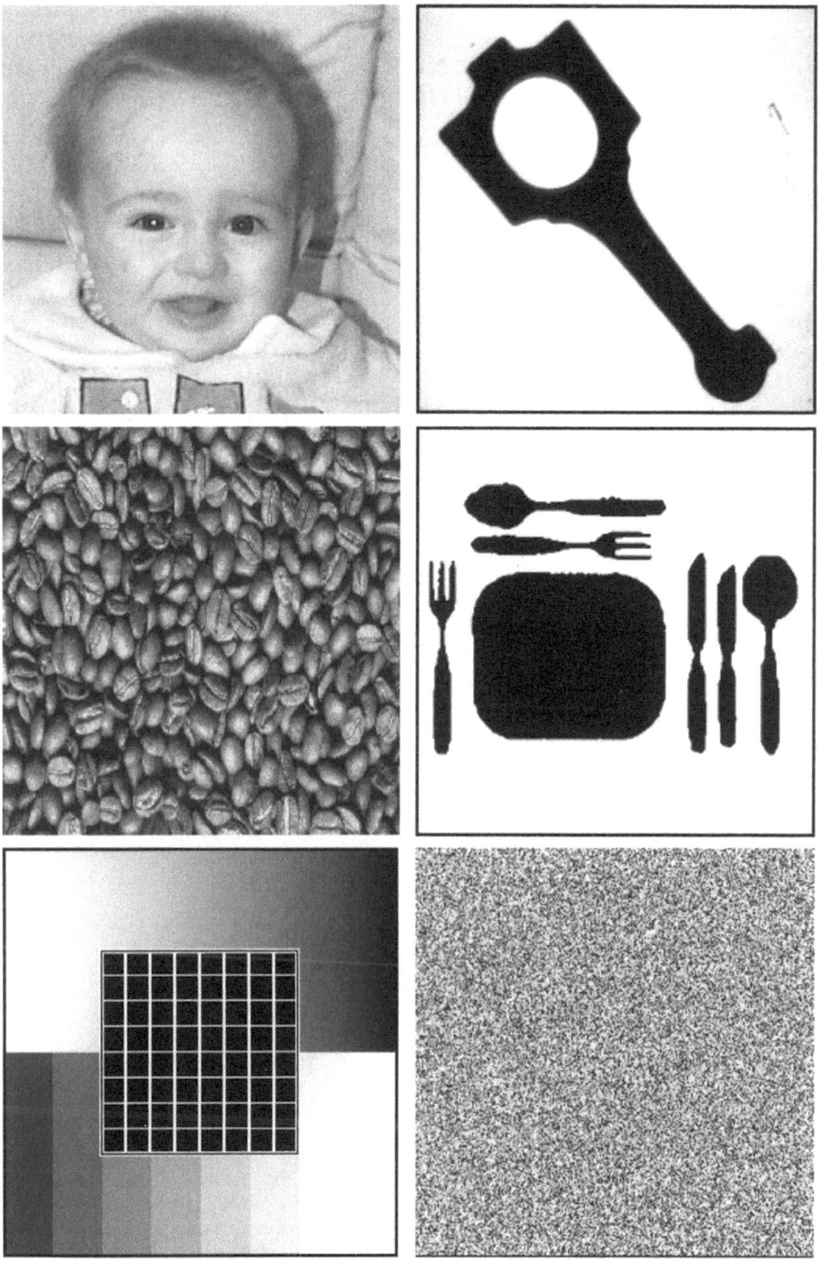

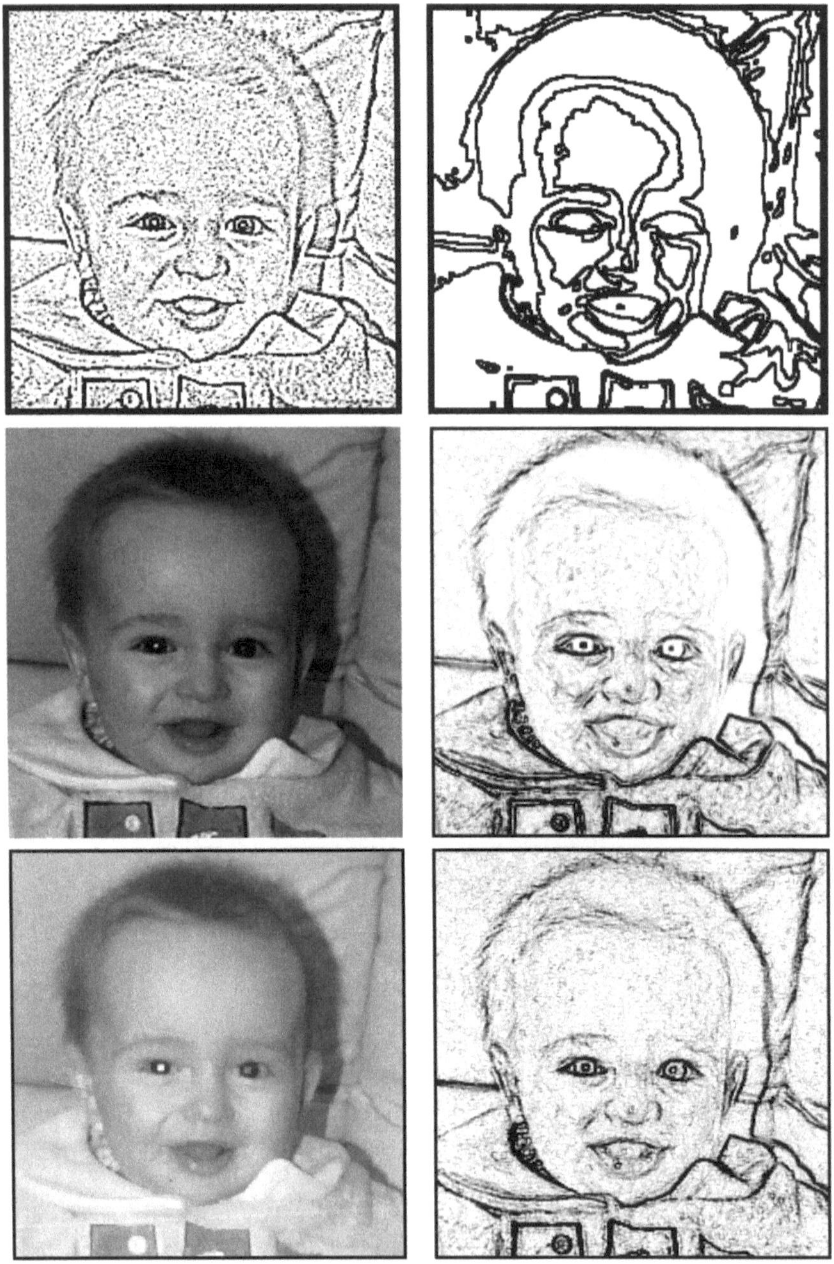

Plate 3

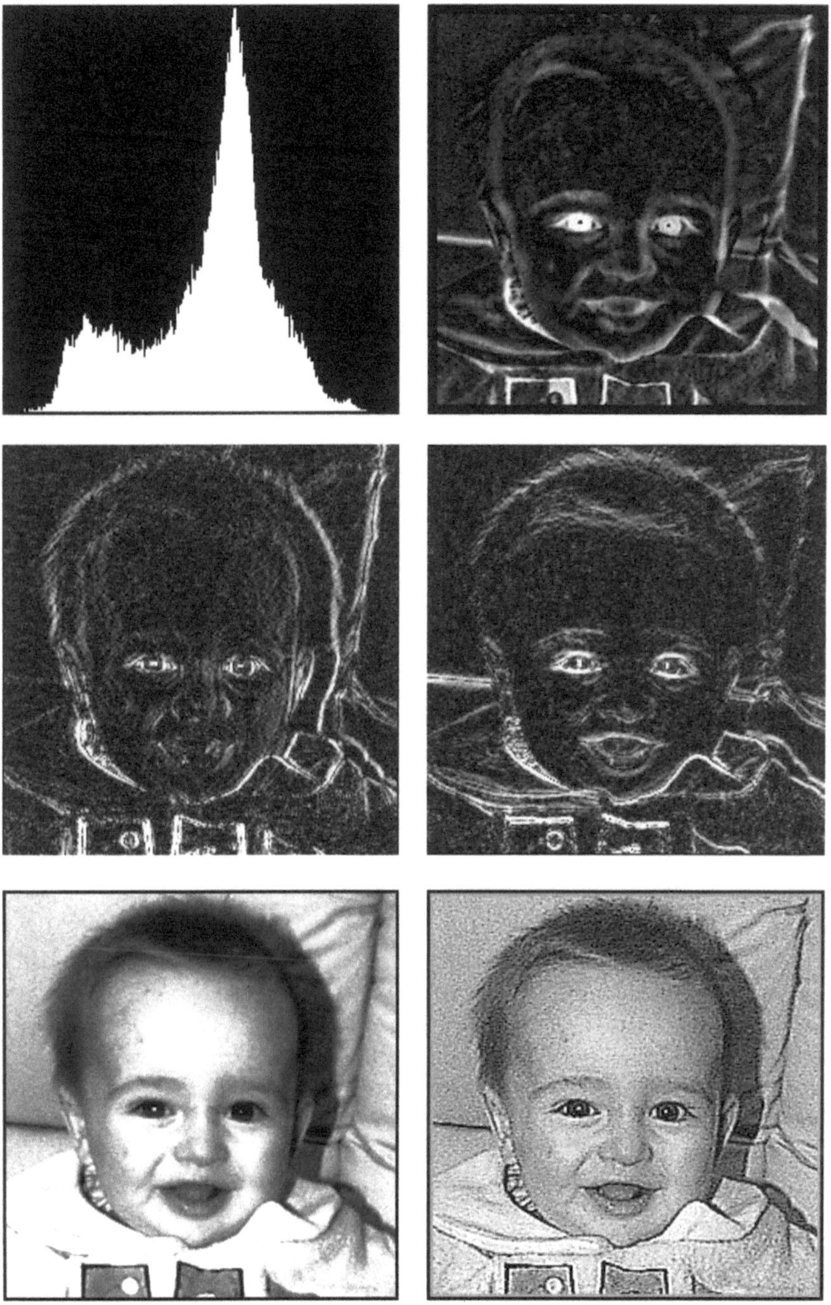

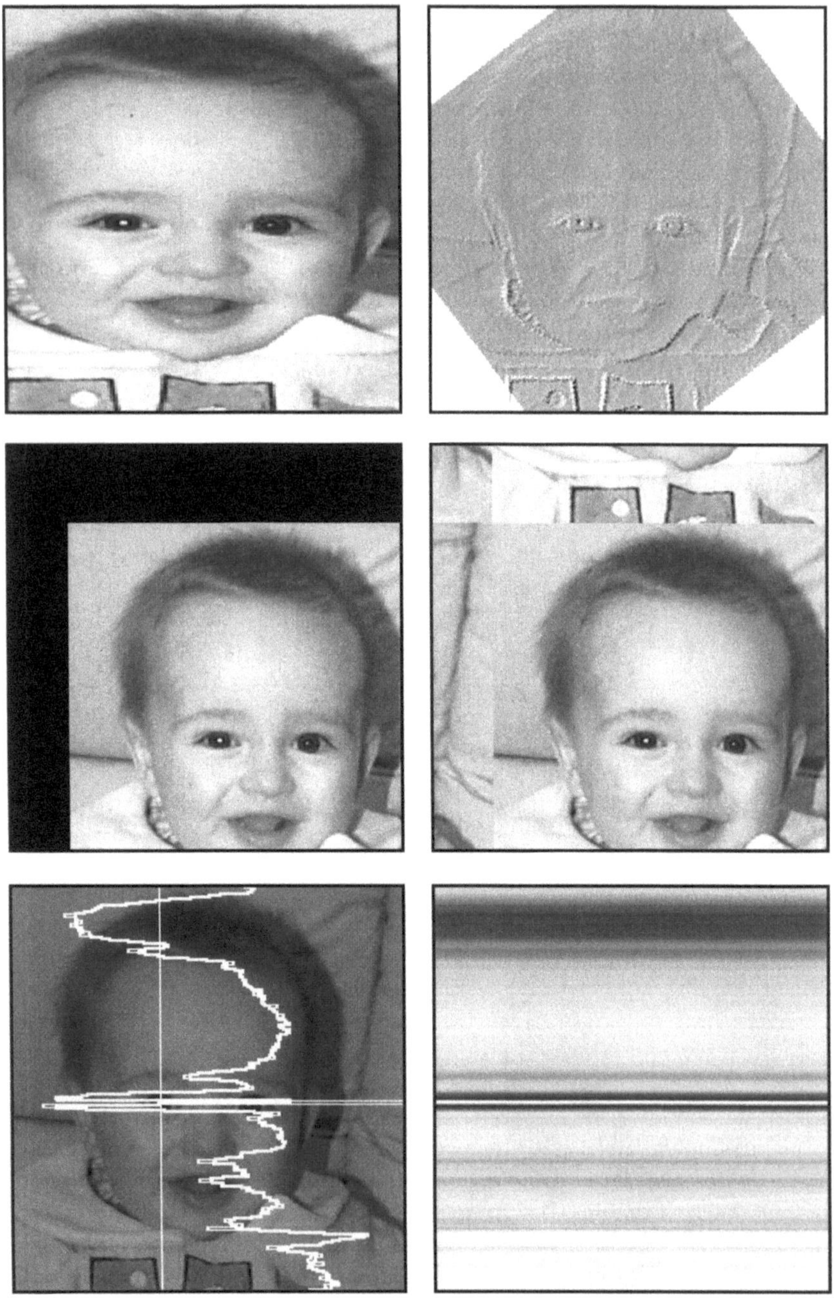

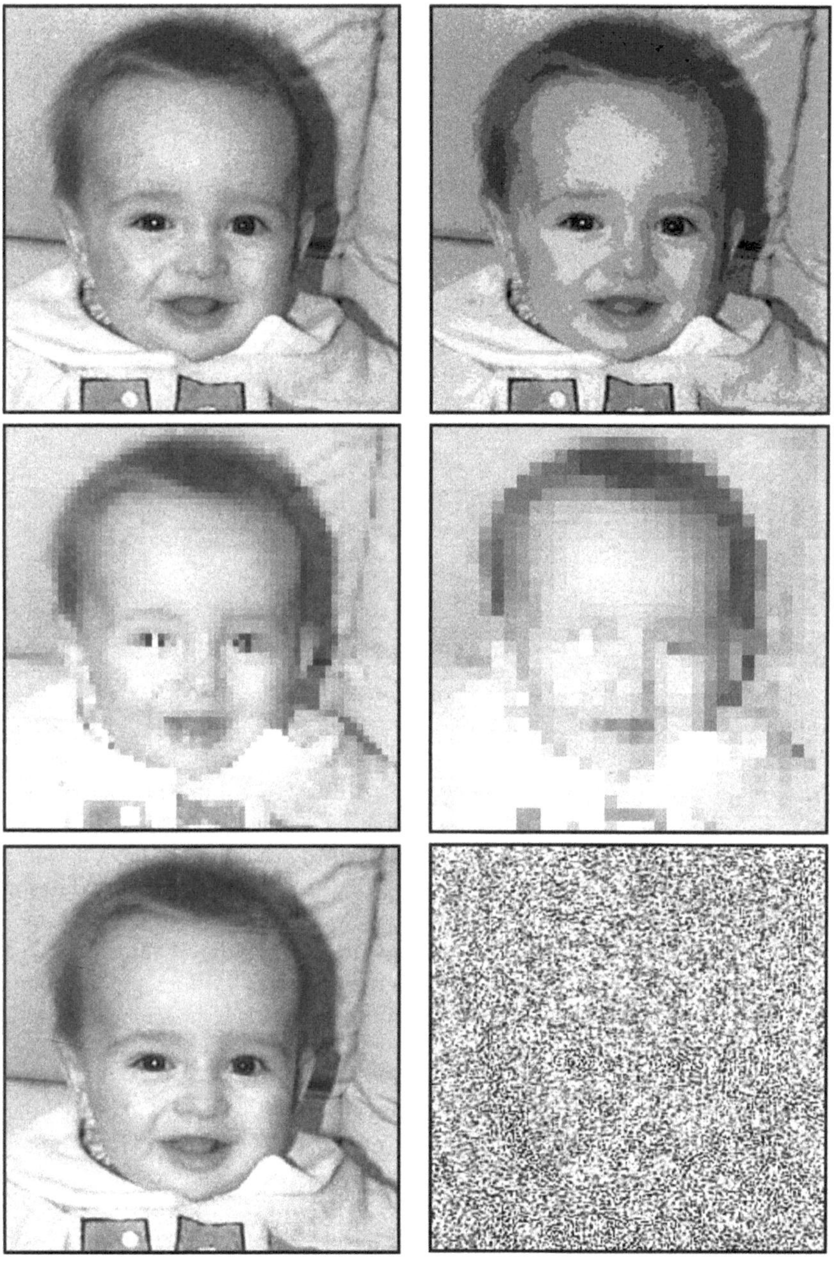

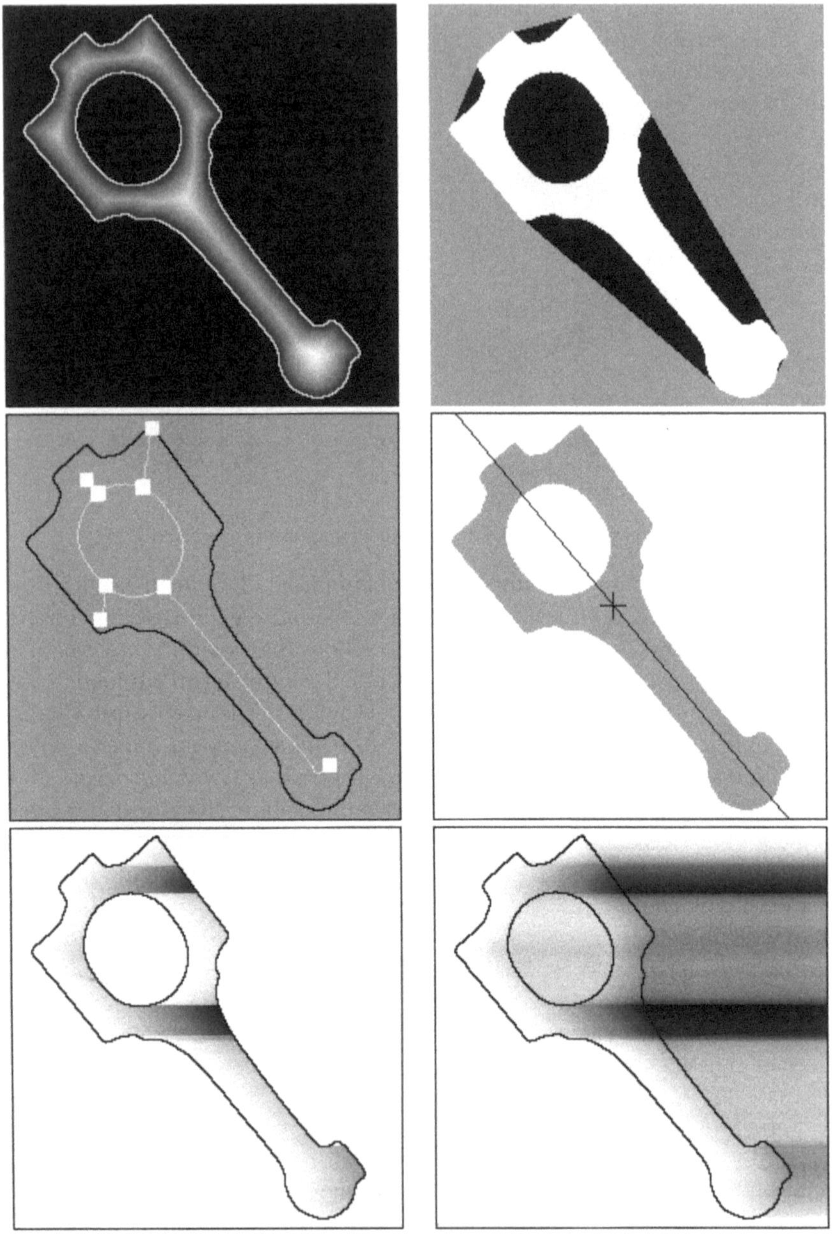

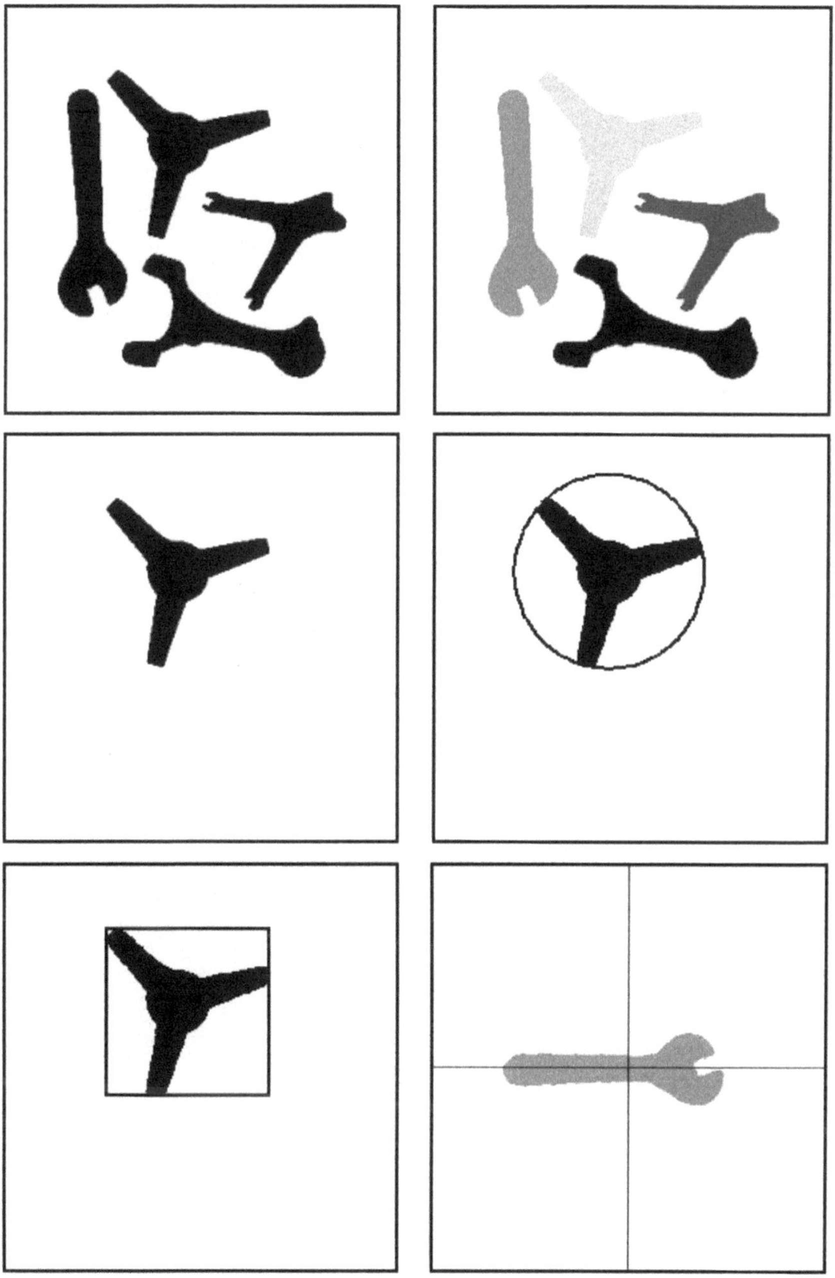

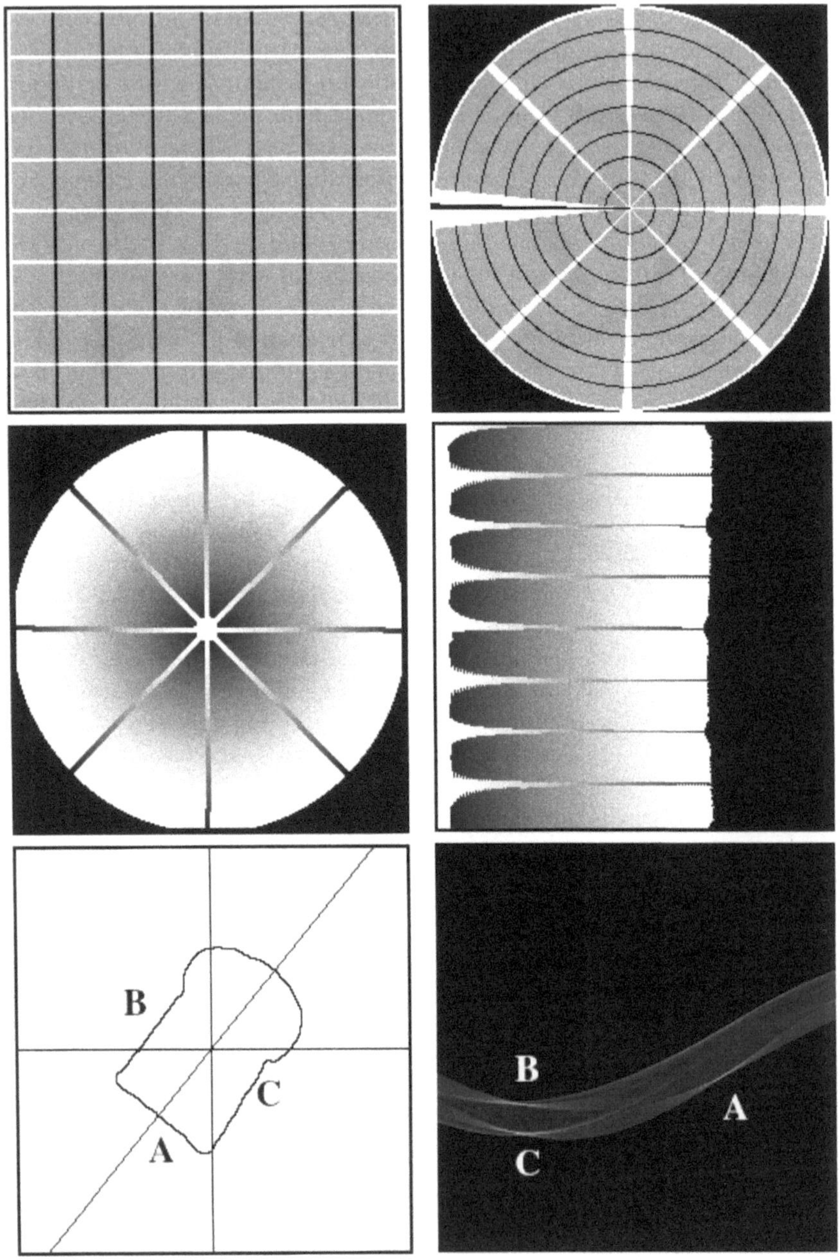
B
C
A
B
A
C

Plate 9

M

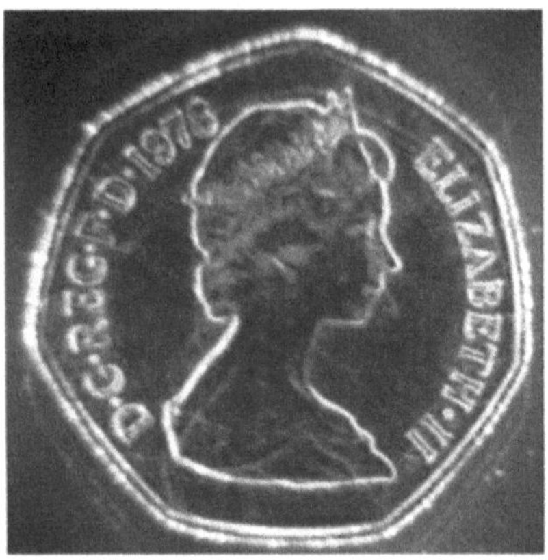
ELIZABETH·II

ELIZABETH II

MILK
BRIDGEN

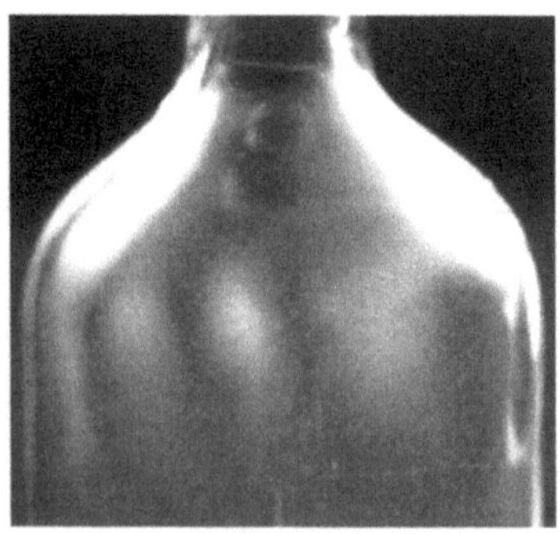

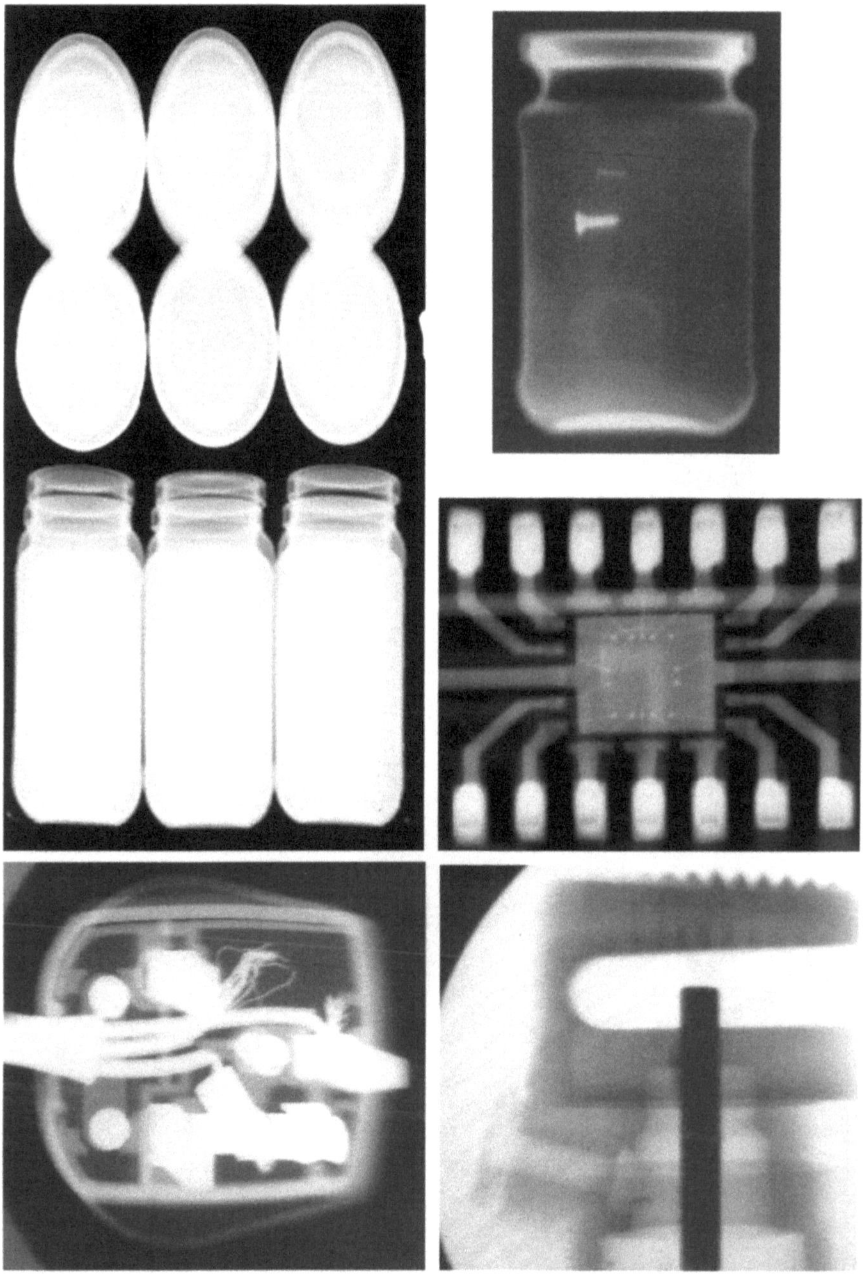

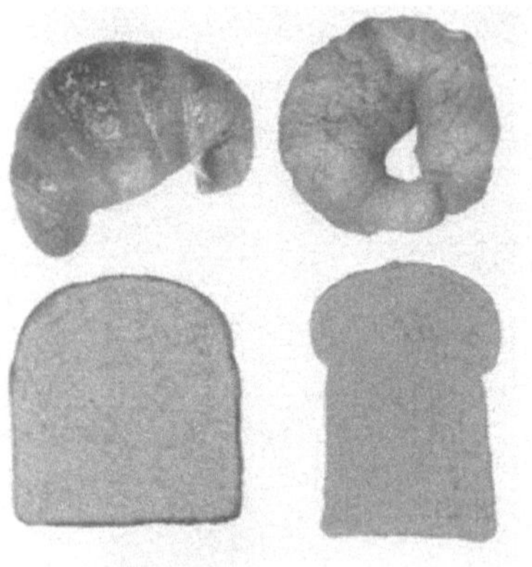

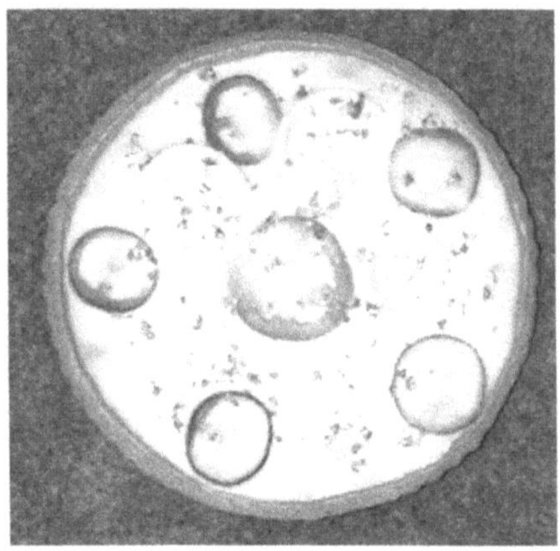
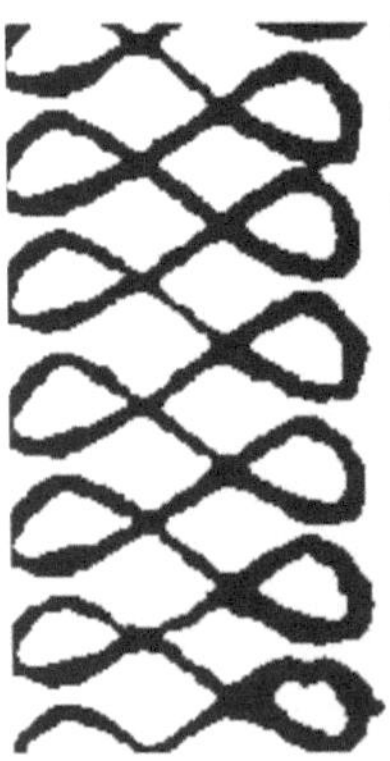

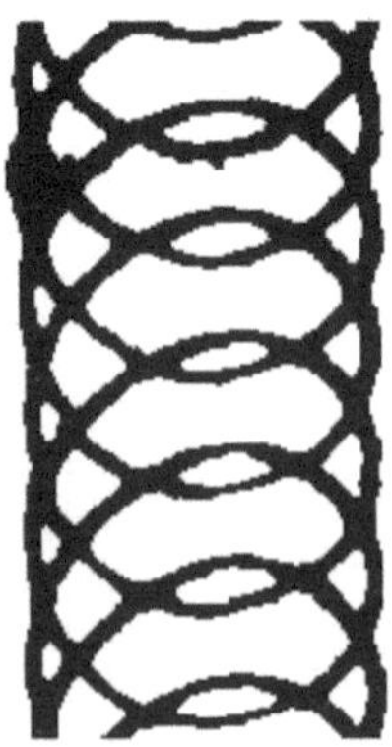

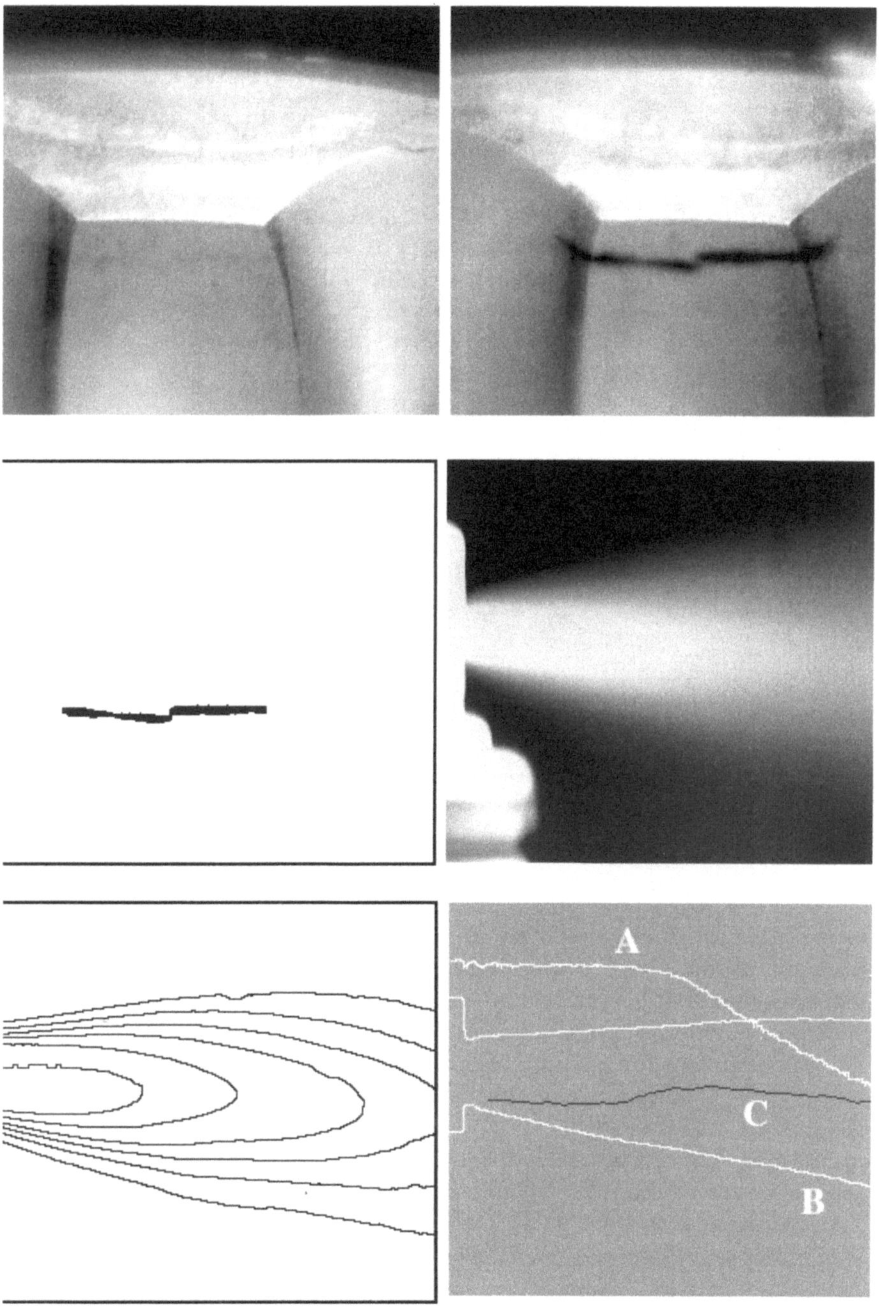
A
C
B

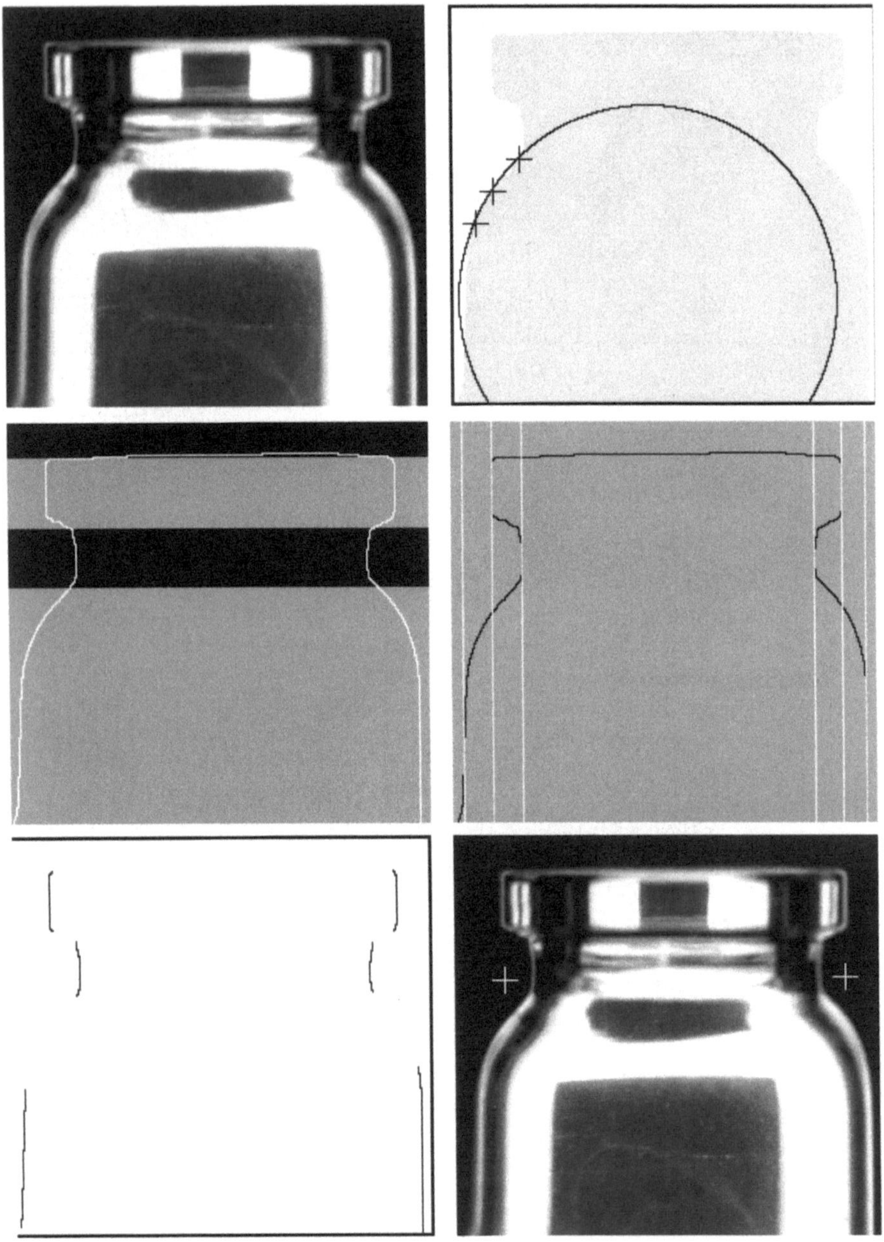

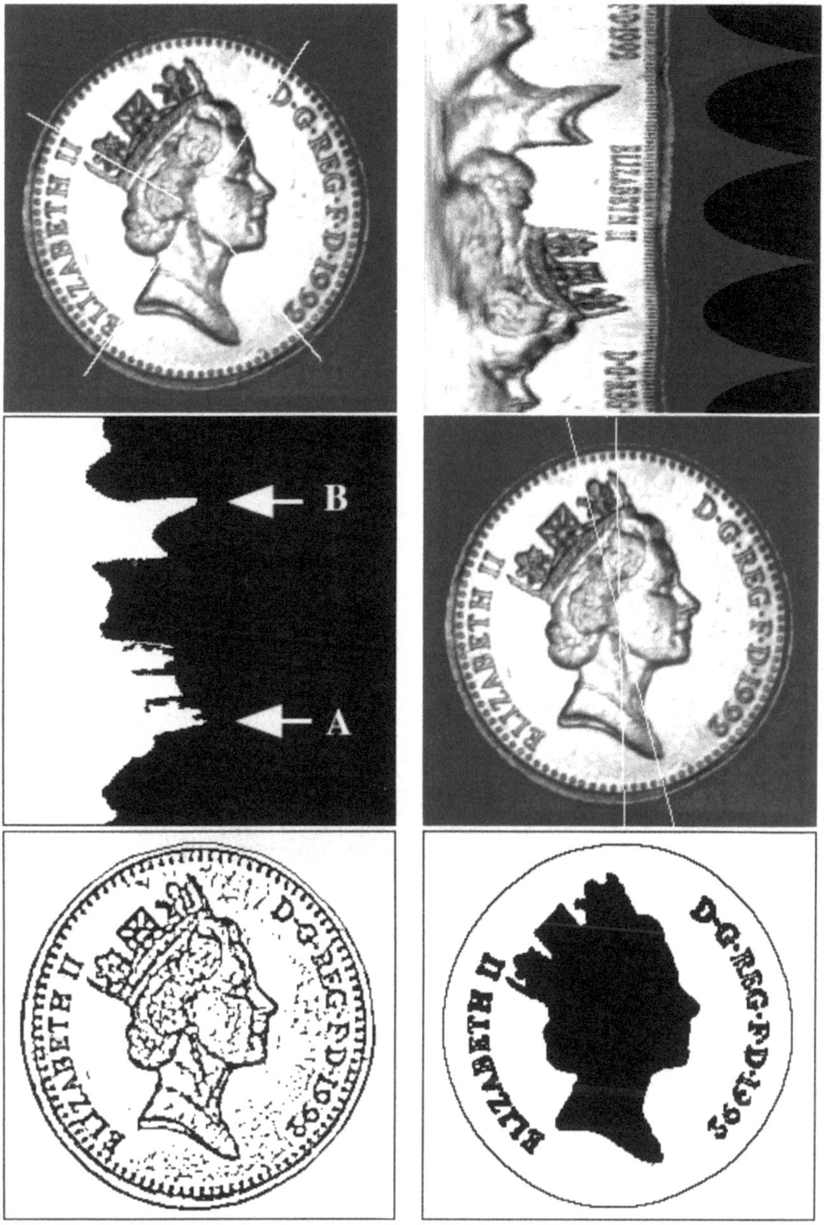
B
A

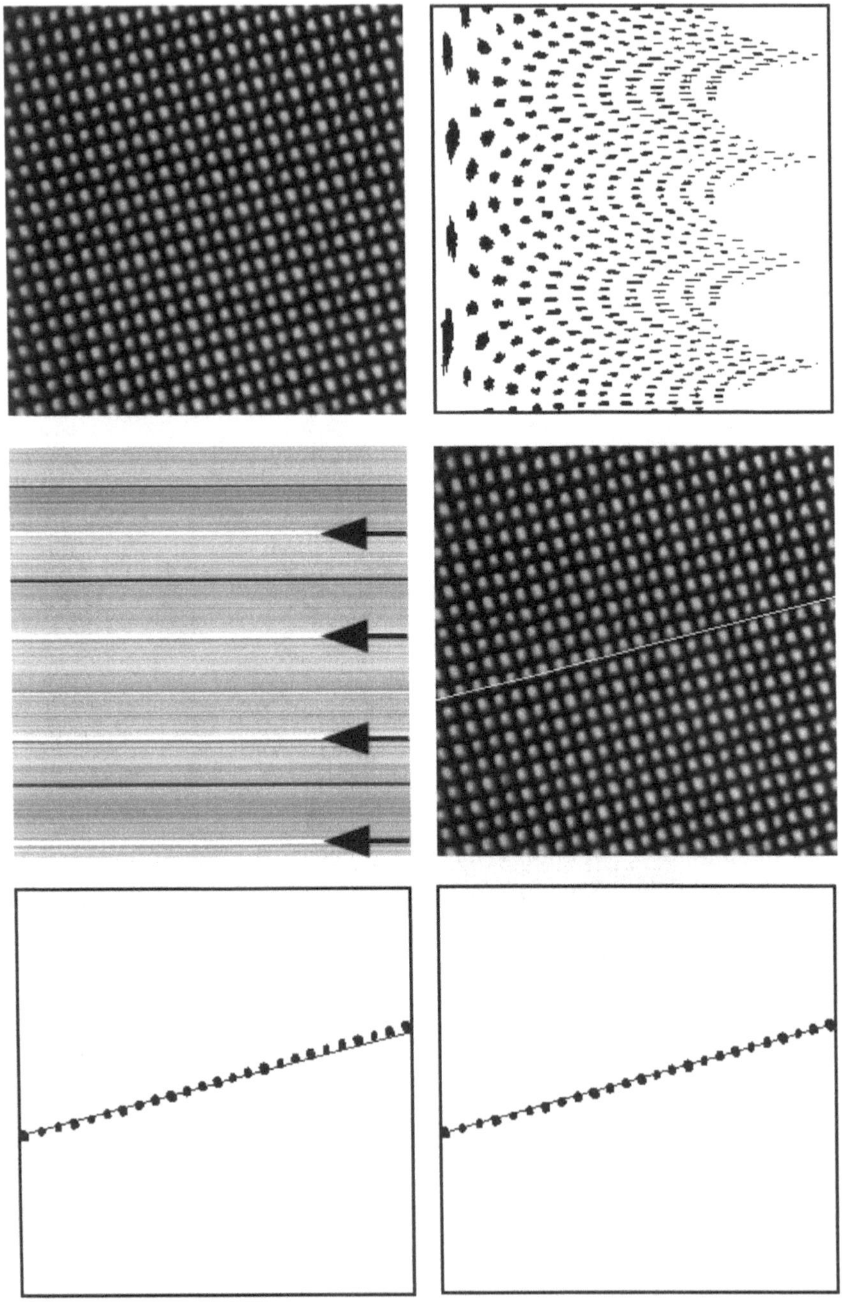

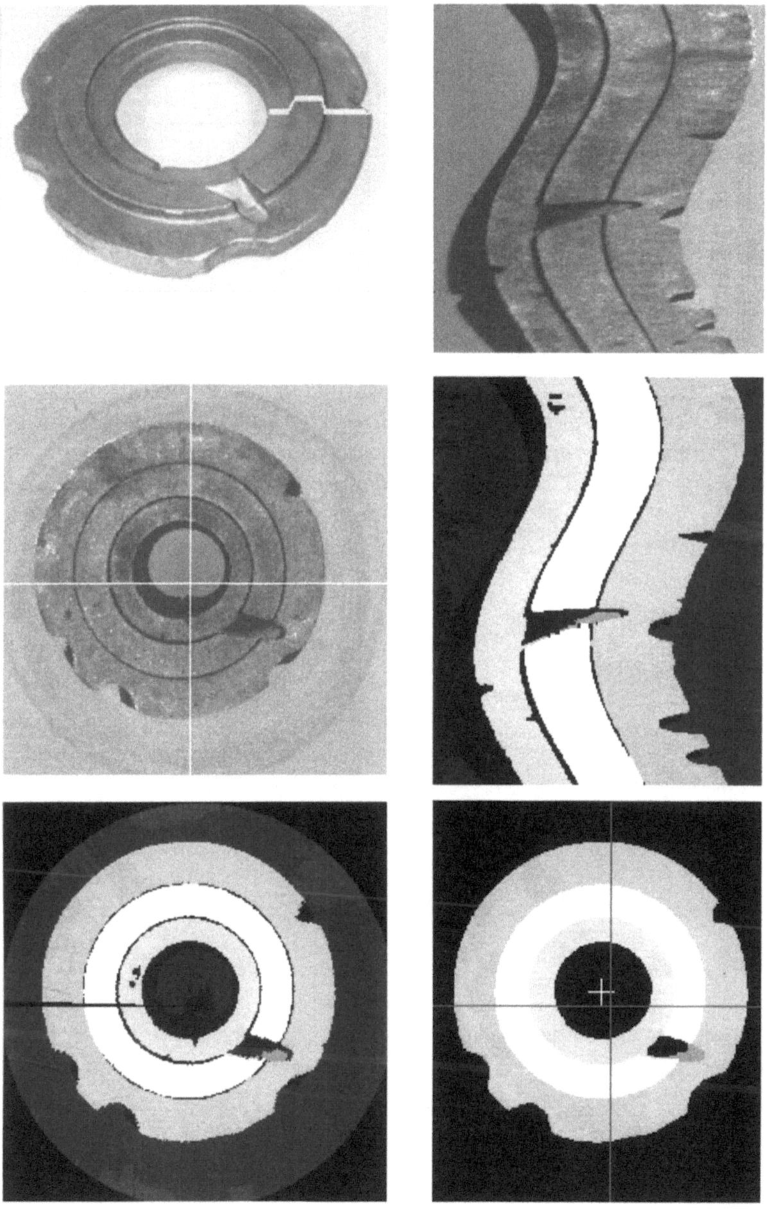

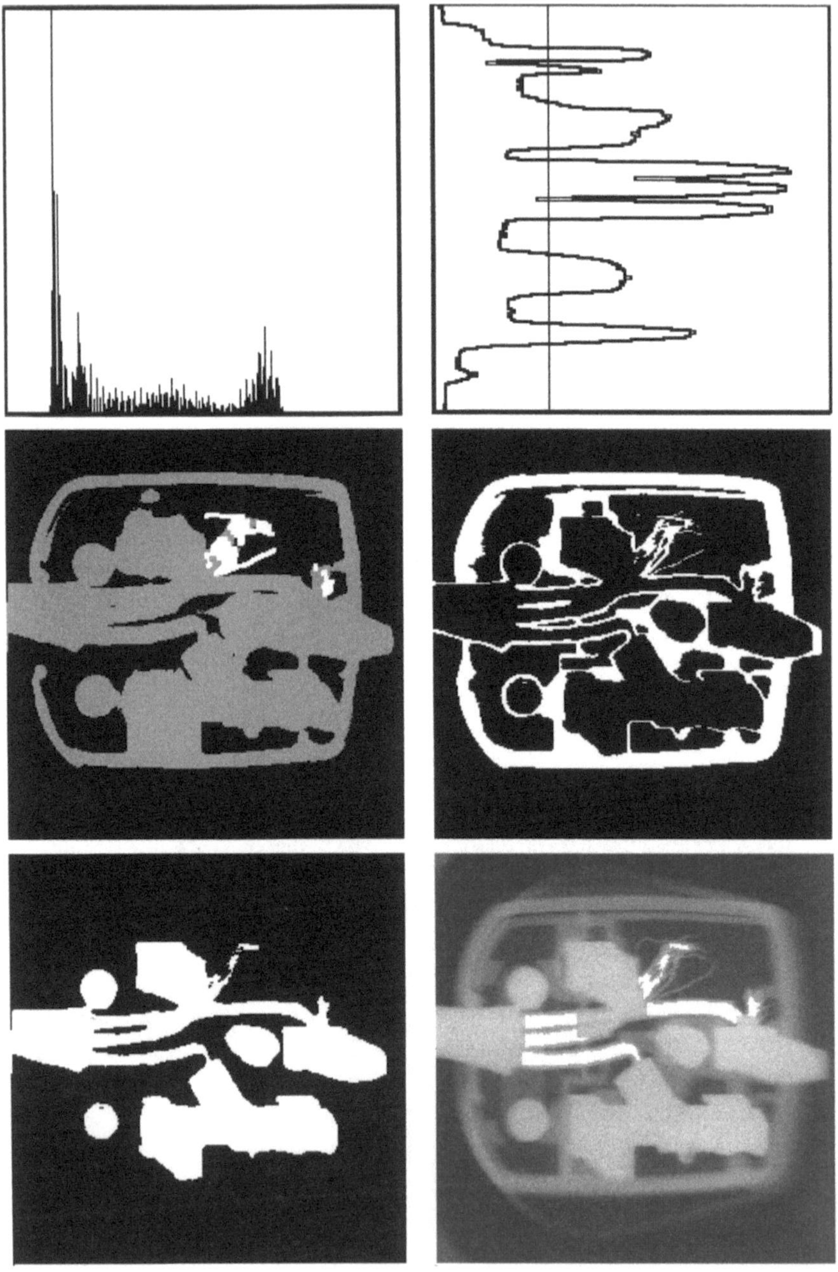

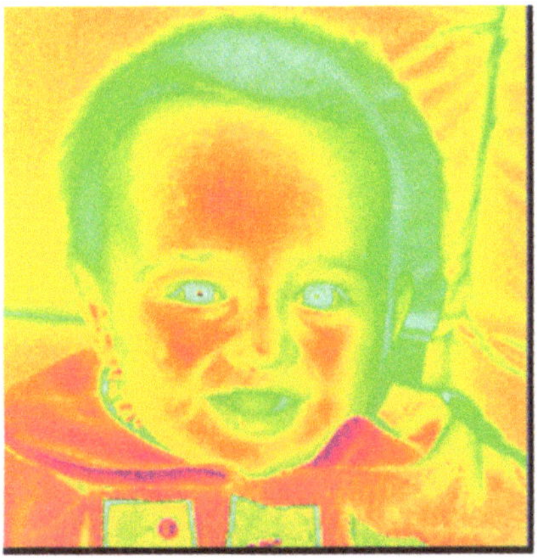

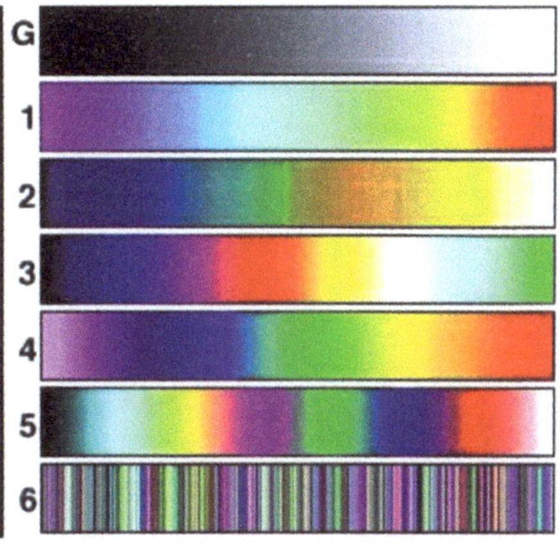
G
1
2
3
4
5
6

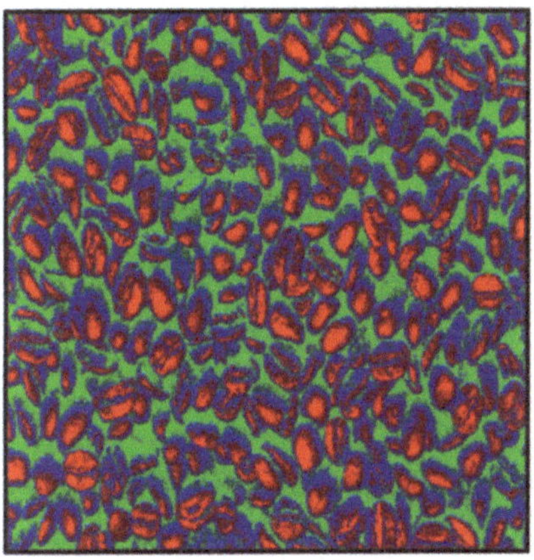

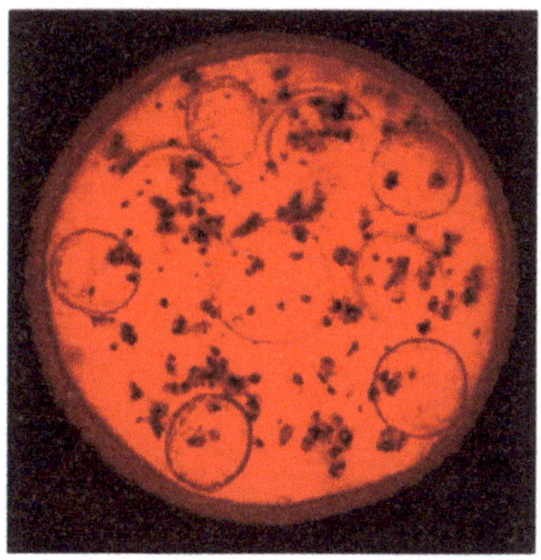

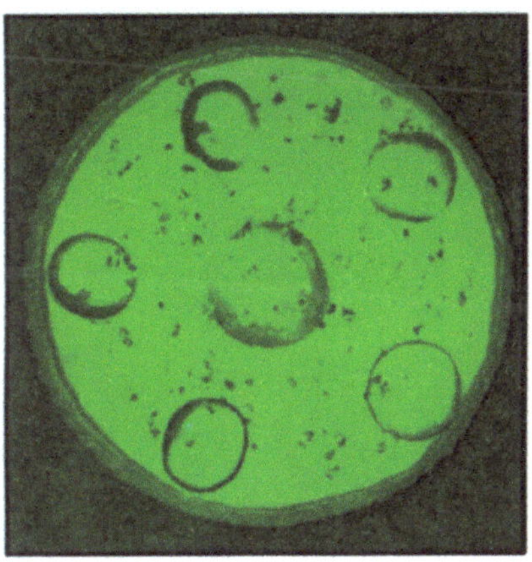

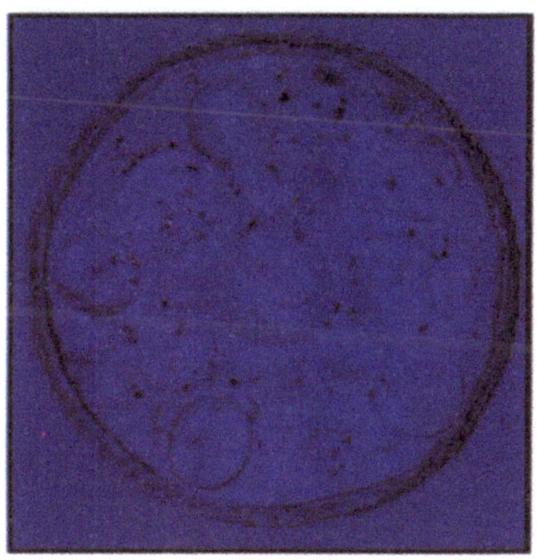

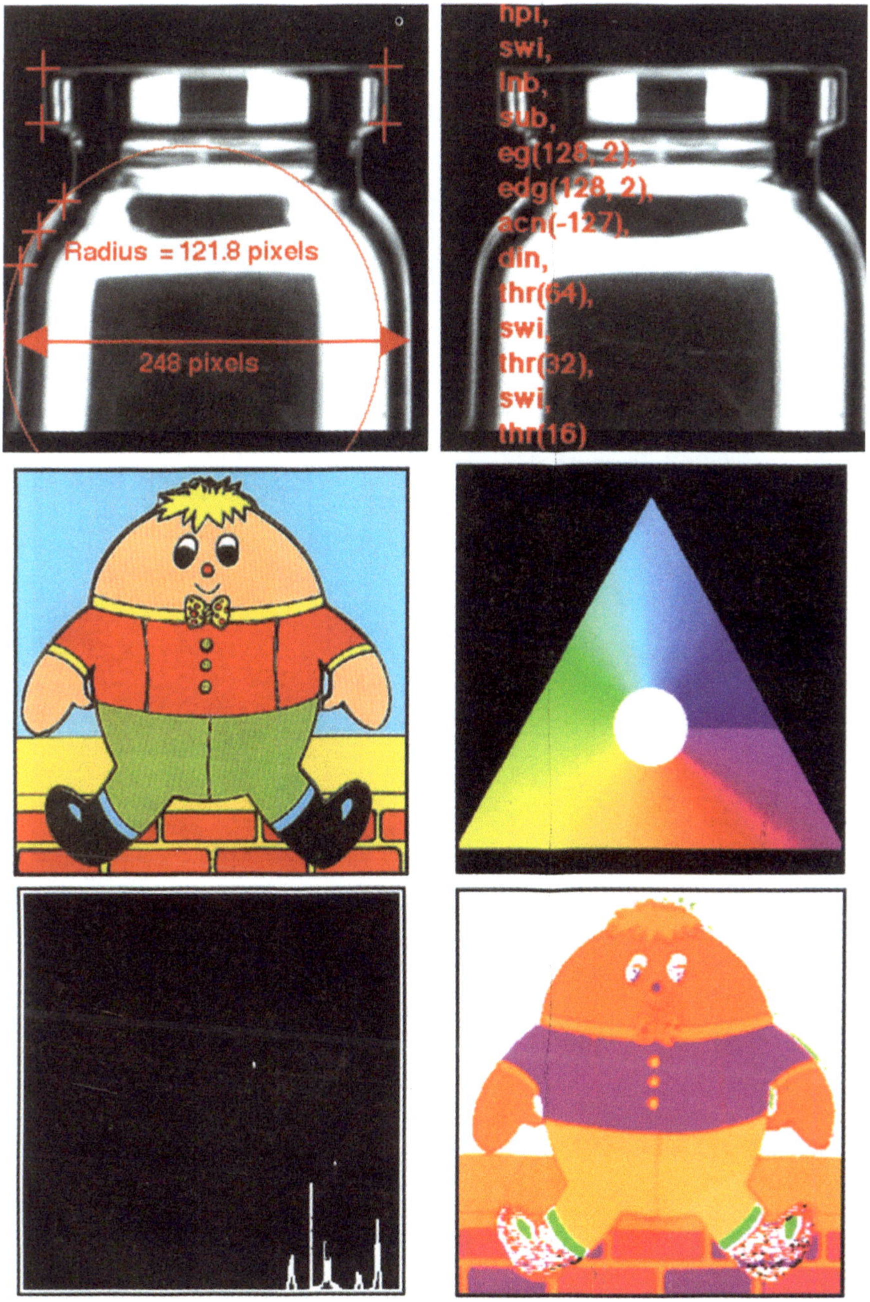
Radius = 121.8 pixels
248 pixels
hpi,
swi,
lnb,
sub,
eg(128, 2),
edg(128, 2),
acn(-127),
din,
thr(64),
swi,
thr(32),
swi,
thr(16)

MIX
Papier aus verantwortungsvollen Quellen
Paper from responsible sources
FSC® C105338

If you have any concerns about our products,
you can contact us on
ProductSafety@springernature.com

In case Publisher is established outside the EU,
the EU authorized representative is:
Springer Nature Customer Service Center GmbH
Europaplatz 3, 69115 Heidelberg, Germany

Printed by Libri Plureos GmbH
in Hamburg, Germany